PRACTICE PROBLEMS

IN

DISCRETE MATHEMATICS

Bojana Obrenić

Queens College, City University of New York

PEARSON

Prentice
Hall

Upper Saddle River, NJ 07458

Editor-in-Chief: Sally Yagan
Executive Editor: George Lobell
Supplement Editor: Jennifer Brady
Assistant Managing Editor: John Matthews
Production Editor: Donna Crilly
Supplement Cover Manager: Paul Gourhan
Supplement Cover Designer: Joanne Alexandris
Manufacturing Buyer: Ilene Kahn

The author and publisher of this book have used their best efforts in preparing this book. These efforts include the development, research, and testing of the theories and programs to determine their effectiveness. The author and publisher make no warranty of any kind, expressed or implied, with regard to these programs or the documentation contained in this book. The author and publisher shall not be liable in any event for incidental or consequential damages in connection with, or arising out of, the furnishing, performance, or use of these programs.

Printed in the United States of America

10 9 8 7 6 5 4 3 2 1

ISBN 0-13-045803-1

Pearson Education Ltd., *London*
Pearson Education Australia Pty. Ltd., *Sydney*
Pearson Education Singapore, Pte. Ltd.
Pearson Education North Asia Ltd., *Hong Kong*
Pearson Education Canada, Inc., *Toronto*
Pearson Educación de Mexico, S.A. de C.V.
Pearson Education—Japan, *Tokyo*
Pearson Education Malaysia, Pte. Ltd.
Pearson Education, *Upper Saddle River, New Jersey*

Preface

This collection of problems with their solutions reflects the content of introductory undergraduate courses in Discrete Mathematics, offered by the author to students of Computer Science at Queens College of the City University of New York over the past nine years.

The background required for studying this book is readily found in any of the many available standard textbooks in introductory discrete mathematics. Most of these textbooks give a fine selection of correctly presented material. However, the transition between knowing a result and being able to employ it in a nontrivial way is usually long and difficult—this book attempts to help on the way.

Students of introductory computer science have to absorb some elementary theoretical techniques quickly. Soon after they encounter such techniques for the first time, they are exposed to nontrivial applications of these techniques in algorithm design and programming. Students are often in a situation where they know enough of a vocabulary to follow a construction or an analysis, but are lacking strength to put it to use independently. Their predicament is typically recognized as a lack of "intuition" and "mathematical maturity". This book attempts to contribute to building this intuition in the early stages of the curriculum.

This book has been written in response to the readiness with which a multitude of the author's students have opted for problem solving exercises as the preferred way of studying this subject. Indeed, whether they aspire to be master programmers (informed pragmatists who keep only a small set of very best tools) or theorists, the students have gained intuition and acquired mathematical sophistication through intensive problem solving exercises, practicing the selected concepts until their correct application becomes habitual. The exercises are designed to be somewhat repetitive and of graduated difficulty. Their solutions are attainable by an individual working alone in the conditions of a closed-book test. The ability to solve these problems under these conditions has served as a faithful practical indicator of the students' mastery of the material and their preparedness for subsequent studies.

A typical student should be able to complete this book during one semester, in parallel with listening to lectures or reading a textbook. The suggested way to work on a problem is to attempt to solve it independently, as if it were an exam problem (with all books closed) and to write down the complete solution. Once completed, the student's solution should be compared with the one given in the book, and the origin of the differences, if any, should be understood. Where a problem has several parts, the parts should be treated as separate problems and studied in the order listed. When a problem is encountered that seems too difficult to solve, the student should study the solution given in the book.

This book contains over 2000 individual questions, with answers whose length ranges from a single symbol to a number of pages. These questions are arranged in 421 problems together with their solutions. The problems are grouped in sixteen sections, which form five chapters. This organization should be helpful to readers in search of problems that address certain topics. However, the grouping is approximate. Even though most of the problems indeed fall under the topic of the section in which they appear, no attempt has been made to confine the problem context within a compartment of a single topic. In fact, the intention is that many later problems bring earlier topics into the focus. In particular, the concept of recursion permeates this problem collection in its entirety. Recursion appears in the context of integer functions, counting, algorithms, and graphs.

Every problem in this book is intended to be self contained—an individual problem statement or solution does not depend on the remainder of the book for completeness, accuracy, or clarity. Exceptions are made in some cases where explicit references are given to definitions or claims presented in earlier problems. However, the order of problems is not arbitrary and, in general, more complex problems follow the simpler ones. The problems within an individual section are intended to be solved in the order in which they appear. The order of sections within an individual chapter should be likewise honored. The listed order of chapters, although preferred, is not mandatory, and after the first two chapters are completed in sequence, the last three may be studied in any order.

George Lobell, Executive Editor for Advanced Mathematics at Prentice Hall, supervised the production of this book. He has been a wise guide, a kind collaborator, and a resourceful manager throughout the project. **Alex Ryba**, **Ruben Michel**, and **Jerry Waxman** have read preliminary versions of the text. Their reliable judgment and technical opinions have influenced this book in many aspects—from general design to points of objective accuracy. The predecessors of this book were a series of informal compilations of practice assignments and actual test questions, published within Queens College for a number of years. It is the many hundreds of students who have worked on "the booklet" with earnest attention that lend credence to this publication. Theirs is a great share of the credit for any accomplishment that this book may facilitate.

Bojana Obrenić
Queens, New York
November, 2002

Contents

Chapter 1

Concepts

1.1 Propositions and Predicates

Problem 1 Let variables p, q, and r denote propositions. Complete the following formulae by writing into each empty box a symbol representing a logical operator or a propositional variable (as appropriate), so that the resulting formula is a tautology (true for any assignment of truth values to its variables.)

Answer:

$(p \wedge q) \iff \left(q \boxed{\wedge} p \right)$

$p \vee (q \vee r) \iff \left(p \boxed{\vee} q \right) \boxed{\vee} r$

$p \boxed{\vee} \neg p$

$(p \wedge p) \boxed{\iff} p$

$(p \wedge p) \boxed{\vee} \boxed{\neg} p$

$\neg \left(p \boxed{\wedge} \neg p \right)$

$(p \wedge q) \iff \neg \left(\neg q \boxed{\vee} \neg p \right)$

$\left(p \boxed{\wedge} \neg p \right) \implies q$

$q \implies \left(p \boxed{\vee} \neg p \right)$

$(p \wedge (q \vee r)) \iff \left(\left(p \boxed{\wedge} q \right) \boxed{\vee} \left(p \boxed{\wedge} r \right) \right)$

$(\neg (p \implies q)) \iff \left(p \boxed{\wedge} \neg q \right)$

$(q \implies p) \iff \left(p \boxed{\vee} \neg q \right)$

$(p \iff q) \iff \left(\left(p \boxed{\implies} q \right) \wedge \left(q \boxed{\implies} p \right) \right)$

$(\neg (p \iff q)) \iff \left(\left(\neg p \wedge \boxed{q} \right) \boxed{\vee} \left(\boxed{p} \wedge \neg q \right) \right)$

$((p \wedge q) \implies r) \iff \neg \left(p \boxed{\wedge} q \boxed{\wedge} \neg r \right)$

Problem 2   Let variables p, q, and r denote propositions. Complete the following formulae by writing into each empty box a symbol representing a logical operator or a propositional variable (as appropriate), so that the resulting formula is a tautology (true for any assignment of truth values to its variables.)

Answer:

$(p \vee q) \iff \left(q \,\boxed{}\, p\right)$ $\qquad$ $(p \vee q) \iff \left(q \,\boxed{\vee}\, p\right)$

$p \wedge (q \wedge r) \iff \left(p \,\boxed{}\, q\right) \,\boxed{}\, r$ $\qquad$ $p \wedge (q \wedge r) \iff \left(p \,\boxed{\wedge}\, q\right) \,\boxed{\wedge}\, r$

$p \vee \neg \,\boxed{}$ $\qquad$ $p \vee \neg \,\boxed{p}$

$(p \vee p) \,\boxed{}\, p$ $\qquad$ $(p \vee p) \,\boxed{\iff}\, p$

$\neg \left(\boxed{} \iff \neg \,\boxed{}\right)$ $\qquad$ $\neg \left(\boxed{p} \iff \neg \,\boxed{p}\right)$

$\neg \left(p \,\boxed{}\, q\right) \iff \left(\neg p \,\boxed{}\, \neg q\right)$ $\qquad$ $\neg \left(p \,\boxed{\vee}\, q\right) \iff \left(\neg p \,\boxed{\wedge}\, \neg q\right)$

$\left(p \,\boxed{}\, \neg p\right) \vee q$ $\qquad$ $\left(p \,\boxed{\vee}\, \neg p\right) \vee q$

$\neg \left(q \wedge \left(p \,\boxed{}\, \neg p\right)\right)$ $\qquad$ $\neg \left(q \wedge \left(p \,\boxed{\wedge}\, \neg p\right)\right)$

$((p \vee q) \wedge (p \vee r)) \iff \left(p \,\boxed{}\, \left(q \,\boxed{}\, r\right)\right)$ $\quad$ $((p \vee q) \wedge (p \vee r)) \iff \left(p \,\boxed{\vee}\, \left(q \,\boxed{\wedge}\, r\right)\right)$

$(\neg p \wedge q) \iff \neg \left(q \,\boxed{}\, p\right)$ $\qquad$ $(\neg p \wedge q) \iff \neg \left(q \,\boxed{\implies}\, p\right)$

$(\neg p \vee q) \iff \left(p \,\boxed{}\, q\right)$ $\qquad$ $(\neg p \vee q) \iff \left(p \,\boxed{\implies}\, q\right)$

$\left((p \implies q) \,\boxed{}\, (q \implies p)\right) \iff \left(p \,\boxed{}\, q\right)$ $\quad$ $\left((p \implies q) \,\boxed{\wedge}\, (q \implies p)\right) \iff \left(p \,\boxed{\iff}\, q\right)$

$\left((\neg p \wedge q) \,\boxed{}\, (p \wedge \neg q)\right) \iff \neg \left(p \,\boxed{}\, q\right)$ $\quad$ $\left((\neg p \wedge q) \,\boxed{\vee}\, (p \wedge \neg q)\right) \iff \neg \left(p \,\boxed{\iff}\, q\right)$

$(p \implies (q \vee r)) \iff \left(\neg p \,\boxed{}\, q \,\boxed{}\, r\right)$ $\quad$ $(p \implies (q \vee r)) \iff \left(\neg p \,\boxed{\vee}\, q \,\boxed{\vee}\, r\right)$

Problem 3   Let variables p, q, and r denote propositions. Complete the following formulae by writing into each empty box a symbol representing a logical operator or a propositional variable (as appropriate), so that the resulting formula is a tautology (true for any assignment of truth values to its variables.)

Answer:

$\neg \left(p \,\boxed{}\, \neg p\right)$ $\qquad$ $\neg \left(p \,\boxed{\wedge}\, \neg p\right)$

$\left(p \,\boxed{}\, p\right) \vee q$ $\qquad$ $\left(p \,\boxed{\iff}\, p\right) \vee q$

$\left(p \wedge \left(q \,\boxed{}\, r\right)\right) \iff \left((p \,\boxed{}\, q) \wedge \,\boxed{}\right)$ $\quad$ $\left(p \wedge \left(q \,\boxed{\wedge}\, r\right)\right) \iff \left((p \,\boxed{\wedge}\, q) \wedge \,\boxed{r}\right)$

$\left((p \vee q) \,\boxed{}\, r\right) \iff \left((q \vee r) \,\boxed{}\, p\right)$ $\quad$ $\left((p \vee q) \,\boxed{\vee}\, r\right) \iff \left((q \vee r) \,\boxed{\vee}\, p\right)$

$\left(p \,\boxed{} \,\boxed{}\, p\right) \vee q$ $\qquad$ $\left(p \,\boxed{\vee} \,\boxed{\neg}\, p\right) \vee q$

$\left((p \vee q) \wedge \left(p \,\boxed{}\, r\right)\right) \iff \left(p \,\boxed{}\, \left(q \,\boxed{}\, r\right)\right)$ $\quad$ $\left((p \vee q) \wedge \left(p \,\boxed{\vee}\, r\right)\right) \iff \left(p \,\boxed{\vee}\, \left(q \,\boxed{\wedge}\, r\right)\right)$

$\neg (p \wedge \neg q) \iff \left(p \,\boxed{}\, q\right)$ $\qquad$ $\neg (p \wedge \neg q) \iff \left(p \,\boxed{\implies}\, q\right)$

Answer:

$$\boxed{}\;(p \wedge q) \Longleftrightarrow \left(\neg p \,\boxed{}\,\boxed{}\, q\right) \qquad \boxed{\neg}\;(p \wedge q) \Longleftrightarrow \left(\neg p \,\boxed{\vee}\,\boxed{\neg}\, q\right)$$

$$\neg\,(\neg p \vee \neg q) \Longleftrightarrow \left(\boxed{}\,\boxed{}\,\boxed{}\right) \qquad \neg\,(\neg p \vee \neg q) \Longleftrightarrow \left(\boxed{p}\,\boxed{\wedge}\,\boxed{q}\right)$$

$$(p \Longrightarrow (q \vee r)) \Longleftrightarrow \left(\left(p\,\boxed{}\,\boxed{}\,q\right)\boxed{}\,r\right) \qquad (p \Longrightarrow (q \vee r)) \Longleftrightarrow \left(\left(p\,\boxed{\wedge}\,\boxed{\neg}\,q\right)\boxed{\Longrightarrow}\,r\right)$$

Problem 4 For each of the following formulae, construct an equivalent propositional formula that contains no operators other than negation, conjunction, and disjunction, and no operands other than p and q.

	Answer:	**equivalently:**
$p \Longrightarrow q$	$\neg(p \wedge \neg q)$	$\neg p \vee q$
$p \Longleftrightarrow q$	$(\neg p \vee q) \wedge (\neg q \vee p)$	$(p \wedge q) \vee (\neg p \wedge \neg q)$
$\neg(p \Longrightarrow q)$	$p \wedge \neg q$	
$\neg(p \Longleftrightarrow q)$	$(p \wedge \neg q) \vee (q \wedge \neg p)$	$(\neg p \vee \neg q) \wedge (p \vee q)$
$p \Longrightarrow (p \Longrightarrow q)$	$\neg(p \wedge \neg q)$	$\neg p \vee q$
$\neg p \Longrightarrow (p \Longrightarrow q)$	$p \vee \neg p$	$q \vee \neg q$
$p \Longrightarrow (q \Longrightarrow p)$	$p \vee \neg p$	$q \vee \neg q$
$p \Longrightarrow (p \Longleftrightarrow q)$	$\neg p \vee q$	
$(p \Longleftrightarrow q) \Longrightarrow p$	$p \vee q$	
$(p \Longrightarrow q) \Longrightarrow p$	p	
$(p \Longrightarrow q) \Longrightarrow q$	$p \vee q$	

Problem 5 For each of the following formulae, construct an equivalent propositional formula that contains no operators other than negation and implication, and no operands other than p and q.

	Answer:	**equivalently:**
$p \wedge q$	$\neg(p \Longrightarrow \neg q)$	$\neg(q \Longrightarrow \neg p)$
$\neg(p \wedge q)$	$p \Longrightarrow \neg q$	$q \Longrightarrow \neg p$
$p \vee q$	$\neg p \Longrightarrow q$	$\neg q \Longrightarrow p$
$\neg p \vee q$	$p \Longrightarrow q$	$\neg q \Longrightarrow \neg p$
$\neg(p \vee q)$	$\neg(\neg p \Longrightarrow q)$	$\neg(\neg q \Longrightarrow p)$
$\neg p \wedge q$	$\neg(q \Longrightarrow p)$	$\neg(\neg p \Longrightarrow \neg q)$
$(p \wedge \neg q) \vee r$	$(p \Longrightarrow q) \Longrightarrow r$	
$\neg p \vee \neg q \vee r$	$p \Longrightarrow (q \Longrightarrow r)$	
$p \Longleftrightarrow q$	$\neg((q \Longrightarrow p) \Longrightarrow \neg(p \Longrightarrow q))$	$\neg((p \Longrightarrow q) \Longrightarrow \neg(q \Longrightarrow p))$
$\neg(p \Longleftrightarrow q)$	$(q \Longrightarrow p) \Longrightarrow \neg(p \Longrightarrow q)$	$(p \Longrightarrow q) \Longrightarrow \neg(q \Longrightarrow p)$
$(p \Longleftrightarrow q) \Longleftrightarrow q$	p	
$(p \Longrightarrow q) \Longleftrightarrow q$	$p \vee q$	

Problem 6   Consider the propositional operators $\longleftarrow$, $\oplus$, $\downarrow$, and $\uparrow$, defined by the following truth tables.

p	q	$p \longleftarrow q$	$p \oplus q$	$p \uparrow q$	$p \downarrow q$
1	1	1	0	0	0
1	0	1	1	1	0
0	1	0	1	1	0
0	0	1	0	1	1

(Variables p and q denote propositions.)

For each of the following formulae, construct an equivalent propositional formula that contains no operators other than negation, conjunction, and disjunction, and no operands other than p and q.

	Answer:	**equivalently:**
$p \longleftarrow q$	$\neg q \vee p$	
$\neg p \longleftarrow \neg q$	$q \vee \neg p$	
$p \oplus q$	$(p \wedge \neg q) \vee (q \wedge \neg p)$	$(\neg p \vee \neg q) \wedge (p \vee q)$
$p \oplus p$	$(p \wedge \neg p)$	
$\neg(p \oplus q)$	$(p \wedge q) \vee (\neg p \wedge \neg q)$	
$\neg(p \longleftarrow q)$	$q \wedge \neg p$	
$p \downarrow q$	$\neg(p \vee q)$	
$q \uparrow p$	$\neg(p \wedge q)$	
$p \downarrow p$	$\neg p$	
$p \uparrow p$	$\neg p$	
$(p \downarrow p) \downarrow (p \downarrow p)$	p	
$(p \uparrow p) \uparrow (p \uparrow p)$	p	
$\neg(p \uparrow q)$	$p \wedge q$	
$\neg(p \downarrow q)$	$p \vee q$	
$p \Longrightarrow (q \longleftarrow p)$	$\neg p \vee q$	
$(p \Longrightarrow q) \longleftarrow p$	$\neg p \vee q$	
$(q \Longrightarrow p) \longleftarrow p$	$\neg p \vee p$	$\neg q \vee q$
$(p \longleftarrow q) \oplus q$	$((\neg q \vee p) \wedge \neg q) \vee (q \wedge \neg(\neg q \vee p))$	$\neg p \vee \neg q$
$(p \oplus q) \longleftarrow p$	$\neg p \vee ((p \wedge \neg q) \vee (q \wedge \neg p))$	$\neg p \vee \neg q$
$q \Longleftrightarrow (p \uparrow q)$	$(q \wedge \neg(p \wedge q)) \vee (\neg q \wedge (p \wedge q))$	$\neg p \wedge q$
$p \Longrightarrow (p \downarrow q)$	$\neg p \vee \neg(p \vee q)$	
$(p \longleftarrow q) \wedge (q \longleftarrow p)$	$(p \wedge q) \vee (\neg p \wedge \neg q)$	

Problem 7   Consider the propositional operators $\longleftarrow$, $\oplus$, $\downarrow$, and $\uparrow$, defined in Problem 6.

For each of the following formulae, construct an equivalent propositional formula that contains no operators other than negation and implication, and no operands other than p and q.

	Answer:	**equivalently:**
$p \longleftarrow q$	$q \Longrightarrow p$	
$\neg p \longleftarrow \neg q$	$\neg q \Longrightarrow \neg p$	
$p \oplus q$	$(p \Longrightarrow q) \Longrightarrow \neg(q \Longrightarrow p)$	$(q \Longrightarrow p) \Longrightarrow \neg(p \Longrightarrow q)$

	Answer:	**equivalently:**
$p \oplus p$	$\neg(p \Longrightarrow p)$	
$\neg(p \oplus q)$	$\neg((p \Longrightarrow q) \Longrightarrow \neg(q \Longrightarrow p))$	$\neg((q \Longrightarrow p) \Longrightarrow \neg(p \Longrightarrow q))$
$\neg(p \longleftarrow q)$	$\neg(q \Longrightarrow p)$	
$p \downarrow q$	$\neg(\neg p \Longrightarrow q)$	$\neg(\neg q \Longrightarrow p)$
$q \uparrow p$	$p \Longrightarrow \neg q$	$q \Longrightarrow \neg p$
$p \downarrow p$	$\neg p$	
$p \uparrow p$	$\neg p$	
$\neg(p \uparrow q)$	$\neg(p \Longrightarrow \neg q)$	$\neg(q \Longrightarrow \neg p)$
$\neg(p \downarrow q)$	$\neg p \Longrightarrow q$	$\neg q \Longrightarrow p$
$p \Longrightarrow (q \longleftarrow p)$	$p \Longrightarrow (p \Longrightarrow q)$	$p \Longrightarrow q$
$(p \Longrightarrow q) \longleftarrow p$	$p \Longrightarrow (p \Longrightarrow q)$	$p \Longrightarrow q$
$(q \Longrightarrow p) \longleftarrow p$	$p \Longrightarrow (q \Longrightarrow p)$	$p \Longrightarrow p$
$(p \longleftarrow q) \oplus q$	$((q \Longrightarrow p) \Longrightarrow q) \Longrightarrow \neg(q \Longrightarrow (q \Longrightarrow p))$	$p \Longrightarrow \neg q$
$(p \oplus q) \longleftarrow p$	$p \Longrightarrow ((q \Longrightarrow p) \Longrightarrow \neg(p \Longrightarrow q))$	$p \Longrightarrow \neg q$
$q \Longleftrightarrow (p \uparrow q)$	$\neg((q \Longrightarrow (q \Longrightarrow \neg p)) \Longrightarrow \neg((q \Longrightarrow \neg p) \Longrightarrow q))$	$\neg(q \Longrightarrow p)$
$p \Longrightarrow (p \downarrow q)$	$p \Longrightarrow \neg(\neg p \Longrightarrow q)$	
$(p \longleftarrow q) \wedge (q \longleftarrow p)$	$\neg((q \Longrightarrow p) \Longrightarrow \neg(p \Longrightarrow q))$	$\neg((p \Longrightarrow q) \Longrightarrow \neg(q \Longrightarrow p))$

Problem 8   Consider the propositional operators $\longleftarrow$, $\oplus$, $\downarrow$, and $\uparrow$, defined in Problem 6.

For each of the following formulae, construct an equivalent propositional formula that contains no operators other than negation and conjunction, and no operands other than p and q.

	Answer:
$p \vee q$	$\neg(\neg p \wedge \neg q)$
$\neg p \vee q$	$\neg(p \wedge \neg q)$
$\neg(p \vee q)$	$\neg p \wedge \neg q$
$p \vee \neg p$	$\neg(p \wedge \neg p)$
$\neg(p \Longrightarrow \neg q)$	$p \wedge q$
$p \Longleftrightarrow q$	$\neg(p \wedge \neg q) \wedge \neg(q \wedge \neg p)$
$\neg(p \Longleftrightarrow q)$	$\neg(p \wedge q) \wedge \neg(\neg p \wedge \neg q)$
$p \Longrightarrow (q \Longrightarrow p)$	$\neg(p \wedge \neg p)$
$p \Longrightarrow (p \Longleftrightarrow q)$	$\neg(p \wedge \neg q)$
$(p \Longleftrightarrow q) \Longrightarrow p$	$\neg(\neg p \wedge \neg q)$
$p \longleftarrow q$	$\neg(q \wedge \neg p)$
$\neg p \longleftarrow \neg q$	$\neg(\neg q \wedge p)$
$p \oplus q$	$\neg(p \wedge q) \wedge \neg(\neg p \wedge \neg q)$
$\neg(p \oplus q)$	$\neg(\neg(p \wedge q) \wedge \neg(\neg p \wedge \neg q))$
$p \downarrow q$	$\neg p \wedge \neg q$
$\neg(p \downarrow q)$	$\neg(\neg p \wedge \neg q)$
$\neg(p \uparrow q)$	$p \wedge q$

Answer:

$$p \implies (q \longleftarrow p) \quad \neg(p \wedge \neg q)$$
$$(p \longleftarrow q) \oplus q \quad \neg(p \wedge q)$$
$$p \implies (p \downarrow q) \quad \neg(p \wedge \neg(\neg p \wedge \neg q))$$

Problem 9   Consider the propositional operators $\longleftarrow$, $\oplus$, $\downarrow$, and $\uparrow$, defined in Problem 6.

For each of the following formulae, construct an equivalent propositional formula that contains no operators other than negation and disjunction, and no operands other than p and q.

Answer:

$p \wedge q$	$\neg(\neg p \vee \neg q)$
$p \wedge \neg p$	$\neg(p \vee \neg p)$
$\neg(p \wedge q)$	$\neg p \vee \neg q$
$p \implies q$	$\neg p \vee q$
$\neg p \implies q$	$p \vee q$
$\neg(\neg p \implies q)$	$\neg(p \vee q)$
$p \iff q$	$\neg(\neg p \vee \neg q) \vee \neg(p \vee q)$
$\neg(p \iff q)$	$\neg(\neg p \vee q) \vee \neg(\neg q \vee p)$
$p \oplus q$	$\neg(\neg p \vee q) \vee \neg(\neg q \vee p)$
$\neg(p \oplus q)$	$\neg(\neg p \vee \neg q) \vee \neg(p \vee q)$
$\neg(p \longleftarrow q)$	$\neg(\neg q \vee p)$
$p \uparrow q$	$\neg p \vee \neg q$
$\neg(p \uparrow q)$	$\neg(\neg p \vee \neg q)$
$\neg(p \downarrow q)$	$p \vee q$

Problem 10   Consider the propositional operators $\longleftarrow$, $\oplus$, $\downarrow$, and $\uparrow$, defined in Problem 6.

For each of the following formulae, construct an equivalent propositional formula that contains no operators other than $\downarrow$, and no operands other than p and q.

Answer:

$\neg p$	$p \downarrow p$
$p \wedge q$	$(p \downarrow p) \downarrow (q \downarrow q)$
$p \vee q$	$(p \downarrow q) \downarrow (p \downarrow q)$
$p \wedge \neg p$	$(p \downarrow p) \downarrow p$
$p \vee \neg p$	$(p \downarrow (p \downarrow p)) \downarrow (p \downarrow (p \downarrow p))$
$p \implies q$	$((p \downarrow p) \downarrow q) \downarrow ((p \downarrow p) \downarrow q)$
$\neg(p \implies q)$	$(p \downarrow p) \downarrow q$
$\neg(\neg p \implies q)$	$p \downarrow q$
$\neg p \wedge \neg q$	$p \downarrow q$
$p \iff q$	$((p \downarrow q) \downarrow ((p \downarrow p) \downarrow (q \downarrow q))) \downarrow ((p \downarrow q) \downarrow ((p \downarrow p) \downarrow (q \downarrow q)))$
$\neg(p \iff q)$	$(p \downarrow q) \downarrow ((p \downarrow p) \downarrow (q \downarrow q))$
$\neg(p \vee \neg q)$	$p \downarrow (q \downarrow q)$

Answer:

$$p \Longrightarrow \neg q \qquad ((p \downarrow p) \downarrow (q \downarrow q)) \downarrow ((p \downarrow p) \downarrow (q \downarrow q))$$

$$(p \wedge q) \vee \neg q \qquad ((q \downarrow q) \downarrow p) \downarrow ((q \downarrow q) \downarrow p)$$

$$p \longleftarrow q \qquad ((q \downarrow q) \downarrow p) \downarrow ((q \downarrow q) \downarrow p)$$

$$p \uparrow q \qquad ((p \downarrow p) \downarrow (q \downarrow q)) \downarrow ((p \downarrow p) \downarrow (q \downarrow q))$$

$$p \oplus q \qquad (p \downarrow q) \downarrow ((p \downarrow p) \downarrow (q \downarrow q))$$

Problem 11   Consider the propositional operators $\longleftarrow$, $\oplus$, $\downarrow$, and $\uparrow$, defined in Problem 6.

For each of the following formulae, construct an equivalent propositional formula that contains no operators other than $\uparrow$, and no operands other than p and q.

Answer:

$$\neg p \qquad p \uparrow p$$

$$p \wedge q \qquad (p \uparrow q) \uparrow (p \uparrow q)$$

$$p \vee q \qquad (p \uparrow p) \uparrow (q \uparrow q)$$

$$p \wedge \neg p \qquad (p \uparrow (p \uparrow p)) \uparrow (p \uparrow (p \uparrow p))$$

$$p \vee \neg p \qquad (p \uparrow p) \uparrow p$$

$$p \Longrightarrow q \qquad p \uparrow (q \uparrow q)$$

$$\neg(p \Longrightarrow q) \qquad (p \uparrow (q \uparrow q)) \uparrow (p \uparrow (q \uparrow q))$$

$$\neg(\neg p \Longrightarrow q) \qquad ((p \uparrow p) \uparrow (q \uparrow q)) \uparrow ((p \uparrow p) \uparrow (q \uparrow q))$$

$$\neg p \vee \neg q \qquad p \uparrow q$$

$$p \Longleftrightarrow q \qquad ((p \uparrow (q \uparrow q)) \uparrow (q \uparrow (p \uparrow p))) \uparrow ((p \uparrow (q \uparrow q)) \uparrow (q \uparrow (p \uparrow p)))$$

$$\neg(p \Longleftrightarrow q) \qquad (p \uparrow (q \uparrow q)) \uparrow (q \uparrow (p \uparrow p))$$

$$\neg(p \vee \neg q) \qquad ((p \uparrow p) \uparrow q) \uparrow ((p \uparrow p) \uparrow q)$$

$$p \Longrightarrow \neg q \qquad p \uparrow q$$

$$p \longleftarrow q \qquad q \uparrow (p \uparrow p)$$

$$p \downarrow q \qquad ((p \uparrow p) \uparrow (q \uparrow q)) \uparrow ((p \uparrow p) \uparrow (q \uparrow q))$$

$$p \oplus q \qquad (p \uparrow (q \uparrow q)) \uparrow (q \uparrow (p \uparrow p))$$

Problem 12   Consider the propositional operators $\longleftarrow$, $\oplus$, $\downarrow$, and $\uparrow$, defined in Problem 6.

For each of the following formulae, construct an equivalent propositional formula that contains no operators other than negation and $\longleftarrow$, and no operands other than propositional variables x, y, z, w.

Answer:

$$x \wedge y \wedge z \qquad \neg(\neg z \longleftarrow (\neg(\neg y \longleftarrow x)))$$

$$x \vee y \vee z \qquad z \longleftarrow \neg(y \longleftarrow \neg x)$$

$$(x \wedge y) \vee z \qquad z \longleftarrow (\neg y \longleftarrow x)$$

$$x \wedge (y \vee z) \qquad \neg(\neg x \longleftarrow (y \longleftarrow \neg z))$$

$$(x \wedge y) \vee (z \wedge w) \qquad (\neg(\neg y \longleftarrow x)) \longleftarrow (\neg z \longleftarrow w)$$

$$(x \vee y) \wedge (z \vee w) \qquad \neg(\neg(x \longleftarrow \neg y) \longleftarrow (z \longleftarrow \neg w))$$

Answer:

$(x \Longrightarrow y) \vee (z \Longrightarrow y)$	$(y \longleftarrow z) \longleftarrow \neg(y \longleftarrow x)$
$(x \Longrightarrow y) \wedge (z \Longleftrightarrow w)$	$\neg(\neg(y \longleftarrow x) \longleftarrow \neg(\neg(z \longleftarrow w) \longleftarrow (w \longleftarrow z)))$
$(x \downarrow y) \vee (x \uparrow y)$	$(\neg x \longleftarrow y) \longleftarrow (y \longleftarrow \neg x)$
$(x \downarrow y) \wedge (x \uparrow y)$	$\neg(\neg(\neg x \longleftarrow y) \longleftarrow \neg(x \longleftarrow \neg y))$
$x \oplus y$	$\neg(x \longleftarrow y) \longleftarrow (y \longleftarrow x)$
$(x \oplus y) \oplus z$	$\neg((\neg(x \longleftarrow y) \longleftarrow (y \longleftarrow x)) \longleftarrow z) \longleftarrow (z \longleftarrow (\neg(x \longleftarrow y) \longleftarrow (y \longleftarrow x)))$

Problem 13   Propositional functions P, Q, R, S are defined by the following truth tables.

a	b	$P(a,b)$	$Q(a,b)$	$R(a,b)$	$S(a,b)$
0	0	0	1	0	1
0	1	1	1	0	0
1	0	0	1	1	1
1	1	0	0	0	0

(The truth values of false and true are represented by 0 and 1, respectively.)

(a) Represent $P(a,b)$, $Q(a,b)$, $R(a,b)$, and $S(a,b)$ in the disjunctive normal form.

Answer:

$$P(a,b) = \neg a \wedge b$$
$$Q(a,b) = (\neg a \wedge \neg b) \vee (\neg a \wedge b) \vee (a \wedge \neg b)$$
$$R(a,b) = a \wedge \neg b$$
$$S(a,b) = (\neg a \wedge \neg b) \vee (a \wedge \neg b)$$

(b) Represent $P(a,b)$, $Q(a,b)$, $R(a,b)$, and $S(a,b)$ in the conjunctive normal form.

Answer:

$$P(a,b) = (a \vee b) \wedge (\neg a \vee b) \wedge (\neg a \vee \neg b)$$
$$Q(a,b) = (\neg a \vee \neg b)$$
$$R(a,b) = (a \vee b) \wedge (a \vee \neg b) \wedge (\neg a \vee \neg b)$$
$$S(a,b) = (a \vee \neg b) \wedge (\neg a \vee \neg b)$$

(c) Represent $P(a,b)$, $Q(a,b)$, $R(a,b)$, and $S(a,b)$ as propositional formulae that contain no operators other than negation and disjunction, and no operands other than a and b.

Answer:

$$P(a,b) = \neg(a \vee \neg b)$$
$$Q(a,b) = (\neg a \vee \neg b)$$
$$R(a,b) = \neg(\neg a \vee b)$$
$$S(a,b) = \neg(a \vee b) \vee \neg(\neg a \vee b)$$

(d) Represent $P(a,b)$, $Q(a,b)$, $R(a,b)$, and $S(a,b)$ as propositional formulae that contain no operators other than negation and conjunction, and no operands other than a and b.

Answer:

$$P(a,b) = \neg a \wedge b$$
$$Q(a,b) = \neg(a \wedge b)$$
$$R(a,b) = a \wedge \neg b$$
$$S(a,b) = \neg(\neg a \wedge b) \wedge \neg(a \wedge b)$$

Problem 14   Propositional functions P and Q are defined by the following truth tables.

a	b	c	$P(a,b,c)$	$Q(a,b,c)$
0	0	0	1	0
0	0	1	1	1
0	1	0	0	0
0	1	1	1	0
1	0	0	1	0
1	0	1	0	0
1	1	0	1	1
1	1	1	0	0

(a) Represent $P(a,b,c)$ as a propositional formula that contains no operators other than negation, conjunction and disjunction, and no operands other than a, b, and c.

Answer: In the truth table of P, those rows in which P assumes the value of 0 are fewer than those in which P assumes the value of 1. Hence, the conjunctive normal form is more convenient than the disjunctive normal form.

$$P(a,b,c) = (a \vee \neg b \vee c) \wedge (\neg a \vee b \vee \neg c) \wedge (\neg a \vee \neg b \vee \neg c)$$

(b) Represent $Q(a,b,c)$ as a propositional formula that contains no operators other than negation, conjunction and disjunction, and no operands other than a, b, and c.

Answer: In the truth table of Q, those rows in which Q assumes the value of 1 are fewer than those in which Q assumes the value of 0. Hence, the disjunctive normal form is more convenient than the conjunctive normal form.

$$Q(a,b,c) = (\neg a \wedge \neg b \wedge c) \vee (a \wedge b \wedge \neg c)$$

Problem 15   For each of the following properties, represent a propositional function $f(a,b,c,d)$ that has that property in the form of a propositional formula that contains no operators other than negation, conjunction and disjunction, and no operands other than a,b,c,d.

(a) The value of a propositional function $f(a,b,c,d)$ is equal to 1 exactly when all those arguments (if any) that have a value of 0 precede in the alphabetical order all those arguments (if any) that have a value of 1.

Answer:
$$f(a,b,c,d) = abcd + \overline{a}bcd + \overline{a}\overline{b}cd + \overline{a}\overline{b}\overline{c}d + \overline{a}\overline{b}\overline{c}\overline{d}$$

(where $xy \equiv x \wedge y$, $\quad x + y \equiv x \vee y$, $\quad$ and $\overline{x} \equiv \neg x$, for any x and y.)

(b) The value of a propositional function $f(a,b,c,d)$ is equal to 0 if and only if the number of those arguments that have a value of 1 is even but no two arguments adjacent in the alphabetical order have a value of 1.

Answer:

$$f(a,b,c,d) = (a + b + c + d)(\overline{a} + b + \overline{c} + d)(\overline{a} + b + c + \overline{d})(a + \overline{b} + c + \overline{d})$$

(where $xy \equiv x \wedge y$, $\quad x + y \equiv x \vee y$, $\quad$ and $\overline{x} \equiv \neg x$, for any x and y.)

Problem 16   For each of the following properties, represent a propositional function $f(a, b, c, d, e)$ that has that property in the form of a propositional formula that contains no operators other than negation, conjunction and disjunction, and no operands other than a, b, c, d, e.

(a) The value of a propositional function $f(a, b, c, d, e)$ is equal to 1 exactly when no more than two of its arguments have a value of 1, provided that those arguments that have a value of 1 are adjacent in the alphabetical order.

Answer:

$$f(a, b, c, d, e) = \overline{a}\overline{b}\overline{c}\overline{d}\overline{e} + a\overline{b}\overline{c}\overline{d}\overline{e} + \overline{a}b\overline{c}\overline{d}\overline{e} + \overline{a}\overline{b}c\overline{d}\overline{e} + \overline{a}\overline{b}\overline{c}d\overline{e} + \overline{a}\overline{b}\overline{c}\overline{d}e + ab\overline{c}\overline{d}\overline{e} + \overline{a}bc\overline{d}\overline{e} + \overline{a}\overline{b}cd\overline{e} + \overline{a}\overline{b}\overline{c}de$$

(where $xy \equiv x \wedge y$, $x + y \equiv x \vee y$, and $\overline{x} \equiv \neg x$, for any x and y.)

(b) The value of a propositional function $f(a, b, c, d, e)$ is equal to 0 exactly when no more than one of its arguments has a value diffferent from the value assumed by the majority of the arguments.

Answer:

$$
\begin{aligned}
f(a, b, c, d, e) = \\
(a + b + c + d + e) \\
\cdot \ (\overline{a} + b + c + d + e)(a + \overline{b} + c + d + e)(a + b + \overline{c} + d + e)(a + b + c + \overline{d} + e)(a + b + c + d + \overline{e}) \\
\cdot \ (\overline{a} + \overline{b} + \overline{c} + \overline{d} + \overline{e}) \\
\cdot \ (a + \overline{b} + \overline{c} + \overline{d} + \overline{e})(\overline{a} + b + \overline{c} + \overline{d} + \overline{e})(\overline{a} + \overline{b} + c + \overline{d} + \overline{e})(\overline{a} + \overline{b} + \overline{c} + d + \overline{e})(\overline{a} + \overline{b} + \overline{c} + \overline{d} + e)
\end{aligned}
$$

(where $xy \equiv x \cdot y \equiv x \wedge y$, $x + y \equiv x \vee y$, and $\overline{x} \equiv \neg x$, for any x and y.)

Problem 17   Let $P(m, n)$ be a predicate which is true exactly when m divides n, where the domain of both variables is the set of positive integers. Determine the truth value of each of the following propositions, and justify your answer briefly.

(a) $P(4, 5)$
    Answer: False.
    5 is not divisible by 4.

(b) $P(2, 4)$
    Answer: True.
    4 is divisible by 2.

(c) $(\forall m)(\forall n)(P(m, n))$
    Answer: False.
    For instance, 3 is not divisible by 2.

(d) $(\exists m)(\forall n)(P(m, n))$
    Answer: True.
    Let $m = 1$, and observe that every integer is divisible by 1.

(e) $(\forall m)(\exists n)(P(m, n))$
    Answer: True.
    For instance, let $n = 2m$.

(f) $(\exists m)(\exists n)(P(m, n))$
    Answer: True.
    For instance, let $n = m = 17$.

(g) $(\exists n)(\forall m)(P(m, n))$
    Answer: False.
    For instance, let $m = n + 1$, and observe that a positive integer cannot be divisible by a greater number.

(h) $(\forall n)(\exists m)(P(m, n))$
    Answer: True.
    For instance, let $m = n$.

(i) $(\forall n)(P(1, n))$

Answer: True.
Every integer is divisible by 1.

(j) $(\exists m)(\exists k)(m \neq k \wedge P(m, 1) \wedge P(k, 1))$

Answer: False.
Number 1 cannot have two different divisors, since it is divisible only by itself.

Problem 18   Sets S, N^+, Z, Q are defined as follows.

$$S = \{0, 1, 2, 3, 4, 5\};$$
$$N^+ = \{1, 2, \ldots\}, \quad \text{(positive integers)} ;$$
$$Z = \{\ldots, -2, -1, 0, 1, 2, \ldots\}, \quad \text{(integers)} ;$$
$$Q = \{(p/q) \mid p \in Z \wedge q \in N^+\}, \quad \text{(rational numbers)}$$

Interpret each of the following formulae with each of these four sets as the domain of the variables x, y, z, and determine if the resulting statement is true or false. For each false statement, expose a counterexample. For each true statement, determine the value(s) of variable(s) bound to existential quantifier(s) for which the statement is true.

Answer:

(a) $(\forall x)(\exists y)(y \geq x)$

True in any of S, N^+, Z, Q; let $y = x$.

(b) $(\forall x)(\exists y)(y \leq x)$

True in any of S, N^+, Z, Q; let $y = x$.

(c) $(\forall x)(\exists y)(y > x)$

False in S; let $x = 5$.
True in any of N^+, Z, Q; let $y = x + 1$.

(d) $(\forall x)(\exists y)(y < x)$

False in S; let $x = 0$.
False in N^+; let $x = 1$.
True in any of Z, Q; let $y = x - 1$.

(e) $(\forall x)(\exists y)(y \neq x \wedge y \geq x)$

False in S; let $x = 5$.
True in any of N^+, Z, Q; let $y = x + 1$.

(f) $(\forall x)(\exists y)(y \neq x \wedge y \leq x)$

False in S; let $x = 0$.
False in N^+; let $x = 1$.
True in any of Z, Q; let $y = x - 1$.

(g) $(\forall x)(\exists y)(y \neq x \wedge y > x)$

False in S; let $x = 5$.
True in any of N^+, Z, Q; let $y = x + 1$.

(h) $(\forall x)(\exists y)(y \neq x \wedge y < x)$

False in S; let $x = 0$.
False in N^+; let $x = 1$.
True in any of Z, Q; let $y = x - 1$.

(i) $(\forall x)(\forall y)(\exists z)(x < z < y)$

False in any of S, N^+, Z, Q; let $x = y$.

(j) $(\forall x)(\forall y)(x < y \implies (\exists z)(x < z < y))$

False in any of S, N^+, Z; let $y = x + 1$.
True in Q; let $z = (x + y)/2$.

(k) $(\forall x)(\forall y)(x < y \implies (\exists z)(x < z \leq y))$

True in any of S, N^+, Z, Q; let $z = y$.

(l) $(\forall x)(\forall y)(x \neq y \implies (\exists z)(x \leq z < y))$

False in any of S, N^+, Z, Q; let $x > y$.

Problem 19　　Complete the following formulae by writing into each empty box either the universal quantifier symbol ($\forall$) or the existential quantifier symbol ($\exists$), as appropriate, so as to obtain a true claim. Explain your answer briefly. (All variables assume values from the set of positive integers $N^+ = \{1, 2, \ldots\}$.)

Answer:

(a)　$(\forall x) \left(\boxed{} y \right) \left(\boxed{} z \right) (x < y \land x < z)$　　　$(\forall x) \left(\boxed{\exists} y \right) \left(\boxed{\exists} z \right) (x < y \land x < z)$

(b)　$\lnot \left((\forall x) \left(\boxed{} y \right) \left(\boxed{} z \right) (x < y \land x < z) \right)$　　$\lnot \left((\forall x) \left(\boxed{\forall} y \right) \left(\boxed{\forall} z \right) (x < y \land x < z) \right)$

(c)　$\left(\boxed{} x \right) (\forall y)(x \ne y \Longrightarrow x < y)$　　　$\left(\boxed{\exists} x \right) (\forall y)(x \ne y \Longrightarrow x < y)$

(d)　$\lnot \left(\left(\boxed{} x \right) (\forall y)(x \ne y \Longrightarrow x < y) \right)$　　$\lnot \left(\left(\boxed{\forall} x \right) (\forall y)(x \ne y \Longrightarrow x < y) \right)$

(e)　$(\forall x) \left(\boxed{} y \right) \left(\boxed{} z \right) (x < z < y)$　　　$(\forall x) \left(\boxed{\exists} y \right) \left(\boxed{\exists} z \right) (x < z < y)$

(f)　$\lnot \left((\forall x) \left(\boxed{} y \right) \left(\boxed{} z \right) (x < z < y) \right)$　　$\lnot \left((\forall x) \left(\boxed{\forall} y \right) \left(\boxed{\exists} z \right) (x < z < y) \right)$

(g)　$(\exists z) \left(\boxed{} y \right) \left(\boxed{} x \right) (x < y \Longrightarrow x < z)$　　$(\exists z) \left(\boxed{\exists} y \right) \left(\boxed{\forall} x \right) (x < y \Longrightarrow x < z)$

(h)　$\lnot \left((\exists z) \left(\boxed{} y \right) \left(\boxed{} x \right) (x < y \Longrightarrow x < z) \right)$　$\lnot \left((\exists z) \left(\boxed{\forall} y \right) \left(\boxed{\forall} x \right) (x < y \Longrightarrow x < z) \right)$

(i)　$(\forall z) \left(\boxed{} y \right) \left(\boxed{} x \right) (z < y + x)$　　$(\forall z) \left(\boxed{\forall} y \right) \left(\boxed{\exists} x \right) (z < y + x)$

(j)　$\lnot \left((\forall z) \left(\boxed{} y \right) \left(\boxed{} x \right) (z < y + x) \right)$　　$\lnot \left((\forall z) \left(\boxed{\forall} y \right) \left(\boxed{\forall} x \right) (z < y + x) \right)$

(k)　$\lnot \left(\left(\boxed{} z \right) \left(\boxed{} y \right) \left(\boxed{} x \right) (x + y + z < 3) \right)$　$\lnot \left(\left(\boxed{\exists} z \right) \left(\boxed{\exists} y \right) \left(\boxed{\exists} x \right) (x + y + z < 3) \right)$

(l)　$(\exists x)(\forall y) \left(\boxed{} z \right) (\forall w)(y - z \le x \le w)$　　$(\exists x)(\forall y) \left(\boxed{\exists} z \right) (\forall w)(y - z \le x \le w)$

For an explanation of the answer, consider the following:

(a)　Let $y = x + 1$ and $z = x + 1$.
(b)　Let $x \ge y$ or $x \ge z$.
(c)　Let $x = 1$.
(d)　Let $x > y$.
(e)　Let $y = x + 2$ and $z = x + 1$.
(f)　Let $y \le x + 1$.
(g)　Let y be arbitrary and $z > y$.
(h)　Let $x \ge z \land x < y$.
(i)　Let $x = z + 1$.
(j)　Let $z = x + y$.
(k)　Let $x \ge 1, y \ge 1, z \ge 1$.
(l)　Let $x = 1, z = y$.

Problem 20 Sets A, B, C, and D are subsets of the set $N^+ = \{1, 2, \ldots\}$ of positive integers, defined as follows.

$$A = \{1, 2, 3, 4, 5, 6, 7\}$$
$$B = \{x \mid x \text{ is odd }\}$$
$$C = \{x \mid (x \in A \wedge x \notin B) \vee (x \in B \wedge x \notin A)\}$$
$$D = \{x \mid (x \in A \wedge x \in B)\}$$

(All variables assume values from the set of positive integers $N^+ = \{1, 2, \ldots\}$.)

Let $P(x)$ and $Q(x)$ be two predicates whose domain is the set $N^+ = \{1, 2, \ldots\}$ of positive integers, such that $P(x)$ is true if and only if $x \in A$, and $Q(x)$ is true if and only if $x \in B$.

Complete the following formulae, by filling in each empty box a quantifier symbol, a logical operator, a variable, or a constant (as appropriate), so as to obtain a true claim.

Answer:

$P(5) \wedge Q\left(\boxed{}\right)$ $\qquad$ $P(5) \wedge Q\left(\boxed{1}\right)$

$P(15) \vee Q\left(\boxed{}\right)$ $\qquad$ $P(15) \vee Q\left(\boxed{1}\right)$

$P(15) \Longleftrightarrow Q\left(\boxed{}\right)$ $\qquad$ $P(15) \Longleftrightarrow Q\left(\boxed{2}\right)$

$P(2) \Longrightarrow Q\left(\boxed{}\right)$ $\qquad$ $P(2) \Longrightarrow Q\left(\boxed{1}\right)$

$P\left(\boxed{}\right) \Longrightarrow Q\left(2 \cdot \boxed{}\right)$ $\qquad$ $P\left(\boxed{12}\right) \Longrightarrow Q\left(2 \cdot \boxed{1}\right)$

$Q(2) \Longrightarrow P\left(\boxed{}\right)$ $\qquad$ $Q(2) \Longrightarrow P\left(\boxed{7}\right)$

$Q(2) \Longleftrightarrow P\left(\boxed{}\right)$ $\qquad$ $Q(2) \Longleftrightarrow P\left(\boxed{17}\right)$

$\left(\boxed{} x\right)(P(x) \Longrightarrow Q(x))$ $\qquad$ $\left(\boxed{\exists} x\right)(P(x) \Longrightarrow Q(x))$

$(x = 11 \wedge P(x)) \boxed{} Q(x)$ $\qquad$ $(x = 11 \wedge P(x)) \boxed{\Longrightarrow} Q(x)$

$(x = 3 \wedge P(x)) \boxed{} Q\left(\boxed{}\right)$ $\qquad$ $(x = 3 \wedge P(x)) \boxed{\Longleftrightarrow} Q\left(\boxed{x}\right)$

$(\forall x)\left(\boxed{} y\right)\left(x < y \wedge Q\left(\boxed{}\right)\right)$ $\qquad$ $(\forall x)\left(\boxed{\exists} y\right)\left(x < y \wedge Q\left(\boxed{y}\right)\right)$

$(\forall x)\left((x \in D) \Longrightarrow \left(P(x) \boxed{} Q(x)\right)\right)$ $\qquad$ $(\forall x)\left((x \in D) \Longrightarrow \left(P(x) \boxed{\wedge} Q(x)\right)\right)$

$(\forall x)\left((x \in C) \Longrightarrow \left(P(x) \boxed{} Q(x)\right)\right)$ $\qquad$ $(\forall x)\left((x \in C) \Longrightarrow \left(P(x) \boxed{\vee} Q(x)\right)\right)$

Problem 21 Sets A and B are subsets of the set $N^+ = \{1, 2, \ldots\}$ of positive integers, defined as follows.

$$A = \{1, 2, 3, 4, 5, 6, 7, 8, 9\}$$
$$B = \{x \mid x \text{ is even }\}$$

(All variables assume values from the set of positive integers $N^+ = \{1, 2, \ldots\}$.)

Let $P(x)$ and $Q(x)$ be two predicates whose domain is the set $N^+ = \{1, 2, \ldots\}$ of positive integers, such that $P(x)$ is true if and only if $x \in A$, and $Q(x)$ is true if and only if $x \in B$.

Complete the following formulae, by filling in each empty box a quantifier symbol, a logical operator, a variable, or a constant (as appropriate), so as to obtain a true claim.

Answer:

$(\forall x)\left(Q\left(2\cdot\boxed{}\right)\right)$ $(\forall x)\left(Q\left(2\cdot\boxed{x}\right)\right)$

$P(15)\Longleftrightarrow Q\left(\boxed{}\right)$ $P(15)\Longleftrightarrow Q\left(\boxed{1}\right)$

$P(9)\Longrightarrow Q\left(\boxed{}\right)$ $P(9)\Longrightarrow Q\left(\boxed{2}\right)$

$(\forall x)\left(\left(x\in B\ \boxed{}\ x\in A\right)\Longrightarrow(P(x)\vee Q(x))\right)$ $(\forall x)\left(\left(x\in B\ \boxed{\vee}\ x\in A\right)\Longrightarrow(P(x)\vee Q(x))\right)$

$(\forall x)\left(\left(x\in B\ \boxed{}\ x\in A\right)\Longrightarrow(P(x)\wedge Q(x))\right)$ $(\forall x)\left(\left(x\in B\ \boxed{\wedge}\ x\in A\right)\Longrightarrow(P(x)\wedge Q(x))\right)$

$(\forall x)\left(\left(x\in B\ \boxed{}\ x\ \boxed{}\ A\right)\Longrightarrow(\neg P(x)\wedge Q(x))\right)$ $(\forall x)\left(\left(x\in B\ \boxed{\wedge}\ x\ \boxed{\notin}\ A\right)\Longrightarrow(\neg P(x)\wedge Q(x))\right)$

$(x=12\wedge P(x))\Longleftrightarrow Q\left(\boxed{}\right)$ $(x=12\wedge P(x))\Longleftrightarrow Q\left(\boxed{1}\right)$

$(x=7\wedge P(x))\Longrightarrow Q\left(\boxed{}\right)$ $(x=7\wedge P(x))\Longrightarrow Q\left(\boxed{2}\right)$

$(x=14\wedge P(x))\vee\neg Q\left(\boxed{}\right)$ $(x=14\wedge P(x))\vee\neg Q\left(\boxed{1}\right)$

$(\forall x)(\forall y)\left((x>y\wedge P(x))\Longrightarrow y\leq\boxed{}\right)$ $(\forall x)(\forall y)\left((x>y\wedge P(x))\Longrightarrow y\leq\boxed{8}\right)$

Problem 22 X, Y, and Z are subsets of the set $N^{+}=\{1,2,\ldots\}$ of positive integers, defined as follows.

$$X=\{1,2,3,4,5\}$$
$$Y=\{y\mid y<10\}$$
$$Z=\{z\mid(\exists n)(z=2n)\}$$

(All variables denote positive integers.)

Let variables x,y,z assume values from the sets $\{X,Y,Z\}$, respectively.

Complete the following formulae by writing into each empty box a symbol representing a quantifier, an operator, a variable, or a constant (as appropriate), so as to obtain a true claim.

Answer:

$(\exists x)(\exists z)\left(x>z+\boxed{}\right)$ $(\exists x)(\exists z)\left(x>z+\boxed{1}\right)$

$(\exists x)(\forall z)\left(x<z+\boxed{}\right)$ $(\exists x)(\forall z)\left(x<z+\boxed{1}\right)$

$(\forall x)(\exists z)\left(x+\boxed{}>z\right)$ $(\forall x)(\exists z)\left(x+\boxed{2}>z\right)$

$(\forall x)(\forall z)\left(x<z+\boxed{}\right)$ $(\forall x)(\forall z)\left(x<z+\boxed{4}\right)$

$\neg(\forall z)\left(z+1<\boxed{}\right)$ $\neg(\forall z)\left(z+1<\boxed{100}\right)$

$\neg\left(\boxed{}y\right)\left(y>\boxed{}\right)$ $\neg\left(\boxed{\forall}y\right)\left(y>\boxed{5}\right)$

$\left(\boxed{}z\right)(\forall x)\left(z\ \boxed{}\ x\right)$ $\left(\boxed{\exists}z\right)(\forall x)\left(z\ \boxed{>}\ x\right)$

$(\forall x)(\forall y)\left(\boxed{}z\right)\left(x+y\ \boxed{}\ z-1\right)$ $(\forall x)(\forall y)\left(\boxed{\exists}z\right)\left(x+y\ \boxed{>}\ z-1\right)$

Answer:

$$\left(\boxed{}\,x\right)\left(\boxed{}\,y\right)\left(\boxed{}\,z\right)\left(x\,\boxed{}\,y\,\boxed{}\,z\right) \qquad \left(\boxed{\exists}\,x\right)\left(\boxed{\exists}\,y\right)\left(\boxed{\exists}\,z\right)\left(x\,\boxed{=}\,y\,\boxed{=}\,z\right)$$

$$\neg\left(\boxed{}\,x\right)\left(\boxed{}\,z\right)\left(x+z>\boxed{}\right) \qquad \neg\left(\boxed{\forall}\,x\right)\left(\boxed{\forall}\,z\right)\left(x+z>\boxed{19}\right)$$

$$\neg\left(\boxed{}\,x\right)\left(\boxed{}\,z\right)\left(x+z<\boxed{}\right) \qquad \neg\left(\boxed{\exists}\,x\right)\left(\boxed{\exists}\,z\right)\left(x+z<\boxed{3}\right)$$

$$(\exists y)\left(\boxed{}\,z\right)\left(y+1\le\boxed{}\right) \qquad (\exists y)\left(\boxed{\forall}\,z\right)\left(y+1\le\boxed{z}\right)$$

Problem 23 $X, Y,$ and Z are subsets of the set $N^+ = \{1, 2, \ldots\}$ of positive integers, defined as follows.

$$X = \{3, 4, 5, 6, 7\}$$
$$Y = \{y \mid y < 16\}$$
$$Z = \{z \mid (\exists n)(z = 4n)\}$$

(All variables denote positive integers.)

Let variables x, y, z assume values from the sets $\{X, Y, Z\}$, respectively.

Complete the following formulae by writing into each empty box a symbol representing a universal quantifier, an existential quantifier, or a positive integer (as appropriate), so as to obtain a true claim.

Answer:

$$\neg\,(\forall z)\left(2z<\boxed{}\right) \qquad \neg\,(\forall z)\left(2z<\boxed{9}\right)$$

$$\neg\left(\boxed{}\,y\right)\left(y>\boxed{}\right) \qquad \neg\left(\boxed{\forall}\,y\right)\left(y>\boxed{8}\right)$$

$$(\exists x)\,(\exists z)\left(2x>z+\boxed{}\right) \qquad (\exists x)\,(\exists z)\left(2x>z+\boxed{3}\right)$$

$$(\forall x)\,(\exists z)\left(x+\boxed{}>2z\right) \qquad (\forall x)\,(\exists z)\left(x+\boxed{6}>2z\right)$$

$$(\exists x)\,(\forall z)\left(2x<z+\boxed{}\right) \qquad (\exists x)\,(\forall z)\left(2x<z+\boxed{3}\right)$$

$$(\forall x)\,(\forall z)\left(2x<z+\boxed{}\right) \qquad (\forall x)\,(\forall z)\left(2x<z+\boxed{11}\right)$$

$$\left(\boxed{}\,z\right)(\forall x)(z>x) \qquad \left(\boxed{\exists}\,z\right)(\forall x)(z>x)$$

$$\left(\boxed{}\,z\right)(\forall x)\,(\forall y)\left(x+y>z-\boxed{}\right) \qquad \left(\boxed{\exists}\,z\right)(\forall x)\,(\forall y)\left(x+y>z-\boxed{1}\right)$$

$$(\forall y)\,(\forall x)\left(\boxed{}\,z\right)\left(2y>z-x-\boxed{}\right) \qquad (\forall y)\,(\forall x)\left(\boxed{\exists}\,z\right)\left(2y>z-x-\boxed{4}\right)$$

$$\neg\left(\boxed{}\,x\right)\left(\boxed{}\,z\right)\left(x+z<\boxed{}\right) \qquad \neg\left(\boxed{\forall}\,x\right)\left(\boxed{\forall}\,z\right)\left(x+z<\boxed{1}\right)$$

$$\neg\left(\boxed{}\,x\right)\left(\boxed{}\,z\right)\left(x+z>\boxed{}\right) \qquad \neg\left(\boxed{\forall}\,x\right)\left(\boxed{\forall}\,z\right)\left(x+z>\boxed{9}\right)$$

$$(\exists y)\left(\boxed{}\,z\right)\left(y+1\le\boxed{}\right) \qquad (\exists y)\left(\boxed{\exists}\,z\right)\left(y+1\le\boxed{4}\right)$$

Problem 24 $X, Y,$ and Z are subsets of the set $N^+ = \{1, 2, \ldots\}$ of positive integers. Let variables x, y, z assume values from the sets X, Y, Z, respectively.

For each of the following formulae, construct the variable domains (X, Y, and Z) so that the formula is true. Your construction has to satisfy an additional constraint that at least one of the domains over which the formula is actually interpreted has to be and infinite set.

Answer:

$(\exists y)(\forall x)(y < x)$	$Y = N^+$, $X = N^+ \setminus \{1\}$;
$\neg((\exists y)(\forall x)(y < x))$	$Y = N^+$, $X = N^+$; (let $x = y$)
$(\exists y)(\forall x)(y > x)$	$Y = N^+$, $X = \{1, 2, 3\}$; (let $y = 4$)
$\neg((\exists y)(\forall x)(y > x))$	$Y = N^+$, $X = N^+$; (let $x = y$)
$(\forall x)(\exists y)(\exists z)(x < y < z)$	$X = N^+$, $Y = N^+$, $Z = N^+$; (let $y = x + 1$, $z = x + 2$)
$\neg((\forall x)(\exists y)(\exists z)(x < y < z))$	$X = N^+$, $Y = N^+$, $Z = \{1, 2, 3\}$; (let $x = 4$)
$(\exists x)(\forall y)(\forall z)(x < y < z)$	$X = N^+$, $Y = \{3, 4, 5\}$, $Z = \{6, 7, 8\}$; (let $x = 1$)
$\neg((\exists x)(\forall y)(\forall z)(x < y < z))$	$X = N^+$, $Y = N^+$, $Z = N^+$. (let $y = z = x$)

Problem 25   Let $Z = \{\ldots, -2, -1, 0, 1, 2, \ldots\}$ be the set of all integers, and let $N^+ = \{1, 2, \ldots\}$ be the set of all positive integers. Let D_x and D_y be nonempty sets such that:

$$D_x \subseteq Z$$
$$D_y \subseteq N^+$$

Let the variables x and y assume values from the sets D_x and D_y, respectively.

For each of the following formulae, construst the sets D_x and D_y, such that the formula is true over these sets as the universes of discourse of the variables x and y.

Answer:

$(\forall x)(\exists y)(y \geq x)$	$D_x = Z$, $D_y = N^+$
$\neg((\forall x)(\exists y)(y \geq x))$	$D_x = Z$, $D_y = \{1, 2\}$
$(\exists y)(\forall x)(y \geq x)$	$D_x = \{1, 2\}$, $D_y = N^+$
$\neg((\exists y)(\forall x)(y \geq x))$	$D_x = Z$, $D_y = N^+$
$(\forall x)(\forall y)(y \geq x)$	$D_x = \{-11\}$, $D_y = N^+$
$\neg((\forall x)(\forall y)(y \geq x))$	$D_x = Z$, $D_y = N^+$
$(\exists x)(\exists y)(y \geq x)$	$D_x = Z$, $D_y = N^+$
$\neg((\exists x)(\exists y)(y \geq x))$	$D_x = \{2\}$, $D_y = \{1\}$
$(\exists x)(\forall y)(y \geq x)$	$D_x = Z$, $D_y = N^+$
$\neg((\exists x)(\forall y)(y \geq x))$	$D_x = \{2\}$, $D_y = N^+$
$(\forall y)(\exists x)(y \geq x)$	$D_x = N^+$, $D_y = N^+$
$\neg((\forall y)(\exists x)(y \geq x))$	$D_x = \{2\}$, $D_y = N^+$
$(\exists y)(\exists x)(y \geq x)$	$D_x = Z$, $D_y = N^+$
$\neg((\exists y)(\exists x)(y \geq x))$	$D_x = \{2\}$, $D_y = \{1\}$

Problem 26   Let $Z = \{\ldots, -2, -1, 0, 1, 2, \ldots\}$ be the set of all integers, and let $N^+ = \{1, 2, \ldots\}$ be the set of all positive integers. Let D_x and D_y be nonempty sets such that:

$$D_x \subseteq Z$$
$$D_y \subseteq N^+$$

Let the variables x and y assume values from the sets D_x and D_y, respectively.

For each of the following formulae, construct the sets D_x and D_y, such that the formula is true over these sets as the universes of discourse of the variables x and y.

Answer:

$(\forall x)(\exists y)(x + y < 0)$	$D_x = Z^{--}$, $D_y = N^+$
$\neg((\forall x)(\exists y)(x + y < 0))$	$D_x = N^+$, $D_y = N^+$
$(\exists y)(\forall x)(x + y < 0)$	$D_x = Z^{--}$, $D_y = \{1\}$
$\neg((\exists y)(\forall x)(x + y < 0))$	$D_x = N^+$, $D_y = N^+$
$(\forall x)(\forall y)(x + y < 0)$	$D_x = Z^{--}$, $D_y = \{1\}$
$\neg((\forall x)(\forall y)(x + y < 0))$	$D_x = Z$, $D_y = N^+$
$(\exists x)(\exists y)(x + y < 0)$	$D_x = Z$, $D_y = N^+$
$\neg((\exists x)(\exists y)(x + y < 0))$	$D_x = N^+$, $D_y = N^+$
$(\exists x)(\forall y)(x + y < 0)$	$D_x = Z$, $D_y = \{1\}$
$\neg((\exists x)(\forall y)(x + y < 0))$	$D_x = Z$, $D_y = N^+$
$(\forall y)(\exists x)(x + y < 0)$	$D_x = Z$, $D_y = N^+$
$\neg((\forall y)(\exists x)(x + y < 0))$	$D_x = N^+$, $D_y = N^+$
$(\exists y)(\exists x)(x + y < 0)$	$D_x = Z$, $D_y = N^+$
$\neg((\exists y)(\exists x)(x + y < 0))$	$D_x = N^+$, $D_y = N^+$

(Z^{--} denotes the set $\{\ldots, -3, -2\}$ of all negative integers excluding -1.)

Problem 27   Let the universe of discourse of the variable n be the set $N^+ = \{1, 2, \ldots\}$ of all positive integers. For each of the following formulae, find a finite, nonempty set Y consisting of positive integers, such that the formula is true over set Y as the universe of discourse of the variable y.

Answer:

(a) $(\exists y)(\exists n)(n < y)$

$Y = \{2\}$
or any $Y \neq \{1\}$

(b) $(\exists y)(\exists n)(y < n)$

$Y = \{100\}$
or any $Y \neq \emptyset$

(c) $(\exists y)(\forall n)(y \leq n)$

$Y = \{1\}$
or any Y such that $1 \in Y$.

(d) $(\exists y)(\forall n)(y \leq n) \wedge \neg(\exists y)(\exists n)(n < y)$

$Y = \{1\}$ only.

(e) $(\forall y)(\exists n)(n < y < 5)$

$Y = \{2, 3, 4\}$
or any $Y \neq \emptyset$ such that $Y \subseteq \{2, 3, 4\}$.

(f) $(\forall y)(\exists n)(y = 2n \wedge y < 5)$

$Y = \{2, 4\}$
or any $Y \neq \emptyset$ such that $Y \subseteq \{2, 4\}$.

(g) $(\forall y)(\exists n)(n + 2 < y)$

$Y = \{4\}$
or any Y such that $1 \notin Y \wedge 2 \notin Y \wedge 3 \notin Y$

(h) $(\exists n)((n \geq 6) \wedge (\forall y)(y \geq n \wedge y < 7))$

$Y = \{6\}$ only.

Answer:

(i) $(\forall n)((n \leq 4) \implies (\exists y)(y = n))$ $Y = \{1, 2, 3, 4, 100\}$
or any Y such that $\{1, 2, 3, 4\} \subseteq Y$.

(j) $(\forall n)((n \leq 4) \iff (\exists y)(y = n))$ $Y = \{1, 2, 3, 4\}$ only.

(k) $(\forall n)(\exists y)(y < 2n)$ $Y = \{1, 100\}$
or any Y such that $1 \in Y$.

(l) $(\forall n)(\exists y)(1 < y < n + 2)$ $Y = \{2, 100\}$
or any Y such that $2 \in Y$.

Problem 28   Let the universe of discourse of the variable n be the set $N^+ = \{1, 2, \ldots\}$ of all positive integers. For each of the following formulae, find a finite, nonempty set Y consisting of positive integers, such that the formula is true over set Y as the universe of discourse of the variable y.

Answer:

(a) $(\exists y)(\exists n)(n + 1 < y)$ $Y = \{3\}$
or any $Y \nsubseteq \{1, 2\}$.

(b) $(\exists y)(\exists n)(y < n + 1)$ $Y = \{100\}$
or any $Y \neq \emptyset$.

(c) $(\exists y)(\forall n)(y \leq n + 1)$ $Y = \{1\}$
or any Y such that $1 \in Y \lor 2 \in Y$.

(d) $(\exists y)(\forall n)(y \leq n) \land \neg(\exists y)(\exists n)(n + 1 < y)$ $Y = \{1, 2\}$
or $Y = \{1\}$.

(e) $(\forall y)(\exists n)(n \leq y \leq 5)$ $Y = \{1, 2, 3, 4, 5\}$
or any $Y \neq \emptyset$ such that $Y \subseteq \{1, 2, 3, 4, 5\}$.

(f) $(\forall y)(\exists n)(y = 2n + 1 \land y < 6)$ $Y = \{3, 5\}$
or any $Y \neq \emptyset$ such that $Y \subseteq \{3, 5\}$.

(g) $(\forall y)(\exists n)(n + 1 < y)$ $Y = \{3\}$
or any Y such that $1 \notin Y \land 2 \notin Y$

(h) $(\exists n)((n \geq 6) \land (\forall y)(y \geq n \land y < 9)$ $Y = \{6, 7, 8\}$
or any $Y \neq \emptyset$ such that $Y \subseteq \{6, 7, 8\}$.

(i) $(\forall n)((n < 4) \implies (\exists y)(y = n))$ $Y = \{1, 2, 3, 100\}$
or any Y such that $\{1, 2, 3\} \subseteq Y$.

(j) $(\forall n)((n < 4) \iff (\exists y)(y = n))$ $Y = \{1, 2, 3\}$ only.

(k) $(\forall n)(\exists y)(y < 3n)$ $Y = \{1, 2, 100\}$
or any Y such that $1 \in Y \lor 2 \in Y$.

(l) $(\forall n)(\exists y)(1 < y \leq n + 1)$ $Y = \{2, 100\}$
or any Y such that $2 \in Y$.

1.2 Sets

Problem 29 Sets A, B, C, and D are defined as follows.

$$A = \{1, 2, 3\}$$
$$B = \{x \mid (\exists y)(y \in A \land x = 2^y)\}$$
$$C = \{x \mid (\exists y)(\exists z)(y \in A \land z \in B \land x = y \cdot z)\}$$
$$D = \{x \mid (\exists y)(\exists z)(y \in A \land z \in B \land x > y + z)\}$$

(All variables assume values from the set of natural numbers $N = \{0, 1, \ldots\}$.)

In each of the following cases, write the required list. If this is impossible, explain why.

(a) List all elements of B.

Answer:

$$2, 4, 8$$

(b) List all elements of C.

Answer:

$$2, 4, 6, 8, 12, 16, 24$$

(c) List all elements of D.

Answer: Impossible, since D is an infinite set, which contains exactly those natural numbers that are greater than 3.

(d) List all elements of $A \cup B$.

Answer:

$$1, 2, 3, 4, 8$$

(e) List the largest element of $A \cap B$.

Answer:

$$2$$

(f) List the smallest element of $A \cap B$.

Answer:

$$2$$

(g) List three elements of $B \cap C$.

Answer:

$$2, 4, 8$$

(h) List all subsets of B.

Answer:

$$\emptyset, \{2\}, \{4\}, \{8\}, \{2, 4\}, \{2, 8\}, \{4, 8\}, \{2, 4, 8\}$$

(i) List all elements of $A \setminus B$.

Answer:

$$1, 3$$

(j) List all elements of $C \cap D$.

Answer:

$$4, 6, 8, 12, 16, 24$$

Problem 30   Sets A, B, C, D, E, F are defined as follows.

$$A = \{0, 1, 2, 3, 4, 5, 6\}$$
$$B = \{x \mid (\exists y \in A)(x = 2y + 1)\}$$
$$C = \{z \mid z \in A \wedge z \in B \wedge z > 1\}$$
$$D = \left\{x \mid (\exists y \in B)\left(x = \sqrt{y}\right)\right\}$$
$$E = \{y \mid (\forall x \in B)(x > y)\}$$
$$F = \{y \mid (\forall x \in B)(x \le y)\}$$

(All variables assume values from the set of natural numbers $N = \{0, 1, \ldots\}$.)

In each of the following cases, write the required list. If this is impossible, explain why.

(a) List all elements of B.

> **Answer:**
> $$1, 3, 5, 7, 9, 11, 13$$

(b) List all elements of C.

> **Answer:**
> $$3, 5$$

(c) List all elements of D.

> **Answer:**
> $$1, 3$$

(d) List the largest element of E.

> **Answer:**
> $$0$$

(e) List all elements of F.

> **Answer:** Impossible, since F is an infinite set, which contains exactly those natural numbers that are greater than 12.

(f) List three elements of $C \cap D$.

> **Answer:** Impossible, since $C \cap D$ is a singleton, containing the natural number 3.

(g) List all elements of $\mathcal{P}(C)$.

> **Answer:**
> $$\emptyset, \{3\}, \{5\}, \{3, 5\}$$

(h) List the smallest element of $F \setminus C$.

> **Answer:**
> $$13$$

(i) List the largest element of $E \cup B$.

> **Answer:**
> $$13$$

(j) List all subsets of $F \cap A$.

Answer:
$$\varnothing$$

($F \cap A$ does not have any elements, but it has one subset.)

(k) List the largest element of $F \cap A$.

Answer: Impossible, since set $F \cap A$ is empty, and there is no largest element.

Problem 31   Sets A, B, C are defined as follows.

$$A = \{0, 1, 2, 3, 4, 5, 6, 7, 8\}$$
$$B = \left\{x \mid (\exists y \in A)\left(x^2 = 2y\right)\right\}$$
$$C = \{z \mid (\forall y \in A)(y < 2z)\}$$

(All variables assume values from the set of natural numbers $N = \{0, 1, \ldots\}$.)

In each of the following cases, write the required list. If this is impossible, explain why.

(a) List all elements of B.

Answer:
$$0, 2, 4$$

(b) List all subsets of B.

Answer:

$$\varnothing, \{0\}, \{2\}, \{4\}, \{0, 2\}, \{0, 4\}, \{2, 4\}, \{0, 2, 4\}$$

(c) List all elements of C.

Answer: Impossible, since C is an infinite set, which consists of those natural numbers that are greater than 4.

(d) List all elements of $A \cap C$.

Answer:
$$5, 6, 7, 8$$

(e) List all elements of $A \setminus C$.

Answer:
$$0, 1, 2, 3, 4$$

(f) List all elements of $B \setminus A$.

Answer: It is impossible to construct a nonempty list, since:

$$B \setminus A = \varnothing$$

(g) List all subsets of $B \setminus A$.

Answer:
$$\varnothing$$

($B \setminus A$ does not have any elements, but it has one subset.)

Problem 32   Sets A, B and C are defined as follows.

$$A = \{1, 2, 3, 4, 5\}$$
$$B = \{y \mid (\exists x \in A)(y = x \vee y = x + 2)\}$$
$$C = \{z \mid (\forall y \in B)(\exists x \in A)(y - x < z)\}$$

(All variables assume values from the set of natural numbers $N = \{0, 1, \ldots\}$.)

In each of the following cases, write the required list. If this is impossible, explain why.

(a) List all elements of B.

> **Answer:**
> $$1, 2, 3, 4, 5, 6, 7$$

(b) List all elements of C.

> **Answer:** Impossible, since C is an infinite set, which contains exactly those natural numbers that are greater than 2.

(c) List all elements of $N \setminus C$.

> **Answer:**
> $$0, 1, 2$$

(d) List all elements of $A \setminus B$.

> **Answer:** It is impossible to construct a nonempty list, since the set $A \setminus B$ is empty.

(e) List all subsets of $A \setminus B$.

> **Answer:**
> $$\emptyset$$

(f) List all elements of $A \cap B$.

> **Answer:**
> $$1, 2, 3, 4, 5$$

(g) List all elements of $A \cup B$.

> **Answer:**
> $$1, 2, 3, 4, 5, 6, 7$$

(h) List the smallest element of C.

> **Answer:**
> $$3$$

(i) List the largest element of $B \cap C$.

> **Answer:**
> $$7$$

(j) List the largest element of $B \setminus C$.

> **Answer:**
> $$2$$

(k) List the smallest element of $A \cap C$.

> **Answer:**
> $$3$$

Problem 33   Sets A, B, C, D, E are defined as follows.

$$A = \{0, 1, 2, 3\}$$
$$B = \{x \mid (\forall y \in A)\,(x > y^2)\}$$
$$C = \{z \mid z \notin A \cup B\}$$
$$D = \{x \mid (\exists y \in C)\,(x^2 = y)\}$$
$$E = \{z \mid z \notin B \wedge (\exists y)\,(z = y^2)\}$$

(All variables assume values from the set of natural numbers $N = \{0, 1, \ldots\}$.)

In each of the following cases, write the required list. If this is impossible, explain why.

(a) List four elements of B.

Answer:
$$18, 19, 20, 21$$

(b) List the smallest element of B.

Answer:
$$10$$

(c) List four elements of $\mathcal{P}(B)$.

Answer:
$$\emptyset, \{10\}, \{11\}, \{10, 11\}$$

(d) List the largest element of B.

Answer: Impossible—B is an infinite set, which contains all natural numbers greater than 9.

(e) List all elements of C.

Answer:
$$4, 5, 6, 7, 8, 9$$

(f) List all elements of D.

Answer:
$$2, 3$$

(g) List all elements of E.

Answer:
$$0, 1, 4, 9$$

(h) List all elements of $C \setminus E$.

Answer:
$$5, 6, 7, 8$$

(i) List the smallest element of $A \cap D$.

Answer:
$$2$$

(j) List the largest element of $A \cap D$.

Answer:
$$3$$

Problem 34   Sets A, B, C, D, and E are defined as follows.

$$A = \{1, 2, 3, 4, 5\}$$
$$B = \{11, 12, 13, 14, 15, 16\}$$
$$C = \left\{ z \mid (\exists x \in A)(\exists y \in B)\left(z = \frac{x+y}{2} \right) \right\}$$
$$D = \{x \mid (\exists z \in A)(\forall y \in C)(x = y - z)\}$$
$$E = \{x \mid (\forall y \in C)(\exists z \in A)(x = y - z)\}$$

(All variables assume values from the set of natural numbers $N = \{0, 1, \ldots\}$.)
In each of the following cases, write the required list. If this is impossible, explain why.

(a) List all elements of C.

> **Answer:**
> $$6, 7, 8, 9, 10$$

(b) List one element of D.

> **Answer:** Impossible, since the set D is empty and there are no elements to list.

(c) List one subset of D.

> **Answer:**
> $$\varnothing$$

(d) List all elements of $\mathcal{P}(D)$.

> **Answer:**
> $$\varnothing$$

(e) List all subsets of $\mathcal{P}(D)$.

> **Answer:**
> $$\varnothing, \{\varnothing\}$$

(f) List all elements of E.

> **Answer:**
> $$5$$

(g) List all elements of $\mathcal{P}(E)$.

> **Answer:**
> $$\varnothing, \{5\}$$

(h) List all subsets of $\mathcal{P}(E)$.

> **Answer:**
> $$\varnothing, \{\varnothing\}, \{\{5\}\}, \{\varnothing, \{5\}\}$$

(i) List all elements of $A \setminus E$.

> **Answer:**
> $$1, 2, 3, 4$$

(j) List all elements of $C \cap A$.

> **Answer:** It is impossible to construct a nonempty list, since the set $C \cap A$ is empty.

(k) List all elements of $A \cap D$.

> **Answer:** It is impossible to construct a nonempty list, since the set $A \cap D$ is empty.

(l) List all elements of $A \cup D$.

> **Answer:**
>
> $$1, 2, 3, 4, 5$$

Problem 35 Sets A, B, and C are defined as follows.

$$A = \{x \mid 0 \le x \le 6 \lor x \ge 16\}$$
$$B = \{x \mid (\exists y \in A)(\exists z \in A)(y < x < z)\}$$
$$C = \{x \mid (\exists y)(y \notin A \land x = 2y)\}$$

(All variables assume values from the set of natural numbers $N = \{0, 1, \ldots\}$.)

For each of the following sets, construct an equivalent set expression, which contains no operators other than the set difference operator, applied to no sets other than explicitly enumerated finite sets and the sets N and $\emptyset$.

Answer:

A	$A = N \setminus \{7, 8, 9, 10, 11, 12, 13, 14, 15\}$
$\overline{A}$	$\overline{A} = \{7, 8, 9, 10, 11, 12, 13, 14, 15\}$
B	$B = N \setminus \{0\}$
$\overline{B}$	$\overline{B} = \{0\}$
$B \setminus N$	$B \setminus N = \emptyset$
$A \cup B$	$A \cup B = N$
$\overline{A \cup B}$	$\overline{A \cup B} = \emptyset$
C	$C = \{14, 16, 18, 20, 22, 24, 26, 28, 30\}$
$B \setminus C$	$B \setminus C = N \setminus \{0, 14, 16, 18, 20, 22, 24, 26, 28, 30\}$
$A \cap C$	$A \cap C = \{16, 18, 20, 22, 24, 26, 28, 30\}$
$A \cup C$	$A \cup C = N \setminus \{7, 8, 9, 10, 11, 12, 13, 15\}$
$B \setminus (A \cup C)$	$B \setminus (A \cup C) = \{7, 8, 9, 10, 11, 12, 13, 15\}$
$C \setminus A$	$C \setminus A = \{14\}$

Problem 36 Sets A, B, and C are subsets of the set N of natural numbers, defined in Problem 35.

Complete the following formulae, by writing into each empty box a symbol representing a set operator or a natural number (as appropriate), so that the resulting formula is true. Where such a completion is impossible, explain why.

Answer:

$$\boxed{} \notin A \qquad\qquad \boxed{7} \notin A$$

$$(\forall x)\left(x \in A \cap C \Longrightarrow x \le \boxed{}\right) \qquad (\forall x)\left(x \in A \cap C \Longrightarrow x \le \boxed{30}\right)$$

$$(\forall x)\left(x \in A \cap C \Longrightarrow x > \boxed{}\right) \qquad (\forall x)\left(x \in A \cap C \Longrightarrow x > \boxed{15}\right)$$

Answer:

$(\exists x)\left(x \notin A \wedge 0 < 3x < 25 \wedge x = \boxed{}\right)$ $(\exists x)\left(x \notin A \wedge 0 < 3x < 25 \wedge x = \boxed{8}\right)$

$(\exists x)\left(x \notin A \wedge 100 > 4x > 50 \wedge x = \boxed{}\right)$ $(\exists x)\left(x \notin A \wedge 100 > 4x > 50 \wedge x = \boxed{13}\right)$

$(\exists x)\left(x \in B \setminus C \wedge 13 < x < 30 \wedge x = \boxed{}\right)$ $(\exists x)\left(x \in B \setminus C \wedge 13 < x < 30 \wedge x = \boxed{15}\right)$

$(\forall x)\left((x \in B \setminus (A \cup C)) \Longrightarrow \left(\boxed{} \leq x\right)\right)$ $(\forall x)\left((x \in B \setminus (A \cup C)) \Longrightarrow \left(\boxed{7} \leq x\right)\right)$

$(\forall x)\left((x \in B \setminus (A \cup C)) \Longrightarrow \left(x < \boxed{}\right)\right)$ $(\forall x)\left((x \in B \setminus (A \cup C)) \Longrightarrow \left(x < \boxed{16}\right)\right)$

$\boxed{} \in N \setminus B$ $\boxed{0} \in N \setminus B$

$(\forall x \in A)\left(2x \leq \boxed{} \vee 2x \geq \boxed{}\right)$ $(\forall x \in A)\left(2x \leq \boxed{12} \vee 2x \geq \boxed{32}\right)$

$\boxed{} \in B \setminus N$ $\boxed{?} \in B \setminus N$ —impossible: $B \setminus N = \emptyset$

$\boxed{} \in C \setminus A$ $\boxed{14} \in C \setminus A$

$(\forall x)(\exists y)\left(y = 10x \wedge \{x,y\} \subseteq A \boxed{} B\right)$ $(\forall x)(\exists y)\left(y = 10x \wedge \{x,y\} \subseteq A \boxed{\cup} B\right)$

$\boxed{} \in \overline{A \cup B}$ $\boxed{?} \in \overline{A \cup B}$ —impossible: $\overline{A \cup B} = \emptyset$

Problem 37   Sets A, B, and C are subsets of the set $N^+ = \{1, 2, \ldots\}$ of positive integers, defined as follows.

$$A = \{1, 2, 3, 4\}$$
$$B = \{x \mid x \geq 3\}$$
$$C = \{x \mid (\exists y)(y \in B \wedge x = 2y)\}$$

(All variables assume values from the set of positive integers $N^+ = \{1, 2, \ldots\}$.)

Complete the following formulae, by writing into each empty box a symbol representing a logical operator, a set operator, or a positive integer (as appropriate), so that the resulting formula is true. Where such a completion is impossible, explain why.

Answer:

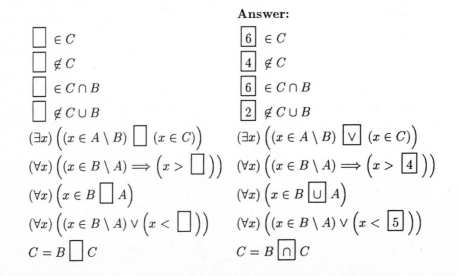

Answer:

$B = B \,\square\, C$ $B = B \,\boxed{\cup}\, C$

$(\forall x)\Big((x \in A \cap B) \Longleftrightarrow \big(\square < x < \square\big)\Big)$ $(\forall x)\Big((x \in A \cap B) \Longleftrightarrow \big(\boxed{2} < x < \boxed{5}\big)\Big)$

$(\forall x)\Big((x \in A \cup C) \Longrightarrow \big(x \neq \square\big)\Big)$ $(\forall x)\Big((x \in A \cup C) \Longrightarrow \big(x \neq \boxed{5}\big)\Big)$

Problem 38   Sets A, B, and C are subsets of the set $N^+ = \{1, 2, \ldots\}$ of positive integers, defined as follows.

$$A = \{x \mid x \leq 6\}$$
$$B = \{4, 5, 6, 7, 8\}$$
$$C = \{x \mid (\exists y)(y \in A \cup B \wedge x = y + 5)\}$$

(All variables assume values from the set of positive integers $N^+ = \{1, 2, \ldots\}$.)

Complete the following formulae, by writing into each empty box a symbol representing a logical operator, a set operator , or a positive integer (as appropriate), so that the resulting formula is true. Where such a completion is impossible, explain why.

Answer:

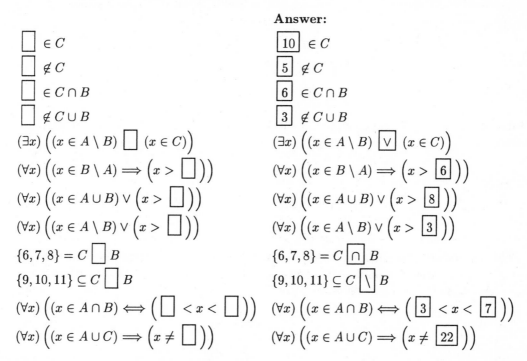

Problem 39   Sets A, B, and C are subsets of the set $N^+ = \{1, 2, \ldots\}$ of positive integers, defined as follows.

$$A = \{x \mid 6 \leq x \leq 10\}$$
$$B = \{4, 5, 6, 7, 8\}$$
$$C = \{x \mid (\exists y)(y \in A \cap B \wedge x = 2y)\}$$

(All variables assume values from the set of positive integers $N^+ = \{1, 2, \ldots\}$.)

Complete the following formulae, by writing into each empty box a symbol representing a logical operator, a set operator, or a positive integer (as appropriate), so that the resulting formula is true. Where such a completion is impossible, explain why.

Answer:

$\square \notin C$ $\qquad$ $\boxed{11} \notin C$

$\square \in C$ $\qquad$ $\boxed{12} \in C$

$\square \notin C \cap B$ $\qquad$ $\boxed{9} \notin C \cap B$

$\square \in C \cup B$ $\qquad$ $\boxed{14} \in C \cup B$

$(\forall x)\Big((x \in B \setminus A) \Longrightarrow \big(\square \leq x \leq \square\big)\Big)$ $\qquad$ $(\forall x)\Big((x \in B \setminus A) \Longrightarrow \big(\boxed{4} \leq x \leq \boxed{5}\big)\Big)$

$(\exists x)\Big((x \in A \setminus B)\,\square\,(x \in C)\Big)$ $\qquad$ $(\exists x)\Big((x \in A \setminus B)\,\boxed{\vee}\,(x \in C)\Big)$

$(\forall x)\Big((x \notin A \setminus B) \Longleftrightarrow \big(x > \square \vee x < \square\big)\Big)$ $\qquad$ $(\forall x)\Big((x \notin A \setminus B) \Longleftrightarrow \big(x > \boxed{10} \vee x < \boxed{9}\big)\Big)$

$(\forall x)\Big((x \notin A \cup C) \Longrightarrow \big(x > \square \vee x < \square\big)\Big)$ $\qquad$ $(\forall x)\Big((x \notin A \cup C) \Longrightarrow \big(x > \boxed{10} \vee x < \boxed{6}\big)\Big)$

$\{8, 12, 16\} \subseteq C \,\square\, B$ $\qquad$ $\{8, 12, 16\} \subseteq C \,\boxed{\cup}\, B$

$\{6, 7, 8\} = A \,\square\, B$ $\qquad$ $\{6, 7, 8\} = A \,\boxed{\cap}\, B$

$(\forall x)\Big((x \in A \cup C) \Longrightarrow \big(x \neq \square\big)\Big)$ $\qquad$ $(\forall x)\Big((x \in A \cup C) \Longrightarrow \big(x \neq \boxed{13}\big)\Big)$

$(\forall x)\Big((x \in A \cap B) \Longleftrightarrow \big(\square < x < \square\big)\Big)$ $\qquad$ $(\forall x)\Big((x \in A \cap B) \Longleftrightarrow \big(\boxed{5} < x < \boxed{9}\big)\Big)$

Problem 40 Sets A, B, and C are subsets of the set $N^+ = \{1, 2, \ldots\}$ of positive integers, defined as follows.

$$A = \{2, 4, 6, 8\}$$
$$B = \{x \mid 2 < x < 19\}$$
$$C = \left\{x \mid (\exists y \in A)(\exists z \in B)\left(x = \frac{y}{2} + z\right)\right\}$$

(All variables assume values from the set of positive integers $N^+ = \{1, 2, \ldots\}$.)

Complete the following formulae, by writing into each empty box a symbol representing a logical operator, a set operator, or a positive integer (as appropriate), so that the resulting formula is true. Where such a completion is impossible, explain why.

Answer:

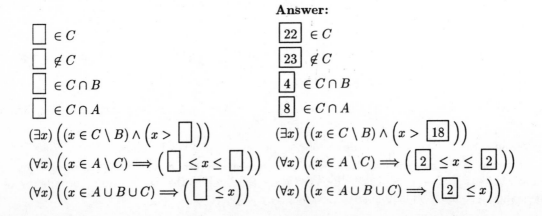

$\square \in C$ $\qquad$ $\boxed{22} \in C$

$\square \notin C$ $\qquad$ $\boxed{23} \notin C$

$\square \in C \cap B$ $\qquad$ $\boxed{4} \in C \cap B$

$\square \in C \cap A$ $\qquad$ $\boxed{8} \in C \cap A$

$(\exists x)\Big((x \in C \setminus B) \wedge \big(x > \square\big)\Big)$ $\qquad$ $(\exists x)\Big((x \in C \setminus B) \wedge \big(x > \boxed{18}\big)\Big)$

$(\forall x)\Big((x \in A \setminus C) \Longrightarrow \big(\square \leq x \leq \square\big)\Big)$ $\qquad$ $(\forall x)\Big((x \in A \setminus C) \Longrightarrow \big(\boxed{2} \leq x \leq \boxed{2}\big)\Big)$

$(\forall x)\Big((x \in A \cup B \cup C) \Longrightarrow \big(\square \leq x\big)\Big)$ $\qquad$ $(\forall x)\Big((x \in A \cup B \cup C) \Longrightarrow \big(\boxed{2} \leq x\big)\Big)$

$$(\forall x)\Big((x \in A \cup B \cup C) \Longrightarrow \big(x \le \Box\big)\Big)$$
$$(\forall x)\Big((x \in B \cap C) \Longrightarrow \big(x > \Box\big)\Big)$$
$$C \subseteq B \,\Box\, C$$
$$B \subseteq B \,\Box\, C$$
$$(\forall x)\Big((x \in A \cap B) \Longrightarrow \big(\Box < x < \Box\big)\Big)$$
$$(\forall x)\Big((x \in C \setminus B) \Longrightarrow \big(x < \Box \land x > \Box\big)\Big)$$

Answer:

$$(\forall x)\Big((x \in A \cup B \cup C) \Longrightarrow \big(x \le \boxed{22}\big)\Big)$$
$$(\forall x)\Big((x \in B \cap C) \Longrightarrow \big(x > \boxed{3}\big)\Big)$$
$$C \subseteq B \,\boxed{\cup}\, C$$
$$B \subseteq B \,\boxed{\cup}\, C$$
$$(\forall x)\Big((x \in A \cap B) \Longrightarrow \big(\boxed{3} < x < \boxed{9}\big)\Big)$$
$$(\forall x)\Big((x \in C \setminus B) \Longrightarrow \big(x < \boxed{23} \land x > \boxed{18}\big)\Big)$$

Problem 41 Sets A and B are subsets of the set $N^+ = \{1, 2, \ldots\}$ of positive integers, defined as follows.

$$A = \{1, 2, 3, 4, 5, 6\}$$
$$B = \{4, 5, 6, 7, 8, 9, 10\}$$

Complete the following formulae, by writing a positive integer into each empty box, so that the resulting formula is true. Where such a completion is impossible, explain why.

Answer:

$\Box \in A \cup B$	$\boxed{1} \in A \cup B$
$\Box \in A \cap B$	$\boxed{4} \in A \cap B$
$\Box \notin A \cap B$	$\boxed{1} \notin A \cap B$
$\Box \in A - B$	$\boxed{1} \in A - B$
$\Box \notin B - A$	$\boxed{1} \notin B - A$
$\Box \in N - A$	$\boxed{7} \in N - A$
$\Box \notin N - B$	$\boxed{4} \notin N - B$
$\Box \in N - (A \cup B)$	$\boxed{11} \in N - (A \cup B)$
$\Box \in N - (A \cap B)$	$\boxed{1} \in N - (A \cap B)$
$\Box \in (N - A) \cap B$	$\boxed{7} \in (N - A) \cap B$
$\Box \notin (N - B) \cap A$	$\boxed{5} \notin (N - B) \cap A$
$\Box \in N$	$\boxed{1} \in N$
$\Box \notin N - (A \cap B)$	$\boxed{4} \notin N - (A \cap B)$
$\Box \notin N - (A \cup B)$	$\boxed{1} \notin N - (A \cup B)$

Problem 42 Sets A, B, and C are subsets of the set $N = \{0, 1, \ldots\}$ of natural numbers, defined as follows.

$$A = \{x \mid x^2 < 9\}$$
$$B = \{x \mid (\exists y)(y \in A \land x = 2^y)\}$$
$$C = \{x \mid (\exists y)(\exists z)(y \in B \land z \in N \land x = y^z)\}$$

(All variables assume values from the set of natural numbers $N = \{0, 1, \ldots\}$.)

Complete the following formulae, by writing a natural number into each empty box so that the resulting formula is true. Where such a completion is impossible, explain why.

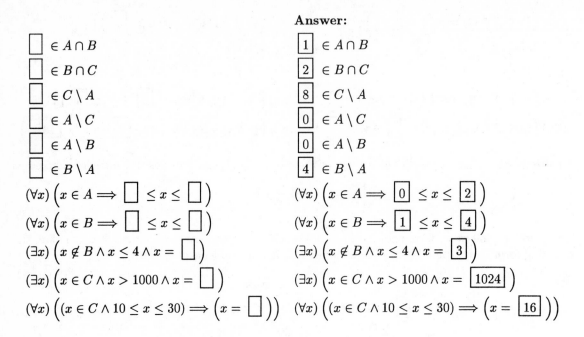

Answer:

$\boxed{} \in A \cap B$	$\boxed{1} \in A \cap B$
$\boxed{} \in B \cap C$	$\boxed{2} \in B \cap C$
$\boxed{} \in C \setminus A$	$\boxed{8} \in C \setminus A$
$\boxed{} \in A \setminus C$	$\boxed{0} \in A \setminus C$
$\boxed{} \in A \setminus B$	$\boxed{0} \in A \setminus B$
$\boxed{} \in B \setminus A$	$\boxed{4} \in B \setminus A$
$(\forall x)\left(x \in A \Longrightarrow \boxed{} \le x \le \boxed{}\right)$	$(\forall x)\left(x \in A \Longrightarrow \boxed{0} \le x \le \boxed{2}\right)$
$(\forall x)\left(x \in B \Longrightarrow \boxed{} \le x \le \boxed{}\right)$	$(\forall x)\left(x \in B \Longrightarrow \boxed{1} \le x \le \boxed{4}\right)$
$(\exists x)\left(x \notin B \wedge x \le 4 \wedge x = \boxed{}\right)$	$(\exists x)\left(x \notin B \wedge x \le 4 \wedge x = \boxed{3}\right)$
$(\exists x)\left(x \in C \wedge x > 1000 \wedge x = \boxed{}\right)$	$(\exists x)\left(x \in C \wedge x > 1000 \wedge x = \boxed{1024}\right)$
$(\forall x)\left((x \in C \wedge 10 \le x \le 30) \Longrightarrow \left(x = \boxed{}\right)\right)$	$(\forall x)\left((x \in C \wedge 10 \le x \le 30) \Longrightarrow \left(x = \boxed{16}\right)\right)$

Problem 43   Sets A, B and C are defined as follows.

$$A = \{a, b, c, d, e, f, g, h\}$$
$$B = \{b, d, f, g, h, i, j, k, l, m\}$$
$$C = \{a, c, d, f, g, h, j, l, m, n, o, p, r\}$$

For each of the following sets, construct an equivalent set expression, which employs only (if at all) the set operations of union, intersection and difference, applied to (some of) the sets A, B, and C:

Answer:

$$S_1 = \{j, l, m, n, o, p, r\} \qquad S_1 = C \setminus A$$
$$S_2 = \{n, o, p, r\} \qquad S_2 = C \setminus (A \cup B)$$
$$S_3 = \{i, k\} \qquad S_3 = B \setminus (A \cup C)$$
$$S_4 = \{a, c, d, f, g, h\} \qquad S_4 = A \cap C$$
$$S_5 = \{d, f, g, h, j, l, m\} \qquad S_5 = B \cap C$$
$$S_6 = \{b, e, i, k\} \qquad S_6 = (A \cup B) \setminus C$$

Problem 44   Sets A, B and C are defined as follows.

$$A = \{1, 3, 4, 7, 8, 9, 11, 12, 13, 14, 15, 18\}$$
$$B = \{2, 3, 4, 8, 11, 15, 17, 19, 21\}$$
$$C = \{5, 6, 10, 11, 14, 15, 16, 17, 18, 19, 20\}$$

For each of the following sets, construct an equivalent set expression, which employs only (if at all) the set operations of union, intersection and difference, applied to (some of) the sets A, B, and C:

Answer:

$S_1 = \{3, 4, 8, 11, 15\}$ $\qquad$ $S_1 = A \cap B$

$S_2 = \{2, 3, 4, 5, 6, 8, 10, 11, 14, 15, 16, 17, 18, 19, 20, 21\}$ $\quad$ $S_2 = B \cup C$

$S_3 = \{17, 19\}$ $\qquad$ $S_3 = (B \cap C) \setminus A$

$S_4 = \{11, 15\}$ $\qquad$ $S_4 = A \cap B \cap C$

$S_5 = \{1, 2, 7, 9, 12, 13, 14, 17, 18, 19, 21\}$ $\qquad$ $S_5 = (B \setminus A) \cup (A \setminus B)$

$S_6 = \{2, 5, 6, 10, 11, 14, 15, 16, 17, 18, 19, 20, 21\}$ $\qquad$ $S_6 = (B \setminus A) \cup C$

$S_7 = \{2, 3, 4, 8, 11, 15, 21\}$ $\qquad$ $S_7 = (A \cap B) \cup (B \setminus C)$

Problem 45 $\quad$ Sets A, B and C are defined as follows.

$$A = \{1, 2, 3, 4, 7, 8, 9, 10, 11, 13, 14, 15, 16\}$$
$$B = \{2, 3, 8, 14, 15\}$$
$$C = \{3, 4, 5, 6, 7, 11, 12, 13, 14, 17, 18\}$$

For each of the following sets, construct an equivalent set expression, which employs only (if at all) the set operations of union, intersection and difference, applied to (some of) the sets A, B, and C:

Answer:

$S_1 = \{2, 3, 8, 14, 15\}$ $\qquad$ $S_1 = B$

$S_2 = \{3, 14\}$ $\qquad$ $S_2 = B \cap C$

$S_3 = \{1, 3, 4, 5, 6, 7, 9, 10, 11, 12, 13, 14, 16, 17, 18\}$ $\quad$ $S_3 = (A \setminus B) \cup C$

$S_4 = \{2, 5, 6, 8, 12, 15, 17, 18\}$ $\qquad$ $S_4 = (B \setminus C) \cup (C \setminus A)$

$S_5 = \{4, 7, 11, 13\}$ $\qquad$ $S_5 = (A \cap C) \setminus B$

$S_6 = \{2, 3, 4, 5, 6, 7, 8, 11, 12, 13, 14, 15, 17, 18\}$ $\qquad$ $S_6 = B \cup C$

$S_7 = \{2, 8, 15\}$ $\qquad$ $S_7 = B \setminus C$

Problem 46 $\quad$ Sets A, B and C are defined as follows.

$$A = \{1, 2, 3, 7, 8, 9, 13, 14\}$$
$$B = \{3, 4, 5, 6, 7, 9, 10, 11, 12, 13, 18\}$$
$$C = \{3, 6, 9, 12, 15, 18, 21, 24\}$$

For each of the following sets, construct an equivalent set expression, which employs only (if at all) the set operations of union, intersection and difference, applied to (some of) the sets A, B, and C:

Answer:

$S_1 = \{7, 13\}$ $\qquad$ $S_1 = (A \cap B) \setminus C$

$S_2 = \{3, 7, 9, 13\}$ $\qquad$ $S_2 = A \cap (B \cup C)$

$S_3 = \{1, 2, 8, 14\}$ $\qquad$ $S_3 = A \setminus B$

$S_4 = \{1, 2, 3, 8, 9, 14\}$ $\qquad$ $S_4 = A \setminus (B \setminus C)$

$S_5 = \{6, 12, 18\}$ $\qquad$ $S_5 = (B \setminus A) \cap C$

$S_6 = \{1, 2, 4, 5, 8, 10, 11, 14\}$ $\qquad$ $S_6 = ((A \setminus B) \cup (B \setminus A)) \setminus C$

$S_7 = \{3, 6, 7, 9, 12, 13, 18\}$ $\qquad$ $S_7 = (A \cap B) \cup (B \cap C)$

Problem 47 A, B, and C are finite sets consisting of positive integers. Find the set C such that all of the following nine claims are true. (All variables assume values from the set of positive integers $N^+ = \{1, 2, \ldots\}$.)

(1) $A = \{x \mid x \le 20 \land (\exists k)\,(x = 2k)\}$

(2) $B = \{x \mid x > 5 \land (\exists k)\,(x = 2k - 1)\}$

(3) $(\forall x)\,(x \in C \Longrightarrow x \in B \cup A)$

(4) $(\forall x)\,(x \in C \Longrightarrow (\exists y)\,(y \in A \land y > x))$

(5) $7 \notin C \cup B$

(6) $(\exists y)\,(y \in A \setminus C \land y < 3)$

(7) $(\exists y)\,(y \in \{7, 8\} \cap C)$

(8) $(\exists y)\,(7 < y < 10 \land y \notin C)$

(9) C has exactly 13 elements

Explain your answer briefly.

Answer: From (3) we infer that:

$$C \subseteq A \cup B$$

while (4) says that all elements of C are less than 20. Combining these two, we see that C is a subset of the following set:

$$\{2, 4, 6, 7, 8, 9, 10, 11, 12, 13, 14, 15, 16, 17, 18, 19\}$$

which has 16 elements, whereas C has only 13 elements, by (9). We have to find the 3 elements that are not in C. By (5), 7 is not in C. By (6), 2 is not in C, since 2 is the only element of A that is less than 3. By (7), 8 must be in C, since 7 is not. Finally, by (8), 9 is not in C, since 8 must be. Hence, $2, 7, 9$ are not in C, while the remaining 13 elements belong to C:

$$C = \{4, 6, 8, 10, 11, 12, 13, 14, 15, 16, 17, 18, 19\}$$

Problem 48 A, B, and C are finite sets consisting of positive integers. Find the set C such that all of the following eight claims are true. (All variables assume values from the set of positive integers $N^+ = \{1, 2, \ldots\}$.)

(1) $A = \{3, 4, 5, 6\}$

(2) $B = \{9, 10, 11, 12\}$

(3) $(\forall x)\,(\forall y)\,((x \in C \land y \in B) \Longrightarrow x \le y)$

(4) $(\forall x)\,(x = 9 \Longleftrightarrow x \in B \cap C)$

(5) $8 \notin C$

(6) $(\exists x)\,(x \in C \setminus B \land (\forall y)\,(y \in A \Longrightarrow y < x))$

(7) $\{5\} = A \cap C$

(8) $\{1, 2\} \subseteq C \cup B$

Answer:

$$C = \{9, 7, 5, 2, 1\}$$

Problem 49 A, B, and C are finite sets consisting of positive integers. Find the set C such that all of the following seven claims are true. (All variables assume values from the set of positive integers $N^+ = \{1, 2, \ldots\}$.)

(1) $A = \{x \mid x \leq 4\}$

(2) $B = \{x \mid x > 15\}$

(3) $(\forall x) ((x \in B \cap C) \iff (19 \leq x < 20))$

(4) $(\exists x) (\exists y) (x \in C \land y \in B \cap C \land y = x + 10)$

(5) $8 \in C \setminus A$

(6) $\{1, 2\} \subseteq A \cap C$

(7) C has exactly 5 elements

Answer:

$$C = \{19, 9, 8, 1, 2\}$$

Problem 50 A, B, and C are finite sets consisting of positive integers. Find the set C such that all of the following seven claims are true. (All variables assume values from the set of positive integers $N^+ = \{1, 2, \ldots\}$.)

(1) $A = \{x \mid 10 \leq x^2 \leq 40\}$

(2) $B = \{x \mid 10 \leq 2^x \leq 40\}$

(3) $(\exists x) (x \in A \cap B \cap C)$

(4) $(\forall x) (x \in C \implies (\exists y \in A \cup B)(x \leq 3y))$

(5) $(\forall x)(\forall y) (((\{x, y\} \subseteq C \land x > y) \implies (\exists k)(x - y = 5k))$

(6) $\{10, 11\} \cap C \neq \emptyset$

(7) $(\exists x) (\exists y) (\{x, y\} \subseteq C \land y = 3x)$

Answer:

$$C = \{5, 10, 15\}$$

Problem 51 A, B, and C are finite sets of positive integers, such that all of the following nine claims are true. (All variables assume values from the set $N^+ = \{1, 2, \ldots\}$ of positive integers.)

(1) $A = \{x \mid x \leq 6\}$

(2) $B = \{x \mid x > 16\}$

(3) $(\exists x) (x \in C \land x > 16)$

(4) $(\forall x) ((x \in B \cap C) \iff (19 \leq x < 20))$

(5) $(\forall x) ((x \in C \setminus B) \iff (x \leq 10))$

(6) $8 \in C \setminus A$

(7) $\{6\} = A \cap C$

(8) $(\forall x) ((x \in C \setminus (A \cup B)) \iff (8 \leq x \leq 10))$

(9) C has exactly 5 elements

Complete the following formulae, by writing a natural number into each empty box so that the resulting formula is true. Where such a completion is impossible, explain why.

Answer:

$\boxed{} \in C \setminus \{x \mid x \geq 15\}$ $\boxed{6} \in C \setminus \{x \mid x \geq 15\}$

$\boxed{} \notin C \setminus \{x \mid x \geq 15\}$ $\boxed{19} \notin C \setminus \{x \mid x \geq 15\}$

Answer:

☐ $\in C \setminus \{x \mid x \leq 7\}$	19	$\in C \setminus \{x \mid x \leq 7\}$
☐ $\notin C \setminus \{x \mid x \leq 7\}$	2	$\notin C \setminus \{x \mid x \leq 7\}$
☐ $\in C \cap \{x \mid 6 < x < 12\}$	8	$\in C \cap \{x \mid 6 < x < 12\}$
☐ $\notin C \cap \{x \mid 6 < x < 12\}$	7	$\notin C \cap \{x \mid 6 < x < 12\}$
☐ $\in C \cap \{x \mid (x/2) \in A\}$	10	$\in C \cap \{x \mid (x/2) \in A\}$
☐ $\notin C \cap \{x \mid (x/2) \in A\}$	9	$\notin C \cap \{x \mid (x/2) \in A\}$
☐ $\in C \cap \{x \mid 2x \in B\}$	9	$\in C \cap \{x \mid 2x \in B\}$
☐ $\notin C \cap \{x \mid 2x \in B\}$	11	$\notin C \cap \{x \mid 2x \in B\}$
☐ $\in C \cap \{x \mid (\exists y)(x = 2y + 1)\}$	9	$\in C \cap \{x \mid (\exists y)(x = 2y + 1)\}$
☐ $\notin C \cap \{x \mid (\exists y)(x = 2y + 1)\}$	10	$\notin C \cap \{x \mid (\exists y)(x = 2y + 1)\}$

Note that: $C = \{6, 8, 9, 10, 19\}$.

Problem 52 A, B, and C are finite sets consisting of positive integers. Find the set C such that all of the following seven claims are true. (All variables assume values from the set of positive integers $N^+ = \{1, 2, \ldots\}$.)

(1) $A = \{x \mid (\exists k)(x = 3k + 2)\}$

(2) $B = \{x \mid (\exists k)(x = 4k + 1)\}$

(3) $(\exists x)(20 \leq x \leq 30 \wedge x \in A \cap B \cap C)$

(4) $\{1, 2, 3, 4\} \subseteq C \cap \{x \mid 1 \leq x \leq 10\}$

(5) $(\exists x)(5 < x \leq 10 \wedge x \in A \cap C)$

(6) $(\exists x)(5 < x \leq 10 \wedge x \in B \cap C)$

(7) C has exactly 8 elements.

(8) $(\exists x)(x \in C \cap \{x \mid 10 < x \leq 20 \wedge (\exists k)(x = k^4)\})$

Answer:

$$C = \{1, 2, 3, 4, 8, 9, 16, 29\}$$

1.3   Products and Functions

Problem 53   Let:

$$A = \{1, 2, 3, 4, 5, 6\}$$

and let subset $\rho \subseteq A \times A$ be defined as follows.

$$\rho = \{(1, 2), (2, 3), (3, 4), (3, 5)\}$$

(a) Is ρ a function from A to A? Explain your answer.

Answer: To see that mapping ρ is not a function, observe that ρ assigns two different values, precisely 4 and 5, to the element 3. Additionally, ρ does not assign any values to elements 4, 5, and 6.

(b) Find a function $\sigma : A \to A$ such that $\rho \subseteq \sigma$. If this is impossible, explain why.

Answer: Since ρ already maps the argument 3 to two distinct values, it is impossible to extend ρ a function.

(c) Find a function $\tau : A \to A$ such that $\tau \subseteq \rho$. If this is impossible, explain why.

Answer: Since ρ already leaves arguments 4, 5, and 6 unassigned, it is impossible to restrict ρ to a function.

(d) Find a set $\mu \subseteq A \times A$ such that $\rho \subseteq \mu$ and the mapping μ^{-1} is a function from A to A. If this is impossible, explain why.

Answer: Let μ be defined as follows.

$$\mu = \rho \cup \{(1,1), (6,6)\} = \{(1,2), (2,3), (3,4), (3,5), (1,1), (6,6)\}$$

yielding the function:

$$\mu^{-1} = \{(1,1), (2,1), (3,2), (4,3), (5,3), (6,6)\}$$

(e) Find a set $\nu \subseteq A \times A$ such that $\rho \subseteq \nu$ and the mapping ν^{-1} is an injective function from A to A. If this is impossible, explain why.

Answer: Since ρ^{-1} already maps two distinct elements, 4 and 5, to the same value 3, it is impossible to extend ρ^{-1} to an injective function.

(f) Find a set $\eta \subseteq A \times A$ such that $\rho \subseteq \eta$ and the mapping η^{-1} is a surjective function from A to A. If this is impossible, explain why.

Answer: Every surjective function from any finite set to a set of equal cardinality (including the set itself) is also injective. However, the analysis given in the answer to the previous part shows that ρ^{-1} cannot be extended to an injection. Hence, ρ^{-1} cannot be extended to a surjection.

Problem 54 Sets A and B are defined as follows.

$$A = \{a, b\}$$
$$B = \{1, 2, 3\}$$

(a) Write all possible functions from the set A to the set B, or explain why it is impossible.
Answer: There are

$$|B|^{|A|} = 3^2 = 9$$

functions from A to B. They are defined as follows.

A	f_1	f_2	f_3	f_4	f_5	f_6	f_7	f_8	f_9
a	1	1	1	2	2	2	3	3	3
b	1	2	3	1	2	3	1	2	3

(b) For each (if any) of the functions which you constructed in the answer to part **(a)**, determine if it is injective or not and if it is surjective or not. Explain your answers briefly.

Answer: There are no surjective functions from A to B, since:

$$|B| = 3 > 2 = |A|$$

Six of the nine functions are injective:

$$f_2, f_3, f_4, f_6, f_7, f_8$$

since a and b have distinct images under each of these functions. The remaining 3 functions:

$$f_1, f_5, f_9$$

are not injective, since each of them sends both a and b to the same element of B.

(c) Which (if any) of the functions which you constructed in the answer to part **(a)** have an inverse function whose domain is the set B?

Answer: There are no functions from the set A to the set B that have such an inverse function. While each of the six injective functions, say f, has a partial inverse function f^{-1}, the domain of f^{-1} is the set $f(A) \subset B$ of images of elements of A under f, which is a proper subset of the set B, since A has fewer elements than B.

Problem 55   Sets A, B, C are defined as follows.

$$A = \{a, b, c\}$$
$$B = \{0, 1\}$$

(a) Write all possible functions from the set A to the set B, or explain why it is impossible.

Answer: There are

$$|B|^{|A|} = 2^3 = 8$$

functions from A to B. They are defined as follows.

A	f_0	f_1	f_2	f_3	f_4	f_5	f_6	f_7
a	0	0	0	0	1	1	1	1
b	0	0	1	1	0	0	1	1
c	0	1	0	1	0	1	0	1

(b) For each (if any) of the functions which you constructed in the answer to part **(a)**, determine if it is injective or not and if it is surjective or not. Explain your answers briefly.

Answer: There are no injective functions from A to B, since:

$$|A| = 3 > 2 = |B|$$

Six of the eight functions are surjective:

$$f_1, f_2, f_3, f_4, f_6$$

since every element of the set B is an image of some element of the set A under each of these functions. The remaining 2 functions:

$$f_0, f_7$$

are not surjective, since each of them leaves one element of the set B unassigned.

Problem 56   Sets A and B are defined as follows.

$$A = \{1, 2, 3\}$$
$$B = \{3, 4, 5, 6, 7\}$$

In each of the following cases, construct a set f that satisfies the stated constraints. If this is impossible, explain why.

(a) $f \subseteq A \times B$ and f is not a function from A to B.

Answer:
$$f = \{(1, 3), (2, 3)\}$$

Set f is not a function because it does not contain a pair in which 3 appears as the first component.

(b) $f : A \to B$.

Answer:

$$f = \{(1, 3), (2, 3), (3, 7)\}$$

(c) $f : A \to B$, and f is injective but not surjective.

Answer:

$$f = \{(1, 4), (2, 5), (3, 6)\}$$

(d) $f : A \to B$ and f is surjective but not injective.

Answer: Impossible, since $|B| > |A|$.

(e) $f : A \to B$ and f is not injective and not surjective.

Answer:

$$f = \{(1, 4), (2, 5), (3, 5)\}$$

(f) f is the partial inverse function of the function constructed in the answer to part **(c)**.

Answer:

$$f = \{(4, 1), (5, 2), (6, 3)\}$$

(g) f is the partial inverse function of the function constructed in the answer to part **(e)**.

Answer: Impossible—the original function is not injective and the partial inverse does not exist.

(h) $f : A \to B$ and $|f(A)| = 1$.

Answer:
$$f(x) = 7$$

(i) $f : A \to B$ and $(\exists x)(f(x) = x)$

Answer:
$$f(x) = 3$$

(j) $f : A \to B$; f is injective, and $(\forall x)(2x < f(x))$.

Answer:
$$f(x) = 2x + 1$$

Problem 57   Sets A and B are defined as follows.

$$A = \{2, 3, 4\}$$
$$B = \{0, 1, 2, 3\}$$

In each of the following cases, construct a set f that satisfies the stated constraints. If this is impossible, explain why.

(a) $f \subseteq A \times B$ and f is not a function from A to B.

> **Answer:** The set f defined as follows is not a function because it contains more than one (precisely two) distinct pairs with 4 as the first component.

$$f = \{(2, 0), (3, 0), (4, 2), (4, 3)\}$$

(b) $f : A \to B$.

> **Answer:**

$$f = \{(2, 1), (3, 1), (4, 3)\}$$

(c) $f : A \to B$, and f is injective but not surjective.

> **Answer:**

$$f = \{(2, 1), (3, 2), (4, 3)\}$$

(d) $f : A \to B$ and f is not surjective and not injective.

> **Answer:**

$$f = \{(2, 3), (3, 2), (4, 3)\}$$

(e) $f : A \to B$ and f is surjective but not injective.

> **Answer:** Impossible, since $|B| > |A|$.

(f) f is the partial inverse function of the function constructed in the answer to part **(c)**.

> **Answer:**

$$f = \{(1, 2), (2, 3), (3, 4)\}$$

(g) f is the partial inverse function of the function constructed in the answer to part **(d)**.

> **Answer:** Impossible—f is not injective.

(h) $f : A \to B$ and $f(A) \subseteq A$.

> **Answer:**

$$f = \{(2, 3), (3, 3), (4, 2)\}$$

(i) $f : A \to B$; f is injective, and $(\forall x)(f(x) < x)$.

> **Answer:**
$$f(x) = x - 1$$

(j) $f : A \to B$ and $(\exists x)(f(x) > x)$.

> **Answer:**
$$f(x) = 3$$

Problem 58   Sets S and W are defined as follows.

$$S = \{1, 2, 3, 4\}$$
$$W = \{A, B, C, D, E\}$$

In each of the following cases, provide the required construction or explain why it is impossible.

(a) Construct a bijection $f : S \to W$.

Answer: Impossible:

$$|S| \neq |W|$$

(b) Construct an injection $g : S \to W$.

Answer:

$$g = \{(1, A), (2, B), (3, C), (4, D)\}$$

(c) Construct a surjection $h : W \to S$.

Answer:

$$h = \{(A, 1), (B, 2), (C, 3), (D, 4), (E, 1)\}$$

(d) Construct an injection $e : S \to S \times W$.

Answer:
$$e(x) = (x, A)$$

(e) Construct an injection $b : W \times W \to S$.

Answer: Impossible:

$$|S| = 4 < 25 = |W \times W|$$

(f) Construct a function $c : W \to S$ such that c is not injective.

Answer:

$$c = \{(A, 1), (B, 2), (C, 3), (D, 4), (E, 1)\}$$

(g) Construct a function $d : W \to S$ such that d is not surjective.

Answer:

$$d = \{(A, 1), (B, 3), (C, 2), (D, 1), (E, 1)\}$$

(h) Construct a relation $\rho \subseteq S \times W$ such that ρ is not a function from S to W.

Answer:
$$\rho = \{(1, A), (3, B), (2, C)\}$$

ρ is not a function because it leaves the element 4 unassigned.

(i) Construct a relation $\sigma \subseteq W \times S$ such that σ is not a function from W to S.

Answer:

$$\sigma = \{(A, 1), (B, 1), (C, 1), (D, 1), (E, 1), (E, 2)\}$$

σ is not a function because it assignes two different values to the element E.

Problem 59   Sets A, B, and C are defined as follows.

$$A = \{1, 2, 3\}$$
$$B = \{2, 3, 4, 5\}$$
$$C = \{3, 4, 5\}$$

In each of the following cases, provide the required construction or explain why it is impossible.

(a) Construct three functions from $A \cup C$ to B.

Answer:

$$f_1(x) = 2$$
$$f_2(x) = 3$$
$$f_3(x) = 4$$

(b) Construct three functions from $A \cap C$ to B.

Answer:

$$f_1(x) = 2$$
$$f_2(x) = 3$$
$$f_3(x) = 4$$

(c) Construct three functions from B to $A \cap C$.

Answer: Impossible—observe that:

$$|B| = |\{2, 3, 4, 5\}| = 4$$

whereas:

$$|A \cap C| = |\{3\}| = 1$$

Hence, the number of functions from B to $A \cap C$ is equal to:

$$(|A \cap C|)^{|B|} = 1^4 = 1$$

Indeed, there exists only one such function:

$$f(x) = 3$$

(d) Construct three injections from C to B.

Answer:

x	$f_1(x)$	$f_2(x)$	$f_3(x)$
3	2	3	2
4	3	4	3
5	4	5	5

(e) Construct three surjections from A to B.

Answer: Impossible—observe that:

$$|B| = |\{2, 3, 4, 5\}| = 4$$

whereas:

$$|A| = |\{1, 2, 3\}| = 3$$

Since $|A| < |B|$, there are no surjections from A to B.

Problem 60   Sets A, B, and C are defined as follows.

$$A = \{1, 2, 3\}$$
$$B = \{2, 3, 4, 5\}$$
$$C = \{3, 4\}$$

In each of the following cases, provide the required construction or explain why it is impossible.

(a) Construct two functions from A to B.

Answer:

A	$f(A)$	$g(A)$
1	2	3
2	2	3
3	2	3

(b) Construct two functions from $A \times C$ to B.

Answer:

$A \times C$	$f(A \times C)$	$g(A \times C)$
$(1, 3)$	2	3
$(1, 4)$	3	4
$(2, 3)$	4	5
$(2, 4)$	5	2
$(3, 3)$	2	3
$(3, 4)$	3	4

(c) Construct two injections from $A \times C$ to B.

Answer: Impossible, since:

$$|A \times C| = 6 > 4 = |B|$$

(d) Construct two surjections from B to C.

Answer:

B	$f(B)$	$g(B)$
2	3	4
3	4	3
4	4	3
5	3	3

(e) Construct two surjections from A to C.

Answer:

A	$f(A)$	$g(A)$
1	3	4
2	4	3
3	3	4

(f) List all elements of $\mathcal{P}(\mathcal{P}(A \cap C))$.

Answer:

$$\emptyset, \{\emptyset\}, \{\{3\}\}, \{\emptyset, \{3\}\}$$

Problem 61   Let sets A, B, and C be defined as follows.

$$A = \{1, 2, 3\}$$
$$B = \{2, 3, 4, 5, 6\}$$
$$C = \{3, 4\}$$

In each of the following cases, provide the required construction or explain why it is impossible.

(a) Construct two functions from A to B.

Answer:

$$f_1(x) = x + 1$$
$$f_2(x) = x + 2$$

(b) Construct two functions from $C \times A$ to B.

Answer:

$C \times A$	$f_1(C \times A)$	$f_2(C \times A)$
$(3, 1)$	2	3
$(3, 2)$	4	3
$(3, 3)$	5	3
$(4, 1)$	2	3
$(4, 2)$	3	3
$(4, 3)$	6	3

(c) Construct two injections from B to $C \times A$.

Answer:

B	$f_1(B)$	$f_2(B)$
2	$(3, 1)$	$(4, 3)$
3	$(3, 2)$	$(3, 1)$
4	$(3, 3)$	$(3, 2)$
5	$(4, 1)$	$(3, 3)$
6	$(4, 2)$	$(4, 1)$

(d) Construct two surjections from B to C.

Answer:

B	$f_1(B)$	$f_2(B)$
2	4	3
3	3	3
4	4	3
5	4	3
6	3	4

(e) Construct two surjections from B to A.

Answer:

$$f_1(x) = \left\lfloor \frac{x}{2} \right\rfloor$$
$$f_2(x) = \left\lceil \frac{x}{2} \right\rceil$$

Problem 62   Sets A, B, and C are defined as follows.

$$A = \{1, 2, 3\}$$
$$B = \{x \mid (\exists y)((y \in A) \wedge (y < x < 2y))\}$$
$$C = \{x \mid (\exists y)((y \in B) \wedge (y < x))\}$$

(All variables assume values from the set of natural numbers $N = \{0, 1, \ldots\}$.)

In each of the following cases, provide the required construction or explain why it is impossible.

(a) List all elements of C.

> **Answer:** It is impossible to list all elements of C, since C is an infinite set, which consists of all natural numbers greater than 3.

(b) List all elements of $A \times B$.

> **Answer:**
>
> $$(1, 3), (2, 3), (3, 3), (1, 4), (2, 4), (3, 4), (1, 5), (2, 5), (3, 5)$$

(c) Construct a surjection $f : A \to B$.

> **Answer:**
>
> $$f(x) = x + 2$$

(d) Construct an bijection $g : A \to B$.

> **Answer:**
>
> $$g(x) = x + 2$$

(e) Construct an injection $h : B \to C$.

> **Answer:**
>
> $$h(x) = 2x$$

(f) Construct an injection $b : A \to (B \times A)$.

> **Answer:**
>
> $$b(x) = (3, x)$$

(g) Construct a relation ρ between A and B such that ρ is not a function, but would be a function if one of its elements was removed, and specify that element.

> **Answer:**
>
> $$\rho = \{(1, 3), (2, 3), (3, 5)(3, 4)\}$$
>
> Relation $\rho \setminus \{(3, 4)\}$ is a function.

(h) Construct a relation σ between A and B such that σ is not a function, but would be a function if one element was added to it, and specify that element.

> **Answer:**
>
> $$\sigma = \{(1, 3), (3, 4)\}$$
>
> Relation $\sigma \cup \{(2, 3)\}$ is a function.

Problem 63   Sets A, B, C, D, E are defined as follows.

$$A = \{1, 2, 3\}$$
$$B = \{14, 15, 16, 17, 18\}$$
$$C = \{z \mid (\exists x \in A)(\exists y \in B)(z = x + y)\}$$
$$D = \{x \mid (\forall y \in A \cup B)(x > y)\}$$
$$E = \{x \mid (\exists y \in A \cup B)(x > y)\}$$

(All variables assume values from the set of natural numbers $N = \{0, 1, \ldots\}$.)
In each of the following cases, provide the required construction or explain why it is impossible.

(a) List all elements of C.

Answer:

$$15, 16, 17, 18, 19, 20, 21$$

(b) List all elements of D.

Answer: It is impossible to list all elements of D, since D is an infinite set, which consists of all natural numbers greater than 18:

$$D = \{x \mid x > 18\}$$

(c) List all elements of E.

Answer: It is impossible to list all elements of E, since E is an infinite set, which consists of all natural numbers greater than 1:

$$E = \{x \mid x > 1\}$$

(d) Construct an injection $f : A \rightarrow C$.

Answer:

$$f = \{(1, 15) ,\ (2, 16) , (3, 17)\}$$

(e) Construct a surjection $g : A \rightarrow C$.

Answer: There are no surjections from A to C, since A has fewer elements that C:

$$|A| = 3 < 7 = |C|$$

(f) Construct a surjection $g_1 : C \rightarrow A$.

Answer:

$$g_1 = \{(15, 1), (16, 2), (17, 3), (18, 1), (19, 2), (20, 1), (21, 1)\}$$

(g) Construct a bijection $h : A \rightarrow B$.

Answer: There are no bijections between A and B, since A and B have different cardinalities.

(h) List all elements of $C \times A$.

Answer:

$$(15, 1), (16, 1), (17, 1), (18, 1), (19, 1), (20, 1), (21, 1),$$
$$(15, 2), (16, 2), (17, 2), (18, 2), (19, 2), (20, 2), (21, 2),$$
$$(15, 3), (16, 3), (17, 3), (18, 3), (19, 3), (20, 3), (21, 3)$$

Problem 64 Sets A, B, and C are defined as follows.

$$A = \{0, 1, 2, 3, 4, 5\}$$
$$B = \{11, 12, 13, 14, 15, 16\}$$
$$C = \{z \mid (\exists x \in A)(\exists y \in B)(y - x = 2^z)\}$$

(All variables assume values from the set of natural numbers $N = \{0, 1, \ldots\}$.)

In each of the following cases, provide the required construction or explain why it is impossible.

(a) List all elements of C.

Answer:

$$3, 4$$

(b) Construct three injective functions from C to A.

Answer:

$$f_1(x) = x$$
$$f_2(x) = x + 1$$
$$f_3(x) = x - 1$$

(c) Construct three surjective functions from C to A.

Answer: Impossible—since $|C| = 2 < 6 = |A|$, there are no surjections from C to A.

(d) Construct three bijective functions from B to A.

Answer:

x	$f_1(x)$	$f_2(x)$	$f_3(x)$
11	0	1	2
12	1	2	3
13	2	3	4
14	3	4	5
15	4	5	0
16	5	0	1

(e) Construct three functions from B to A which are not bijective.

Answer:

x	$f_1(x)$	$f_2(x)$	$f_3(x)$
11	0	1	2
12	1	2	3
13	2	3	4
14	0	4	5
15	4	5	5
16	5	4	1

(f) Construct three functions from C to A which are not injective.

Answer:

$$f_1(x) = 0$$
$$f_2(x) = 1$$
$$f_3(x) = 2$$

Problem 65   Sets A, B, C, D, E, F are defined as follows.

$$A = \{3, 4, 5, 6\}$$
$$B = \{11, 12, 13, 14, 15, 16, 17, 18, 19, 20\}$$
$$C = \{z \mid (\exists x \in A)(\exists y \in B)(y = xz)\}$$
$$D = \{y \in B \mid (\forall x \in A)(y > 2x)\}$$
$$E = \{x \mid 1 \le x \le 5\}$$
$$F = E \setminus A$$

(All variables assume values from the set of natural numbers $N = \{0, 1, \ldots\}$.)

In each of the following cases, provide the required construction or explain why it is impossible.

(a) List all elements of C.

Answer:

$$2, 3, 4, 5, 6$$

(b) List all elements of D.

Answer:

$$13, 14, 15, 16, 17, 18, 19, 20$$

(c) Construct two functions from B to A.

Answer:

$$f_1(x) = 3$$
$$f_2(x) = 4$$

(d) Construct two injective functions from A to B.

Answer:

$$f_1(x) = x + 8$$
$$f_2(x) = x + 9$$

(e) List all elements of $F \times A$.

Answer:

$$(1, 3), \ (1, 4), \ (1, 5), \ (1, 6),$$
$$(2, 3), \ (2, 4), \ (2, 5), \ (2, 6)$$

(f) Construct two injective functions from $F \cup A$ to D.

Answer:

$$f_1(x) = x + 12$$
$$f_2(x) = x + 14$$

(g) Construct two functions from C to $F \cap A$.

Answer: Impossible—the set $F \cap A$ is empty and cannot be the codomain of any function.

(h) Construct two functions from $C \cap D$ to D.

Answer: Impossible—the set $C \cap D$ is empty and cannot be the domain of any function.

Problem 66   Sets A, B, and C are defined as follows.

$$A = \{1, 2, 3\}$$
$$B = \{5, 6, 7\}$$
$$C = \{x \mid (\exists y)((y \in A \lor y \in B) \land x = 2y)\}$$

(All variables assume values from the set of natural numbers $N = \{0, 1, \ldots\}$.)

In each of the following cases, provide the required construction or explain why it is impossible.

(a) Construct a surjection $g : A \to C$.

> **Answer:** Impossible, since:
>
> $$C = \{2, 4, 6, 10, 12, 14\}.$$
>
> Hence, $|C| = 6 > 3 = |A|$, while the domain of a surjection cannot have fewer elements than the codomain.

(b) Construct an injection $f : A \to C$.

> **Answer:**
>
> | A | $f(A)$ |
> |-----|--------|
> | 1 | 2 |
> | 2 | 4 |
> | 3 | 6 |

(c) Construct an injection $h : A \to B$, such that h is not surjective.

> **Answer:** Impossible. Since $|A| = |B| = 3$, every injection from A to B is also surjective.

(d) Construct an injection $e : (A \cup B) \to C$.

> **Answer:**
>
> $$e(x) = 2x$$

(e) Construct an injection $a : (A \cup C) \to B$.

> **Answer:** Impossible, since $|A \cup C| = 8$, while $|B| = 3$. The domain of an injection cannot have more elements than the codomain.

(f) List all elements of $C \times A$.

> **Answer:**
>
> $(2, 1)$, $(2, 2)$, $(2, 3)$, $(4, 1)$, $(4, 2)$, $(4, 3)$,
> $(6, 1)$, $(6, 2)$, $(6, 3)$, $(10, 1)$, $(10, 2)$, $(10, 3)$,
> $(12, 1)$, $(12, 2)$, $(12, 3)$, $(14, 1)$, $(14, 2)$, $(14, 3)$

(g) Construct a relation ρ between C and B such that ρ is not a function.

> **Answer:**
>
> $$\rho = \{(14, 6)\}$$
>
> ρ is not a function, since it leaves some elements of the set C unassigned.

(h) Construct a function $d : (A \cup C) \to B$.

Answer:
$$d(x) = 5$$

(i) Construct a surjection $b : (A \cup C) \to B$.

Answer:
$$b(x) = \left\lfloor \frac{x}{7} \right\rfloor + 5$$

(j) Calculate the value of $b(1)$, $b(10)$, and $b(14)$, for the function b of the previous part.

Answer:
$$b(1) = \left\lfloor \frac{1}{7} \right\rfloor + 5 = 0 + 5 = 5$$
$$b(10) = \left\lfloor \frac{10}{7} \right\rfloor + 5 = 1 + 5 = 6$$
$$b(14) = \left\lfloor \frac{14}{7} \right\rfloor + 5 = 2 + 5 = 7$$

Problem 67   Sets A, B, C, D, E are defined as follows.

$$A = \{1, 2, 3\}$$
$$B = \{3, 4, 5\}$$
$$C = \left\{ z \mid (\exists x \in B)(\exists y \in B)\left(x^2 + 3 < z < y^2 - 3\right) \right\}$$
$$D = \{x \mid (\exists y \in A)(\exists z \in N)((y < 2) \wedge (x = 3z + y))\}$$
$$E = \{x \mid (\exists y \in A)(\exists z \in N)((y \geq 2) \wedge (x = 3z + y))\}$$

(All variables assume values from the set of natural numbers $N = \{0, 1, \ldots\}$.)

In each of the following cases, provide the required construction or explain why it is impossible.

(a) List all elements of C.

Answer:
$$13, 14, 15, 16, 17, 18, 19, 20, 21$$

(b) List all elements of D.

Answer: This is impossible, since D is an infinite set, which contains all natural numbers whose remainder after the division by 3 is equal to 1.

(c) List all elements of E.

Answer: This is impossible, since E is an infinite set, which contains all positive natural numbers whose remainder after the division by 3 is equal to 0 or 2. (Observe that $0 \notin E$.)

(d) List all elements of $E \cap D$.

Answer: There are no elements to list:

$$E \cap D = \emptyset$$

(e) List all elements of $\mathcal{P}(\mathcal{P}(E \cap D))$.

$$\varnothing, \{\varnothing\}$$

(f) List all elements of $N \setminus (D \cup E)$.

Answer:

$$0$$

(g) List all subsets of $N \setminus (D \cup E)$.

Answer:

$$\varnothing, \{0\}$$

(h) Construct an injection $f : (A \cup B) \to C$.

Answer:

x	1	2	3	4	5
$f(x)$	13	14	15	16	17

(i) Construct an injection $g : (A \times B) \to C$.

Answer:

$$g\left((x, y)\right) = 7 + 3x + y$$

(j) Calculate $g\left((2, 5)\right)$, where g is the function constructed in the answer to the previous part.

Answer:

$$g\left((2, 5)\right) = 7 + 3 \cdot 2 + 5 = 18$$

(k) Construct a function $h : A \to B$, such that h is not injective.

Answer:

$$h(1) = h(2) = h(3) = 4$$

(l) Construct a function $e : A \to B$, such that e is not surjective.

Answer:

$$e(1) = e(2) = e(3) = 4$$

(m) Construct a surjection $\ell : (A \times B) \to C$, such that ℓ is not injective.

Answer: This is impossible, since the domain and the codomain are finite sets that have equal cardinalities:

$$|A \times B| = |C| = 9$$

Hence, every surjection (or injection) is also a bijection.

Problem 68   Let S be an arbitrary set of natural numbers: $S \subseteq N$. The characteristic function $f_S : N \to \{0, 1\}$ of the set S is defined as follows.

$$f_S(x) = \begin{cases} 1 & \text{if } x \in S \\ 0 & \text{if } x \notin S \end{cases}$$

Sets A, B, C are defined as follows.

$$A = \{1, 2, 3, 4, 5\}$$
$$B = \{3, 4, 5, 6, 7\}$$
$$C = \{0, 2, 4, 6, 8\}$$

Complete each of the following formulae, by writing a number into the empty box, so as to obtain a true claim.

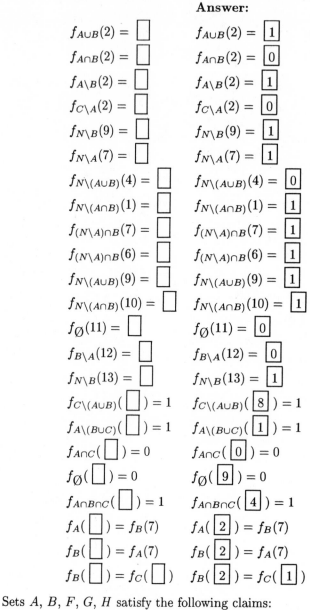

Answer:

$f_{A \cup B}(2) = \boxed{}$ $f_{A \cup B}(2) = \boxed{1}$

$f_{A \cap B}(2) = \boxed{}$ $f_{A \cap B}(2) = \boxed{0}$

$f_{A \setminus B}(2) = \boxed{}$ $f_{A \setminus B}(2) = \boxed{1}$

$f_{C \setminus A}(2) = \boxed{}$ $f_{C \setminus A}(2) = \boxed{0}$

$f_{N \setminus B}(9) = \boxed{}$ $f_{N \setminus B}(9) = \boxed{1}$

$f_{N \setminus A}(7) = \boxed{}$ $f_{N \setminus A}(7) = \boxed{1}$

$f_{N \setminus (A \cup B)}(4) = \boxed{}$ $f_{N \setminus (A \cup B)}(4) = \boxed{0}$

$f_{N \setminus (A \cap B)}(1) = \boxed{}$ $f_{N \setminus (A \cap B)}(1) = \boxed{1}$

$f_{(N \setminus A) \cap B}(7) = \boxed{}$ $f_{(N \setminus A) \cap B}(7) = \boxed{1}$

$f_{(N \setminus A) \cap B}(6) = \boxed{}$ $f_{(N \setminus A) \cap B}(6) = \boxed{1}$

$f_{N \setminus (A \cup B)}(9) = \boxed{}$ $f_{N \setminus (A \cup B)}(9) = \boxed{1}$

$f_{N \setminus (A \cap B)}(10) = \boxed{}$ $f_{N \setminus (A \cap B)}(10) = \boxed{1}$

$f_{\emptyset}(11) = \boxed{}$ $f_{\emptyset}(11) = \boxed{0}$

$f_{B \setminus A}(12) = \boxed{}$ $f_{B \setminus A}(12) = \boxed{0}$

$f_{N \setminus B}(13) = \boxed{}$ $f_{N \setminus B}(13) = \boxed{1}$

$f_{C \setminus (A \cup B)}(\boxed{}) = 1$ $f_{C \setminus (A \cup B)}(\boxed{8}) = 1$

$f_{A \setminus (B \cup C)}(\boxed{}) = 1$ $f_{A \setminus (B \cup C)}(\boxed{1}) = 1$

$f_{A \cap C}(\boxed{}) = 0$ $f_{A \cap C}(\boxed{0}) = 0$

$f_{\emptyset}(\boxed{}) = 0$ $f_{\emptyset}(\boxed{9}) = 0$

$f_{A \cap B \cap C}(\boxed{}) = 1$ $f_{A \cap B \cap C}(\boxed{4}) = 1$

$f_A(\boxed{}) = f_B(7)$ $f_A(\boxed{2}) = f_B(7)$

$f_B(\boxed{}) = f_A(7)$ $f_B(\boxed{2}) = f_A(7)$

$f_B(\boxed{}) = f_C(\boxed{})$ $f_B(\boxed{2}) = f_C(\boxed{1})$

Problem 69   Sets A, B, F, G, H satisfy the following claims:

$$A = \{1, 2, 3, 4, 5\}$$
$$B = \{a, b, c, d, e\}$$
$$F = \{(1, e), (2, d), (3, a), (4, b)\} \subseteq A \times B$$
$$G = \{(1, b), (3, a), (5, d)\} \subseteq A \times B$$
$$H \subseteq A \times B \text{ and } H \neq \emptyset$$

In each of the following cases, construct a set H that satisfies the stated additional constraints, or explain why it is impossible.

	Answer:
$F \cup H$ is a function.	$H = \{(5, b)\}$
$F \cup H$ is not a function.	$H = \{(4, a)\}$
$G \cup H$ is a function.	$H = \{(2, a), (4, a)\}$
$G \cup H$ is not a function.	$H = \{(2, a)\}$
$F \cup H$ is an injection.	$H = \{(5, c)\}$
$F \cup H$ is a function but is not a surjection.	$H = \{(5, b)\}$
$F \cup H$ is a bijection.	$H = \{(5, c)\}$
$F \cup H$ is a surjection.	$H = \{(5, c)\}$
$G \cup H$ is an injection.	$H = \{(2, c), (4, e)\}$
$G \cup H$ is a surjection.	$H = \{(2, e), (4, c)\}$
$G \cup H$ is a bijection.	$H = \{(2, c), (4, e)\}$
$G \cup H$ is a function but is not an injection.	$H = \{(2, c), (4, c)\}$
$G \cup H$ is a function but is not a surjection.	$H = \{(2, a), (4, b)\}$
$G \cup H$ is a function but is not a bijection.	$H = \{(2, a), (4, e)\}$

Problem 70   Let $N^+ = \{1, 2, \ldots\}$ be the set of positive integers, and let $f : N^+ \to N^+$ and $g : N^+ \to N^+$ be two functions such that:

$$\{(1, 2), (3, 5), (4, 4), (2, 5), (5, 3)\} \subseteq f$$

and

$$(\forall x \in N^+)(g(x) = 2x)$$

For each of the following claims, determine if the claim is true or false and explain your answer briefly.

(a) g is a surjective.

Answer: False. Odd numbers are not in the range of g.

(b) g is injective.

Answer: True. Indeed, by the definition of g:

$$g(x) = g(y) \iff 2x = 2y \iff x = y$$

(c) f is injective.

Answer: False. We conclude, by inspecton of the given subset of f, that f maps distinct elements of its domain into the same value:
$$f(2) = f(3) = 5$$

(d) $(\exists x)(f(x) = g(x))$

Answer: True. Indeed:

$$f(1) = g(1) = 2$$

(e) if $(\forall x > 5)(f(x) = x + 5)$ then f is surjective.

Answer: False. For instance, 1 is not in the range of f.

(f) if $(\forall x > 5)(f(x) = x)$ then f is injective.

> **Answer: False.** The analysis given in the answer to the part **(c)** shows that f cannot be extended into an injective function.

(g) $(\forall x)(\exists y)(f(x) < g(y))$

> **Answer: True.** Let $y = f(x)$:

$$(\forall x)(f(x) < 2f(x) = g(f(x)))$$

Problem 71   For any set A and a function $f : A \to A$, the set $\mathcal{F}(f)$ of fixed points of f is defined as follows.

$$\mathcal{F}(f) = \{a \mid a \in A \land f(a) = a\}$$

For each of the following functions, find the set of fixed points and calculate the cardinality of this set. Assume that the set of natural numbers $N = \{0, 1, \dots\}$ is the domain and the codomain of all the functions.

(a) $g_1(n) = 3n$

> **Answer:** $\mathcal{F}(g_1)$ is the set of points such that:

$$3n = n$$

> which yields a singleton set:

$$\mathcal{F}(g_1) = \{0\}$$

(b) $g_2(n) = 2n + 1$

> **Answer:** $\mathcal{F}(g_2)$ is the set of points such that:

$$2n + 1 = n$$

> This claim is not satisfied by any natural numbers. hence:

$$\mathcal{F}(g_2) = \emptyset$$

(c) $g_3(n) = n^3$

> **Answer:** $\mathcal{F}(g_3)$ is the set of points such that:

$$n^3 = n$$

> which yields the two-element set:

$$\mathcal{F}(g_3) = \{0, 1\}$$

(d) $g_4(n) = n^3 - 24$

> **Answer:** $\mathcal{F}(g_4)$ is the set of points such that:

$$n^3 - 24 = n$$

> which yields a singleton set:

$$\mathcal{F}(g_4) = \{3\}$$

(e) $g_5(n) = 2^n - 12$

Answer: $\mathcal{F}(g_5)$ is the set of points such that:

$$2^n - 12 = n$$

which yields a singleton set:

$$\mathcal{F}(g_5) = \{4\}$$

(f) $g_6(n) = (-1)^n \cdot n$

Answer: $\mathcal{F}(g_6)$ is the set of points such that:

$$(-1)^n \cdot n = n$$

which yields an infinite set of all even natural numbers:

$$\mathcal{F}(g_6) = \{n \in N \mid (\exists k \in N)(n = 2k)\}$$

(g) $g_7(n) = n^2 - 3n + 3$

Answer: $\mathcal{F}(g_7)$ is the set of points such that:

$$n^2 - 3n + 3 = n \iff n^2 - 4n + 3 = 0 \iff (n-1)(n-3) = 0$$

which yields a set with two elements:

$$\mathcal{F}(g_7) = \{1, 3\}$$

Problem 72   Sets N^+, Z, Q are defined as follows.

$$N^+ = \{1, 2, \ldots\}, \quad \text{(positive integers)} ;$$
$$Z = \{\ldots, -2, -1, 0, 1, 2, \ldots\}, \quad \text{(integers)} ;$$
$$Q = \{(p/q) \mid p \in Z \wedge q \in N^+\}, \quad \text{(rational numbers)}$$

For each of the following functions, determine the range of the function and construct the (partial) inverse function or explain why it is impossible.

(a) $f_1 : N^+ \to N^+$ such that $f_1(x) = 3x$.

Answer: The range of f_1 is the set of all positive integers divisible by 3:

$$f_1(N^+) = \{y \in N^+ \mid (\exists n \in N^+)(y = 3n)\}$$

and the partial inverse function of f_1 is:

$$f_1^{-1} : y \mapsto \frac{y}{3} \text{ for } y \in f_1(N^+)$$

(b) $f_2 : Q \to Q$ such that $f_2(x) = 3x$.

Answer: The range of f_2 is the set of all rational numbers

$$f_2(Q) = Q$$

and the inverse function of f_2 is:

$$f_2^{-1} : y \mapsto \frac{y}{3} \text{ for } y \in Q$$

(c) $f_3 : N^+ \to N^+$ such that $f_3(x) = x + 5$.

> **Answer:** The range of f_3 is the set of all positive integers greater than 5:
>
> $$f_3(N^+) = \{y \in N^+ \mid y > 5\} = N^+ \setminus \{1, 2, 3, 4, 5\}$$
>
> and the partial inverse function of f_3 is:
>
> $$f_3^{-1} : y \mapsto y - 5 \text{ for } y \in N^+ \setminus \{1, 2, 3, 4, 5\}$$

(d) $f_4 : Z \to Z$ such that $f_4(x) = x + 5$.

> **Answer:** The range of f_4 is the set of all integers:
>
> $$f_4(Z) = Z$$
>
> and the inverse function of f_4 is:
>
> $$f_4^{-1} : y \mapsto y - 5 \text{ for } y \in Z$$

(e) $f_5 : N^+ \to N^+$ such that $f_5(x) = x^2$.

> **Answer:** The range of f_5 is the set of all positive integers that are perfect squares:
>
> $$f_5(N^+) = \{y \in N^+ \mid (\exists n \in N^+)(y = n^2)\}$$
>
> and the partial inverse function of f_5 is:
>
> $$f_5^{-1} : y \mapsto \sqrt{y} \text{ for } y \in f_5(N^+)$$

(f) $f_6 : Z \to Z$ such that $f_6(x) = x^2$.

> **Answer:** The range of f_6 is the set of all non-negative integers that are perfect squares:
>
> $$f_6(Z) = \{y \in Z \mid (\exists n \in Z)(y = n^2)\} = f_5(N^+) \cup \{0\}$$
>
> The inverse function of f_6 does not exist, since f_6 sends two distinct arguments to every non-zero value in its range:
>
> $$f_6(x) = f_6(-x) = x^2$$

(g) $f_7 : N^+ \to N^+$ such that $f_7(x) = \left\lfloor \dfrac{x + 5}{5} \right\rfloor$.

> **Answer:** The range of f_7 is the set of all positive integers.
>
> $$f_7(N^+) = N^+$$
>
> The inverse function of f_7 does not exist, since f_7 sends five distinct elements of its domain to each value in its range. Precisely, for any r such that: $0 \leq r < 5$:
>
> $$f_7(5k + r) = \left\lfloor \frac{(5k + r) + 5}{5} \right\rfloor = \left\lfloor \frac{5(k + 1) + r}{5} \right\rfloor = k + 1 + \left\lfloor \frac{r}{5} \right\rfloor = k + 1 + 0 = k + 1$$

Chapter 2

Recursion

2.1 Induction

Problem 73   Use mathematical induction to prove that:

(handwritten: $(n-5)(n-3) \geq 0$)

$$n^2 - 8n + 15 \geq 0$$

for every natural number n greater than 4.

Answer: Let:

$$L(n) = n^2 - 8n + 15$$

The claim is that $L(n) \geq 0$ whenever $n \geq 5$.

The base case occurs for $n = 5$, and requires a proof that $L(5) \geq 0$.

$$L(5) = 5^2 - 8 \cdot 5 + 15 = 25 - 40 + 15 = 0 \geq 0$$

whence the base case.

For the inductive step, assume:

$$L(n) \geq 0$$

for some $n \geq 5$. We have to prove that:

$$L(n + 1) \geq 0$$

Indeed:

$$
\begin{aligned}
L(n+1) \quad &= \\
\triangle \qquad &= (n+1)^2 - 8(n+1) + 15 = (n^2 + 2n + 1) - 8n - 8 + 15 \\
&= (n^2 - 8n + 15) + (2n - 7) \\
\bullet \qquad &\geq 0 + 2n - 7 = 2n - 7 = 2(n - 4) + 1 \\
\star \qquad &> 0 + 1 > 0
\end{aligned}
$$

The inequality $\bullet$ follows from the inductive hypothesis, equality $\triangle$ follows from the definition of L, the inequality $\star$ follows from the assumption that $n \geq 5$, while the other (in)equalities are obtained by arithmetic transformations.

Problem 74   Use mathematical induction to prove that:

$$3^n > 3n + 50$$

for every natural number n greater than 3.

Answer: Let:

$$L(n) = 3^n$$
$$R(n) = 3n + 50$$

The claim is that $L(n) > R(n)$ whenever $n \geq 4$.

The base case occurs for $n = 4$:

$$L(4) = 3^4 = 81$$
$$R(4) = 3 \cdot 4 + 50 = 62$$

Since $81 > 62$, we conclude that: $L(4) > R(4)$, verifying the base case.

For the inductive step, assume that:

$$L(n) > R(n)$$

for some $n \geq 4$. We have to prove that:

$$L(n + 1) > R(n + 1)$$

Indeed:

$$
\begin{aligned}
L(n+1) \quad &= \\
\triangle \qquad &= 3^{n+1} = 3 \cdot 3^n \\
\triangle \qquad &= 3L(n) \\
\bullet \qquad &> 3R(n) = R(n) + R(n) + R(n) \\
\nabla \qquad &= (3n + 50) + (3n + 50) + (3n + 50) = 3n + 3 + 50 + 6n + 97 \\
&= (3(n + 1) + 50) + (6n + 97) \\
\nabla \qquad &= R(n + 1) + (6n + 97) > R(n + 1) + 0 = R(n + 1)
\end{aligned}
$$

The inequality $\bullet$ follows from the inductive hypothesis, equalities $\triangle$ follow from the definition of L, equalities ∇ follow from the definition of R, while the other(in)equalities are obtained by arithmetic transformations.

Problem 75   Use mathematical induction to prove that:

$$n^3 - 49 > 3n$$

for every natural number $n \geq 4$.

Answer: The base case occurs for $n = 4$, and requires a proof that $4^3 - 49 > 3 \cdot 4$.

$$4^3 - 49 = 64 - 49 = 15 > 12 = 3 \cdot 4$$

Inductively, assume that:

$$n^3 - 49 > 3n$$

for some natural number $n \geq 4$. We need to prove that:

$$(n + 1)^3 - 49 > 3(n + 1)$$

$$
\begin{aligned}
(n+1)^3 - 49 \quad &= n^3 + 3n^2 + 3n + 1 - 49 = (n^3 - 49) + (3n^2 + 3n + 1) \\
\bullet \quad &> 3n + (3n^2 + 3n + 1) = (3n + 3n) + (3n^2 + 1) \\
* \quad &\geq (3n + 3 \cdot 4) + (3n^2 + 1) > (3n + 3) + (3n^2 + 1) = 3(n+1) + (3n^2 + 1) > 3(n+1)
\end{aligned}
$$

The inequality • is obtained from the inductive hypothesis; * follows from the assumption that $n \geq 4$; and the other (in)equalities are obtained by arithmetic transformations.

Problem 76   Use mathematical induction to prove:

$$3^{n+6} - 10 < (n+6)!$$

for every natural number n.

Answer: The base case occurs for $n = 0$, and requires a proof that $3^{0+6} - 10 < (0+6)!$.

$$3^{0+6} - 10 = 3^6 - 10 = 729 - 10 = 719 < 720 = 6 \cdot 5 \cdot 4 \cdot 3 \cdot 2 = 6!$$

Inductively, assume that:

$$3^{n+6} - 10 < (n+6)!$$

for some natural number n. We need to prove that:

$$3^{(n+1)+6} - 10 < ((n+1)+6)!$$

or, equivalently:

$$3^{n+7} - 10 < (n+7)!$$

Indeed:

$$
\begin{aligned}
3^{n+7} - 10 \quad &= 3 \cdot 3^{n+6} - 10 = 3 \cdot 3^{n+6} - 30 + 20 = 3 \left(3^{n+6} - 10\right) + 20 \\
\bullet \quad &< 3 \cdot (n+6)! + 20 < 3 \cdot (n+6)! + 720 = 3 \cdot (n+6)! + 6! \\
* \quad &\leq 3 \cdot (n+6)! + (n+6)! = 4 \cdot (n+6)! < 7 \cdot (n+6)! = (n+7)!
\end{aligned}
$$

The inequality • follows from the inductive hypothesis; * is based on the fact that $n!$ is an increasing function of n—hence, $(n+6)! \geq 6!$ since $n+6 \geq 6$; and the other (in)equalities are obtained by arithmetic transformations.

Problem 77   Use mathematical induction to prove that every natural number $n \geq 110$ can be represented in the form:

$$n = \alpha \cdot 6 + \beta \cdot 23$$

for some non-negative integers α and β.

Answer: In the base case:

$$
\begin{aligned}
110 &= 3 \cdot 6 + 4 \cdot 23 \\
111 &= 7 \cdot 6 + 3 \cdot 23 \\
112 &= 11 \cdot 6 + 2 \cdot 23 \\
113 &= 15 \cdot 6 + 1 \cdot 23 \\
114 &= 19 \cdot 6 + 0 \cdot 23 \\
115 &= 0 \cdot 6 + 5 \cdot 23
\end{aligned}
$$

Inductively, assume that for an arbitrary $n \geq 115$:

$$k = \alpha_k \cdot 6 + \beta_k \cdot 23$$

for all k such that $110 \leq k \leq n$, where α_k and β_k are non-negative integers (that depend on k.)

We have to show that:

$$n + 1 = \alpha' \cdot 6 + \beta' \cdot 23$$

for some non-negative integers α' and β'.

The inductive hypothesis applies to $n - 5$, since:

$$n - 5 < n$$

and also:

$$110 = 115 - 5 \leq n - 5$$

Hence:

$$n - 5 = a \cdot 6 + b \cdot 23$$

for some non-negative integers a and b, yielding:

$$n + 1 = (n - 5) + 6 = a \cdot 6 + b \cdot 23 + 6 = (a + 1) \cdot 6 + b \cdot 23$$

Problem 78   Use mathematical induction to prove that every amount of money greater than $ 1.43 can be paid using only 9-cent and 19-cent coins.

Answer: The claim is that every natural number $n \geq 144$ can be represented in the form:

$$n = \alpha \cdot 9 + \beta \cdot 19$$

for some non-negative integers α and β.

In the base case:

$$
\begin{array}{rclclcl}
144 & = & 16 \cdot 9 & + & 0 \cdot 19 \\
145 & = & 14 \cdot 9 & + & 1 \cdot 19 \\
146 & = & 12 \cdot 9 & + & 2 \cdot 19 \\
147 & = & 10 \cdot 9 & + & 3 \cdot 19 \\
148 & = & 8 \cdot 9 & + & 4 \cdot 19 \\
149 & = & 6 \cdot 9 & + & 5 \cdot 19 \\
150 & = & 4 \cdot 9 & + & 6 \cdot 19 \\
151 & = & 2 \cdot 9 & + & 7 \cdot 19 \\
152 & = & 0 \cdot 9 & + & 8 \cdot 19 \\
\end{array}
$$

Inductively, assume that for an arbitrary $n \geq 152$:

$$k = \alpha_k \cdot 9 + \beta_k \cdot 19$$

for all k such that $144 \leq k \leq n$, where α_k and β_k are non-negative integers (that depend on k.) We have to show that:

$$n + 1 = \alpha' \cdot 9 + \beta' \cdot 19$$

for some non-negative integers α' and β'.

The inductive hypothesis applies to $n - 8$, since:

$$n > n - 8 \geq 152 - 8 = 144$$

Hence:

$$n - 8 = a \cdot 9 + b \cdot 19$$

for some non-negative integers a and b, yielding:

$$n + 1 = (n - 8) + 9 = a \cdot 9 + b \cdot 19 + 9 = (a + 1) \cdot 9 + b \cdot 19$$

Problem 79   Use mathematical induction to prove that every toll of at least $ 1.60 can be paid using only 11-cent and 17-cent tokens.

Answer: The claim is that every natural number $n \geq 160$ can be represented in the form:

$$n = \alpha \cdot 11 + \beta \cdot 17$$

for some non-negative integers α and β.

In the base case:

$$
\begin{aligned}
160 &= 13 \cdot 11 &+& \quad 1 \cdot 17 \\
161 &= 10 \cdot 11 &+& \quad 3 \cdot 17 \\
162 &= 7 \cdot 11 &+& \quad 5 \cdot 17 \\
163 &= 4 \cdot 11 &+& \quad 7 \cdot 17 \\
164 &= 1 \cdot 11 &+& \quad 9 \cdot 17 \\
165 &= 15 \cdot 11 &+& \quad 0 \cdot 17 \\
166 &= 12 \cdot 11 &+& \quad 2 \cdot 17 \\
167 &= 9 \cdot 11 &+& \quad 4 \cdot 17 \\
168 &= 6 \cdot 11 &+& \quad 6 \cdot 17 \\
169 &= 3 \cdot 11 &+& \quad 8 \cdot 17 \\
170 &= 0 \cdot 11 &+& \quad 11 \cdot 17
\end{aligned}
$$

Inductively, assume that for an arbitrary $n \geq 170$:

$$k = \alpha_k \cdot 11 + \beta_k \cdot 17$$

for all k such that $160 \leq k \leq n$, where α_k and β_k are non-negative integers (that depend on k.) We have to show that:

$$n + 1 = \alpha' \cdot 11 + \beta' \cdot 17$$

for some non-negative integers α' and β'.

The inductive hypothesis applies to $n - 10$, since:

$$n > n - 10 \geq 170 - 10 = 160$$

Hence:

$$n - 10 = a \cdot 11 + b \cdot 17$$

for some non-negative integers a and b, yielding:

$$n + 1 = (n - 10) + 11 = a \cdot 11 + b \cdot 17 + 11 = (a + 1) \cdot 11 + b \cdot 17$$

Problem 80   Use mathematical induction to prove that a wall of any height exceeding 17 inches can be built using only three types of blocks, whose height is equal to 5 inches, 9 inches and 11 inches.

Answer: The claim is that every natural number $n \geq 18$ can be represented in the form:

$$n = \alpha \cdot 5 + \beta \cdot 9 + \gamma \cdot 11$$

for some non-negative integers α, β, and γ.

In the base case:

$$
\begin{aligned}
18 &= 0 \cdot 5 &+& \quad 2 \cdot 9 &+& \quad 0 \cdot 11 \\
19 &= 2 \cdot 5 &+& \quad 1 \cdot 9 &+& \quad 0 \cdot 11 \\
20 &= 0 \cdot 5 &+& \quad 1 \cdot 9 &+& \quad 1 \cdot 11 \\
21 &= 2 \cdot 5 &+& \quad 0 \cdot 9 &+& \quad 1 \cdot 11 \\
22 &= 0 \cdot 5 &+& \quad 0 \cdot 9 &+& \quad 2 \cdot 11
\end{aligned}
$$

Inductively, assume that for an arbitrary $n \geq 22$:

$$k = \alpha_k \cdot 5 + \beta_k \cdot 9 + \gamma_k \cdot 11$$

for all k such that $18 \leq k \leq n$, where α_k, β_k and γ_k are non-negative integers (that depend on k.) We have to show that:

$$n + 1 = \alpha' \cdot 5 + \beta' \cdot 9 + \gamma' \cdot 11$$

for some non-negative integers α', β', and γ'.

The inductive hypothesis applies to $n - 4$, since:

$$n > n - 4 \geq 22 - 4 = 18$$

Hence:

$$n - 4 = a \cdot 5 + b \cdot 9 + c \cdot 11$$

for some non-negative integers a, b, and c, yielding:

$$n + 1 = (n - 4) + 5 = (a \cdot 5 + b \cdot 9 + c \cdot 11) + 5 = (a + 1) \cdot 5 + b \cdot 9 + c \cdot 11$$

Problem 81   Use mathematical induction to prove that every postage of at least 25 cents can be paid using only four types of stamps, which cost $7, 9, 19$, and 22 cents.

Answer: The claim is that every natural number $n \geq 25$ can be represented in the form:

$$n = \alpha \cdot 7 + \beta \cdot 9 + \gamma \cdot 19 + \delta \cdot 22$$

for some non-negative integers α, β, γ and δ.

In the base case:

$$
\begin{aligned}
25 &= 1 \cdot 7 + 2 \cdot 9 + 0 \cdot 19 + 0 \cdot 22 \\
26 &= 1 \cdot 7 + 0 \cdot 9 + 1 \cdot 19 + 0 \cdot 22 \\
27 &= 0 \cdot 7 + 3 \cdot 9 + 0 \cdot 19 + 0 \cdot 22 \\
28 &= 0 \cdot 7 + 1 \cdot 9 + 1 \cdot 19 + 0 \cdot 22 \\
29 &= 1 \cdot 7 + 0 \cdot 9 + 0 \cdot 19 + 1 \cdot 22 \\
30 &= 3 \cdot 7 + 1 \cdot 9 + 0 \cdot 19 + 0 \cdot 22 \\
31 &= 0 \cdot 7 + 1 \cdot 9 + 0 \cdot 19 + 1 \cdot 22
\end{aligned}
$$

Inductively, assume that for an arbitrary $n \geq 31$:

$$k = \alpha_k \cdot 7 + \beta_k \cdot 9 + \gamma_k \cdot 19 + \delta_k \cdot 22$$

for all k such that $25 \leq k \leq n$, where α_k, β_k, γ_k and δ_k are non-negative integers (that depend on k.) We have to show that:

$$n + 1 = \alpha' \cdot 7 + \beta' \cdot 9 + \gamma' \cdot 19 + \delta' \cdot 22$$

for some non-negative integers α', β', γ', and δ'.

The inductive hypothesis applies to $n - 6$, since:

$$n > n - 6 \geq 31 - 6 = 25$$

Hence:

$$n - 6 = a \cdot 7 + b \cdot 9 + c \cdot 19 + d \cdot 22$$

for some non-negative integers a, b, c, and d, yielding:

$$n + 1 = (n - 6) + 7 = (a \cdot 7 + b \cdot 9 + c \cdot 19 + d \cdot 22) + 7 = (a + 1) \cdot 7 + b \cdot 9 + c \cdot 19 + d \cdot 22$$

Problem 82   Use mathematical induction to prove that

$$3^{4n+1} + 5^{2n+1}$$

is divisible by 8 for every natural number n.

Answer: In the base case, for $n = 0$:

$$3^{4 \cdot 0 + 1} + 5^{2 \cdot 0 + 1} = 3^1 + 5^1 = 8 = 1 \cdot 8$$

whence the base case.

Inductively, assume that for some natural number n:

$$3^{4n+1} + 5^{2n+1} = 8k$$

for some integer k. We have to show that:

$$3^{4(n+1)+1} + 5^{2(n+1)+1} = 8\ell$$

for some integer ℓ.

Indeed:

$$
\begin{aligned}
3^{4(n+1)+1} + 5^{2(n+1)+1} \ &= 3^{4n+4+1} + 5^{2n+2+1} = 3^{4n+1} \cdot 3^4 + 5^{2n+1} \cdot 5^2 \\
&= 3^{4n+1} \cdot 81 + 5^{2n+1} \cdot 25 = 3^{4n+1} \cdot (56 + 25) + 5^{2n+1} \cdot 25 \\
&= 56 \cdot 3^{4n+1} + 25(3^{4n+1} + 5^{2n+1}) \\
\spadesuit \quad &= 56 \cdot 3^{4n+1} + 25 \cdot 8k = 8(7 \cdot 3^{4n+1} + 25k)
\end{aligned}
$$

The equality ♠ is obtained from the inductive hypothesis, while the other equalities are obtained by arithmetic transformations.

Problem 83   Use mathematical induction to prove that

$$16 \cdot 7^n - 6n + 2$$

is divisible by 18 for every natural number n.

Answer: In the base case, for $n = 0$:

$$16 \cdot 7^0 - 6 \cdot 0 + 2 = 16 \cdot 1 + 2 = 18$$

whence the base case.

Inductively, assume that for some natural number n:

$$16 \cdot 7^n - 6n + 2 = 18k$$

for some integer k. As a consequence, it follows that:

$$16 \cdot 7^n = 6n - 2 + 18k$$

We have to show that:

$$16 \cdot 7^{n+1} - 6(n+1) + 2 = 18\ell$$

for some integer ℓ.

Indeed:

$$
\begin{aligned}
16 \cdot 7^{n+1} - 6(n+1) + 2 \ &= 16 \cdot 7^n \cdot 7 - 6n - 6 + 2 \\
\clubsuit \quad &= (6n - 2 + 18k) \cdot 7 - 6n - 6 + 2 = 42n - 14 + 18 \cdot 7k - 6n - 4 \\
&= 36n + 18 \cdot 7k - 18 = 18 \cdot (2n + 7k - 1)
\end{aligned}
$$

The equality ♣ follows from the consequence of the inductive hypothesis, while the other equalities are obtained by arithmetic transformations.

Problem 84   Use mathematical induction to prove that

$$5 \cdot 6^n + 2^{5n+3}$$

is divisible by 13 for every natural number n.

Answer: In the base case, for $n = 0$:

$$5 \cdot 6^0 + 2^{5 \cdot 0 + 3} = 5 \cdot 1 + 2^3 = 5 + 8 = 13 = 1 \cdot 13$$

whence the base case.

Inductively, assume that for some natural number n:

$$5 \cdot 6^n + 2^{5n+3} = 13k$$

for some integer k. We have to show that:

$$5 \cdot 6^{n+1} + 2^{5(n+1)+3} = 13\ell$$

for some integer ℓ.

Indeed:

$$
\begin{aligned}
5 \cdot 6^{n+1} + 2^{5(n+1)+3} &= 5 \cdot 6^n \cdot 6 + 2^{5n+5+3} = 5 \cdot 6^n \cdot 6 + 2^{5n+3} \cdot 2^5 \\
&= 5 \cdot 6^n \cdot (32 - 26) + 2^{5n+3} \cdot 32 = 32 \left(5 \cdot 6^n + 2^{5n+3}\right) - 26 \cdot 5 \cdot 6^n \\
\spadesuit \quad &= 32 \cdot 13k - 13 \cdot 2 \cdot 5 \cdot 6^n = 13 \left(32k - 10 \cdot 6^n\right)
\end{aligned}
$$

The equality $\spadesuit$ is obtained from the inductive hypothesis, while the other equalities are obtained by arithmetic transformations.

Problem 85   Use mathematical induction to prove that

$$(n - 6)(n^2 + 1) + 10n$$

is divisible by 3 for every natural number $n \geq 0$.

Answer: In the base case, for $n = 0$:

$$(0 - 6)(0^2 + 1) + 10 \cdot 0 = (-6) \cdot 1 + 0 = -6 = 3 \cdot (-2)$$

Hence, the base case holds.

For the inductive step, assume that for some $n \geq 0$:

$$(n - 6)(n^2 + 1) + 10n = 3k$$

for some integer k. We have to prove that:

$$((n + 1) - 6)((n + 1)^2 + 1) + 10(n + 1) = 3\ell$$

for some integer ℓ.

$$
\begin{aligned}
&\quad ((n + 1) - 6)((n + 1)^2 + 1) + 10(n + 1) \\
&= (n - 5)((n^2 + 2n + 1) + 1) + (10n + 10) \\
&= ((n - 6) + 1)((n^2 + 1) + (2n + 1)) + (10n + 10) \\
&= (n - 6)(n^2 + 1) + (n^2 + 1) + (n - 5)(2n + 1) + (10n + 10) \\
&= (n - 6)(n^2 + 1) + (n^2 + 1) + (2n^2 + n - 10n - 5) + (10n + 10) \\
&= (n - 6)(n^2 + 1) + 10n + 3n^2 - 9n + 6 = (n - 6)(n^2 + 1) + 10n + 3(n^2 - 3n + 2) \\
\bullet \quad &= 3k + 3(n^2 - 3n + 2) = 3(k + n^2 - 3n + 2)
\end{aligned}
$$

The equality $\bullet$ follows from the inductive hypothesis, while the other equalities are obtained by arithmetic transformations.

Problem 86   Use mathematical induction to prove that

$$5 \cdot 7^{n+2} + 5 \cdot 4^n - 10^{n+1}$$

is divisible by 3 for every natural number n.

Answer: In the base case, for $n = 0$:

$$5 \cdot 7^{0+2} + 5 \cdot 4^0 - 10^{0+1} = 5 \cdot 49 + 5 \cdot 1 - 10 = 240 = 3 \cdot 80$$

whence the base case.

Inductively, assume that for some natural number n:

$$5 \cdot 7^{n+2} + 5 \cdot 4^n - 10^{n+1} = 3k$$

for some integer k. We have to show that:

$$5 \cdot 7^{(n+1)+2} + 5 \cdot 4^{n+1} - 10^{(n+1)+1} = 3\ell$$

for some integer ℓ.

Indeed:

$$
\begin{aligned}
5 \cdot 7^{(n+1)+2} + 5 \cdot 4^{n+1} - 10^{(n+1)+1} &= 5 \cdot 7^{n+3} + 5 \cdot 4^{n+1} - 10^{n+2} \\
&= 7 \cdot 5 \cdot 7^{n+2} + 4 \cdot 5 \cdot 4^n - 10 \cdot 10^{n+1} = (4+3) \cdot 5 \cdot 7^{n+2} + 4 \cdot 5 \cdot 4^n - (4+6) \cdot 10^{n+1} \\
&= 4 \left(5 \cdot 7^{n+2} + 5 \cdot 4^n - 10^{n+1} \right) + 3 \cdot 5 \cdot 7^{n+2} - 6 \cdot 10^{n+1} \\
\spadesuit \quad &= 4 \cdot 3k + 3 \cdot 5 \cdot 7^{n+2} - 6 \cdot 10^{n+1} = 3 \left(4k + 5 \cdot 7^{n+2} - 2 \cdot 10^{n+1} \right)
\end{aligned}
$$

The equality $\spadesuit$ is obtained from the inductive hypothesis, while the other equalities are obtained by arithmetic transformations.

Problem 87   Use mathematical induction to prove that

$$19 \cdot 9^n - 8n - 3$$

is divisible by 16 for every natural number n.

Answer: In the base case, for $n = 0$:

$$19 \cdot 9^0 - 8 \cdot 0 - 3 = 19 \cdot 1 - 3 = 16$$

whence the base case.

Inductively, assume that for some natural number n:

$$19 \cdot 9^n - 8n - 3 = 16k$$

for some integer k. As a consequence, it follows that:

$$19 \cdot 9^n = 8n + 3 + 16k$$

We have to show that:

$$19 \cdot 9^{n+1} - 8(n+1) - 3 = 16\ell$$

for some integer ℓ.

Indeed:

$$
\begin{aligned}
19 \cdot 9^{n+1} - 8(n+1) - 3 \quad &= 19 \cdot 9^n \cdot 9 - 8n - 8 - 3 \\
\clubsuit \quad &= (8n + 3 + 16k) \cdot 9 - 8n - 8 - 3 = 72n + 27 + 16 \cdot 9k - 8n - 11 \\
&= 64n + 16 \cdot 9k + 16 = 16 \cdot (4n + 9k + 1)
\end{aligned}
$$

The equality $\clubsuit$ follows from the consequence of the inductive hypothesis, while the other equalities are obtained by arithmetic transformations.

Problem 88   Use mathematical induction to prove that

$$5 \cdot 8^{n+1} + 7 \cdot 5^n + 2^{7n+4}$$

is divisible by 3 for every natural number n.

Answer: In the base case, for $n = 0$:

$$5 \cdot 8^{0+1} + 7 \cdot 5^0 + 2^{7 \cdot 0 + 4} = 5 \cdot 8 + 7 \cdot 1 + 16 = 63 = 3 \cdot 21$$

whence the base case.

Inductively, assume that for some natural number n:

$$5 \cdot 8^{n+1} + 7 \cdot 5^n + 2^{7n+4} = 3k$$

for some integer k. We have to show that:

$$5 \cdot 8^{(n+1)+1} + 7 \cdot 5^{(n+1)} + 2^{7(n+1)+4} = 3\ell$$

for some integer ℓ.

Indeed:

$$
\begin{aligned}
5 \cdot 8^{(n+1)+1} + 7 \cdot 5^{n+1} + 2^{7(n+1)+4} &= 5 \cdot 8^{n+2} + 7 \cdot 5^{n+1} + 2^{7n+4+7} \\
&= 8 \cdot 5 \cdot 8^{n+1} + 5 \cdot 7 \cdot 5^n + 2^7 \cdot 2^{7n+4} = (5+3) \cdot 5 \cdot 8^{n+1} + 5 \cdot 7 \cdot 5^n + (5+123) \cdot 2^{7n+4} \\
&= 5 \left(5 \cdot 8^{n+1} + 7 \cdot 5^n + 2^{7n+4} \right) + 3 \cdot 5 \cdot 8^{n+1} + 123 \cdot 2^{7n+4} \\
\spadesuit \quad &= 5 \cdot 3k + 3 \left(5 \cdot 8^{n+1} + 41 \cdot 2^{7n+4} \right) = 3 \left(5k + 5 \cdot 8^{n+1} + 41 \cdot 2^{7n+4} \right)
\end{aligned}
$$

The equality $\spadesuit$ is obtained from the inductive hypothesis, while the other equalities are obtained by arithmetic transformations.

Problem 89   Use mathematical induction to prove that

$$n^4 + 2n^3 + 3n^2 - 2n$$

is divisible by 4 for every natural number n.

Answer: In the base case, for $n = 0$:

$$0^4 + 2 \cdot 0^3 + 3 \cdot 0^2 - 2 \cdot 0 = 0 = 0 \cdot 4$$

Hence, the base case holds.

For the inductive step, assume that for some natural number n:

$$n^4 + 2n^3 + 3n^2 - 2n = 4k$$

for some integer k. We have to prove that:

$$(n+1)^4 + 2(n+1)^3 + 3(n+1)^2 - 2(n+1) = 4\ell$$

for some integer ℓ.

$$
\begin{aligned}
&(n+1)^4 + 2(n+1)^3 + 3(n+1)^2 - 2(n+1) \\
&= (n^2 + 2n + 1)^2 + 2(n+1)(n^2 + 2n + 1) + 3(n^2 + 2n + 1) - 2(n+1) \\
&= (n^4 + 4n^3 + 6n^2 + 4n + 1) + 2(n^3 + 3n^2 + 3n + 1) + (3n^2 + 6n + 3) - 2n - 2 \\
&= (n^4 + 2n^3 + 3n^2 - 2n) + (4n^3 + 12n^2 + 16n + 4) \\
&= (n^4 + 2n^3 + 3n^2 - 2n) + 4(n^3 + 3n^2 + 4n + 1) \\
\bullet \quad &= 4k + 4(n^3 + 3n^2 + 4n + 1) = 4 \cdot (k + n^3 + 3n^2 + 4n + 1)
\end{aligned}
$$

The equality $\bullet$ follows from the inductive hypothesis, while the other equalities are obtained by arithmetic transformations.

Problem 90 Use mathematical induction to prove that

$$2 \cdot 11^n + 3 \cdot 4^{2n+3} - 9^{2n+1}$$

is divisible by 5 for every natural number n.

Answer: In the base case, for $n = 0$:

$$2 \cdot 11^0 + 3 \cdot 4^{2 \cdot 0 + 3} - 9^{2 \cdot 0 + 1} = 2 \cdot 1 + 3 \cdot 64 - 9 = 185 = 5 \cdot 37$$

whence the base case.

Inductively, assume that for some natural number n:

$$2 \cdot 11^n + 3 \cdot 4^{2n+3} - 9^{2n+1} = 5k$$

for some integer k. We have to show that:

$$2 \cdot 11^{n+1} + 3 \cdot 4^{2(n+1)+3} - 9^{2(n+1)+1} = 5\ell$$

for some integer ℓ.

$$2 \cdot 11^{n+1} + 3 \cdot 4^{2(n+1)+3} - 9^{2(n+1)+1} = 2 \cdot 11^{n+1} + 3 \cdot 4^{2n+3+2} - 9^{2n+1+2}$$
$$= 11 \cdot 2 \cdot 11^n + 4^2 \cdot 3 \cdot 4^{2n+3} - 9^2 \cdot 9^{2n+1}$$
$$= 11 \cdot 2 \cdot 11^n + (11 + 5) \cdot 3 \cdot 4^{2n+3} - (11 + 70) \cdot 9^{2n+1}$$
$$= 11 \left(2 \cdot 11^n + 3 \cdot 4^{2n+3} - 9^{2n+1}\right) + 5 \cdot 3 \cdot 4^{2n+3} - 70 \cdot 9^{2n+1}$$
$$\spadesuit \quad = 11 \cdot 5k + 5 \cdot 3 \cdot 4^{2n+3} - 70 \cdot 9^{2n+1} = 5\left(11k + 3 \cdot 4^{2n+3} - 14 \cdot 9^{2n+1}\right)$$

The equality $\spadesuit$ is obtained from the inductive hypothesis, while the other equalities are obtained by arithmetic transformations.

Problem 91 Use mathematical induction to prove that

$$9 \cdot 13^n - 4n^2 - 1$$

is divisible by 8 for every natural number n.

Answer: In the base case, for $n = 0$:

$$9 \cdot 13^0 - 4 \cdot 0^2 - 1 = 9 - 1 = 8$$

whence the base case.

Inductively, assume that for some natural number n:

$$9 \cdot 13^n - 4n^2 - 1 = 8k$$

for some integer k. As a consequence, it follows that:

$$9 \cdot 13^n = 4n^2 + 1 + 8k$$

We have to show that:

$$9 \cdot 13^{n+1} - 4(n+1)^2 - 1 = 8\ell$$

for some integer ℓ.

$$9 \cdot 13^{n+1} - 4(n+1)^2 - 1 = 9 \cdot 13^n \cdot 13 - 4(n^2 + 2n + 1) - 1$$
$$\clubsuit \quad = (4n^2 + 1 + 8k) \cdot 13 - 4n^2 - 8n - 5 = 52n^2 + 13 + 8 \cdot 13k - 4n^2 - 8n - 5$$
$$= 48n^2 + 8 \cdot 13k - 8n + 8 = 8 \cdot (6n^2 + 13k - n + 1)$$

The equality $\clubsuit$ follows from the consequence of the inductive hypothesis, while the other equalities are obtained by arithmetic transformations.

Problem 92   Use mathematical induction to prove that

$$2 \cdot 5^{3n+2} - 3 \cdot 6^{n+1} + 4 \cdot 3^{3n+1} + 2 \cdot 13^{n+1}$$

is divisible by 7 for every natural number n.

Answer: In the base case, for $n = 0$:

$$2 \cdot 5^{3 \cdot 0 + 2} - 3 \cdot 6^{0+1} + 4 \cdot 3^{3 \cdot 0 + 1} + 2 \cdot 13^{0+1} = 2 \cdot 25 - 3 \cdot 6 + 4 \cdot 3 + 2 \cdot 13 = 70 = 7 \cdot 10$$

whence the base case.

Inductively, assume that for some natural number n:

$$2 \cdot 5^{3n+2} - 3 \cdot 6^{n+1} + 4 \cdot 3^{3n+1} + 2 \cdot 13^{n+1} = 7k$$

for some integer k. We have to show that:

$$2 \cdot 5^{3(n+1)+2} - 3 \cdot 6^{(n+1)+1} + 4 \cdot 3^{3(n+1)+1} + 2 \cdot 13^{(n+1)+1} = 7\ell$$

for some integer ℓ.

Indeed:

$$
\begin{aligned}
& 2 \cdot 5^{3(n+1)+2} - 3 \cdot 6^{(n+1)+1} + 4 \cdot 3^{3(n+1)+1} + 2 \cdot 13^{(n+1)+1} \\
={} & 2 \cdot 5^{3n+2+3} - 3 \cdot 6^{n+1+1} + 4 \cdot 3^{3n+1+3} + 2 \cdot 13^{n+1+1} \\
={} & 5^3 \cdot 2 \cdot 5^{3n+2} - 6 \cdot 3 \cdot 6^{n+1} + 3^3 \cdot 4 \cdot 3^{3n+1} + 13 \cdot 2 \cdot 13^{n+1} \\
={} & (6 + 119) \cdot 2 \cdot 5^{3n+2} - 6 \cdot 3 \cdot 6^{n+1} + (6 + 21) \cdot 4 \cdot 3^{3n+1} + (6 + 7) \cdot 2 \cdot 13^{n+1} \\
={} & 6 \left(2 \cdot 5^{3n+2} - 3 \cdot 6^{n+1} + 4 \cdot 3^{3n+1} + 2 \cdot 13^{n+1} \right) + 119 \cdot 2 \cdot 5^{3n+2} + 21 \cdot 4 \cdot 3^{3n+1} + 7 \cdot 2 \cdot 13^{n+1} \\
\bullet={} & 6 \cdot 7k + 119 \cdot 2 \cdot 5^{3n+2} + 21 \cdot 4 \cdot 3^{3n+1} + 7 \cdot 2 \cdot 13^{n+1} \\
={} & 7 \left(6k + 17 \cdot 2 \cdot 5^{3n+2} + 3 \cdot 4 \cdot 3^{3n+1} + 2 \cdot 13^{n+1} \right)
\end{aligned}
$$

The equality $\bullet$ is obtained from the inductive hypothesis, while the other equalities are obtained by arithmetic transformations.

Problem 93   Use mathematical induction to prove that

$$3 \cdot 5^n + 4^{2n+1} + 3^{3n+2} - 5 \cdot 7^{2n}$$

is divisible by 11 for every natural number n.

Answer: In the base case, for $n = 0$:

$$3 \cdot 5^0 + 4^{2 \cdot 0 + 1} + 3^{3 \cdot 0 + 2} - 5 \cdot 7^{2 \cdot 0} = 3 + 4 + 9 - 5 = 11$$

whence the base case.

Inductively, assume that for some natural number n:

$$3 \cdot 5^n + 4^{2n+1} + 3^{3n+2} - 5 \cdot 7^{2n} = 11k$$

for some integer k. We have to show that:

$$3 \cdot 5^{n+1} + 4^{2(n+1)+1} + 3^{3(n+1)+2} - 5 \cdot 7^{2(n+1)} = 11\ell$$

for some integer ℓ.

Indeed:

$$3 \cdot 5^{n+1} + 4^{2(n+1)+1} + 3^{3(n+1)+2} - 5 \cdot 7^{2(n+1)} = 3 \cdot 5^{n+1} + 4^{2n+1+2} + 3^{3n+2+3} - 5 \cdot 7^{2n+2}$$
$$= 5 \cdot 3 \cdot 5^n + 4^2 \cdot 4^{2n+1} + 3^3 \cdot 3^{3n+2} - 7^2 \cdot 5 \cdot 7^{2n}$$
$$= 5 \cdot 3 \cdot 5^n + (5+11) \cdot 4^{2n+1} + (5+22) \cdot 3^{3n+2} - (5+44) \cdot 5 \cdot 7^{2n}$$
$$= 5 \left(3 \cdot 5^n + 4^{2n+1} + 3^{3n+2} - 5 \cdot 7^{2n} \right) + 11 \cdot 4^{2n+1} + 22 \cdot 3^{3n+2} - 44 \cdot 5 \cdot 7^{2n}$$
$$\bullet = 5 \cdot 11k + 11 \cdot 4^{2n+1} + 22 \cdot 3^{3n+2} - 44 \cdot 5 \cdot 7^{2n} = 11 \left(5k + 4^{2n+1} + 2 \cdot 3^{3n+2} - 4 \cdot 5 \cdot 7^{2n} \right)$$

The equality • is obtained from the inductive hypothesis, while the other equalities are obtained by arithmetic transformations.

Problem 94   Use mathematical induction to prove that

$$17^{n+1} - 2 \cdot 5^{n+3} + 11^{n+3}$$

is divisible by 3 for every natural number n.

Answer: In the base case, for $n = 0$:

$$17^{0+1} - 2 \cdot 5^{0+3} + 11^{0+3} = 17 - 2 \cdot 125 + 1331 = 1098 = 3 \cdot 366$$

whence the base case.

Inductively, assume that for some natural number n:

$$17^{n+1} - 2 \cdot 5^{n+3} + 11^{n+3} = 3k$$

for some integer k. We have to show that:

$$17^{(n+1)+1} - 2 \cdot 5^{(n+1)+3} + 11^{(n+1)+3} = 3\ell$$

for some integer ℓ.

Indeed:

$$17^{(n+1)+1} - 2 \cdot 5^{(n+1)+3} + 11^{(n+1)+3} = 17^{n+1+1} - 2 \cdot 5^{n+3+1} + 11^{n+3+1}$$
$$= (5+12) \cdot 17^{n+1} - 5 \cdot 2 \cdot 5^{n+3} + (5+6) \cdot 11^{n+3}$$
$$= 5 \left(17^{n+1} - 2 \cdot 5^{n+3} + 11^{n+3} \right) + 12 \cdot 17^{n+1} + 6 \cdot 11^{n+3}$$
$$\spadesuit = 5 \cdot 3k + 12 \cdot 17^{n+1} + 6 \cdot 11^{n+3} = 3 \left(5k + 4 \cdot 17^{n+1} + 2 \cdot 11^{n+3} \right)$$

The equality ♠ is obtained from the inductive hypothesis, while the other equalities are obtained by arithmetic transformations.

Problem 95   Use mathematical induction to prove that

$$2 \cdot 3^{2n} + 5^{2n} + 7^{4n} + 9^{2n+1} + 11^{2n+1}$$

is divisible by 8 for every natural number n.

Answer: In the base case, for $n = 0$:

$$2 \cdot 3^{2 \cdot 0} + 5^{2 \cdot 0} + 7^{4 \cdot 0} + 9^{2 \cdot 0+1} + 11^{2 \cdot 0+1} = 2 + 1 + 1 + 9 + 11 = 24 = 3 \cdot 8$$

whence the base case.

Inductively, assume that for some natural number n:

$$2 \cdot 3^{2n} + 5^{2n} + 7^{4n} + 9^{2n+1} + 11^{2n+1} = 8k$$

for some integer k. We have to show that:

$$2 \cdot 3^{2(n+1)} + 5^{2(n+1)} + 7^{4(n+1)} + 9^{2(n+1)+1} + 11^{2(n+1)+1} = 8\ell$$

for some integer ℓ.

Indeed:

$$
\begin{aligned}
& 2 \cdot 3^{2(n+1)} + 5^{2(n+1)} + 7^{4(n+1)} + 9^{2(n+1)+1} + 11^{2(n+1)+1} \\
=~& 2 \cdot 3^{2n+2} + 5^{2n+2} + 7^{4n+4} + 9^{2n+1+2} + 11^{2n+1+2} \\
=~& 3^2 \cdot 2 \cdot 3^{2n} + 5^2 \cdot 5^{2n} + 7^4 \cdot 7^{4n} + 9^2 \cdot 9^{2n+1} + 11^2 \cdot 11^{2n+1} \\
=~& 9 \cdot 2 \cdot 3^{2n} + (9+16) \cdot 5^{2n} + (9+2392) \cdot 7^{4n} + (9+72) \cdot 9^{2n+1} + (9+112) \cdot 11^{2n+1} \\
=~& 9 \left(2 \cdot 3^{2n} + 5^{2n} + 7^{4n} + 9^{2n+1} + 11^{2n+1} \right) + 16 \cdot 5^{2n} + 2392 \cdot 7^{4n} + 72 \cdot 9^{2n+1} + 112 \cdot 11^{2n+1} \\
\bullet =~& 9 \cdot 8k + 16 \cdot 5^{2n} + 2392 \cdot 7^{4n} + 72 \cdot 9^{2n+1} + 112 \cdot 11^{2n+1} \\
=~& 8 \left(9k + 2 \cdot 5^{2n} + 299 \cdot 7^{4n} + 9 \cdot 9^{2n+1} + 14 \cdot 11^{2n+1} \right)
\end{aligned}
$$

The equality $\bullet$ is obtained from the inductive hypothesis; the other equalities are obtained by arithmetic transformations.

2.2   Recursive Definitions

Problem 96   Sequence b is defined recursively as follows.

$$
\begin{aligned}
b_0 &= 1 \\
b_1 &= 2 \\
b_2 &= 3 \\
b_n &= \frac{b_{n-2}^2 + n - 3}{b_{n-3}}, \text{ for } n \geq 3
\end{aligned}
$$

Sets A, B, C, D, E are defined as follows.

$$
\begin{aligned}
A &= \{1, 2, 3, 4, 5\} \\
B &= \{x \mid (\exists y \in A)(x = b_y)\} \\
C &= \{z \mid (\exists x \in B)(z \leq x) \wedge (\forall x \in A)(z > x)\} \\
D &= \{x \mid (\exists y \in B)(y = x^2)\} \\
E &= \{3, 5\}
\end{aligned}
$$

(All variables assume values from the set of natural numbers $N = \{0, 1, \ldots\}$.)

In each of the following cases, write the required list. If this is impossible, explain why.

(a) List all elements of B.

Answer: Since $B = b(A)$ (where $b(A)$ denotes the set of images under b of elements of the set A), we need to calculate b_3, b_4, b_5:

$$b_3 = \frac{b_1^2 + 3 - 3}{b_0} = \frac{2^2 + 3 - 3}{1} = 4$$

$$b_4 = \frac{b_2^2 + 4 - 3}{b_1} = \frac{3^2 + 4 - 3}{2} = 5$$

$$b_5 = \frac{b_3^2 + 5 - 3}{b_2} = \frac{4^2 + 5 - 3}{3} = 6$$

whence the listing of the elements of B:

$$2, 3, 4, 5, 6$$

(b) List all elements of C.

Answer:

$$6$$

(c) List all elements of D.

Answer:

$$2$$

(d) Construct an injection $f : A \to B$.

Answer:

$$f(x) = x + 1$$

(e) Construct an injection $f : C \to B$.

Answer:

$$f(x) = x - 1$$

(f) Construct an injection $f : D \to B$.

Answer:

$$f(x) = x + 2$$

(g) Construct an injection $f : E \to B$.

Answer:

$$f(x) = x$$

(h) Construct an injection $f : B \to D$.

Answer: Impossible, since:

$$|B| = 5 > 1 = |D|$$

(i) Construct a surjection $f : C \to D$.

Answer:

$$f(x) = x - 4$$

(j) Construct a surjection $f : D \to E$.

Answer: Impossible, since:

$$|D| = 1 < 2 = |E|$$

Problem 97   Sequence g is defined recursively as follows.

$$g_0 = 1$$
$$g_1 = 2$$
$$g_2 = 4$$
$$g_n = \frac{g_{n-1}g_{n-2}}{g_{n-3}}, \text{ for } n \geq 3$$

Sets A, B, C, D, E are defined as follows.

$$A = \{1, 2, 3, 4, 5\}$$
$$B = \{x \mid (\exists y \in A)(x = g_y)\}$$
$$C = \{z \mid ((\exists x \in B)(3z \leq x)) \wedge ((\forall x \in A)(z > x))\}$$
$$D = \{x \mid (\exists y \in B)(y = x^2)\}$$
$$E = \{3, 5\}$$

(All variables assume values from the set of natural numbers $N = \{0, 1, \ldots\}$.)

In each of the following cases, write the required list. If this is impossible, explain why.

(a) List all elements of B.

Answer: Since $B = g(A)$, we calculate g_3, g_4, g_5:

$$g_3 = \frac{g_2 g_1}{g_0} = \frac{4 \cdot 2}{1} = 8$$
$$g_4 = \frac{g_3 g_2}{g_1} = \frac{8 \cdot 4}{2} = 16$$
$$g_5 = \frac{g_4 g_3}{g_2} = \frac{16 \cdot 8}{4} = 32$$

Hence:

$$g(A) = \{2, 4, 8, 16, 32\}$$

whence the listing of B:

$$2, 4, 8, 16, 32$$

(b) List all elements of $A \setminus B$.

Answer:

$$1, 3, 5$$

(c) List all elements of $B \setminus A$.

Answer:

$$8, 16, 32$$

(d) List all elements of $B \cup A$.

Answer:

$$1, 2, 3, 4, 5, 8, 16, 32$$

(e) List all elements of C.

Answer:

$$6, 7, 8, 9, 10$$

(f) List all elements of $B \cap C$.

Answer:

8

(g) List all elements of $D \times E$.

Answer:

$(2,3), (2,5), (4,3), (4,5)$

Problem 98   Sequence g is defined recursively as follows.

$$g_0 = 2$$
$$g_1 = 5$$
$$g_n = 5g_{n-1} - 6g_{n-2}, \text{ for } n \geq 2$$

Sets A, B, C, D, E are defined as follows.

$$A = \{0, 1, 2, 3, 4\}$$
$$B = \{x \mid (\exists y \in A)(x = g_y)\}$$
$$C = \{z \mid z \in B \wedge 5 \mid z\}$$
$$D = \{x \mid (\exists y \in B)(y > 3^x)\}$$
$$E = \{4, 5\}$$

(All variables assume values from the set of natural numbers $N = \{0, 1, \ldots\}$.)

In each of the following cases, provide the required construction or explain why it is impossible.

(a) List all elements of B.

Answer: Observe that:
$$B = g(A) = \{g_0, g_1, g_2, g_3, g_4\}$$

We employ the recursive definition of the sequence g to calculate its first five values:

$$g(A) = \{2, 5, 13, 35, 97\}$$

whence the listing of B:

$$2, 5, 13, 35, 97$$

(b) List all elements of $A \setminus B$.

Answer:

$0, 1, 3, 4$

(c) List all elements of $B \cap A$.

Answer:

2

(d) List all elements of $B \cup A$.

Answer:

$0, 1, 2, 3, 4, 5, 13, 35, 97$

(e) List all elements of C.

$$\text{Answer:}$$
$$5, 35$$

(f) List all elements of $\mathcal{P}(C)$.

$$\text{Answer:}$$
$$\emptyset, \{5\}, \{35\}, \{5, 35\}$$

(g) List all elements of D.

$$\text{Answer:}$$
$$0, 1, 2, 3, 4$$

(h) List all elements of $C \times E$.

$$\text{Answer:}$$
$$(5, 4), (5, 5), (35, 4), (35, 5)$$

(i) List all elements of $E \times C$.

$$\text{Answer:}$$
$$(4, 5), (4, 35), (5, 5), (5, 35)$$

Problem 99   Sequence g is defined recursively as follows.

$$g_0 = 3$$
$$g_1 = 7$$
$$g_n = 5g_{n-1} - 6g_{n-2}, \text{ for } n \geq 2$$

Sets A, B, C, D, E are defined as follows.

$$A = \{0, 1, 2, 3\}$$
$$B = \{x \mid (\forall y \in A)(x > g_y)\}$$
$$C = \{z \mid z \notin B \wedge 8 \mid z\}$$
$$D = \{x \mid (\forall y \in C)(x^2 < y + 10)\}$$
$$E = \{4, 5\}$$

(All variables assume values from the set of natural numbers $N = \{0, 1, \ldots\}$.)
In each of the following cases, provide the required construction or explain why it is impossible.

(a) List all elements of B.

Answer: First, calculate the value of g for all elements of A:

$$g_2 = 5g_1 - 6g_0 = 5 \cdot 7 - 6 \cdot 3 = 17$$
$$g_3 = 5g_2 - 6g_1 = 5 \cdot 17 - 6 \cdot 7 = 43$$

Since the largest number in $g(A)$ is $g_3 = 43$, the set B consists of all numbers greater than **43**. It is impossible to list all the elements of B, since B is an infinite set.

(b) Construct a function from A to B that is not injective.

$$\text{Answer:}$$
$$f(x) = 44$$

(c) List all elements of $A \setminus B$.

Answer:
$$0, 1, 2, 3$$

(d) List one element of $A \cap B$.

Answer: Impossible, since:

$$A \cap B = \emptyset$$

and there are no elements to list.

(e) List all elements of C.

Answer:
$$0, 8, 16, 24, 32, 40$$

(f) Construct a function from C to E that is not surjective.

Answer:
$$f(x) = 4$$

(g) List all elements of D.

Answer: Since 0 is the smallest number in the set C, the definition of the set D becomes:

$$D = \{x \mid (x^2 < 10)\}$$

whence the listing:

$$0, 1, 2, 3$$

(h) List all elements of $D \times E$.

Answer:
$$(0, 4), (1, 4), (2, 4), (3, 4), (0, 5), (1, 5), (2, 5), (3, 5)$$

(i) List all elements of $\mathcal{P}(E)$.

Answer:
$$\emptyset, \{4\}, \{5\}, \{4, 5\}$$

Problem 100 Sequence b is defined recursively as follows.

$$b_0 = 0$$
$$b_1 = 1$$
$$b_n = 5b_{n-1} - 6b_{n-2}, \text{ for } n \geq 2$$

Sets A, B, C, D, E are defined as follows.

$$A = \{0, 1, 2, 3, 4\}$$
$$B = \{x \mid (\exists y \in A)(x = b_y)\}$$
$$C = \{z \mid (\forall x \in B)(z \geq x) \wedge z < 70\}$$
$$D = \{x \mid (\exists y \in B)(y > x^2)\}$$
$$E = \{4, 5\}$$

(All variables assume values from the set of natural numbers $N = \{0, 1, \ldots\}$.)

In each of the following cases, write the required list. If this is impossible, explain why.

(a) List all elements of B.

  Answer: Observe that:

$$B = b(A) = \{b_0, b_1, b_2, b_3, b_4\}$$

We employ the recursive definition of the sequence b to calculate its first five values:

$$b(A) = \{0, 1, 5, 19, 65\}$$

whence the listing of B:

$$0, 1, 5, 19, 65$$

(b) List all elements of C.

      Answer:

$$65, 66, 67, 68, 69$$

(c) List all elements of D.

      Answer:

$$0, 1, 2, 3, 4, 5, 6, 7, 8$$

(d) List one element of $C \cap D$.

      Answer: There are no elements to list, since:

$$C \cap D = \emptyset$$

(e) List one subset of $C \cap D$.

      Answer:

$$\emptyset$$

(f) List all elements of $N \setminus D$.

      Answer: Impossible, since $N \setminus D$ is an infinite set, defined as follows.

$$N \setminus D = \{x \in N \mid x > 8\}$$

(g) List one element of $N \setminus D$.

      Answer:

$$100$$

(h) List one subset of $N \setminus D$.

      Answer:

$$\{100\}$$

Problem 101 Sequence a is defined recursively as follows.

$$
\begin{aligned}
a_0 &= 5 \\
a_1 &= 8 \\
a_2 &= 11 \\
a_n &= a_{n-3} + 9 \text{ for } n \geq 3
\end{aligned}
$$

Sets A, B, C are defined as follows.

$$A = \{0, 1, 2, 3, 4, 5, 6, 7, 8\}$$
$$B = \{x \mid x \in A \wedge a_x < 25\}$$
$$C = \{z \mid z > a_5\}$$

(All variables assume values from the set of natural numbers $N = \{0, 1, \ldots\}$.)

In each of the following cases, write the required list. If this is impossible, explain why.

(a) List the largest element of B.

Answer:

6

(b) List the smallest element of B.

Answer:

0

(c) List the largest element of C.

Answer: Impossible—set C is infinite; it contains all positive integers greater than 20.

(d) List the smallest element of C.

Answer:

21

(e) List all elements of $A \cap C$.

Answer: There are no elements to list, since:

$$A \cap C = \emptyset$$

(f) List all subsets of $A \cap C$.

Answer:

$\emptyset$

Problem 102   Sequence a is defined recursively as follows.

$$a_0 = 1$$
$$a_1 = 3$$
$$a_n = 2a_{n-1} + 4a_{n-2} + 5 \text{ for } n \geq 2$$

Sets A, B, C are defined as follows.

$$A = \{0, 1, 2, 3, 4, 5, 6, 7, 8, 9, 10, 11, 12, 13, 14, 15, 16\}$$
$$B = \{x \mid x \in A \wedge (\exists y)(x = a_y)\}$$
$$C = \{y \mid y < 64 \wedge (\exists x \in A)(a_x = y)\}$$

(All variables assume values from the set of natural numbers $N = \{0, 1, \ldots\}$.)

In each of the following cases, write the required list. If this is impossible, explain why.

(a) List all elements of B.

Answer:

$1, 3, 15$

(b) List all elements of C.

Answer:

$1, 3, 15, 47$

Problem 103   Sequence a is defined recursively as follows.

$$a_0 = 0$$
$$a_1 = 1$$
$$a_n = 8a_{n-1} - 15a_{n-2}, \text{ for } n \geq 2$$

Sets A, B, C, D are defined as follows.

$$A = \{0, 1, 2, 3, 4, 5, 6\}$$
$$B = \{x \mid x = a_2 \lor x = a_3\}$$
$$C = \{z \mid (\exists y \in B)(z^2 = y \lor z^3 = y)\}$$
$$D = \{z \mid (\forall y \in A)(\exists x \in B)(x + y + z \leq 16)\}$$

(All variables assume values from the set of natural numbers $N = \{0, 1, \ldots\}$.)

In each of the following cases, provide the required constructon or explain why it is impossible.

(a) List the largest element of B.

Answer:

49

(b) List the smallest element of B.

Answer:

8

(c) List all elements of C.

Answer:

$2, 7$

(d) List all elements of D.

Answer:

$0, 1, 2$

(e) Construct a function from A to B that is not injective.

Answer:

$f(x) = 8, \text{ for all } x \in A$

(f) Construct two injective functions from C to D.

Answer:

$f_1 = \{(2, 0), (7, 1)\}$
$f_2 = \{(2, 0), (7, 2)\}$

(g) Construct a function from B to A that is not injective and not surjective.

Answer:

$f(x) = 4, \text{ for all } x \in B$

(h) Construct two surjective functions from D to C.

Answer:

$$f_1 = \{(0,7),(1,2),(2,2)\}$$
$$f_2 = \{(0,7),(1,7),(2,2)\}$$

Problem 104   Sequence a is defined recursively as follows.

$$a_0 = 0$$
$$a_1 = 1$$
$$a_n = 8a_{n-1} - 15a_{n-2}, \text{ for } n \geq 2$$

Sets A, B, C, D are defined as follows.

$$A = \{0,1,2,3,4\}$$
$$B = \{x \mid x < a_3\}$$
$$C = \{z \mid (\exists y \in B)(5z^3 = y)\}$$
$$D = \{x \mid (\exists y \in B)(10x = y)\}$$

(All variables assume values from the set of natural numbers $N = \{0,1,\ldots\}$.)

In each of the following cases, provide the required consruction or explain why it is impossible.

(a) List the largest element of B.

Answer:

48

(b) List the smallest element of B.

Answer:

0

(c) List all elements of C.

Answer:

$0,1,2$

(d) List the largest element of D.

Answer:

4

(e) Construct a function from A to B

Answer:

$$f(0) = f(1) = f(2) = f(3) = f(4) = 37$$

(f) Construct an injection from A to D.

Answer:

$$f(x) = x$$

(g) Construct a surjection from A to C.

Answer:

$$f(x) = \left\lfloor \frac{x}{2} \right\rfloor$$

(h) Construct a function from C to B that is not injective and not surjective.

Answer:

$$f(0) = f(1) = f(2) = 38$$

Problem 105   Set S is defined recursively as follows.

$$5 \in S$$
$$x \in S \implies (x + 3) \in S$$

Furthermore, S has no elements other than those obtained by finitely many applications of the recursive rule given in its definition.

Sets A, B, C are defined as follows.

$$A = \{1, 2, 3, 4, 5, 6, 7, 8, 9, 10\}$$
$$B = \{x \mid (\exists k)(x = 2k + 1)\}$$
$$C = \{11, 12, 13, 14, 15\}$$

(All variables assume values from the set of natural numbers $N = \{0, 1, \ldots\}$.)

In each of the following cases, write the required list. If this is impossible, explain why.

(a) List nine elements of S.

Answer:

$$5, 8, 11, 14, 17, 20, 23, 26, 29$$

(b) List nine elements of $N \setminus S$.

Answer:

$$30, 31, 33, 34, 36, 37, 39, 40, 42$$

(c) List all elements of $A \setminus S$.

Answer:

$$1, 2, 3, 4, 6, 7, 9, 10$$

(d) List all elements of $A \setminus (B \cup S)$.

Answer:

$$2, 4, 6, 10$$

(e) List all alements of $A \setminus (B \cap S)$.

Answer:

$$1, 2, 3, 4, 6, 7, 8, 9, 10$$

(f) List all elements of $A \cap (B \setminus S)$.

Answer:

$$1, 3, 7, 9$$

(g) List all elements of $A \setminus (S \setminus B)$.

Answer:

$$1, 2, 3, 4, 5, 6, 7, 9, 10$$

(h) List all elements of $A \cup (C \setminus S)$.

Answer:

$1, 2, 3, 4, 5, 6, 7, 8, 9, 10, 12, 13, 15$

(i) List all elements of $C \cup (A \cap (B \cup S))$.

Answer:

$1, 3, 5, 7, 8, 9, 11, 12, 13, 14, 15$

Problem 106   Consider two arbitrary sets, A and B. Let F be a set of sets, defined recursively as follows.

$$A \in F \wedge B \in F$$
$$(S_1 \in F \wedge S_2 \in F) \implies (S_1 \cap S_2 \in F \wedge S_1 \cup S_2 \in F)$$

Furthermore, F has no elements other than those obtained by finitely many applications of the recursive rule given in its definition.

(a) Prove that all sets that are elements of F are empty whenever both of the two sets A and B are empty. Precisely:

$$(A = \emptyset \wedge B = \emptyset) \implies F = \{\emptyset\}$$

Answer: Assuming that the sets A and B are empty, we use mathematical induction to prove the claim that every set which has been obtained by n applications of the recursive rule of the definition of F is empty.

The base case occurs for $n = 0$. The only sets in F that can be obtained by zero applications of the recursive rule are the sets A and B, obtained by the base case. Since the sets A and B are indeed empty, the base case holds.

For the inductive step, assume that every set in F obtained by k applications of the recursive rule is empty, for all k such that $0 \le k \le n$, for some $n \ge 0$. We have to prove that every set in F obtained by $n + 1$ applications of the recursive rule is empty.

Consider a set $Y \in F$, obtained by $n + 1$ applications of the recursive rule. The last of these applications requires the existence of two sets $S_1, S_2 \in F$ such that:

$$Y = S_1 \cap S_2 \ \text{ or } \ Y = S_1 \cup S_2$$

and S_1 and S_2 are obtained by no more than n applications of the recursive rule, performed prior to the last, $(n + 1)$st application, which generates Y either as the intersection or as the union of two already generated elements of F. Since S_1 and S_2 are obtained by n applications of the recursive rule, the inductive hypothesis applies to S_1 and S_2, guaranteeing that both S_1 and S_2 are empty:

$$S_1 = S_2 = \emptyset$$

However, if S_1 and S_2 are empty, both of the two possible forms of the last application guarantee that Y is also empty:

$$Y = S_1 \cap S_2 = \emptyset \cap \emptyset = \emptyset$$
$$Y = S_1 \cup S_2 = \emptyset \cap \emptyset = \emptyset$$

(b) Assume that the sets A and B are instantiated as follows.

$$A = \{n \in N \mid 2^n = 15\}$$
$$B = \{n \in N \mid 2n = 15\}$$

Prove that the set of natural numbers N does not belong to F. Precisely: $N \notin F$.

Answer: The definition of A and B implies that:

$$A = B = \emptyset$$

By the claim proved in part **(a)**, all elements of the set F are empty:

$$F = \{\emptyset\}$$

Since $N \neq \emptyset$ we conclude that $N \notin F$.

Problem 107   Consider two arbitrary propositions, p and q. Let F be a set of propositions, defined recursively as follows.

$$p \in F \wedge q \in F$$
$$(w_1 \in F \wedge w_2 \in F) \Longrightarrow ((w_1 \wedge w_2) \in F \bigwedge (w_1 \vee w_2) \in F)$$

Furthermore, F has no elements other than those obtained by finitely many applications of the recursive rule given in its definition.

(a) Prove that all propositions that are elements of F are false whenever both of the two propositions p and q are false. Precisely, let $\tau(x)$ denote the truth value of a propositional formula x, which may be equal to 0 (corresponding to false) or to 1 (corresponding to true.) The claim is that:

$$(\tau(p) = 0 \wedge \tau(q) = 0) \Longrightarrow (\forall x \in F)(\tau(x) = 0)$$

Answer: Assuming that the propositions p and q are false, we use mathematical induction to prove the claim that every proposition which has been obtained by n applications of the recursive rule of the definition of F is false.

The base case occurs for $n = 0$. The only propositions in F that can be obtained by zero applications of the recursive rule are the propositions p and q, obtained by the base case. Since the propositions p and q are indeed false, the base case holds.

For the inductive step, assume that every proposition in F obtained by k applications of the recursive rule is false, for all k such that $0 \leq k \leq n$, for some $n \geq 0$. We have to prove that every proposition in F obtained by $n + 1$ applications of the recursive rule is false.

Consider a proposition $t \in F$, obtained by $n + 1$ applications of the recursive rule. The last of these applications requires the existence of two propositions $w_1, w_2 \in F$ such that:

$$t = w_1 \wedge w_2 \ \text{ or } t = w_1 \vee w_2$$

and w_1 and w_2 are obtained by no more than n applications of the recursive rule, performed prior to the last, $(n + 1)$st application, which generates t either as the conjunction or as the disjunction of two already generated elements of F. Since w_1 and w_2 are obtained by n applications of the recursive rule, the inductive hypothesis applies to w_1 and w_2, guaranteeing that both w_1 and w_2 are false:

$$\tau(w_1) = \tau(w_2) = 0$$

However, if w_1 and w_2 are false, both $w_1 \wedge w_2$ and $w_1 \vee w_2$ are false, guaranteeing that t is also false.

(b) Assume that p and q are false. Prove that the formula $p \Longrightarrow q$ is not logically equivalent to any formula in F. Precisely:

$$(\forall x \in F)(\tau(p \Longrightarrow q) \neq \tau(x))$$

Answer: If there exists a formula $x \in F$ equivalent to $p \Longrightarrow q$, then x and $p \Longrightarrow q$ must have identical truth values. However, the claim proved in part **(a)** guarantees that all propositions in F are false:

$$(\tau(p) = \tau(q) = 0) \Longrightarrow (\forall x \in F)(\tau(x) = 0)$$

However, in this case, the implication $p \Longrightarrow q$ is true:

$$(\tau(p) = \tau(q) = 0) \Longrightarrow (\tau(p \Longrightarrow q) = 1)$$

whence the claim.

Problem 108   Consider a set S of strings over alphabet $A = \{a, b\}$, defined recursively as follows.

$$a \in S$$
$$x \in S \Longrightarrow bx \in S$$

Furthermore, S has no elements other than those obtained by finitely many applications of the recursive rule given in its definition.

(a) List four elements of S.

Answer:

$$a,\ ba,\ bba,\ bbba$$

(b) Use mathematical induction to prove that every string that belongs to the set S contains exactly one occurrence of the letter a.

Answer: We use mathematical induction to prove the claim that every string which has been obtained by n applications of the recursive rule of the definition of S contains exactly one occurrence of the letter a, for every $n \geq 0$.

The base case occurs for $n = 0$. The only string in S that can be obtained by zero applications of the recursive rule is the string a, obtained by the base case. Since a evidently contains exactly one occurrence of the letter a, the base case holds.

For the inductive step, assume that every string in S obtained by n applications of the recursive rule contains exactly one occurrence of the letter a, for some $n \geq 0$. We have to prove that every string in S obtained by $n + 1$ applications of the recursive rule contains exactly one occurrence of the letter a.

Consider a string $y \in S$, obtained by $n + 1$ applications of the recursive rule. The last of these applications requires the existence of a string $x \in S$ such that:

$$y = bx$$

and x has been obtained by the previous n applications of the recursive rule (performed prior to the last, $(n + 1)$st application, which generates y.) Since x is obtained by n applications of the recursive rule, the inductive hypothesis applies to x, guaranteeing that x contains exactly one occurrence of the letter a. However, y has no occurrences of a that are outside x. Hence, a occurs in y exactly as many times as it occurs in x, which is indeed once.

Problem 109   Consider a set S of strings over alphabet $A = \{a, b\}$, defined recursively as follows.

$$b \in S$$
$$x \in S \Longrightarrow abx \in S$$

Furthermore, S has no elements other than those obtained by finitely many applications of the recursive rule given in its definition.

(a) List four elements of S.

Answer:

$$b, \; abb, \; ababb, \; abababb$$

(b) Use mathematical induction to prove that there are no strings in S that end with the letter a.

Answer: We use mathematical induction to prove the claim that every string which has been obtained by n applications of the recursive rule of the definition of S ends with the letter b, for every $n \geq 0$.

The base case occurs for $n = 0$. The only string in S that can be obtained by zero applications of the recursive rule is the string b, obtained by the base case. Since b evidently ends with the letter b, the base case holds.

For the inductive step, assume that every string in S obtained by n applications of the recursive rule ends with the letter b, for some $n \geq 0$. We have to prove that every string in S obtained by $n + 1$ applications of the recursive rule ends with the letter b.

Consider a string $y \in S$, obtained by $n + 1$ applications of the recursive rule. The last of these applications requires the existence of a string $x \in S$ such that:

$$y = abx$$

and x has been obtained by the previous n applications of the recursive rule (performed prior to the last, $(n + 1)$st application, which generates y.) Since x is obtained by n applications of the recursive rule, the inductive hypothesis applies to x, guaranteeing that x ends with the letter b. However, the last letter of y is identical to the last letter of x. Hence, the string y indeed ends with the letter b.

Problem 110   Consider a set S of strings over alphabet $A = \{a, b\}$, defined recursively as follows.

$$a \in S$$
$$x \in S \implies abx \in S$$

Furthermore, S has no elements other than those obtained by finitely many applications of the recursive rule given in its definition.

For an arbitrary string $y \in S$, let $\alpha(y)$ be the number of occurrences of the letter a in the string y, and let $\beta(y)$ be the number of occurrences of the letter b in the string y.

Use mathematical induction to prove that:

$$\beta(y) = \alpha(y) - 1$$

for every string $y \in S$.

Answer: Our claim is:

$$(\forall y \in S)(\alpha(y) = n \implies \beta(y) = n - 1)$$

We use mathematical induction to prove that the claim holds for every positive integer n.

The base case occurs for $n = 1$. In this case, we have to prove that:

$$(\forall y \in S)(\alpha(y) = 1 \implies \beta(y) = 0)$$

In other words, we have to prove that if a string $y \in S$ contains exactly one occurrence of the letter a, then y does not contain any occurrences of the letter b.

Indeed, if $\alpha(y) = 1$, then y must by obtained by the base case of the definition of S, since applications of the recursive rule would introduce additional occurrences of the letter a. Hence, we conclude that $y = a$. String a indeed has one occurrence of the letter a and zero occurrences of the letter b, whence the claim.

For the inductive step, assume that every string in S that has n occurrences of the letter a has one fewer, namely $n-1$, occurrences of the letter b. Precisely:

$$(\forall y \in S)(\alpha(y) = n \implies \beta(y) = n - 1)$$

for some $n \geq 1$. We have to prove that every string that has $n+1$ occurrences of the letter a has n occurrences of letter b. Precisely:

$$(\forall y \in S)(\alpha(y) = n + 1 \implies \beta(y) = n)$$

Consider a string $y \in S$, such that:
$$\alpha(y) = n + 1$$

Since y has more than one occurrence of the letter a, it must be obtained by a certain number of applications of the recursive rule of the definition of S. The last of these applications requires the existence of a string $x \in S$ such that:
$$y = abx$$

Since the letter a in the string y occurs exactly once outside x, and $\alpha(y) = n + 1$, it must be that:

$$\alpha(x) = \alpha(y) - 1 = n + 1 - 1 = n$$

Hence, the inductive hypothesis applies to x, yielding:

$$\beta(x) = n - 1$$

However, the letter b in the string y occurs exactly once outside x, hence:

$$\beta(y) = 1 + \beta(x) = 1 + (n - 1) = n$$

Problem 111   Consider a set S of strings over alphabet $A = \{0, 1, 2\}$, defined recursively as follows.
$$1012 \in S$$
$$x \in S \implies 10x \in S$$

Furthermore, S has no elements other than those obtained by finitely many applications of the recursive rule given in its definition.

Use mathematical induction to prove that every string that belongs to the set S begins with the substring 101.

Answer: We use mathematical induction to prove the claim that every string which has been obtained by n applications of the recursive rule of the definition of S begins with the substring 101, for every $n \geq 0$.

The base case occurs for $n = 0$. The only string in S that can be obtained by zero applications of the recursive rule is 1012, obtained by the base case. Since 1012 evidently begins with the substring 101, the base case holds.

For the inductive step, assume that every string in S obtained by n applications of the recursive rule begins with the substring 101, for some $n \geq 0$. We have to prove that every string in S obtained by $n + 1$ applications of the recursive rule begins with the substring 101.

Consider a string $y \in S$, obtained by $n + 1$ applications of the recursive rule. The last of these applications requires the existence of a string $x \in S$ such that:

$$y = 10x$$

and x has been obtained by the previous n applications of the recursive rule (performed prior to the last, $(n + 1)$st application, which generates y.) Since x is obtained by n applications of the recursive rule, the inductive hypothesis applies to x, guaranteeing that x begins with the substring 101. Precisely:

$$x = 101z$$

for some string z. However, the last application of the recursive rule yields:

$$y = 10x = 10101z$$

Hence, y indeed begins with the substring 101.

Problem 112   Consider a set S of strings over alphabet $A = \{a, b\}$, defined recursively as follows.

$$bab \in S$$
$$x \in S \implies baxb \in S$$

Furthermore, S has no elements other than those obtained by finitely many applications of the recursive rule given in its definition.

Use mathematical induction to prove that every string that belongs to the set S contains an even number of occurrences of the letter b.

Answer: We use mathematical induction to prove the claim that every string which has been obtained by n applications of the recursive rule of the definition of S contains an even number of occurrences of the letter b, for every $n \geq 0$.

The base case occurs for $n = 0$. The only string in S that can be obtained by zero applications of the recursive rule is bab, obtained by the base case. Since bab evidently contains an even number of occurrences of the letter b, the base case holds.

For the inductive step, assume that every string in S obtained by n applications of the recursive rule contains an even number of occurrences of the letter b, for some $n \geq 0$. We have to prove that every string in S obtained by $n + 1$ applications of the recursive rule contains an even number (precisely, two) of occurrences of the letter b.

For an arbitrary string z, let $\beta(z)$ be the number of occurrences of the letter b in the string z. Consider a string $y \in S$, obtained by $n + 1$ applications of the recursive rule. The last of these applications requires the existence of a string $x \in S$ such that:

$$y = baxb$$

and x has been obtained by the previous n applications of the recursive rule (performed prior to the last, $(n + 1)$st application, which generates y.) Since x is obtained by n applications of the recursive rule, the inductive hypothesis applies to x, guaranteeing that x contains an even number of occurrences of the letter b. Precisely:

$$\beta(x) = 2k$$

for some integer k. However, y has exactly two occurrences of b that are outside x. Hence:

$$\beta(y) = 2 + \beta(x) = 2 + 2k = 2(k + 1)$$

Indeed, the number of occurrences of the letter b in the string y is even.

Problem 113   Sets S and T contain strings over alphabet $A = \{a, b, c\}$. Set S is defined as follows.

$$S = \{a, b, c, ab, bc, abb, abc, acc, bac, bab\};$$

Set T is defined recursively as follows.

$$a \in T \wedge b \in T \wedge c \in T$$
$$(x \in T) \implies (axa \in T \wedge bxb \in T \wedge cxc \in T)$$

Furthermore, T has no elements other than those obtained by finitely many applications of the recursive rule given in its definition.

Let $|x|$ denote the length of string x and let $N^+ = \{1, 2, \ldots\}$ be the set of positive integers.

Complete the following formulae by writing into each empty box a symbol representing a quantifier, logical operator, set operator, integer operator, set name, or one alphabet letter (as appropriate, if at all), so as to obtain a true claim.

Answer :

$a \boxed{} \, a \in T$ $a \boxed{b} \, a \in T$

$\left(ab \boxed{} \in S\right) \wedge \left(ab \boxed{} \notin T\right)$ $\left(ab \boxed{c} \in S\right) \wedge \left(ab \boxed{c} \notin T\right)$

$ab \boxed{} \in T$ $ab \boxed{a} \in T$

$bb \boxed{} \in T$ $bb \boxed{b} \in T$

$ca \boxed{} \in T$ $ca \boxed{c} \in T$

$accc \boxed{} \in T$ $accc \boxed{a} \in T$

$abc \boxed{} \boxed{} \in T$ $abc \boxed{b} \boxed{a} \in T$

$abaca \boxed{} \boxed{} \in T$ $abaca \boxed{b} \boxed{a} \in T$

$(bac \in S) \boxed{} (bac \in T)$ $(bac \in S) \boxed{\vee} (bac \in T)$

$\left(\forall \beta \in \boxed{}\right) \left(b \boxed{} \beta c \boxed{} \in T\right)$ $\left(\forall \beta \in \boxed{T}\right) \left(b \boxed{c} \beta c \boxed{b} \in T\right)$

$(d \in A) \boxed{} (ddd \in T)$ $(d \in A) \boxed{\implies} (ddd \in T)$

$(\forall n \in N^+) (\exists x \in T) \left(|x| \boxed{} n\right)$ $(\forall n \in N^+) (\exists x \in T) \left(|x| \boxed{>} n\right)$

$(\forall x_1 \in S) \left(\boxed{} x_2 \in T\right) \left(|x_1| \boxed{} |x_2|\right)$ $(\forall x_1 \in S) \left(\boxed{\exists} x_2 \in T\right) \left(|x_1| \boxed{\leq} |x_2|\right)$

$\left(\boxed{} x\right) \left(x \in \left(S \boxed{} T\right)\right)$ $\left(\boxed{\exists} x\right) \left(x \in \left(S \boxed{\cap} T\right)\right)$

$\left(\boxed{} x\right) \left(\left(|x| = \boxed{}\right) \implies \left(x \in S \boxed{} T\right)\right)$ $\left(\boxed{\forall} x\right) \left(\left(|x| = \boxed{1}\right) \implies \left(x \in S \boxed{\cap} T\right)\right)$

$\boxed{} \in S \cap T$ $\boxed{a} \in S \cap T$

$(\forall x \in S \cap T) \left((|x| > 1) \implies \left(x = \boxed{}\boxed{}\boxed{}\right)\right)$ $(\forall x \in S \cap T) \left((|x| > 1) \implies \left(x = \boxed{b}\boxed{a}\boxed{b}\right)\right)$

$(\forall x \in T \setminus S) \left(|x| \geq \boxed{}\right)$ $(\forall x \in T \setminus S) \left(|x| \geq \boxed{3}\right)$

Problem 114   Sequence g is defined recursively as follows.

$$g_0 = 1$$
$$g_1 = 2$$
$$g_2 = 4$$
$$g_n = g_{n-1} + g_{n-2} + \frac{g_{n-1}g_{n-3}}{g_{n-2}} \quad \text{for } n \geq 3$$

(a) Calculate g_3, g_4, g_5, g_6, g_7 and show your work.

Answer:

$$g_3 = g_{3-1} + g_{3-2} + \frac{g_{3-1}g_{3-3}}{g_{3-2}} = g_2 + g_1 + \frac{g_2 g_0}{g_1} = 4 + 2 + \frac{4 \cdot 1}{2} = 8$$

$$g_4 = g_{4-1} + g_{4-2} + \frac{g_{4-1}g_{4-3}}{g_{4-2}} = g_3 + g_2 + \frac{g_3 g_1}{g_2} = 8 + 4 + \frac{8 \cdot 2}{4} = 16$$

$$g_5 = g_{5-1} + g_{5-2} + \frac{g_{5-1}g_{5-3}}{g_{5-2}} = g_4 + g_3 + \frac{g_4 g_2}{g_3} = 16 + 8 + \frac{16 \cdot 4}{8} = 32$$

$$g_6 = g_{6-1} + g_{6-2} + \frac{g_{6-1}g_{6-3}}{g_{6-2}} = g_5 + g_4 + \frac{g_5 g_3}{g_4} = 32 + 16 + \frac{32 \cdot 8}{16} = 64$$

$$g_7 = g_{7-1} + g_{7-2} + \frac{g_{7-1}g_{7-3}}{g_{7-2}} = g_6 + g_5 + \frac{g_6 g_4}{g_5} = 64 + 32 + \frac{64 \cdot 16}{32} = 128$$

(b) Guess a closed form (non-recursive definition) for g_n.

Answer:

$$g_n = 2^n \quad \text{for } n \geq 0$$

(c) Use mathematical induction to prove your answer in part **(b)**. That is, prove that the recursive definition of g indeed defines the function which you have proposed in part **(b)**.

Answer: The base case occurs for $n = 0, 1, 2$ and follows from the base case of the recursive definition of the sequence g. Indeed:

$$g_0 = 1 = 2^0$$
$$g_1 = 2 = 2^1$$
$$g_2 = 4 = 2^2$$

For the inductive step, assume that:

$$g_k = 2^k$$

for every k such that $0 \leq k \leq n$, for some $n \geq 2$. We have to prove that:

$$g_{n+1} = 2^{n+1}$$

Indeed:

$$g_{n+1} =$$

$$\bullet \qquad = g_{(n+1)-1} + g_{(n+1)-2} + \frac{g_{(n+1)-1}g_{(n+1)-3}}{g_{(n+1)-2}} = g_n + g_{n-1} + \frac{g_n g_{n-2}}{g_{n-1}}$$

$$* \qquad = 2^n + 2^{n-1} + \frac{2^n \cdot 2^{n-2}}{2^{n-1}} = 2^n + 2^{n-1} + \frac{2^{2n-2}}{2^{n-1}} = 2^n + 2^{n-1} + 2^{2n-2-(n-1)}$$

$$= 2^n + 2^{n-1} + 2^{n-1} = 2^{n-1}(2 + 1 + 1) = 2^{n-1} \cdot 4 = 2^{n-1} \cdot 2^2 = 2^{n-1+2} = 2^{n+1}$$

The equality $\bullet$ follows from the recursive definition of g_{n+1} for $n+1 \geq 3$; the equality $*$ is obtained by three applications of the inductive hypothesis: to g_n, to g_{n-1}, and to g_{n-2}; and the other equalities are obtained by arithmetic transformations.

Problem 115   Sequence g is defined recursively as follows.

$$g_0 = 0$$
$$g_1 = 1$$
$$g_2 = 2$$
$$g_n = \frac{g_{n-1}}{2} + \frac{g_{n-2}}{3} + \frac{g_{n-3}}{6} + \frac{5}{3} \quad \text{for } n \geq 3$$

(a) Calculate g_3, g_4, g_5, g_6 and show your work.

Answer:

$$g_3 = \frac{g_{3-1}}{2} + \frac{g_{3-2}}{3} + \frac{g_{3-3}}{6} + \frac{5}{3} = \frac{g_2}{2} + \frac{g_1}{3} + \frac{g_0}{6} + \frac{5}{3} = \frac{2}{2} + \frac{1}{3} + \frac{0}{6} + \frac{5}{3} = \frac{6 + 2 + 0 + 10}{6} = 3$$

$$g_4 = \frac{g_{4-1}}{2} + \frac{g_{4-2}}{3} + \frac{g_{4-3}}{6} + \frac{5}{3} = \frac{g_3}{2} + \frac{g_2}{3} + \frac{g_1}{6} + \frac{5}{3} = \frac{3}{2} + \frac{2}{3} + \frac{1}{6} + \frac{5}{3} = \frac{9 + 4 + 1 + 10}{6} = 4$$

$$g_5 = \frac{g_{5-1}}{2} + \frac{g_{5-2}}{3} + \frac{g_{5-3}}{6} + \frac{5}{3} = \frac{g_4}{2} + \frac{g_3}{3} + \frac{g_2}{6} + \frac{5}{3} = \frac{4}{2} + \frac{3}{3} + \frac{2}{6} + \frac{5}{3} = \frac{12 + 6 + 2 + 10}{6} = 5$$

$$g_6 = \frac{g_{6-1}}{2} + \frac{g_{6-2}}{3} + \frac{g_{6-3}}{6} + \frac{5}{3} = \frac{g_5}{2} + \frac{g_4}{3} + \frac{g_3}{6} + \frac{5}{3} = \frac{5}{2} + \frac{4}{3} + \frac{3}{6} + \frac{5}{3} = \frac{15 + 8 + 3 + 10}{6} = 6$$

(b) Guess a closed form (non-recursive definition) for g_n.

Answer:

$$g_n = n \quad \text{for } n \geq 0$$

(c) Use mathematical induction to prove your answer in part **(b)**. That is, prove that the recursive definition of g indeed defines the function which you have proposed in part **(b)**.

Answer: The base case occurs for $n = 0, 1, 2$ and follows from the base case of the recursive definition of the sequence g. Indeed:

$$g_0 = 0$$
$$g_1 = 1$$
$$g_2 = 2$$

For the inductive step, assume that:

$$g_k = k$$

for every k such that $0 \leq k \leq n$, for some $n \geq 2$. We have to prove that:

$$g_{n+1} = n + 1$$

Indeed:

$$g_{n+1} =$$

$$\bullet \qquad = \frac{g_{(n+1)-1}}{2} + \frac{g_{(n+1)-2}}{3} + \frac{g_{(n+1)-3}}{6} + \frac{5}{3} = \frac{g_n}{2} + \frac{g_{n-1}}{3} + \frac{g_{n-2}}{6} + \frac{5}{3}$$

$$* \qquad = \frac{n}{2} + \frac{n-1}{3} + \frac{n-2}{6} + \frac{5}{3} = \frac{3n + 2(n-1) + (n-2) + 10}{6} = \frac{6n + 6}{6} = n + 1$$

The equality $\bullet$ follows from the recursive definition of g_{n+1} for $n+1 \geq 3$; the equality $*$ is obtained by three applications of the inductive hypothesis: to g_n, to g_{n-1}, and to g_{n-2}; and the other equalities are obtained by arithmetic transformations.

Problem 116   Sequence g is defined recursively as follows.

$$g_0 = 0$$
$$g_n = g_{n-1} + 2\sqrt{g_{n-1}} + 1 \quad \text{for } n \geq 1$$

(a) Calculate $g_1, g_2, g_3, g_4, g_5, g_6, g_7$ and show your work.

Answer:

$$g_1 = g_{1-1} + 2\sqrt{g_{1-1}} + 1 = g_0 + 2\sqrt{g_0} + 1 = 0 + 2 \cdot \sqrt{0} + 1 = 0 + 0 + 1 = 1$$
$$g_2 = g_{2-1} + 2\sqrt{g_{2-1}} + 1 = g_1 + 2\sqrt{g_1} + 1 = 1 + 2 \cdot \sqrt{1} + 1 = 1 + 2 + 1 = 4$$
$$g_3 = g_{3-1} + 2\sqrt{g_{3-1}} + 1 = g_2 + 2\sqrt{g_2} + 1 = 4 + 2 \cdot \sqrt{4} + 1 = 4 + 4 + 1 = 9$$
$$g_4 = g_{4-1} + 2\sqrt{g_{4-1}} + 1 = g_3 + 2\sqrt{g_3} + 1 = 9 + 2 \cdot \sqrt{9} + 1 = 9 + 6 + 1 = 16$$
$$g_5 = g_{5-1} + 2\sqrt{g_{5-1}} + 1 = g_4 + 2\sqrt{g_4} + 1 = 16 + 2 \cdot \sqrt{16} + 1 = 16 + 8 + 1 = 25$$
$$g_6 = g_{6-1} + 2\sqrt{g_{6-1}} + 1 = g_5 + 2\sqrt{g_5} + 1 = 25 + 2 \cdot \sqrt{25} + 1 = 25 + 10 + 1 = 36$$
$$g_7 = g_{7-1} + 2\sqrt{g_{7-1}} + 1 = g_6 + 2\sqrt{g_6} + 1 = 36 + 2 \cdot \sqrt{36} + 1 = 36 + 12 + 1 = 49$$

(b) Guess a closed form (non-recursive definition) for g_n.

Answer:
$$g_n = n^2 \quad \text{for } n \geq 0$$

(c) Use mathematical induction to prove your answer in part **(b)**. That is, prove that the recursive definition of g indeed defines the function which you have proposed in part **(b)**.

Answer: The base case occurs for $n = 0$ and follows from the base case of the recursive definition of the sequence g. Indeed:

$$g_0 = 0 = 0^2$$

For the inductive step, assume that:

$$g_n = n^2$$

for some $n \geq 0$. We have to prove that:

$$g_{n+1} = (n+1)^2$$

Indeed:

$$
\begin{aligned}
g_{n+1} &= \\
\bullet \quad &= g_{(n+1)-1} + 2\sqrt{g_{(n+1)-1}} + 1 = g_n + 2\sqrt{g_n} + 1 \\
* \quad &= n^2 + 2\sqrt{n^2} + 1 = n^2 + 2n + 1 = (n+1)^2
\end{aligned}
$$

The equality $\bullet$ follows from the recursive definition of g_{n+1} for $n+1 \geq 1$; the equality $*$ is obtained by an application of the inductive hypothesis to g_n; and the other equalities are obtained by arithmetic transformations.

Problem 117   Sequence g is defined recursively as follows.

$$g_0 = 2$$
$$g_1 = 3$$
$$g_2 = 5$$
$$g_n = 12g_{n-3} - 2g_{n-2} - 9 \quad \text{for } n \geq 3$$

(a) Calculate g_3, g_4, g_5, g_6, g_7 and show your work.

Answer:

$$g_3 = 12g_0 - 2g_1 - 9 = 12 \cdot 2 - 2 \cdot 3 - 9 = 9$$
$$g_4 = 12g_1 - 2g_2 - 9 = 12 \cdot 3 - 2 \cdot 5 - 9 = 17$$
$$g_5 = 12g_2 - 2g_3 - 9 = 12 \cdot 5 - 2 \cdot 9 - 9 = 33$$
$$g_6 = 12g_3 - 2g_4 - 9 = 12 \cdot 9 - 2 \cdot 17 - 9 = 65$$
$$g_7 = 12g_4 - 2g_5 - 9 = 12 \cdot 17 - 2 \cdot 33 - 9 = 129$$

(b) Guess a closed form (non-recursive definition) for g_n.

Answer:

$$g_n = 2^n + 1 \quad \text{for } n \geq 0$$

(c) Use mathematical induction to prove your answer in part **(b)**. That is, prove that the recursive definition of g indeed defines the function which you have proposed in part **(b)**.

Answer: The base case occurs for $n = 0, 1, 2$ and follows from the base case of the recursive definition of the sequence g. Indeed:

$$g_0 = 2 = 2^0 + 1$$
$$g_1 = 3 = 2^1 + 1$$
$$g_2 = 5 = 2^2 + 1$$

For the inductive step, assume that:

$$g_k = 2^k + 1$$

for every k such that $0 \leq k \leq n$, for some $n \geq 2$. We have to prove that:

$$g_{n+1} = 2^{n+1} + 1$$

Indeed:

$$
\begin{aligned}
g_{n+1} &= \\
\bullet \quad &= 12g_{n-2} - 2g_{n-1} - 9 \\
* \quad &= 12(2^{n-2} + 1) - 2(2^{n-1} + 1) - 9 = 3 \cdot 4 \cdot 2^{n-2} + 12 - 2 \cdot 2^{n-1} - 2 - 9 \\
&= 3 \cdot 2^n - 2^n + 1 = 2 \cdot 2^n + 1 = 2^{n+1} + 1
\end{aligned}
$$

The equality $\bullet$ follows from the recursive definition of g_{n+1} for $n + 1 \geq 3$; the equality $*$ is obtained by two applications of the inductive hypothesis: to g_{n-1} and to g_{n-2}; and the other equalities are obtained by arithmetic transformations.

Problem 118   Sequence g is defined recursively as follows.

$$
\begin{aligned}
g_0 &= 5 \\
g_1 &= 5 \\
g_n &= g_{n-1}^2 - 4g_{n-2} \quad \text{for } n \geq 2
\end{aligned}
$$

(a) Calculate g_2, g_3, g_4, g_5 and show your work.

Answer:

$$
\begin{aligned}
g_2 &= g_{2-1}^2 - 4g_{2-2} = g_1^2 - 4g_0 = 5^2 - 4 \cdot 5 = 25 - 20 = 5 \\
g_3 &= g_{3-1}^2 - 4g_{3-2} = g_2^2 - 4g_1 = 5^2 - 4 \cdot 5 = 25 - 20 = 5 \\
g_4 &= g_{4-1}^2 - 4g_{4-2} = g_3^2 - 4g_2 = 5^2 - 4 \cdot 5 = 25 - 20 = 5 \\
g_5 &= g_{5-1}^2 - 4g_{5-2} = g_4^2 - 4g_3 = 5^2 - 4 \cdot 5 = 25 - 20 = 5
\end{aligned}
$$

(b) Guess a closed form (non-recursive definition) for g_n.

Answer:

$$g_n = 5 \quad \text{for } n \geq 0$$

(c) Use mathematical induction to prove your answer in part **(b)**. That is, prove that the recursive definition of g indeed defines the function which you have proposed in part **(b)**.

Answer: The base case occurs for $n = 0, 1$ and follows from the base case of the recursive definition of the sequence g. Indeed:

$$g_0 = 5$$
$$g_1 = 5$$

For the inductive step, assume that:

$$g_k = 5$$

for every k such that $0 \leq k \leq n$, for some $n \geq 1$. We have to prove that:

$$g_{n+1} = 5$$

Indeed:

$$
\begin{aligned}
g_{n+1} &= \\
\bullet \quad &= g_{(n+1)-1}^2 - 4g_{(n+1)-2} = g_n^2 - 4g_{n-1} \\
* \quad &= 5^2 - 4 \cdot 5 = 25 - 20 = 5
\end{aligned}
$$

The equality $\bullet$ follows from the recursive definition of g_{n+1} for $n + 1 \geq 2$; the equality $*$ is obtained by two applications of the inductive hypothesis: to g_n and to g_{n-1}; and the other equalities are obtained by arithmetic transformations.

Problem 119   Sequence g is defined recursively as follows.

$$g_2 = 3$$
$$g_n = g_{n-1}\left(\frac{n}{n-2}\right) - 1 \quad \text{for } n \geq 3$$

(a) Calculate g_3, g_4, g_5, g_6, g_7 and show your work.

Answer:

$$g_3 = g_{3-1}\left(\frac{3}{3-2}\right) - 1 = g_2\left(\frac{3}{1}\right) - 1 = 3 \cdot \frac{3}{1} - 1 = \frac{9}{1} - 1 = 8$$

$$g_4 = g_{4-1}\left(\frac{4}{4-2}\right) - 1 = g_3\left(\frac{4}{2}\right) - 1 = 8 \cdot \frac{4}{2} - 1 = \frac{32}{2} - 1 = 15$$

$$g_5 = g_{5-1}\left(\frac{5}{5-2}\right) - 1 = g_4\left(\frac{5}{3}\right) - 1 = 15 \cdot \frac{5}{3} - 1 = \frac{75}{3} - 1 = 24$$

$$g_6 = g_{6-1}\left(\frac{6}{6-2}\right) - 1 = g_5\left(\frac{6}{4}\right) - 1 = 24 \cdot \frac{6}{4} - 1 = \frac{144}{4} - 1 = 35$$

$$g_7 = g_{7-1}\left(\frac{7}{7-2}\right) - 1 = g_6\left(\frac{7}{5}\right) - 1 = 35 \cdot \frac{7}{5} - 1 = \frac{245}{5} - 1 = 48$$

(b) Guess a closed form (non-recursive definition) for g_n.

Answer:

$$g_n = n^2 - 1 \quad \text{for } n \geq 2$$

(c) Use mathematical induction to prove your answer in part (b). That is, prove that the recursive definition of g indeed defines the function which you have proposed in part (b).

Answer: The base case occurs for $n = 2$ and follows from the base case of the recursive definition of the sequence g. Indeed:

$$g_2 = 3 = 2^2 - 1$$

For the inductive step, assume that:

$$g_n = n^2 - 1$$

for some $n \geq 2$. We have to prove that:

$$g_{n+1} = (n+1)^2 - 1$$

Indeed:

$$
\begin{aligned}
g_{n+1} \quad &= \\
\bullet \quad &= g_{(n+1)-1}\left(\frac{(n+1)}{(n+1)-2}\right) - 1 = g_n\left(\frac{n+1}{n-1}\right) - 1 \\
* \quad &= (n^2 - 1)\left(\frac{n+1}{n-1}\right) - 1 = (n-1)(n+1) \cdot \frac{(n+1)}{(n-1)} - 1 = (n+1)^2 - 1
\end{aligned}
$$

The equality $\bullet$ follows from the recursive definition of g_{n+1} for $n+1 \geq 3$; the equality $*$ is obtained by an application of the inductive hypothesis to g_n; and the other equalities are obtained by arithmetic transformations.

Problem 120 Sequence g is defined recursively as follows.

$$g_1 = 3$$
$$g_n = g_{n-1}\left(1 + \frac{1}{n-1}\right) \quad \text{for } n \geq 2$$

(a) Calculate $g_2, g_3, g_4, g_5, g_6, g_7$ and show your work.

Answer:

$$
\begin{aligned}
g_2 &= g_{2-1}\left(1 + \frac{1}{2-1}\right) = g_1\left(1 + \frac{1}{1}\right) = 3 \cdot 2 = 6 \\
g_3 &= g_{3-1}\left(1 + \frac{1}{3-1}\right) = g_2\left(1 + \frac{1}{2}\right) = 6 \cdot \frac{3}{2} = 9 \\
g_4 &= g_{4-1}\left(1 + \frac{1}{4-1}\right) = g_3\left(1 + \frac{1}{3}\right) = 9 \cdot \frac{4}{3} = 12 \\
g_5 &= g_{5-1}\left(1 + \frac{1}{5-1}\right) = g_4\left(1 + \frac{1}{4}\right) = 12 \cdot \frac{5}{4} = 15 \\
g_6 &= g_{6-1}\left(1 + \frac{1}{6-1}\right) = g_5\left(1 + \frac{1}{5}\right) = 15 \cdot \frac{6}{5} = 18 \\
g_7 &= g_{7-1}\left(1 + \frac{1}{7-1}\right) = g_6\left(1 + \frac{1}{6}\right) = 18 \cdot \frac{7}{6} = 21
\end{aligned}
$$

(b) Guess a closed form (non-recursive definition) for g_n.

Answer:

$$g_n = 3n \quad \text{for } n \geq 1$$

(c) Use mathematical induction to prove your answer in part **(b)**. That is, prove that the recursive definition of g indeed defines the function which you have proposed in part **(b)**.

Answer: The base case occurs for $n = 1$ and follows from the base case of the recursive definition of the sequence g. Indeed:

$$g_1 = 3 = 3 \cdot 1$$

For the inductive step, assume that:

$$g_n = 3n$$

for some $n \geq 1$. We have to prove that:

$$g_{n+1} = 3(n + 1)$$

Indeed:

$$g_{n+1} =$$

$$\bullet \qquad = g_{(n+1)-1}\left(1 + \frac{1}{(n + 1) - 1}\right) = g_n\left(1 + \frac{1}{n}\right)$$

$$* \qquad = 3n\left(1 + \frac{1}{n}\right) = 3n \cdot \frac{n + 1}{n} = 3(n + 1)$$

The equality $\bullet$ follows from the recursive definition of g_{n+1} for $n+1 \geq 2$; the equality $*$ is obtained by an application of the inductive hypothesis to g_n; and the other equalities are obtained by arithmetic transformations.

Problem 121   Sequence g is defined recursively as follows.

$$g_0 = 1$$
$$g_1 = 3$$
$$g_2 = 9$$
$$g_n = g_{n-1} + 2g_{n-2} + \frac{4}{3} \cdot \frac{g_{n-2}^2}{g_{n-3}} \quad \text{for } n \geq 3$$

(a) Calculate g_3, g_4, g_5, g_6 and show your work.

Answer:

$$g_3 = g_{3-1} + 2g_{3-2} + \frac{4}{3} \cdot \frac{g_{3-2}^2}{g_{3-3}} = g_2 + 2g_1 + \frac{4}{3} \cdot \frac{g_1^2}{g_0} = 9 + 2 \cdot 3 + \frac{4}{3} \cdot \frac{3^2}{1} = 27$$

$$g_4 = g_{4-1} + 2g_{4-2} + \frac{4}{3} \cdot \frac{g_{4-2}^2}{g_{4-3}} = g_3 + 2g_2 + \frac{4}{3} \cdot \frac{g_2^2}{g_1} = 27 + 2 \cdot 9 + \frac{4}{3} \cdot \frac{9^2}{3} = 81$$

$$g_5 = g_{5-1} + 2g_{5-2} + \frac{4}{3} \cdot \frac{g_{5-2}^2}{g_{5-3}} = g_4 + 2g_3 + \frac{4}{3} \cdot \frac{g_3^2}{g_2} = 81 + 2 \cdot 27 + \frac{4}{3} \cdot \frac{27^2}{9} = 243$$

$$g_6 = g_{6-1} + 2g_{6-2} + \frac{4}{3} \cdot \frac{g_{6-2}^2}{g_{6-3}} = g_5 + 2g_4 + \frac{4}{3} \cdot \frac{g_4^2}{g_3} = 243 + 2 \cdot 81 + \frac{4}{3} \cdot \frac{81^2}{27} = 729$$

(b) Guess a closed form (non-recursive definition) for g_n.

Answer:

$$g_n = 3^n \quad \text{for } n \geq 0$$

(c) Use mathematical induction to prove your answer in part **(b)**. That is, prove that the recursive definition of g indeed defines the function which you have proposed in part **(b)**.

Answer: The base case occurs for $n = 0, 1, 2$ and follows from the base case of the recursive definition of the sequence g. Indeed:

$$g_0 = 1 = 3^0$$
$$g_1 = 3 = 3^1$$
$$g_2 = 9 = 3^2$$

For the inductive step, assume that:

$$g_k = 3^k$$

for every k such that $0 \leq k \leq n$, for some $n \geq 2$. We have to prove that:

$$g_{n+1} = 3^{n+1}$$

Indeed:

$$g_{n+1} =$$

$$\bullet \quad = g_{(n+1)-1} + 2g_{(n+1)-2} + \frac{4}{3} \cdot \frac{g^2_{(n+1)-2}}{g_{(n+1)-3}} = g_n + 2g_{n-1} + \frac{4}{3} \cdot \frac{g^2_{n-1}}{g_{n-2}}$$

$$* \quad = 3^n + 2 \cdot 3^{n-1} + \frac{4}{3} \cdot \frac{\left(3^{n-1}\right)^2}{3^{n-2}} = 3^n + 2 \cdot 3^{n-1} + \frac{4}{3} \cdot 3^{2(n-1)-(n-2)}$$

$$= 3^n + 2 \cdot 3^{n-1} + \frac{4}{3} \cdot 3^n = 3^{n-1}\left(3 + 2 + \frac{4}{3} \cdot 3\right) = 3^{n-1} \cdot 9 = 3^{n-1} \cdot 3^2 = 3^{n+1}$$

The equality $\bullet$ follows from the recursive definition of g_{n+1} for $n+1 \geq 3$; the equality $*$ is obtained by three applications of the inductive hypothesis: to g_n, to g_{n-1}, and to g_{n-2}; and the other equalities are obtained by arithmetic transformations.

Problem 122   Sequence g is defined recursively as follows.

$$g_0 = 6$$
$$g_1 = 6$$
$$g_n = \frac{g_{n-1}}{2} + \frac{3\sqrt{6g_{n-2}}}{g_{n-1}} \quad \text{for } n \geq 2$$

(a) Calculate g_2, g_3, g_4, g_5 and show your work.

Answer:

$$g_2 = \frac{g_{2-1}}{2} + \frac{3\sqrt{6g_{2-2}}}{g_{2-1}} = \frac{g_1}{2} + \frac{3\sqrt{6g_0}}{g_1} = \frac{6}{2} + \frac{3\sqrt{6 \cdot 6}}{6} = 3 + \frac{3 \cdot 6}{6} = 6$$

$$g_3 = \frac{g_{3-1}}{2} + \frac{3\sqrt{6g_{3-2}}}{g_{3-1}} = \frac{g_2}{2} + \frac{3\sqrt{6g_1}}{g_2} = \frac{6}{2} + \frac{3\sqrt{6 \cdot 6}}{6} = 3 + \frac{3 \cdot 6}{6} = 6$$

$$g_4 = \frac{g_{4-1}}{2} + \frac{3\sqrt{6g_{4-2}}}{g_{4-1}} = \frac{g_3}{2} + \frac{3\sqrt{6g_2}}{g_3} = \frac{6}{2} + \frac{3\sqrt{6 \cdot 6}}{6} = 3 + \frac{3 \cdot 6}{6} = 6$$

$$g_5 = \frac{g_{5-1}}{2} + \frac{3\sqrt{6g_{5-2}}}{g_{5-1}} = \frac{g_4}{2} + \frac{3\sqrt{6g_3}}{g_4} = \frac{6}{2} + \frac{3\sqrt{6 \cdot 6}}{6} = 3 + \frac{3 \cdot 6}{6} = 6$$

(b) Guess a closed form (non-recursive definition) for g_n.

Answer:

$$g_n = 6 \quad \text{for } n \geq 0$$

(c) Use mathematical induction to prove your answer in part **(b)**. That is, prove that the recursive definition of g indeed defines the function which you have proposed in part **(b)**.

Answer: The base case occurs for $n = 0, 1$ and follows from the base case of the recursive definition of the sequence g. Indeed:

$$g_0 = 6$$
$$g_1 = 6$$

For the inductive step, assume that:

$$g_k = 6$$

for every k such that $0 \leq k \leq n$, for some $n \geq 1$. We have to prove that:

$$g_{n+1} = 6$$

Indeed:

$$g_{n+1} =$$

$$\bullet \quad = \frac{g_{(n+1)-1}}{2} + \frac{3\sqrt{6g_{(n+1)-2}}}{g_{(n+1)-1}} = \frac{g_n}{2} + \frac{3\sqrt{6g_{n-1}}}{g_n}$$

$$* \quad = \frac{6}{2} + \frac{3\sqrt{6 \cdot 6}}{6} = 3 + \frac{3 \cdot 6}{6} = 6$$

The equality $\bullet$ follows from the recursive definition of g_{n+1} for $n+1 \geq 2$; the equality $*$ is obtained by two applications of the inductive hypothesis: to g_n and to g_{n-1}; and the other equalities are obtained by arithmetic transformations.

Problem 123   Sequence g is defined recursively as follows.

$$g_0 = \frac{4}{3}$$

$$g_1 = 2$$

$$g_n = g_{n-2}\left(\frac{g_{n-1}-1}{g_{n-2}-1}\right)^2 - 8 \quad \text{for } n \geq 2$$

(a) Calculate g_2, g_3, g_4, g_5, g_6 and show your work.

Answer:

$$g_2 = g_{2-2}\left(\frac{g_{2-1}-1}{g_{2-2}-1}\right)^2 - 8 = g_0\left(\frac{g_1-1}{g_0-1}\right)^2 - 8 = \frac{4}{3}\cdot\left(\frac{2-1}{(4/3)-1}\right)^2 - 8 = \frac{4}{3}\cdot 3^2 - 8 = 4$$

$$g_3 = g_{3-2}\left(\frac{g_{3-1}-1}{g_{3-2}-1}\right)^2 - 8 = g_1\left(\frac{g_2-1}{g_1-1}\right)^2 - 8 = 2\cdot\left(\frac{4-1}{2-1}\right)^2 - 8 = 2\cdot 3^2 - 8 = 10$$

$$g_4 = g_{4-2}\left(\frac{g_{4-1}-1}{g_{4-2}-1}\right)^2 - 8 = g_2\left(\frac{g_3-1}{g_2-1}\right)^2 - 8 = 4\cdot\left(\frac{10-1}{4-1}\right)^2 - 8 = 4\cdot 3^2 - 8 = 28$$

$$g_5 = g_{5-2}\left(\frac{g_{5-1}-1}{g_{5-2}-1}\right)^2 - 8 = g_3\left(\frac{g_4-1}{g_3-1}\right)^2 - 8 = 10\cdot\left(\frac{28-1}{10-1}\right)^2 - 8 = 10\cdot 3^2 - 8 = 82$$

$$g_6 = g_{6-2}\left(\frac{g_{6-1}-1}{g_{6-2}-1}\right)^2 - 8 = g_4\left(\frac{g_5-1}{g_4-1}\right)^2 - 8 = 28\cdot\left(\frac{82-1}{28-1}\right)^2 - 8 = 28\cdot 3^2 - 8 = 244$$

(b) Guess a closed form (non-recursive definition) for g_n.

Answer:

$$g_n = 3^{n-1} + 1 \quad \text{for } n \geq 0$$

(c) Use mathematical induction to prove your answer in part **(b)**. That is, prove that the recursive definition of g indeed defines the function which you have proposed in part **(b)**.

Answer: The base case occurs for $n = 0, 1$ and follows from the base case of the recursive definition of the sequence g. Indeed:

$$g_0 = \frac{4}{3} = \frac{1+3}{3} = \frac{1}{3} + 1 = 3^{-1} + 1 = 3^{0-1} + 1$$
$$g_1 = 2 = 1 + 1 = 3^0 + 1 = 3^{1-1} + 1$$

For the inductive step, assume that:

$$g_k = 3^{k-1} + 1$$

for every k such that $0 \leq k \leq n$, for some $n \geq 1$. We have to prove that:

$$g_{n+1} = 3^{(n+1)-1} + 1$$

Indeed:

$$g_{n+1} =$$
$$\bullet \qquad = g_{(n+1)-2} \left(\frac{g_{(n+1)-1} - 1}{g_{(n+1)-2} - 1} \right)^2 - 8 = g_{n-1} \left(\frac{g_n - 1}{g_{n-1} - 1} \right)^2 - 8$$
$$* \qquad = (3^{n-2} + 1) \left(\frac{(3^{n-1} + 1) - 1}{(3^{(n-2)} + 1) - 1} \right)^2 - 8 = (3^{n-2} + 1) \left(\frac{3^{n-1}}{3^{n-2}} \right)^2 - 8$$
$$= (3^{n-2} + 1) \cdot 3^2 - 8 = 3^n + 9 - 8 = 3^n + 1 = 3^{(n+1)-1} + 1$$

The equality $\bullet$ follows from the recursive definition of g_{n+1} for $n + 1 \geq 2$; the equality $*$ is obtained by two applications of the inductive hypothesis: to g_n and to g_{n-1}; and the other equalities are obtained by arithmetic transformations.

Problem 124   Sequence g is defined recursively as follows.

$$g_0 = 0$$
$$g_1 = 2$$
$$g_n = \frac{g_{n-1}^2 - g_{n-2}^2}{(g_{n-1} - g_{n-2})^2} + 3 \quad \text{for } n \geq 2$$

(a) Calculate g_2, g_3, g_4, g_5 and show your work.

Answer:

$$g_2 = \frac{g_{2-1}^2 - g_{2-2}^2}{(g_{2-1} - g_{2-2})^2} + 3 = \frac{g_1^2 - g_0^2}{(g_1 - g_0)^2} + 3 = \frac{2^2 - 0^2}{(2 - 0)^2} + 3 = \frac{4 - 0}{(2)^2} + 3 = \frac{4}{4} + 3 = 4$$

$$g_3 = \frac{g_{3-1}^2 - g_{3-2}^2}{(g_{3-1} - g_{3-2})^2} + 3 = \frac{g_2^2 - g_1^2}{(g_2 - g_1)^2} + 3 = \frac{4^2 - 2^2}{(4 - 2)^2} + 3 = \frac{16 - 4}{(2)^2} + 3 = \frac{12}{4} + 3 = 6$$

$$g_4 = \frac{g_{4-1}^2 - g_{4-2}^2}{(g_{4-1} - g_{4-2})^2} + 3 = \frac{g_3^2 - g_2^2}{(g_3 - g_2)^2} + 3 = \frac{6^2 - 4^2}{(6 - 4)^2} + 3 = \frac{36 - 16}{(2)^2} + 3 = \frac{20}{4} + 3 = 8$$

$$g_5 = \frac{g_{5-1}^2 - g_{5-2}^2}{(g_{5-1} - g_{5-2})^2} + 3 = \frac{g_4^2 - g_3^2}{(g_4 - g_3)^2} + 3 = \frac{8^2 - 6^2}{(8 - 6)^2} + 3 = \frac{64 - 36}{(2)^2} + 3 = \frac{28}{4} + 3 = 10$$

(b) Guess a closed form (non-recursive definition) for g_n.

Answer:

$$g_n = 2n \quad \text{for } n \geq 0$$

(c) Use mathematical induction to prove your answer in part **(b)**. That is, prove that the recursive definition of g indeed defines the function which you have proposed in part **(b)**.

Answer: The base case occurs for $n = 0, 1$ and follows from the base case of the recursive definition of the sequence g. Indeed:

$$g_0 = 0 = 2 \cdot 0$$
$$g_1 = 2 = 2 \cdot 1$$

For the inductive step, assume that:

$$g_k = 2k$$

for every k such that $0 \leq k \leq n$, for some $n \geq 1$. We have to prove that:

$$g_{n+1} = 2(n + 1)$$

Indeed:

$$g_{n+1} \quad =$$

$$\bullet \quad = \frac{g_{(n+1)-1}^2 - g_{(n+1)-2}^2}{(g_{(n+1)-1} - g_{(n+1)-2})^2} + 3 = \frac{g_n^2 - g_{n-1}^2}{(g_n - g_{n-1})^2} + 3$$

$$* \quad = \frac{(2n)^2 - (2(n-1))^2}{(2n - 2(n-1))^2} + 3 = \frac{4n^2 - 4(n-1)^2}{2^2} + 3 = \frac{4n^2 - 4n^2 + 8n - 4}{4} + 3$$

$$= (2n - 1) + 3 = 2n + 2 = 2(n + 1)$$

The equality $\bullet$ follows from the recursive definition of g_{n+1} for $n + 1 \geq 2$; the equality $*$ is obtained by two applications of the inductive hypothesis: to g_n and to g_{n-1}; and the other equalities are obtained by arithmetic transformations.

Problem 125   Sequence g is defined recursively as follows.

$$g_0 = 0$$
$$g_1 = 1$$
$$g_n = \left(\frac{g_{n-1} - g_{n-2} + 9n - 7}{3} \right) \cdot \left(\frac{g_{n-2} - g_{n-1} + 3n^2 + 7}{9} \right) \quad \text{for } n \geq 2$$

(a) Calculate g_2, g_3, g_4, g_5 and show your work.

Answer:

$$g_2 = \left(\frac{g_{2-1} - g_{2-2} + 9 \cdot 2 - 7}{3} \right) \cdot \left(\frac{g_{2-2} - g_{2-1} + 3 \cdot 2^2 + 7}{9} \right)$$

$$= \left(\frac{g_1 - g_0 + 9 \cdot 2 - 7}{3} \right) \cdot \left(\frac{g_0 - g_1 + 3 \cdot 2^2 + 7}{9} \right)$$

$$= \left(\frac{1 - 0 + 9 \cdot 2 - 7}{3} \right) \cdot \left(\frac{0 - 1 + 3 \cdot 2^2 + 7}{9} \right) = \left(\frac{12}{3} \right) \cdot \left(\frac{18}{9} \right) = 8$$

$$g_3 = \left(\frac{g_{3-1} - g_{3-2} + 9 \cdot 3 - 7}{3} \right) \cdot \left(\frac{g_{3-2} - g_{3-1} + 3 \cdot 3^2 + 7}{9} \right)$$

$$= \left(\frac{g_2 - g_1 + 9 \cdot 3 - 7}{3} \right) \cdot \left(\frac{g_1 - g_2 + 3 \cdot 3^2 + 7}{9} \right)$$

$$= \left(\frac{8 - 1 + 9 \cdot 3 - 7}{3} \right) \cdot \left(\frac{1 - 8 + 3 \cdot 3^2 + 7}{9} \right) = \left(\frac{27}{3} \right) \cdot \left(\frac{27}{9} \right) = 27$$

$$g_4 = \left(\frac{g_{4-1} - g_{4-2} + 9 \cdot 4 - 7}{3}\right) \cdot \left(\frac{g_{4-2} - g_{4-1} + 3 \cdot 4^2 + 7}{9}\right)$$

$$= \left(\frac{g_3 - g_2 + 9 \cdot 4 - 7}{3}\right) \cdot \left(\frac{g_2 - g_3 + 3 \cdot 4^2 + 7}{9}\right)$$

$$= \left(\frac{27 - 8 + 9 \cdot 4 - 7}{3}\right) \cdot \left(\frac{8 - 27 + 3 \cdot 4^2 + 7}{9}\right) = \left(\frac{48}{3}\right) \cdot \left(\frac{36}{9}\right) = 64$$

$$g_5 = \left(\frac{g_{5-1} - g_{5-2} + 9 \cdot 5 - 7}{3}\right) \cdot \left(\frac{g_{5-2} - g_{5-1} + 3 \cdot 5^2 + 7}{9}\right)$$

$$= \left(\frac{g_4 - g_3 + 9 \cdot 5 - 7}{3}\right) \cdot \left(\frac{g_3 - g_4 + 3 \cdot 5^2 + 7}{9}\right)$$

$$= \left(\frac{64 - 27 + 9 \cdot 5 - 7}{3}\right) \cdot \left(\frac{27 - 64 + 3 \cdot 5^2 + 7}{9}\right) = \left(\frac{75}{3}\right) \cdot \left(\frac{45}{9}\right) = 125$$

(b) Guess a closed form (non-recursive definition) for g_n.

Answer:

$$g_n = n^3 \quad \text{for } n \geq 0$$

(c) Use mathematical induction to prove your answer in part **(b)**. That is, prove that the recursive definition of g indeed defines the function which you have proposed in part **(b)**.

Answer: The base case occurs for $n = 0, 1$ and follows from the base case of the recursive definition of the sequence g. Indeed:

$$g_0 = 0 = 0^3$$
$$g_1 = 1 = 1^3$$

For the inductive step, assume that:

$$g_k = k^3$$

for every k such that $0 \leq k \leq n$, for some $n \geq 1$. We have to prove that:

$$g_{n+1} = (n+1)^3$$

Indeed:

$$g_{n+1} =$$

$$\bullet \quad = \left(\frac{g_{(n+1)-1} - g_{(n+1)-2} + 9(n+1) - 7}{3}\right) \cdot \left(\frac{g_{(n+1)-2} - g_{(n+1)-1} + 3(n+1)^2 + 7}{9}\right)$$

$$= \left(\frac{g_n - g_{n-1} + 9n + 2}{3}\right) \cdot \left(\frac{g_{n-1} - g_n + 3(n+1)^2 + 7}{9}\right)$$

$$* \quad = \left(\frac{n^3 - (n-1)^3 + 9n + 2}{3}\right) \cdot \left(\frac{(n-1)^3 - n^3 + 3(n+1)^2 + 7}{9}\right)$$

$$= \left(\frac{n^3 - (n^3 - 3n^2 + 3n - 1) + 9n + 2}{3}\right) \cdot \left(\frac{(n^3 - 3n^2 + 3n - 1) - n^3 + 3(n+1)^2 + 7}{9}\right)$$

$$= \left(\frac{3n^2 + 6n + 3}{3}\right) \cdot \left(\frac{9n + 9}{9}\right) = \left(\frac{3(n^2 + 2n + 1)}{3}\right) \cdot \left(\frac{9(n+1)}{9}\right)$$

$$= \left(\frac{3(n+1)^2}{3}\right) \cdot \left(\frac{9(n+1)}{9}\right) = (n+1)^2(n+1) = (n+1)^3$$

The equality $\bullet$ follows from the recursive definition of g_{n+1} for $n+1 \geq 2$; the equality $*$ is obtained by two applications of the inductive hypothesis: to g_n and to g_{n-1}; and the other equalities are obtained by arithmetic transformations.

Problem 126   Sequence g is defined recursively as follows.

$$g_0 = 0$$
$$g_1 = 1$$
$$g_2 = 3$$
$$g_n = g_{n-1} + g_{n-2} + 2g_{n-3} + 3 \quad \text{for } n \geq 3$$

(a) Calculate g_3, g_4, g_5, g_6 and show your work.

Answer:

$$g_3 = g_{3-1} + g_{3-2} + 2g_{3-3} + 3 = g_2 + g_1 + 2g_0 + 3 = 3 + 1 + 2 \cdot 0 + 3 = 7$$
$$g_4 = g_{4-1} + g_{4-2} + 2g_{4-3} + 3 = g_3 + g_2 + 2g_1 + 3 = 7 + 3 + 2 \cdot 1 + 3 = 15$$
$$g_5 = g_{5-1} + g_{5-2} + 2g_{5-3} + 3 = g_4 + g_3 + 2g_2 + 3 = 15 + 7 + 2 \cdot 3 + 3 = 31$$
$$g_6 = g_{6-1} + g_{6-2} + 2g_{6-3} + 3 = g_5 + g_4 + 2g_3 + 3 = 31 + 15 + 2 \cdot 7 + 3 = 63$$

(b) Guess a closed form (non-recursive definition) for g_n.

Answer:

$$g_n = 2^n - 1 \quad \text{for } n \geq 0$$

(c) Use mathematical induction to prove your answer in part **(b)**. That is, prove that the recursive definition of g indeed defines the function which you have proposed in part **(b)**.

Answer: The base case occurs for $n = 0, 1, 2$ and follows from the base case of the recursive definition of the sequence g. Indeed:

$$g_0 = 0 = 2^0 - 1$$
$$g_1 = 1 = 2^1 - 1$$
$$g_2 = 3 = 2^2 - 1$$

For the inductive step, assume that:

$$g_k = 2^k - 1$$

for every k such that $0 \leq k \leq n$, for some $n \geq 2$. We have to prove that:

$$g_{n+1} = 2^{n+1} - 1$$

Indeed:

$$g_{n+1} =$$
$$\bullet \quad = g_{(n+1)-1} + g_{(n+1)-2} + 2g_{(n+1)-3} + 3 = g_n + g_{n-1} + 2g_{n-2} + 3$$
$$* \quad = (2^n - 1) + (2^{n-1} - 1) + 2(2^{n-2} - 1) + 3 = 2^n - 1 + 2^{n-1} - 1 + 2^{n-1} - 2 + 3$$
$$= 2^n - 1 + 2^{n-1} - 1 + 2^{n-1} - 2 + 3 = 2^{n-1}(2 + 1 + 1) - 1 = 2^{n-1} \cdot 2^2 - 1 = 2^{n+1} - 1$$

The equality $\bullet$ follows from the recursive definition of g_{n+1} for $n+1 \geq 3$; the equality $*$ is obtained by three applications of the inductive hypothesis: to g_n, to g_{n-1}, and to g_{n-2}; and the other equalities are obtained by arithmetic transformations.

Problem 127   Sequence g is defined recursively as follows.

$$g_0 = -1$$
$$g_1 = 1$$
$$g_n = \frac{5}{2} + \frac{g_{n-1}}{2} + \frac{1}{2}\sqrt{g_{n-2}^2 + 4n - 9} \quad \text{for } n \geq 2$$

(a) Calculate g_2, g_3, g_4, g_5 and show your work.

Answer:

$$
\begin{aligned}
g_2 &= \frac{5}{2} + \frac{g_{2-1}}{2} + \frac{1}{2}\sqrt{g_{2-2}^2 + 4\cdot 2 - 9} = \frac{5}{2} + \frac{g_1}{2} + \frac{1}{2}\sqrt{g_0^2 + 4\cdot 2 - 9} \\
&= \frac{5}{2} + \frac{1}{2} + \frac{1}{2}\sqrt{(-1)^2 + 8 - 9} = \frac{5+1}{2} + \frac{1}{2}\cdot 0 = 3 \\
g_3 &= \frac{5}{2} + \frac{g_{3-1}}{2} + \frac{1}{2}\sqrt{g_{3-2}^2 + 4\cdot 3 - 9} = \frac{5}{2} + \frac{g_2}{2} + \frac{1}{2}\sqrt{g_1^2 + 4\cdot 3 - 9} \\
&= \frac{5}{2} + \frac{3}{2} + \frac{1}{2}\sqrt{1^2 + 12 - 9} = \frac{5+3}{2} + \frac{1}{2}\cdot 2 = 5 \\
g_4 &= \frac{5}{2} + \frac{g_{4-1}}{2} + \frac{1}{2}\sqrt{g_{4-2}^2 + 4\cdot 4 - 9} = \frac{5}{2} + \frac{g_3}{2} + \frac{1}{2}\sqrt{g_2^2 + 4\cdot 4 - 9} \\
&= \frac{5}{2} + \frac{5}{2} + \frac{1}{2}\sqrt{3^2 + 16 - 9} = \frac{5+5}{2} + \frac{1}{2}\cdot 4 = 7 \\
g_5 &= \frac{5}{2} + \frac{g_{5-1}}{2} + \frac{1}{2}\sqrt{g_{5-2}^2 + 4\cdot 5 - 9} = \frac{5}{2} + \frac{g_4}{2} + \frac{1}{2}\sqrt{g_3^2 + 4\cdot 5 - 9} \\
&= \frac{5}{2} + \frac{7}{2} + \frac{1}{2}\sqrt{5^2 + 20 - 9} = \frac{5+7}{2} + \frac{1}{2}\cdot 6 = 9
\end{aligned}
$$

(b) Guess a closed form (non-recursive definition) for g_n.

Answer:

$$g_n = 2n - 1 \quad \text{for } n \geq 0$$

(c) Use mathematical induction to prove your answer in part **(b)**. That is, prove that the recursive definition of g indeed defines the function which you have proposed in part **(b)**.

Answer: The base case occurs for $n = 0, 1$ and follows from the base case of the recursive definition of the sequence g. Indeed:

$$
\begin{aligned}
g_0 &= -1 = 2\cdot 0 - 1 \\
g_1 &= 1 = 2\cdot 1 - 1
\end{aligned}
$$

For the inductive step, assume that:

$$g_k = 2k - 1$$

for every k such that $0 \leq k \leq n$, for some $n \geq 1$. We have to prove that:

$$g_{n+1} = 2(n+1) - 1$$

Indeed:

$$
\begin{aligned}
g_{n+1} &= \\
\bullet \quad &= \frac{5}{2} + \frac{g_{(n+1)-1}}{2} + \frac{1}{2}\sqrt{g_{(n+1)-2}^2 + 4(n+1) - 9} = \frac{5}{2} + \frac{g_n}{2} + \frac{1}{2}\sqrt{g_{n-1}^2 + 4(n+1) - 9} \\
* \quad &= \frac{5}{2} + \frac{2n-1}{2} + \frac{1}{2}\sqrt{(2(n-1)-1)^2 + 4(n+1) - 9} \\
&= \frac{5+2n-1}{2} + \frac{1}{2}\sqrt{(2n-3)^2 + 4n - 5} = \frac{2n+4}{2} + \frac{1}{2}\sqrt{4n^2 - 12n + 9 + 4n - 5} \\
&= \frac{2n+4}{2} + \frac{1}{2}\sqrt{4n^2 - 8n + 4} = \frac{2n+4}{2} + \frac{1}{2}\sqrt{(2n-2)^2} = \frac{(2n+4) + (2n-2)}{2} \\
&= \frac{4n+2}{2} = 2n+1 = 2(n+1) - 1
\end{aligned}
$$

The equality $\bullet$ follows from the recursive definition of g_{n+1} for $n+1 \geq 2$; the equality $*$ is obtained by two applications of the inductive hypothesis: to g_n and to g_{n-1}; and the other equalities are obtained by arithmetic transformations.

Problem 128   Sequence g is defined recursively as follows.

$$g_0 = 1$$
$$g_1 = 2$$
$$g_2 = 3$$
$$g_n = 3(g_{n-1} - g_{n-2}) + g_{n-3} \quad \text{for } n \geq 3$$

(a) Calculate g_3, g_4, g_5, g_6 and show your work.

Answer:

$$g_3 = 3(g_{3-1} - g_{3-2}) + g_{3-3} = 3(g_2 - g_1) + g_0 = 3(3 - 2) + 1 = 4$$
$$g_4 = 3(g_{4-1} - g_{4-2}) + g_{4-3} = 3(g_3 - g_2) + g_1 = 3(4 - 3) + 2 = 5$$
$$g_5 = 3(g_{5-1} - g_{5-2}) + g_{5-3} = 3(g_4 - g_3) + g_2 = 3(5 - 4) + 3 = 6$$
$$g_6 = 3(g_{6-1} - g_{6-2}) + g_{6-3} = 3(g_5 - g_4) + g_3 = 3(6 - 5) + 4 = 7$$

(b) Guess a closed form (non-recursive definition) for g_n.

Answer:

$$g_n = n + 1 \quad \text{for } n \geq 0$$

(c) Use mathematical induction to prove your answer in part **(b)**. That is, prove that the recursive definition of g indeed defines the function which you have proposed in part **(b)**.

Answer: The base case occurs for $n = 0, 1, 2$ and follows from the base case of the recursive definition of the sequence g. Indeed:

$$g_0 = 1 = 0 + 1$$
$$g_1 = 2 = 1 + 1$$
$$g_2 = 3 = 2 + 1$$

For the inductive step, assume that:

$$g_k = k + 1$$

for every k such that $0 \leq k \leq n$, for some $n \geq 2$. We have to prove that:

$$g_{n+1} = (n + 1) + 1$$

Indeed:

$$
\begin{array}{ll}
g_{n+1} & = \\
\bullet & = 3(g_{(n+1)-1} - g_{(n+1)-2}) + g_{(n+1)-3} = 3(g_n - g_{n-1}) + g_{n-2} \\
* & = 3((n + 1) - ((n - 1) + 1)) + ((n - 2) + 1) = 3 \cdot 1 + (n - 1) = n + 2 = (n + 1) + 1
\end{array}
$$

The equality $\bullet$ follows from the recursive definition of g_{n+1} for $n+1 \geq 3$; the equality $*$ is obtained by three applications of the inductive hypothesis: to g_n, to g_{n-1}, and to g_{n-2}; and the other equalities are obtained by arithmetic transformations.

Problem 129   Sequence g is defined recursively as follows.

$$g_0 = 7$$
$$g_n = 2 \lg(g_{n-1} + 1) + 1 \quad \text{for } n \geq 1$$

(where $\lg x = \log_2 x$.)

(a) Calculate g_1, g_2, g_3, g_4, g_5 and show your work.

Answer:

$$\begin{aligned}
g_1 &= 2\lg(g_{1-1}+1)+1 = 2\lg(g_0+1)+1 = 2\lg(7+1)+1 = 2\lg 8+1 = 2\cdot 3+1 = 7 \\
g_2 &= 2\lg(g_{2-1}+1)+1 = 2\lg(g_1+1)+1 = 2\lg(7+1)+1 = 2\lg 8+1 = 2\cdot 3+1 = 7 \\
g_3 &= 2\lg(g_{3-1}+1)+1 = 2\lg(g_2+1)+1 = 2\lg(7+1)+1 = 2\lg 8+1 = 2\cdot 3+1 = 7 \\
g_4 &= 2\lg(g_{4-1}+1)+1 = 2\lg(g_3+1)+1 = 2\lg(7+1)+1 = 2\lg 8+1 = 2\cdot 3+1 = 7 \\
g_5 &= 2\lg(g_{5-1}+1)+1 = 2\lg(g_4+1)+1 = 2\lg(7+1)+1 = 2\lg 8+1 = 2\cdot 3+1 = 7
\end{aligned}$$

(b) Guess a closed form (non-recursive definition) for g_n.

Answer:

$$g_n = 7 \quad \text{for } n \geq 0$$

(c) Use mathematical induction to prove your answer in part **(b)**. That is, prove that the recursive definition of g indeed defines the function which you have proposed in part **(b)**.

Answer: The base case occurs for $n = 0$ and follows from the base case of the recursive definition of the sequence g. Indeed:

$$g_0 = 7$$

For the inductive step, assume that:

$$g_n = 7$$

for some $n \geq 0$. We have to prove that:

$$g_{n+1} = 7$$

Indeed:

$$\begin{aligned}
g_{n+1} &= \\
\bullet \quad &= 2\lg(g_{(n+1)-1}+1)+1 = 2\lg(g_n+1)+1 \\
* \quad &= 2\lg(7+1)+1 = 2\lg 8+1 = 2\cdot 3+1 = 7
\end{aligned}$$

The equality $\bullet$ follows from the recursive definition of g_{n+1} for $n+1 \geq 1$; the equality $*$ is obtained by an application of the inductive hypothesis to g_n; and the other equalities are obtained by arithmetic transformations.

Problem 130   Sequence g is defined recursively as follows.

$$\begin{aligned}
g_0 &= 1 \\
g_1 &= 2 \\
g_2 &= 5 \\
g_n &= g_{n-1} + g_{n-2} - g_{n-3} + 4 \quad \text{for } n \geq 3
\end{aligned}$$

(a) Calculate g_3, g_4, g_5, g_6, g_7 and show your work.

Answer:

$$\begin{aligned}
g_3 &= g_{3-1}+g_{3-2}-g_{3-3}+4 = g_2+g_1-g_0+4 = 5+2-1+4 = 10 \\
g_4 &= g_{4-1}+g_{4-2}-g_{4-3}+4 = g_3+g_2-g_1+4 = 10+5-2+4 = 17 \\
g_5 &= g_{5-1}+g_{5-2}-g_{5-3}+4 = g_4+g_3-g_2+4 = 17+10-5+4 = 26 \\
g_6 &= g_{6-1}+g_{6-2}-g_{6-3}+4 = g_5+g_4-g_3+4 = 26+17-10+4 = 37 \\
g_7 &= g_{7-1}+g_{7-2}-g_{7-3}+4 = g_6+g_5-g_4+4 = 37+26-17+4 = 50
\end{aligned}$$

(b) Guess a closed form (non-recursive definition) for g_n.

Answer:

$$g_n = n^2 + 1 \quad \text{for } n \geq 0$$

(c) Use mathematical induction to prove your answer in part **(b)**. That is, prove that the recursive definition of g indeed defines the function which you have proposed in part **(b)**.

Answer: The base case occurs for $n = 0, 1, 2$ and follows from the base case of the recursive definition of the sequence g. Indeed:

$$g_0 = 1 = 0^2 + 1$$
$$g_1 = 2 = 1^2 + 1$$
$$g_2 = 5 = 2^2 + 1$$

For the inductive step, assume that:

$$g_k = k^2 + 1$$

for every k such that $0 \leq k \leq n$, for some $n \geq 2$. We have to prove that:

$$g_{n+1} = (n + 1)^2 + 1$$

Indeed:

$$
\begin{aligned}
g_{n+1} \quad &= \\
\bullet \quad &= g_{(n+1)-1} + g_{(n+1)-2} - g_{(n+1)-3} + 4 = g_n + g_{n-1} - g_{n-2} + 4 \\
* \quad &= (n^2 + 1) + ((n - 1)^2 + 1) - ((n - 2)^2 + 1) + 4 \\
&= (n^2 + 1) + (n^2 - 2n + 1 + 1) - (n^2 - 4n + 4 + 1) + 4 = n^2 + 2n + 2 = (n + 1)^2 + 1
\end{aligned}
$$

The equality $\bullet$ follows from the recursive definition of g_{n+1} for $n+1 \geq 3$; the equality $*$ is obtained by three applications of the inductive hypothesis: to g_n, to g_{n-1}, and to g_{n-2}; and the other equalities are obtained by arithmetic transformations.

Problem 131   Sequence g is defined recursively as follows.

$$
\begin{aligned}
g_0 &= 1 \\
g_1 &= 4 \\
g_2 &= 7 \\
g_n &= g_{n-3} + \left(\frac{g_{n-1} - 1}{n - 1} \right)^2 \quad \text{for } n \geq 3
\end{aligned}
$$

(a) Calculate g_3, g_4, g_5, g_6 and show your work.

Answer:

$$g_3 = g_{3-3} + \left(\frac{g_{3-1} - 1}{3 - 1} \right)^2 = g_0 + \left(\frac{g_2 - 1}{2} \right)^2 = 1 + \left(\frac{7 - 1}{2} \right)^2 = 1 + 3^2 = 10$$

$$g_4 = g_{4-3} + \left(\frac{g_{4-1} - 1}{4 - 1} \right)^2 = g_1 + \left(\frac{g_3 - 1}{3} \right)^2 = 4 + \left(\frac{10 - 1}{3} \right)^2 = 4 + 3^2 = 13$$

$$g_5 = g_{5-3} + \left(\frac{g_{5-1} - 1}{5 - 1} \right)^2 = g_2 + \left(\frac{g_4 - 1}{4} \right)^2 = 7 + \left(\frac{13 - 1}{4} \right)^2 = 7 + 3^2 = 16$$

$$g_6 = g_{6-3} + \left(\frac{g_{6-1} - 1}{6 - 1} \right)^2 = g_3 + \left(\frac{g_5 - 1}{5} \right)^2 = 10 + \left(\frac{16 - 1}{5} \right)^2 = 10 + 3^2 = 19$$

(b) Guess a closed form (non-recursive definition) for g_n.

Answer:

$$g_n = 3n + 1 \quad \text{for } n \geq 0$$

(c) Use mathematical induction to prove your answer in part **(b)**. That is, prove that the recursive definition of g indeed defines the function which you have proposed in part **(b)**.

Answer: The base case occurs for $n = 0, 1, 2$ and follows from the base case of the recursive definition of the sequence g. Indeed:

$$g_0 = 1 = 3 \cdot 0 + 1$$
$$g_1 = 4 = 3 \cdot 1 + 1$$
$$g_2 = 7 = 3 \cdot 2 + 1$$

For the inductive step, assume that:

$$g_k = 3k + 1$$

for every k such that $0 \leq k \leq n$, for some $n \geq 2$. We have to prove that:

$$g_{n+1} = 3(n + 1) + 1$$

Indeed:

$$g_{n+1} =$$

$$\bullet \quad = g_{(n+1)-3} + \left(\frac{g_{(n+1)-1} - 1}{(n + 1) - 1} \right)^2 = g_{n-2} + \left(\frac{g_n - 1}{n} \right)^2$$

$$* \quad = (3(n - 2) + 1) + \left(\frac{(3n + 1) - 1}{n} \right)^2 = 3n - 5 + \left(\frac{3n}{n} \right)^2 = 3n - 5 + 3^2$$

$$= 3n + 4 = 3(n + 1) + 1$$

The equality $\bullet$ follows from the recursive definition of g_{n+1} for $n + 1 \geq 3$; the equality $*$ is obtained by two applications of the inductive hypothesis: to g_n and to g_{n-2}; and the other equalities are obtained by arithmetic transformations.

Problem 132 Sequence g is defined recursively as follows.

$$g_0 = -1$$
$$g_1 = 0$$
$$g_n = 1 + \frac{g_{n-1}^2 + 2g_{n-1}}{g_{n-2} + 3} \quad \text{for } n \geq 2$$

(a) Calculate g_2, g_3, g_4, g_5, g_6 and show your work.

Answer:

$$g_2 = 1 + \frac{g_{2-1}^2 + 2g_{2-1}}{g_{2-2} + 3} = 1 + \frac{g_1^2 + 2g_1}{g_0 + 3} = 1 + \frac{0^2 + 2 \cdot 0}{-1 + 3} = 1 + \frac{0}{2} = 1$$

$$g_3 = 1 + \frac{g_{3-1}^2 + 2g_{3-1}}{g_{3-2} + 3} = 1 + \frac{g_2^2 + 2g_2}{g_1 + 3} = 1 + \frac{1^2 + 2 \cdot 1}{0 + 3} = 1 + \frac{3}{3} = 2$$

$$g_4 = 1 + \frac{g_{4-1}^2 + 2g_{4-1}}{g_{4-2} + 3} = 1 + \frac{g_3^2 + 2g_3}{g_2 + 3} = 1 + \frac{2^2 + 2 \cdot 2}{1 + 3} = 1 + \frac{8}{4} = 3$$

$$g_5 = 1 + \frac{g_{5-1}^2 + 2g_{5-1}}{g_{5-2} + 3} = 1 + \frac{g_4^2 + 2g_4}{g_3 + 3} = 1 + \frac{3^2 + 2 \cdot 3}{2 + 3} = 1 + \frac{15}{5} = 4$$

$$g_6 = 1 + \frac{g_{6-1}^2 + 2g_{6-1}}{g_{6-2} + 3} = 1 + \frac{g_5^2 + 2g_5}{g_4 + 3} = 1 + \frac{4^2 + 2 \cdot 4}{3 + 3} = 1 + \frac{24}{6} = 5$$

(b) Guess a closed form (non-recursive definition) for g_n.

Answer:

$$g_n = n - 1 \quad \text{for } n \geq 0$$

(c) Use mathematical induction to prove your answer in part **(b)**. That is, prove that the recursive definition of g indeed defines the function which you have proposed in part **(b)**.

Answer: The base case occurs for $n = 0, 1$ and follows from the base case of the recursive definition of the sequence g. Indeed:

$$g_0 = -1 = 0 - 1$$
$$g_1 = 0 = 1 - 1$$

For the inductive step, assume that:

$$g_k = k - 1$$

for every k such that $0 \leq k \leq n$, for some $n \geq 1$. We have to prove that:

$$g_{n+1} = (n + 1) - 1$$

Indeed:

$$g_{n+1} =$$

• $$= 1 + \frac{g_{(n+1)-1}^2 + 2g_{(n+1)-1}}{g_{(n+1)-2} + 3} = 1 + \frac{g_n^2 + 2g_n}{g_{n-1} + 3}$$

* $$= 1 + \frac{(n-1)^2 + 2(n-1)}{((n-1)-1) + 3} = 1 + \frac{(n^2 - 2n + 1) + (2n - 2)}{n - 2 + 3} = 1 + \frac{n^2 - 1}{n + 1}$$

$$= 1 + \frac{(n-1)(n+1)}{n+1} = 1 + (n - 1) = n = (n + 1) - 1$$

The equality • follows from the recursive definition of g_{n+1} for $n + 1 \geq 2$; the equality * is obtained by two applications of the inductive hypothesis: to g_n and to g_{n-1}; and the other equalities are obtained by arithmetic transformations.

Problem 133   Sequence g is defined recursively as follows.

$$g_0 = 2$$
$$g_1 = 8$$
$$g_n = g_{n-1} + 2g_{n-2}\left(\frac{g_{n-1} + 1}{g_{n-2} + 1}\right) + 6 \quad \text{for } n \geq 2$$

(a) Calculate g_2, g_3, g_4, g_5 and show your work.

Answer:

$$g_2 = g_{2-1} + 2g_{2-2}\left(\frac{g_{2-1} + 1}{g_{2-2} + 1}\right) + 6 = g_1 + 2g_0\left(\frac{g_1 + 1}{g_0 + 1}\right) + 6 = 8 + 2 \cdot 2 \cdot \left(\frac{8 + 1}{2 + 1}\right) + 6$$
$$= 8 + 2 \cdot 2 \cdot 3 + 6 = 26$$

$$g_3 = g_{3-1} + 2g_{3-2}\left(\frac{g_{3-1} + 1}{g_{3-2} + 1}\right) + 6 = g_2 + 2g_1\left(\frac{g_2 + 1}{g_1 + 1}\right) + 6 = 26 + 2 \cdot 8 \cdot \left(\frac{26 + 1}{8 + 1}\right) + 6$$
$$= 26 + 2 \cdot 8 \cdot 3 + 6 = 80$$

$$g_4 = g_{4-1} + 2g_{4-2}\left(\frac{g_{4-1} + 1}{g_{4-2} + 1}\right) + 6 = g_3 + 2g_2\left(\frac{g_3 + 1}{g_2 + 1}\right) + 6 = 80 + 2 \cdot 26 \cdot \left(\frac{80 + 1}{26 + 1}\right) + 6$$
$$= 80 + 2 \cdot 26 \cdot 3 + 6 = 242$$

$$g_5 = g_{5-1} + 2g_{5-2}\left(\frac{g_{5-1}+1}{g_{5-2}+1}\right) + 6 = g_4 + 2g_3\left(\frac{g_4+1}{g_3+1}\right) + 6 = 242 + 2\cdot 80\cdot\left(\frac{242+1}{80+1}\right) + 6$$

$$= 242 + 2\cdot 80\cdot 3 + 6 = 728$$

(b) Guess a closed form (non-recursive definition) for g_n.

Answer:

$$g_n = 3^{n+1} - 1 \quad \text{for } n \geq 0$$

(c) Use mathematical induction to prove your answer in part **(b)**. That is, prove that the recursive definition of g indeed defines the function which you have proposed in part **(b)**.

Answer: The base case occurs for $n = 0, 1$ and follows from the base case of the recursive definition of the sequence g. Indeed:

$$g_0 = 2 = 3^{0+1} - 1$$
$$g_1 = 8 = 3^{1+1} - 1$$

For the inductive step, assume that:

$$g_k = 3^{k+1} - 1$$

for every k such that $0 \leq k \leq n$, for some $n \geq 1$. We have to prove that:

$$g_{n+1} = 3^{(n+1)+1} - 1$$

Indeed:

$$g_{n+1} =$$

$$\bullet \qquad = g_{(n+1)-1} + 2g_{(n+1)-2}\left(\frac{g_{(n+1)-1}+1}{g_{(n+1)-2}+1}\right) + 6 = g_n + 2g_{n-1}\left(\frac{g_n+1}{g_{n-1}+1}\right) + 6$$

$$* \qquad = (3^{n+1} - 1) + 2(3^{(n-1)+1} - 1)\left(\frac{(3^{n+1}-1)+1}{(3^{(n-1)+1}-1)+1}\right) + 6$$

$$= (3^{n+1} - 1) + 2(3^n - 1)\left(\frac{3^{n+1}-1+1}{3^n-1+1}\right) + 6 = (3^{n+1} - 1) + 2(3^n - 1)\left(\frac{3^{n+1}}{3^n}\right) + 6$$

$$= (3^{n+1} - 1) + 2(3^n - 1)\cdot 3 + 6 = 3^{n+1} - 1 + 2\cdot 3^n\cdot 3 - 2\cdot 3 + 6$$

$$= 3^{n+1} - 1 + 2\cdot 3^{n+1} - 6 + 6 = 3^{n+1}(1 + 2) - 1 = 3^{n+1}\cdot 3 - 1 = 3^{(n+1)+1} - 1$$

The equality $\bullet$ follows from the recursive definition of g_{n+1} for $n + 1 \geq 2$; the equality $*$ is obtained by two applications of the inductive hypothesis: to g_n and to g_{n-1}; and the other equalities are obtained by arithmetic transformations.

Problem 134 Sequence g is defined recursively as follows.

$$g_0 = 1$$
$$g_n = \frac{g_{n-1}^2 + 10n + 1}{g_{n-1} + 3} \quad \text{for } n \geq 1$$

(a) Calculate g_1, g_2, g_3, g_4, g_5 and show your work.

Answer:

$$g_1 = \frac{g_{1-1}^2 + 10\cdot 1 + 1}{g_{1-1} + 3} = \frac{g_0^2 + 10\cdot 1 + 1}{g_0 + 3} = \frac{1^2 + 10\cdot 1 + 1}{1 + 3} = \frac{12}{4} = 3$$

$$g_2 = \frac{g_{2-1}^2 + 10\cdot 2 + 1}{g_{2-1} + 3} = \frac{g_1^2 + 10\cdot 2 + 1}{g_1 + 3} = \frac{3^2 + 10\cdot 2 + 1}{3 + 3} = \frac{30}{6} = 5$$

$$g_3 = \frac{g_{3-1}^2 + 10\cdot 3 + 1}{g_{3-1} + 3} = \frac{g_2^2 + 10\cdot 3 + 1}{g_2 + 3} = \frac{5^2 + 10\cdot 3 + 1}{5 + 3} = \frac{56}{8} = 7$$

$$g_4 = \frac{g_{4-1}^2 + 10 \cdot 4 + 1}{g_{4-1} + 3} = \frac{g_3^2 + 10 \cdot 4 + 1}{g_3 + 3} = \frac{7^2 + 10 \cdot 4 + 1}{7 + 3} = \frac{90}{10} = 9$$

$$g_5 = \frac{g_{5-1}^2 + 10 \cdot 5 + 1}{g_{5-1} + 3} = \frac{g_4^2 + 10 \cdot 5 + 1}{g_4 + 3} = \frac{9^2 + 10 \cdot 5 + 1}{9 + 3} = \frac{132}{12} = 11$$

(b) Guess a closed form (non-recursive definition) for g_n.

Answer:

$$g_n = 2n + 1 \quad \text{for } n \geq 0$$

(c) Use mathematical induction to prove your answer in part **(b)**. That is, prove that the recursive definition of g indeed defines the function which you have proposed in part **(b)**.

Answer: The base case occurs for $n = 0$ and follows from the base case of the recursive definition of the sequence g. Indeed:

$$g_0 = 1 = 2 \cdot 0 + 1$$

For the inductive step, assume that:

$$g_n = 2n + 1$$

for some $n \geq 0$. We have to prove that:

$$g_{n+1} = 2(n + 1) + 1$$

Indeed:

$$g_{n+1} =$$

$$\bullet \quad = \frac{g_{(n+1)-1}^2 + 10(n + 1) + 1}{g_{(n+1)-1} + 3} = \frac{g_n^2 + 10(n + 1) + 1}{g_n + 3}$$

$$* \quad = \frac{(2n + 1)^2 + 10(n + 1) + 1}{(2n + 1) + 3} = \frac{4n^2 + 4n + 1 + 10n + 10 + 1}{2n + 4}$$

$$= \frac{4n^2 + 14n + 12}{2n + 4} = \frac{(2n + 3)(2n + 4)}{2n + 4} = 2n + 3 = 2(n + 1) + 1$$

The equality $\bullet$ follows from the recursive definition of g_{n+1} for $n+1 \geq 1$; the equality $*$ is obtained by an application of the inductive hypothesis to g_n; and the other equalities are obtained by arithmetic transformations.

Problem 135   Sequence g is defined recursively as follows.

$$g_1 = 2$$
$$g_n = \sqrt{g_{n-1}^2 + 6g_{n-1} + \left(\frac{g_{n-1} + 1}{n - 1}\right)^2} \quad \text{for } n \geq 2$$

(a) Calculate g_2, g_3, g_4, g_5 and show your work.

Answer:

$$g_2 = \sqrt{g_{2-1}^2 + 6g_{2-1} + \left(\frac{g_{2-1} + 1}{2 - 1}\right)^2} = \sqrt{g_1^2 + 6g_1 + \left(\frac{g_1 + 1}{2 - 1}\right)^2} = \sqrt{2^2 + 6 \cdot 2 + \left(\frac{2 + 1}{1}\right)^2}$$

$$= \sqrt{4 + 12 + 3^2} = \sqrt{25} = 5$$

$$g_3 = \sqrt{g_{3-1}^2 + 6g_{3-1} + \left(\frac{g_{3-1} + 1}{3 - 1}\right)^2} = \sqrt{g_2^2 + 6g_2 + \left(\frac{g_2 + 1}{3 - 1}\right)^2} = \sqrt{5^2 + 6 \cdot 5 + \left(\frac{5 + 1}{2}\right)^2}$$

$$= \sqrt{25 + 30 + 3^2} = \sqrt{64} = 8$$

$$g_4 = \sqrt{g_{4-1}^2 + 6g_{4-1} + \left(\frac{g_{4-1}+1}{4-1}\right)^2} = \sqrt{g_3^2 + 6g_3 + \left(\frac{g_3+1}{4-1}\right)^2} = \sqrt{8^2 + 6\cdot 8 + \left(\frac{8+1}{3}\right)^2}$$

$$= \sqrt{64 + 48 + 3^2} = \sqrt{121} = 11$$

$$g_5 = \sqrt{g_{5-1}^2 + 6g_{5-1} + \left(\frac{g_{5-1}+1}{5-1}\right)^2} = \sqrt{g_4^2 + 6g_4 + \left(\frac{g_4+1}{5-1}\right)^2} = \sqrt{11^2 + 6\cdot 11 + \left(\frac{11+1}{4}\right)^2}$$

$$= \sqrt{121 + 66 + 3^2} = \sqrt{196} = 14$$

(b) Guess a closed form (non-recursive definition) for g_n.

Answer:

$$g_n = 3n - 1 \quad \text{for } n \geq 1$$

(c) Use mathematical induction to prove your answer in part **(b)**. That is, prove that the recursive definition of g indeed defines the function which you have proposed in part **(b)**.

Answer: The base case occurs for $n = 1$ and follows from the base case of the recursive definition of the sequence g. Indeed:

$$g_1 = 2 = 3 \cdot 1 - 1$$

For the inductive step, assume that:

$$g_n = 3n - 1$$

for some $n \geq 1$. We have to prove that:

$$g_{n+1} = 3(n+1) - 1$$

$g_{n+1} =$

$$\bullet \quad = \sqrt{g_{(n+1)-1}^2 + 6g_{(n+1)-1} + \left(\frac{g_{(n+1)-1}+1}{(n+1)-1}\right)^2} = \sqrt{g_n^2 + 6g_n + \left(\frac{g_n+1}{n}\right)^2}$$

$$* \quad = \sqrt{(3n-1)^2 + 6(3n-1) + \left(\frac{(3n-1)+1}{n}\right)^2} = \sqrt{9n^2 - 6n + 1 + 18n - 6 + \left(\frac{3n}{n}\right)^2}$$

$$= \sqrt{9n^2 + 12n - 5 + 3^2} = \sqrt{9n^2 + 12n + 4} = \sqrt{(3n+2)^2} = 3n + 2 = 3(n+1) - 1$$

The equality $\bullet$ follows from the recursive definition of g_{n+1} for $n + 1 \geq 2$; the equality $*$ is obtained by two applications of the inductive hypothesis: to g_n and to g_{n-1}; and the other equalities are obtained by arithmetic transformations.

2.3 Sums and Sequence Operators

Problem 136 Let p be an arbitrary proposition and let q_n be an arbitrary sequence of propositions such that:

$$q_0 = (p \wedge \neg p)$$

Using mathematical induction, prove that

$$\bigwedge_{i=0}^{n} q_i$$

is false for every natural number n.

Answer: The base case occurs for $n = 0$. To verify that the claim is true in the base case, observe that:

$$\bigwedge_{i=0}^{0} q_i = q_0 = (p \wedge \neg p) = 0$$

by definition of conjunction (where 0 denotes the truth value of false.)

For the inductive step, assume that:

$$\bigwedge_{i=0}^{n} q_i = 0$$

for some $n \geq 0$. We have to prove that:

$$\bigwedge_{i=0}^{n+1} q_i = 0$$

Indeed:

$$\bigwedge_{i=0}^{n+1} q_i \quad =$$

$$\diamond \qquad = \left(\bigwedge_{i=0}^{n} q_i \right) \wedge q_{n+1} =$$

$$\star \qquad = 0 \wedge q_{n+1} =$$

$$\bullet \qquad = 0$$

The equality $\diamond$ follows from the definition of the operator $\bigwedge$ for sequences, $\star$ follows from the inductive hypothesis, and the equality $\bullet$ follows from the definition of conjunction.

Problem 137   Let p be an arbitrary proposition and let q_n be a sequence of propositions such that:

$$q_0 = p$$
$$q_n = \neg p, \text{ for } n \geq 1$$

Using mathematical induction, prove that

$$\bigvee_{i=0}^{n} q_i$$

is true for every positive integer $n \geq 1$.

Answer: The base case occurs for $n = 1$. To verify that the claim is true in the base case, observe that:

$$\bigvee_{i=0}^{1} q_i = q_0 \vee q_1 = p \vee \neg p = 1$$

The first equality follows from the definition of the operator $\bigvee$ for sequences; the second equality is obtained from the definition of the sequence q_n for $n \geq 1$; and the last equality follows from the definition of disjunction (where 1 denotes the truth value of true).

For the inductive step, assume that:

$$\bigvee_{i=0}^{n} q_i = 1$$

for some $n \geq 1$. We have to prove that:

$$\bigvee_{i=0}^{n+1} q_i = 1$$

Indeed:

$$\bigvee_{i=0}^{n+1} q_i \quad =$$

$$\diamond \qquad = \left(\bigvee_{i=0}^{n} q_i \right) \vee q_{n+1} =$$

$$\star \qquad = 1 \vee q_{n+1}$$

$$\spadesuit \qquad = 1 \vee \neg p$$

$$\bullet \qquad = 1$$

The equality $\diamond$ follows from the definition of the operator $\bigvee$ for sequences, $\star$ follows from the inductive hypothesis, $\spadesuit$ follows from the definition of q_{n+1} for $n + 1 \geq 1$, and the equality $\bullet$ follows from the definition of disjunction.

Problem 138   Let A and B be two sets of natural numbers defined as follows.

$$A = \{1, 2, 3\}$$
$$B = \{x \mid x \leq 10\}$$

Let X_n be an arbitrary sequence of sets such that $X_0 = A \setminus B$.

Using mathematical induction, prove that:

$$\bigcap_{i=0}^{n} X_i = \emptyset$$

for every natural number n.

Answer: The base case occurs for $n = 0$. To verify that the claim is true in the base case, observe that:

$$\bigcap_{i=0}^{0} X_i = X_0 = A \setminus B = \emptyset$$

by definition of A and B.

For the inductive step, assume:

$$\bigcap_{i=0}^{n} X_i = \emptyset$$

for some $n \geq 0$. We have to prove that:

$$\bigcap_{i=0}^{n+1} X_i = \emptyset$$

Indeed:

$$\bigcap_{i=0}^{n+1} X_i \quad =$$

$$\diamond \qquad = \left(\bigcap_{i=0}^{n} X_i \right) \cap X_{n+1}$$

$$\star \qquad = \emptyset \cap X_{n+1}$$

$$\bullet \qquad = \emptyset$$

The equality $\diamond$ follows from the definition of the operator $\bigcap$ for sequences, $\star$ follows from the inductive hypothesis, and the equality $\bullet$ follows from the definition of conjunction.

Problem 139   Use mathematical induction to prove that:

$$\sum_{k=0}^{4n-1} 2^k = 16^n - 1$$

for every positive integer $n \geq 1$.

Answer: Let:

$$L(n) = \sum_{k=0}^{4n-1} 2^k$$

$$R(n) = 16^n - 1$$

The claim is that $L(n) = R(n)$ for all $n \geq 1$.

The base case occurs for $n = 1$:

$$L(1) = \sum_{k=0}^{4 \cdot 1 - 1} 2^k = \sum_{k=0}^{3} 2^k = 2^0 + 2^1 + 2^2 + 2^3 = 1 + 2 + 4 + 8 = 15$$

$$R(1) = 16^1 - 1 = 15$$

Since $L(1) = R(1)$, the base case is verified.

For the inductive step, assume that:

$$L(n) = R(n)$$

for some $n \geq 1$. We need to prove that:

$$L(n+1) = R(n+1)$$

Indeed:

$$
\begin{aligned}
L(n+1) \ &= \\
\triangle \qquad &= \sum_{k=0}^{4(n+1)-1} 2^k = \sum_{k=0}^{4n+3} 2^k \\
\clubsuit \qquad &= \left(\sum_{k=0}^{4n-1} 2^k \right) + 2^{4n} + 2^{4n+1} + 2^{4n+2} + 2^{4n+3} \\
\triangle \qquad &= L(n) + 2^{4n} + 2^{4n+1} + 2^{4n+2} + 2^{4n+3} \\
\bullet \qquad &= R(n) + 2^{4n} + 2^{4n+1} + 2^{4n+2} + 2^{4n+3} \\
\triangledown \qquad &= 16^n - 1 + 2^{4n} + 2^{4n+1} + 2^{4n+2} + 2^{4n+3} = 16^n - 1 + 2^{4n}(1 + 2 + 2^2 + 2^3) = \\
&= 16^n - 1 + 2^{4n} \cdot 15 = 16^n - 1 + (2^4)^n \cdot 15 = \\
&= 16^n - 1 + 16^n \cdot 15 = 16^n(1 + 15) - 1 = 16^n \cdot 16 - 1 = 16^{n+1} - 1 \\
\triangledown \qquad &= R(n+1)
\end{aligned}
$$

The equalities $\triangle$ follow from the definition of L; the equalities $\triangledown$ follow from the definition of R; the equality $\clubsuit$ follows from the definition of $\sum$; the equality $\bullet$ follows from the inductive hypothesis; while the other equalities are obtained by arithmetic transformations.

Problem 140   Use mathematical induction to prove that:

$$\sum_{k=0}^{n} 2k < (n+1)^2$$

for every natural number n.

Answer: Let:

$$L(n) = \sum_{k=0}^{n} 2k$$

$$R(n) = (n+1)^2$$

The claim is that $L(n) < R(n)$ for all $n \geq 0$.

The base case occurs for $n = 0$:

$$L(0) = \sum_{k=0}^{0} 2k = 2 \cdot 0 = 0$$

$$R(0) = (0+1)^2 = 1$$

Since $L(0) < R(0)$, the base case is verified.

For the inductive step, assume that:

$$L(n) < R(n)$$

for some $n \geq 0$. We need to prove that:

$$L(n+1) < R(n+1)$$

Indeed:

$$
\begin{array}{ll}
L(n+1) & = \\[4pt]
\triangle & = \displaystyle\sum_{k=0}^{n+1} 2k \\[6pt]
\clubsuit & = \left(\displaystyle\sum_{k=0}^{n} 2k\right) + 2(n+1) \\[6pt]
\triangle & = L(n) + 2(n+1) \\[4pt]
\bullet & < R(n) + 2(n+1) \\[4pt]
\triangledown & = (n+1)^2 + 2(n+1) = (n+1)(n+3) \\[4pt]
& = n^2 + 4n + 3 \\[4pt]
& < n^2 + 4n + 4 = (n+2)^2 = ((n+1)+1)^2 \\[4pt]
\triangledown & = R(n+1)
\end{array}
$$

The equalities $\triangle$ follow from the definition of L; the equalities $\triangledown$ follow from the definition of R; the equality $\clubsuit$ follows from the definition of $\sum$; the inequality $\bullet$ follows from the inductive hypothesis; while the other (in)equalities are obtained by arithmetic transformations.

Problem 141   Use mathematical induction to prove that:

$$\sum_{k=0}^{n} 3k^2 < (n+1)^3$$

for every natural number n.

Answer: Let:

$$L(n) = \sum_{k=0}^{n} 3k^2$$

$$R(n) = (n+1)^3$$

The claim is that $L(n) < R(n)$ for all $n \geq 0$.

The base case occurs for $n = 0$:

$$L(0) = \sum_{k=0}^{0} 3k^2 = 3 \cdot 0 = 0$$

$$R(0) = (0+1)^3 = 1$$

Since $L(0) < R(0)$, the base case is verified.

For the inductive step, assume that:

$$L(n) < R(n)$$

for some $n \geq 0$. We need to prove that:

$$L(n+1) < R(n+1)$$

Indeed:

$$
\begin{aligned}
L(n+1) \quad &= \\[4pt]
\triangle \qquad &= \sum_{k=0}^{n+1} 3k^2 \\
\clubsuit \qquad &= \left(\sum_{k=0}^{n} 3k^2 \right) + 3(n+1)^2 \\
\triangle \qquad &= L(n) + 3(n+1)^2 \\
\bullet \qquad &< R(n) + 3(n+1)^2 \\
\triangledown \qquad &= (n+1)^3 + 3(n+1)^2 = (n^3 + 3n^2 + 3n + 1) + 3(n^2 + 2n + 1) \\
&= n^3 + 6n^2 + 9n + 4 \\
&< n^3 + 6n^2 + 12n + 8 = (n+2)^3 = ((n+1)+1)^3 \\
\triangledown \qquad &= R(n+1)
\end{aligned}
$$

The equalities $\triangle$ follow from the definition of L; the equalities $\triangledown$ follow from the definition of R; the equality $\clubsuit$ follows from the definition of $\sum$; the inequality $\bullet$ follows from the inductive hypothesis; while the other (in)equalities are obtained by arithmetic transformations.

Problem 142   Use mathematical induction to prove that:

$$\sum_{k=0}^{n} 4k^3 < (n+1)^4$$

for every natural number n.

Answer: Let:

$$L(n) = \sum_{k=0}^{n} 4k^3$$

$$R(n) = (n+1)^4$$

The claim is that $L(n) < R(n)$ for all $n \geq 0$.

The base case occurs for $n = 0$:

$$L(0) = \sum_{k=0}^{0} 4k^3 = 4 \cdot 0 = 0$$

$$R(0) = (0+1)^4 = 1$$

Since $L(0) < R(0)$, the base case is verified.

For the inductive step, assume that:

$$L(n) < R(n)$$

for some $n \geq 0$. We need to prove that:

$$L(n+1) < R(n+1)$$

Indeed:

$$
\begin{aligned}
L(n+1) \quad &= \\
\triangle \qquad &= \sum_{k=0}^{n+1} 4k^3 \\
\clubsuit \qquad &= \left(\sum_{k=0}^{n} 4k^3 \right) + 4(n+1)^3 \\
\triangle \qquad &= L(n) + 4(n+1)^3 \\
\bullet \qquad &< R(n) + 4(n+1)^3 \\
\triangledown \qquad &= (n+1)^4 + 4(n+1)^3 = (n+1)^3(n+5) \\
&= (n^3 + 3n^2 + 3n + 1)(n+5) = n^4 + 3n^3 + 3n^2 + n + 5n^3 + 15n^2 + 15n + 5 \\
&= n^4 + 8n^3 + 18n^2 + 16n + 5 \\
&< n^4 + 8n^3 + 24n^2 + 32n + 16 = (n+2)^4 = ((n+1)+1)^4 \\
\triangledown \qquad &= R(n+1)
\end{aligned}
$$

The equalities $\triangle$ follow from the definition of L; the equalities $\triangledown$ follow from the definition of R; the equality $\clubsuit$ follows from the definition of $\sum$; the inequality $\bullet$ follows from the inductive hypothesis; while the other (in)equalities are obtained by arithmetic transformations.

Problem 143   Use mathematical induction to prove that:

$$\sum_{k=0}^{n} k^2 \cdot 2^k < n^2 \cdot 2^{n+1}$$

for every positive integer $n \geq 1$.

Answer: Let:

$$L(n) = \sum_{k=0}^{n} k^2 \cdot 2^k$$
$$R(n) = n^2 \cdot 2^{n+1}$$

The claim is that $L(n) < R(n)$ for all $n \geq 1$.

The base case occurs for $n = 1$:

$$L(1) = \sum_{k=0}^{1} k^2 \cdot 2^k = 0^2 \cdot 2^0 + 1^2 \cdot 2^1 = 0 + 1 \cdot 2 = 2$$
$$R(1) = 1^2 \cdot 2^{1+1} = 1 \cdot 2^2 = 4$$

Since $L(1) < R(1)$, the base case is verified.

For the inductive step, assume that:

$$L(n) < R(n)$$

for some $n \geq 1$. We need to prove that:

$$L(n+1) < R(n+1)$$

Indeed:

$$L(n+1) \quad =$$

$$\triangle \qquad = \sum_{k=0}^{n+1} k^2 \cdot 2^k$$

$$\clubsuit \qquad = \left(\sum_{k=0}^{n} k^2 \cdot 2^k \right) + (n+1)^2 \cdot 2^{n+1}$$

$$\triangle \qquad = L(n) + (n+1)^2 \cdot 2^{n+1}$$

$$\bullet \qquad < R(n) + (n+1)^2 \cdot 2^{n+1}$$

$$\triangledown \qquad = n^2 \cdot 2^{n+1} + (n+1)^2 \cdot 2^{n+1} = 2^{n+1} \left(n^2 + (n+1)^2 \right)$$

$$\qquad \quad < 2^{n+1} \left((n+1)^2 + (n+1)^2 \right) = 2^{n+1} \cdot 2(n+1)^2 = 2^{n+2}(n+1)^2$$

$$\triangledown \qquad = R(n+1)$$

The equalities $\triangle$ follow from the definition of L; the equalities $\triangledown$ follow from the definition of R; the equality $\clubsuit$ follows from the definition of $\sum$; the inequality $\bullet$ follows from the inductive hypothesis; while the other (in)equalities are obtained by arithmetic transformations.

Problem 144   Use mathematical induction to prove that:

$$\sum_{k=0}^{n} k^3 \cdot 3^k < \left(\frac{1}{2} \right) \cdot n^3 \cdot 3^{n+1}$$

for every positive integer $n \geq 1$.

Answer: Let:

$$L(n) = \sum_{k=0}^{n} k^3 \cdot 3^k$$

$$R(n) = \left(\frac{1}{2} \right) \cdot n^3 \cdot 3^{n+1}$$

The claim is that $L(n) < R(n)$ for all $n \geq 1$.

The base case occurs for $n = 1$:

$$L(1) = \sum_{k=0}^{1} k^3 \cdot 3^k = 0^3 \cdot 3^0 + 1^3 \cdot 3^1 = 0 + 1 \cdot 3 = 3$$

$$R(1) = \left(\frac{1}{2} \right) \cdot 1^3 \cdot 3^{1+1} = \left(\frac{1}{2} \right) \cdot 1 \cdot 9 = \frac{9}{2}$$

Since $L(1) < R(1)$, the base case is verified.

For the inductive step, assume that:

$$L(n) < R(n)$$

for some $n \geq 1$. We need to prove that:

$$L(n+1) < R(n+1)$$

Indeed:

$$L(n+1) \quad =$$

$\triangle \qquad \displaystyle = \sum_{k=0}^{n+1} k^3 \cdot 3^k$

$\clubsuit \qquad \displaystyle = \left(\sum_{k=0}^{n} k^3 \cdot 3^k \right) + (n+1)^3 \cdot 3^{n+1}$

$\triangle \qquad = L(n) + (n+1)^3 \cdot 3^{n+1}$

$\bullet \qquad < R(n) + (n+1)^3 \cdot 3^{n+1}$

$\triangledown \qquad \displaystyle = \left(\frac{1}{2} \right) \cdot n^3 \cdot 3^{n+1} + (n+1)^3 \cdot 3^{n+1} = \left(\frac{3^{n+1}}{2} \right) \left(n^3 + 2(n+1)^3 \right)$

$\qquad \displaystyle < \left(\frac{3^{n+1}}{2} \right) \left((n+1)^3 + 2(n+1)^3 \right) = \left(\frac{3^{n+1}}{2} \right) \cdot 3(n+1)^3$

$\qquad \displaystyle = \left(\frac{1}{2} \right) \cdot 3^{n+1} \cdot 3 \cdot (n+1)^3 = \left(\frac{1}{2} \right) \cdot (n+1)^3 \cdot 3^{(n+1)+1}$

$\triangledown \qquad = R(n+1)$

The equalities $\triangle$ follow from the definition of L; the equalities $\triangledown$ follow from the definition of R; the equality $\clubsuit$ follows from the definition of $\sum$; the inequality $\bullet$ follows from the inductive hypothesis; while the other (in)equalities are obtained by arithmetic transformations.

Problem 145   Use mathematical induction to prove that:

$$\sum_{k=1}^{2n+1} \left(k + \frac{1}{2} \right) \cdot 3^k = \left(n + \frac{1}{2} \right) \cdot 9^{n+1}$$

for every natural number $n \geq 0$.

Answer: Let:

$$L(n) = \sum_{k=1}^{2n+1} \left(k + \frac{1}{2} \right) \cdot 3^k$$

$$R(n) = \left(n + \frac{1}{2} \right) \cdot 9^{n+1}$$

The claim is that $L(n) = R(n)$ for all $n \geq 0$.

The base case occurs for $n = 0$:

$$L(0) \; = \sum_{k=1}^{2 \cdot 0+1} \left(k + \frac{1}{2} \right) \cdot 3^k = \sum_{k=1}^{1} \left(k + \frac{1}{2} \right) \cdot 3^k = \left(1 + \frac{1}{2} \right) \cdot 3^1 = \left(\frac{3}{2} \right) \cdot 3 = \frac{9}{2}$$

$$R(0) \; = \left(0 + \frac{1}{2} \right) \cdot 9^{0+1} = \left(\frac{1}{2} \right) \cdot 9 = \frac{9}{2}$$

Since $L(0) = R(0)$, the base case is verified.

For the inductive step, assume that:

$$L(n) = R(n)$$

for some $n \geq 0$.

We need to prove that:

$$L(n+1) = R(n+1)$$

Indeed:

$$L(n+1) \quad =$$

$\triangle$
$$= \sum_{k=1}^{2(n+1)+1} \left(k + \frac{1}{2}\right) \cdot 3^k = \sum_{k=1}^{2n+3} \left(k + \frac{1}{2}\right) \cdot 3^k =$$

$\clubsuit$
$$= \left(\sum_{k=1}^{2n+1} \left(k + \frac{1}{2}\right) \cdot 3^k\right) + \left(2n + 2 + \frac{1}{2}\right) \cdot 3^{2n+2} + \left(2n + 3 + \frac{1}{2}\right) \cdot 3^{2n+3}$$

$\triangle$
$$= L(n) + \left(2n + 2 + \frac{1}{2}\right) \cdot 3^{2n+2} + \left(2n + 3 + \frac{1}{2}\right) \cdot 3^{2n+3}$$

$\bullet$
$$= R(n) + \left(2n + 2 + \frac{1}{2}\right) \cdot 3^{2n+2} + \left(2n + 3 + \frac{1}{2}\right) \cdot 3^{2n+3}$$

$\triangledown$
$$= \left(n + \frac{1}{2}\right) \cdot 9^{n+1} + \left(2n + 2 + \frac{1}{2}\right) \cdot 3^{2n+2} + \left(2n + 3 + \frac{1}{2}\right) \cdot 3^{2n+3}$$

$$= \left(n + \frac{1}{2}\right) \cdot \left(3^2\right)^{n+1} + \left(\frac{3^{2n+2}}{2}\right) \cdot ((4n + 4 + 1) + 3(4n + 6 + 1))$$

$$= \left(\frac{2n+1}{2}\right) \cdot 3^{2n+2} + \left(\frac{3^{2n+2}}{2}\right) \cdot (16n + 26) = \left(\frac{3^{2n+2}}{2}\right) \cdot (18n + 27)$$

$$= \left(\frac{3^{2n+2}}{2}\right) \cdot 9(2n + 3) = \left(\frac{3^{2n+2} \cdot 3^2}{2}\right) \cdot (2n + 3) = \left(\frac{3^{2n+4}}{2}\right) \cdot (2n + 3)$$

$$= 3^{2(n+2)} \cdot \left(\frac{2n+3}{2}\right) = 9^{n+2} \cdot \left(\frac{2n+2}{2} + \frac{1}{2}\right) = \left((n+1) + \frac{1}{2}\right) \cdot 9^{(n+1)+1}$$

$\triangledown$
$$= R(n+1)$$

The equalities $\triangle$ follow from the definition of L; the equalities $\triangledown$ follow from the definition of R; the equality $\clubsuit$ follows from the definition of $\sum$; the equality $\bullet$ follows from the inductive hypothesis; while the other equalities are obtained by arithmetic transformations.

Problem 146   Use mathematical induction to prove that:

$$\sum_{k=0}^{n} (3 - 2k) \cdot 3^{n-k} = 3^{n+1} + n$$

for every natural number $n \geq 0$.

Answer: Let:

$$L(n) = \sum_{k=0}^{n} (3 - 2k) \cdot 3^{n-k}$$
$$R(n) = 3^{n+1} + n$$

The claim is that $L(n) = R(n)$ for all $n \geq 0$.
The base case occurs for $n = 0$:

$$L(0) = \sum_{k=0}^{0} (3 - 2k) \cdot 3^{0-k} = (3 - 2 \cdot 0) \cdot 3^{0-0} = 3 \cdot 1 = 3$$
$$R(0) = 3^{0+1} + 0 = 3^1 = 3$$

Since $L(0) = R(0)$, the base case is verified.

For the inductive step, assume that:

$$L(n) = R(n)$$

for some $n \geq 0$. We need to prove that:

$$L(n + 1) = R(n + 1)$$

Indeed:

$L(n + 1) =$

$\triangle \quad = \displaystyle\sum_{k=0}^{n+1}(3 - 2k) \cdot 3^{(n+1)-k} = \sum_{k=0}^{n+1}\left((3 - 2k) \cdot 3^{n-k}\right) \cdot 3 = 3\left(\sum_{k=0}^{n+1}(3 - 2k) \cdot 3^{n-k}\right)$

$\clubsuit \quad = 3\left(\left(\displaystyle\sum_{k=0}^{n}(3 - 2k) \cdot 3^{n-k}\right) + (3 - 2(n + 1)) \cdot 3^{n-(n+1)}\right)$

$\triangle \quad = 3\left(L(n) + (3 - 2(n + 1)) \cdot 3^{n-(n+1)}\right)$

$\bullet \quad = 3\left(R(n) + (3 - 2(n + 1)) \cdot 3^{n-(n+1)}\right)$

$\triangledown \quad = 3\left((3^{n+1} + n) + (3 - 2(n + 1)) \cdot 3^{n-(n+1)}\right) = 3\left(3^{n+1} + n + (3 - 2n - 2) \cdot 3^{-1}\right)$

$\quad = 3\left(3^{n+1} + n + \dfrac{1 - 2n}{3}\right) = 3^{n+2} + 3n + (1 - 2n) = 3^{(n+1)+1} + (n + 1)$

$\triangledown \quad = R(n + 1)$

The equalities $\triangle$ follow from the definition of L; the equalities $\triangledown$ follow from the definition of R; the equality $\clubsuit$ follows from the definition of $\sum$; the equality $\bullet$ follows from the inductive hypothesis; while the other equalities are obtained by arithmetic transformations.

Problem 147 Use mathematical induction to prove that:

$$\sum_{k=1}^{n} k \cdot k! < (n + 1)!$$

for every positive integer $n \geq 1$.

Answer: Let:

$$L(n) = \sum_{k=1}^{n} k \cdot k!$$
$$R(n) = (n + 1)!$$

The claim is that $L(n) < R(n)$ for all $n \geq 1$.

The base case occurs for $n = 1$:

$$L(0) = \sum_{k=1}^{1} k \cdot k! = 1 \cdot 1! = 1 \cdot 1 = 1$$

$$R(0) = (1 + 1)! = 2! = 2 \cdot 1 = 2$$

Since $L(0) < R(0)$, the base case is verified.
For the inductive step, assume that:

$$L(n) < R(n)$$

for some $n \geq 1$.

We need to prove that:

$$L(n+1) < R(n+1)$$

Indeed:

$$L(n+1) =$$

$$\triangle \qquad = \sum_{k=1}^{n+1} k \cdot k!$$

$$\clubsuit \qquad = \left(\sum_{k=1}^{n} k \cdot k!\right) + (n+1)(n+1)!$$

$$\triangle \qquad = L(n) + (n+1)(n+1)!$$

$$\bullet \qquad < R(n) + (n+1)(n+1)!$$

$$\triangledown \qquad = (n+1)! + (n+1)(n+1)! = ((n+1)!)\,(1+n+1) = ((n+1)!)\,(n+2)$$

$$\spadesuit \qquad = (n+2)! = ((n+1)+1)!$$

$$\triangledown \qquad = R(n+1)$$

The equalities $\triangle$ follow from the definition of L; the equalities $\triangledown$ follow from the definition of R; the equality $\clubsuit$ follows from the definition of $\sum$; the inequality $\bullet$ follows from the inductive hypothesis; the equality $\spadesuit$ follows from the definition of the factorial function; while the other equalities are obtained by arithmetic transformations.

Problem 148   Use mathematical induction to prove that:

$$\sum_{k=0}^{2n-1} (k+2) \cdot 2^{k+1} = n \cdot 4^{n+1}$$

for every positive integer $n \geq 1$.

Answer: Let:

$$L(n) = \sum_{k=0}^{2n-1} (k+2) \cdot 2^{k+1}$$

$$R(n) = n \cdot 4^{n+1}$$

The claim is that $L(n) = R(n)$ for all $n \geq 1$.

The base case occurs for $n = 1$:

$$L(1) = \sum_{k=0}^{2\cdot 1-1} (k+2) \cdot 2^{k+1} = \sum_{k=0}^{1}(k+2) \cdot 2^{k+1} = (0+2) \cdot 2^{0+1} + (1+2) \cdot 2^{1+1} = 2 \cdot 2 + 3 \cdot 4 = 16$$

$$R(1) = 1 \cdot 4^{1+1} = 1 \cdot 16 = 16$$

Since $L(1) = R(1)$, the base case is verified.

For the inductive step, assume that:

$$L(n) = R(n)$$

for some $n \geq 1$. We need to prove that:

$$L(n+1) = R(n+1)$$

Indeed:

$$L(n+1) =$$

$\triangle \qquad = \sum_{k=0}^{2(n+1)-1} (k+2) \cdot 2^{k+1} = \sum_{k=0}^{2n+1} (k+2) \cdot 2^{k+1}$

$\clubsuit \qquad = \left(\sum_{k=0}^{2n-1} (k+2) \cdot 2^{k+1} \right) + (2n+2) \cdot 2^{2n+1} + ((2n+1)+2) \cdot 2^{(2n+1)+1}$

$\qquad = \left(\sum_{k=0}^{2n-1} (k+2) \cdot 2^{k+1} \right) + (2n+2) \cdot 2^{2n+1} + (2n+3) \cdot 2^{2n+2}$

$\triangle \qquad = L(n) + (2n+2) \cdot 2^{2n+1} + (2n+3) \cdot 2^{2n+2}$

$\bullet \qquad = R(n) + (2n+2) \cdot 2^{2n+1} + (2n+3) \cdot 2^{2n+2}$

$\triangledown \qquad = n \cdot 4^{n+1} + (2n+2) \cdot 2^{2n+1} + (2n+3) \cdot 2^{2n+2}$

$\qquad = n \cdot 4^{n+1} + (n+1) \cdot 2 \cdot 2^{2n+1} + (2n+3) \cdot 2^{2n+2}$

$\qquad = n \cdot 4^{n+1} + (n+1) \cdot 2^{2n+2} + (2n+3) \cdot 2^{2n+2} = n \cdot 4^{n+1} + 2^{2n+2} \cdot (n+1+2n+3)$

$\qquad = n \cdot 4^{n+1} + 2^{2(n+1)} \cdot (3n+4) = n \cdot 4^{n+1} + 4^{n+1}(3n+4) = 4^{n+1} \cdot (n+3n+4)$

$\qquad = 4^{n+1}(4n+4) = 4^{n+1} \cdot 4(n+1) = (n+1) \cdot 4^{n+2} = (n+1) \cdot 4^{(n+1)+1}$

$\triangledown \qquad = R(n+1)$

The equalities $\triangle$ follow from the definition of L; the equalities $\triangledown$ follow from the definition of R; the equality $\clubsuit$ follows from the definition of $\sum$; the equality $\bullet$ follows from the inductive hypothesis; while the other equalities are obtained by arithmetic transformations.

Problem 149 Use mathematical induction to prove that:

$$\sum_{k=0}^{n} \frac{2-k}{2^k} = 2 + \frac{n}{2^n}$$

for every natural number $n \geq 0$.

Answer: Let:

$$L(n) = \sum_{k=0}^{n} \frac{2-k}{2^k}$$

$$R(n) = 2 + \frac{n}{2^n}$$

The claim is that $L(n) = R(n)$ for all $n \geq 0$.

The base case occurs for $n = 0$:

$$L(0) = \sum_{k=0}^{0} \frac{2-k}{2^k} = \frac{2-0}{2^0} = \frac{2}{1} = 2$$

$$R(0) = 2 + \frac{n}{2^n} = 2 + \frac{0}{2^0} = 2$$

Since $L(0) = R(0)$, the base case is verified.

For the inductive step, assume that:

$$L(n) = R(n)$$

for some $n \geq 0$.

We need to prove that:

$$L(n + 1) = R(n + 1)$$

Indeed:

$$L(n + 1) \quad =$$

$$\triangle \qquad = \sum_{k=0}^{n+1} \frac{2 - k}{2^k}$$

$$\clubsuit \qquad = \left(\sum_{k=0}^{n} \frac{2 - k}{2^k} \right) + \frac{2 - (n + 1)}{2^{n+1}}$$

$$\triangle \qquad = L(n) + \frac{2 - (n + 1)}{2^{n+1}}$$

$$\bullet \qquad = R(n) + \frac{2 - (n + 1)}{2^{n+1}}$$

$$\triangledown \qquad = 2 + \frac{n}{2^n} + \frac{2 - (n + 1)}{2^{n+1}} = 2 + \frac{2n + 2 - n - 1}{2^{n+1}} = 2 + \frac{n + 1}{2^{n+1}}$$

$$\triangledown \qquad = R(n + 1)$$

The equalities $\triangle$ follow from the definition of L; the equalities $\triangledown$ follow from the definition of R; the equality $\clubsuit$ follows from the definition of $\sum$; the equality $\bullet$ follows from the inductive hypothesis; while the other equalities are obtained by arithmetic transformations.

Problem 150   Use mathematical induction to prove that:

$$\sum_{k=0}^{3n} \frac{3 - k}{2^k} = 4 + \frac{3n - 1}{8^n}$$

for every natural number $n \geq 0$.

Answer: Let:

$$L(n) = \sum_{k=0}^{3n} \frac{3 - k}{2^k}$$

$$R(n) = 4 + \frac{3n - 1}{8^n}$$

The claim is that $L(n) = R(n)$ for all $n \geq 0$.

The base case occurs for $n = 0$:

$$L(0) = \sum_{k=0}^{3n} \frac{3 - k}{2^k} = \sum_{k=0}^{3 \cdot 0} \frac{3 - k}{2^k} = \sum_{k=0}^{0} \frac{3 - k}{2^k} = \frac{3 - 0}{2^0} = \frac{3}{1} = 3$$

$$R(0) \doteq 4 + \frac{3n - 1}{8^n} = 4 + \frac{3 \cdot 0 - 1}{8^0} = 4 + \frac{-1}{1} = 3$$

Since $L(0) = R(0)$, the base case is verified.

For the inductive step, assume that:

$$L(n) = R(n)$$

for some $n \geq 0$. We need to prove that:

$$L(n + 1) = R(n + 1)$$

Indeed:

$$L(n+1) \quad =$$

$$\triangle \qquad = \sum_{k=0}^{3(n+1)} \frac{3-k}{2^k} = \sum_{k=0}^{3n+3} \frac{3-k}{2^k}$$

$$\clubsuit \qquad = \left(\sum_{k=0}^{3n} \frac{3-k}{2^k}\right) + \frac{3-(3n+1)}{2^{3n+1}} + \frac{3-(3n+2)}{2^{3n+2}} + \frac{3-(3n+3)}{2^{3n+3}}$$

$$\triangle \qquad = L(n) + \frac{3-(3n+1)}{2^{3n+1}} + \frac{3-(3n+2)}{2^{3n+2}} + \frac{3-(3n+3)}{2^{3n+3}}$$

$$\bullet \qquad = R(n) + \frac{3-(3n+1)}{2^{3n+1}} + \frac{3-(3n+2)}{2^{3n+2}} + \frac{3-(3n+3)}{2^{3n+3}}$$

$$\triangledown \qquad = 4 + \frac{3n-1}{8^n} + \frac{3-(3n+1)}{2^{3n+1}} + \frac{3-(3n+2)}{2^{3n+2}} + \frac{3-(3n+3)}{2^{3n+3}}$$

$$\qquad \qquad = 4 + \frac{3n-1}{2^{3n}} + \frac{2-3n}{2^{3n+1}} + \frac{1-3n}{2^{3n+2}} + \frac{-3n}{2^{3n+3}}$$

$$\qquad \qquad = 4 + \frac{2^3(3n-1) + 2^2(2-3n) + 2(1-3n) + (-3n)}{2^{3n+3}}$$

$$\qquad \qquad = 4 + \frac{(24n-8) + (8-12n) + (2-6n) + (-3n)}{2^{3n+3}} = 4 + \frac{3n+2}{2^{3(n+1)}} = 4 + \frac{3(n+1)-1}{8^{n+1}}$$

$$\triangledown \qquad = R(n+1)$$

The equalities $\triangle$ follow from the definition of L; the equalities $\triangledown$ follow from the definition of R; the equality $\clubsuit$ follows from the definition of $\sum$; the equality $\bullet$ follows from the inductive hypothesis; while the other equalities are obtained by arithmetic transformations.

Problem 151   Use mathematical induction to prove that:

$$\sum_{k=0}^{3n+1} (6k+7) = (3n+2)(9n+10)$$

for every natural number $n \geq 0$.

Answer: Let:

$$L(n) = \sum_{k=0}^{3n+1} (6k+7)$$

$$R(n) = (3n+2)(9n+10)$$

The claim is that $L(n) = R(n)$ for all $n \geq 0$.

The base case occurs for $n = 0$:

$$L(0) \quad = \sum_{k=0}^{3 \cdot 0+1} (6k+7) = \sum_{k=0}^{1}(6k+7)$$

$$\qquad = (6 \cdot 0 + 7) + (6 \cdot 1 + 7) = 7 + 13 = 20$$

$$R(0) \quad = (3 \cdot 0 + 2)(9 \cdot 0 + 10) = 2 \cdot 10 = 20$$

Since $L(0) = R(0)$, the base case is verified.

For the inductive step, assume that:

$$L(n) = R(n)$$

for some $n \geq 0$.

We need to prove that:

$$L(n + 1) = R(n + 1)$$

Indeed:

$$L(n + 1) \quad =$$

$$\triangle \qquad = \sum_{k=0}^{3(n+1)+1} (6k + 7) = \sum_{k=0}^{3n+4} (6k + 7)$$

$$\clubsuit \qquad = \left(\sum_{k=0}^{3n+1} (6k + 7) \right) + (6 \cdot (3n + 2) + 7) + (6 \cdot (3n + 3) + 7) + (6 \cdot (3n + 4) + 7)$$

$$\triangle \qquad = L(n) + (6 \cdot (3n + 2) + 7) + (6 \cdot (3n + 3) + 7) + (6 \cdot (3n + 4) + 7)$$

$$\bullet \qquad = R(n) + (6 \cdot (3n + 2) + 7) + (6 \cdot (3n + 3) + 7) + (6 \cdot (3n + 4) + 7)$$

$$\triangledown \qquad = (3n + 2)(9n + 10) + (6 \cdot (3n + 2) + 7) + (6 \cdot (3n + 3) + 7) + (6 \cdot (3n + 4) + 7)$$

$$\qquad = (3n + 2)(9n + 10) + (18n + 19) + (18n + 25) + (18n + 31)$$

$$\qquad = (27n^2 + 48n + 20) + (54n + 75) = 27n^2 + 102n + 95$$

$$R(n + 1) \quad =$$

$$\triangledown \qquad = (3(n + 1) + 2)(9(n + 1) + 10) = (3n + 5)(9n + 19) = 27n^2 + 45n + 57n + 95$$

$$\qquad = 27n^2 + 102n + 95$$

The equalities $\triangle$ follow from the definition of L; the equalities $\triangledown$ follow from the definition of R; the equality $\clubsuit$ follows from the definition of $\sum$; the inequality $\bullet$ follows from the inductive hypothesis; while the other equalities are obtained by arithmetic transformations. To complete the proof, observe that $L(n + 1) = R(n + 1)$.

Problem 152   Use mathematical induction to prove that:

$$\sum_{k=1}^{n} \frac{1}{(2k - 1)^2} < 2 - \frac{1}{2n - 1}$$

for every positive integer n such that $n > 1$.

Answer: Let:

$$L(n) = \sum_{k=1}^{n} \frac{1}{(2k - 1)^2}$$

$$R(n) = 2 - \frac{1}{2n - 1}$$

The claim is that $L(n) < R(n)$ for all $n \geq 2$.

The base case occurs for $n = 2$:

$$L(2) \ = \sum_{k=1}^{2} \frac{1}{(2k - 1)^2} = \frac{1}{(2 \cdot 1 - 1)^2} + \frac{1}{(2 \cdot 2 - 1)^2} = \frac{1}{1} + \frac{1}{9} = 1 + \frac{1}{9}$$

$$R(2) \ = 2 - \frac{1}{2 \cdot 2 - 1} = 2 - \frac{1}{3} = 1 + \frac{2}{3} = 1 + \frac{6}{9}$$

Since $L(2) < R(2)$, the base case is verified.

For the inductive step, assume that:

$$L(n) < R(n)$$

for some $n \geq 2$. We need to prove that:

$$L(n+1) < R(n+1)$$

Indeed:

$$L(n+1) \quad =$$

$$\triangle \qquad = \sum_{k=1}^{n+1} \frac{1}{(2k-1)^2}$$

$$\clubsuit \qquad = \left(\sum_{k=1}^{n} \frac{1}{(2k-1)^2} \right) + \frac{1}{(2(n+1)-1)^2}$$

$$\triangle \qquad = L(n) + \frac{1}{(2(n+1)-1)^2}$$

$$\bullet \qquad < R(n) + \frac{1}{(2(n+1)-1)^2}$$

$$\triangledown \qquad = 2 - \frac{1}{2n-1} + \frac{1}{(2(n+1)-1)^2} = 2 - \frac{1}{2n-1} + \frac{1}{(2n+1)^2}$$

$$= 2 - \frac{(2n+1)^2 - (2n-1)}{(2n-1)(2n+1)^2} = 2 - \frac{4n^2 + 4n + 1 - 2n + 1}{(2n-1)(2n+1)^2} = 2 - \frac{4n^2 + 2n + 2}{(2n-1)(2n+1)^2}$$

$$= 2 - \frac{4n^2 - 1}{(2n-1)(2n+1)^2} - \frac{2n+3}{(2n-1)(2n+1)^2}$$

$$\heartsuit \qquad < 2 - \frac{4n^2 - 1}{(2n-1)(2n+1)^2}$$

$$R(n+1) \quad =$$

$$\triangledown \qquad = 2 - \frac{1}{2(n+1)-1} = 2 - \frac{1}{2n+1} = 2 - \frac{1}{(2n+1)} \cdot \left(\frac{(2n-1)(2n+1)}{(2n-1)(2n+1)} \right)$$

$$= 2 - \frac{4n^2 - 1}{(2n-1)(2n+1)^2}$$

The equalities $\triangle$ follow from the definition of L; the equalities $\triangledown$ follow from the definition of R; the equality $\clubsuit$ follows from the definition of $\sum$; the inequality $\bullet$ follows from the inductive hypothesis; the inequality $\heartsuit$ holds because the assumption that $n \geq 2$ guarantees that $2n - 1 > 0$; while the other (in)equalities are obtained by arithmetic transformations. To complete the proof, observe that $L(n+1) < R(n+1)$.

Problem 153   Use mathematical induction to prove that:

$$\sum_{k=1}^{n} (2k-1)^2 < \frac{4}{3} \cdot n^3$$

for every positive integer $n \geq 1$.

Answer: Let:

$$L(n) = \sum_{k=1}^{n} (2k-1)^2$$

$$R(n) = \frac{4}{3} \cdot n^3$$

The claim is that $L(n) < R(n)$ for all $n \geq 1$.

The base case occurs for $n = 1$:

$$L(1) = \sum_{k=1}^{1}(2k-1)^2 = (2 \cdot 1 - 1)^2 = 1$$

$$R(1) = \frac{4}{3} \cdot 1^3 = \frac{4}{3}$$

Since $L(1) < R(1)$, the base case is verified.

For the inductive step, assume that:

$$L(n) < R(n)$$

for some $n \geq 1$. We need to prove that:

$$L(n+1) < R(n+1)$$

Indeed:

$$
\begin{aligned}
L(n+1) \quad &= \\
\triangle \qquad &= \sum_{k=1}^{n+1}(2k-1)^2 \\
\clubsuit \qquad &= \left(\sum_{k=1}^{n}(2k-1)^2\right) + (2(n+1)-1)^2 \\
\triangle \qquad &= L(n) + (2(n+1)-1)^2 \\
\bullet \qquad &< R(n) + (2(n+1)-1)^2 \\
\triangledown \qquad &= \frac{4}{3} \cdot n^3 + (2(n+1)-1)^2 = \frac{4}{3} \cdot n^3 + (2n+1)^2 \\
&= \frac{4}{3} \cdot n^3 + 4n^2 + 4n + 1 \\
&< \frac{4}{3} \cdot n^3 + 4n^2 + 4n + \frac{4}{3} = \frac{4}{3} \cdot (n^3 + 3n^2 + 3n + 1) = \frac{4}{3} \cdot (n+1)^3 \\
\triangledown \qquad &= R(n+1)
\end{aligned}
$$

The equalities $\triangle$ follow from the definition of L; the equalities $\triangledown$ follow from the definition of R; the equality $\clubsuit$ follows from the definition of $\sum$; the inequality $\bullet$ follows from the inductive hypothesis; while the other (in)equalities are obtained by arithmetic transformations.

Problem 154   Use mathematical induction to prove that:

$$\sum_{k=1}^{2n+1}(3k^2 + 4k) = (2n+1)(n+1)(4n+7)$$

for every natural number n.

Answer: Let:

$$L(n) = \sum_{k=1}^{2n+1}(3k^2 + 4k)$$

$$R(n) = (2n+1)(n+1)(4n+7)$$

The claim is that $L(n) = R(n)$ for all $n \geq 0$.

The base case occurs for $n = 0$:

$$L(0) = \sum_{k=1}^{2 \cdot 0 + 1} (3k^2 + 4k) = \sum_{k=1}^{1} (3k^2 + 4k) = 3 \cdot 1^2 + 4 \cdot 1 = 7$$
$$R(0) = (2 \cdot 0 + 1)(0 + 1)(4 \cdot 0 + 7) = 1 \cdot 1 \cdot 7 = 7$$

Since $L(0) = R(0)$, the base case is verified.

For the inductive step, assume that:

$$L(n) = R(n)$$

for some $n \geq 0$. We need to prove that:

$$L(n + 1) = R(n + 1)$$

Indeed:

$$L(n + 1) =$$

$$\triangle \qquad = \sum_{k=1}^{2(n+1)+1} (3k^2 + 4k) = \sum_{k=1}^{2n+3} (3k^2 + 4k)$$

$$\clubsuit \qquad = \left(\sum_{k=1}^{2n+1} (3k^2 + 4k) \right) + \left(3(2n+2)^2 + 4(2n+2) \right) + \left(3(2n+3)^2 + 4(2n+3) \right)$$

$$\triangle \qquad = L(n) + \left(3(2n+2)^2 + 4(2n+2) \right) + \left(3(2n+3)^2 + 4(2n+3) \right)$$

$$\bullet \qquad = R(n) + \left(3(2n+2)^2 + 4(2n+2) \right) + \left(3(2n+3)^2 + 4(2n+3) \right)$$

$$\triangledown \qquad = (2n+1)(n+1)(4n+7)$$
$$\qquad\quad + \left(3(2n+2)^2 + 4(2n+2) \right) + \left(3(2n+3)^2 + 4(2n+3) \right)$$
$$= (2n+1)(4n^2 + 11n + 7)$$
$$\qquad\quad + \left(3(4n^2 + 8n + 4) + (8n+8) \right) + \left(3(4n^2 + 12n + 9) + (8n+12) \right)$$
$$= \left(8n^3 + 22n^2 + 14n + 4n^2 + 11n + 7 \right)$$
$$\qquad\quad + \left(12n^2 + 24n + 12 + 8n + 8 \right) + \left(12n^2 + 36n + 27 + 8n + 12 \right)$$
$$= 8n^3 + 50n^2 + 101n + 66$$

$$R(n + 1) =$$

$$\triangledown \qquad = (2(n+1)+1)((n+1)+1)(4(n+1)+7) = (2n+3)(n+2)(4n+11)$$
$$= (2n+3)(4n^2 + 19n + 22) = (8n^3 + 38n^2 + 44n + 12n^2 + 57n + 66)$$
$$= 8n^3 + 50n^2 + 101n + 66$$

The equalities $\triangle$ follow from the definition of L; the equalities $\triangledown$ follow from the definition of R; the equality $\clubsuit$ follows from the definition of $\sum$; the equality $\bullet$ follows from the inductive hypothesis; while the other equalities are obtained by arithmetic transformations. To complete the proof, observe that $L(n+1) = R(n+1)$.

Problem 155 Use mathematical induction to prove that:

$$\sum_{k=1}^{3n} (4k + 3) = 3n(6n + 5)$$

for every positive integer $n \geq 1$.

Answer: Let:

$$L(n) = \sum_{k=1}^{3n}(4k+3)$$

$$R(n) = 3n(6n+5)$$

The claim is that $L(n) = R(n)$ for all $n \geq 1$.

The base case occurs for $n = 1$:

$$L(1) = \sum_{k=1}^{3\cdot 1}(4k+3) = \sum_{k=1}^{3}(4k+3) = (4\cdot 1+3) + (4\cdot 2+3) + (4\cdot 3+3) = 7+11+15 = 33$$

$$R(1) = 3\cdot 1 \cdot (6\cdot 1+5) = 3\cdot 11 = 33$$

Since $L(1) = R(1)$, the base case is verified.

For the inductive step, assume that:

$$L(n) = R(n)$$

for some $n \geq 1$. We need to prove that:

$$L(n+1) = R(n+1)$$

Indeed:

$$
\begin{aligned}
L(n+1) \quad &= \\
\triangle \qquad &= \sum_{k=1}^{3(n+1)}(4k+3) = \sum_{k=1}^{3n+3}(4k+3) \\
\clubsuit \qquad &= \left(\sum_{k=1}^{3n}(4k+3)\right) + (4(3n+1)+3) + (4(3n+2)+3) + (4(3n+3)+3) \\
\triangle \qquad &= L(n) + (4(3n+1)+3) + (4(3n+2)+3) + (4(3n+3)+3) \\
\bullet \qquad &= R(n) + (4(3n+1)+3) + (4(3n+2)+3) + (4(3n+3)+3) \\
\triangledown \qquad &= 3n(6n+5) + (4(3n+1)+3) + (4(3n+2)+3) + (4(3n+3)+3) \\
&= (18n^2+15n) + (12n+4+3) + (12n+8+3) + (12n+12+3) \\
&= 18n^2+51n+33 = 3(6n^2+17n+11) = 3(6n^2+6n+11n+11) \\
&= 3(6n(n+1)+11(n+1)) = 3(n+1)(6n+11) = 3(n+1)(6(n+1)+5) \\
\triangledown \qquad &= R(n+1)
\end{aligned}
$$

The equalities $\triangle$ follow from the definition of L; the equalities $\triangledown$ follow from the definition of R; the equality $\clubsuit$ follows from the definition of $\sum$; the inequality $\bullet$ follows from the inductive hypothesis; while other equalities are obtained by arithmetic transformations.

Problem 156　　Use mathematical induction to prove that:

$$\sum_{k=1}^{n}(2k-1)^3 < 2n^4$$

for every positive integer n.

Answer: Let:

$$L(n) = \sum_{k=1}^{n}(2k-1)^3$$

$$R(n) = 2n^4$$

The claim is that $L(n) < R(n)$ for all $n \geq 1$.

The base case occurs for $n = 1$:

$$L(1) = \sum_{k=1}^{1} (2k-1)^3 = (2 \cdot 1 - 1)^3 = 1$$
$$R(1) = 2 \cdot 1^4 = 2$$

Since $L(1) < R(1)$, the base case is verified.

For the inductive step, assume that:

$$L(n) < R(n)$$

for some $n \geq 1$. We need to prove that:

$$L(n+1) < R(n+1)$$

Indeed:

$$
\begin{aligned}
L(n+1) &= \\
\triangle \quad &= \sum_{k=1}^{n+1} (2k-1)^3 \\
\clubsuit \quad &= \left(\sum_{k=1}^{n} (2k-1)^3 \right) + (2(n+1)-1)^3 \\
\triangle \quad &= L(n) + (2(n+1)-1)^3 \\
\bullet \quad &< R(n) + (2(n+1)-1)^3 \\
\triangledown \quad &= 2n^4 + (2(n+1)-1)^3 = 2n^4 + (2n+1)^3 = 2n^4 + 8n^3 + 12n^2 + 6n + 1
\end{aligned}
$$

$$
\begin{aligned}
R(n+1) &= \\
\triangledown \quad &= 2(n+1)^4 = 2(n^4 + 4n^3 + 6n^2 + 4n + 1) = 2n^4 + 8n^3 + 12n^2 + 8n + 2
\end{aligned}
$$

The equalities $\triangle$ follow from the definition of L; the equalities $\triangledown$ follow from the definition of R; the equality $\clubsuit$ follows from the definition of $\sum$; the inequality $\bullet$ follows from the inductive hypothesis; while the other (in)equalities are obtained by arithmetic transformations. To complete the proof, observe that $L(n+1) < R(n+1)$.

Problem 157 Use mathematical induction to prove that:

$$\sum_{k=1}^{n} \frac{1}{(2k-1)^3} < 2 - \frac{1}{(2n-1)^2}$$

for every positive integer n such that $n > 1$.

Answer: Let:

$$L(n) = \sum_{k=1}^{n} \frac{1}{(2k-1)^3}$$
$$R(n) = 2 - \frac{1}{(2n-1)^2}$$

The claim is that $L(n) < R(n)$ for all $n \geq 2$.

The base case occurs for $n = 2$:

$$L(2) = \sum_{k=1}^{2} \frac{1}{(2k-1)^3} = \frac{1}{(2 \cdot 1 - 1)^3} + \frac{1}{(2 \cdot 2 - 1)^3} = \frac{1}{1} + \frac{1}{3^3} = 1 + \frac{1}{27}$$

$$R(2) = 2 - \frac{1}{(2 \cdot 2 - 1)^2} = 2 - \frac{1}{3^2} = 2 - \frac{1}{9} = 1 + \frac{8}{9} = 1 + \frac{24}{27}$$

Since $L(2) < R(2)$, the base case is verified.

For the inductive step, assume that:

$$L(n) < R(n)$$

for some $n \geq 2$. We need to prove that:

$$L(n+1) < R(n+1)$$

Indeed:

$$L(n+1) =$$

$$\triangle \qquad = \sum_{k=1}^{n+1} \frac{1}{(2k-1)^3}$$

$$\clubsuit \qquad = \left(\sum_{k=1}^{n} \frac{1}{(2k-1)^3} \right) + \frac{1}{(2(n+1)-1)^3}$$

$$\triangle \qquad = L(n) + \frac{1}{(2(n+1)-1)^3}$$

$$\bullet \qquad < R(n) + \frac{1}{(2(n+1)-1)^3}$$

$$\triangledown \qquad = 2 - \frac{1}{(2n-1)^2} + \frac{1}{(2(n+1)-1)^3} = 2 - \frac{1}{(2n-1)^2} + \frac{1}{(2n+1)^3}$$

$$= 2 - \frac{(2n+1)^3 - (2n-1)^2}{(2n-1)^2(2n+1)^3} = 2 - \frac{(8n^3 + 12n^2 + 6n + 1) - (4n^2 - 4n + 1)}{(2n-1)^2(2n+1)^3}$$

$$= 2 - \frac{8n^3 + 8n^2 + 10n}{(2n-1)^2(2n+1)^3} = 2 - \frac{8n^3 - 4n^2 - 2n + 1}{(2n-1)^2(2n+1)^3} - \frac{12n^2 + 12n - 1}{(2n-1)^2(2n+1)^3}$$

$$\heartsuit \qquad < 2 - \frac{8n^3 - 4n^2 - 2n + 1}{(2n-1)^2(2n+1)^3} - 0 = 2 - \frac{8n^3 - 4n^2 - 2n + 1}{(2n-1)^2(2n+1)^3}$$

$$R(n+1) =$$

$$\triangledown \qquad = 2 - \frac{1}{(2(n+1)-1)^2} = 2 - \frac{1}{(2n+1)^2} = 2 - \frac{1}{(2n+1)^2} \cdot \left(\frac{(2n-1)^2(2n+1)}{(2n-1)^2(2n+1)} \right)$$

$$= 2 - \frac{(4n^2 - 1)(2n - 1)}{(2n-1)^2(2n+1)^3} = 2 - \frac{8n^3 - 4n^2 - 2n + 1}{(2n-1)^2(2n+1)^3}$$

The equalities $\triangle$ follow from the definition of L; the equalities $\triangledown$ follow from the definition of R; the equality $\clubsuit$ follows from the definition of $\sum$; the inequality $\bullet$ follows from the inductive hypothesis; the inequality $\heartsuit$ holds because the assumption that $n \geq 2$ guarantees that $12n^2 + 12n - 1 > 0$; while the other (in)equalities are obtained by arithmetic transformations. To complete the proof, observe that $L(n+1) < R(n+1)$.

Problem 158   Use mathematical induction to prove that:

$$\sum_{k=1}^{n} \frac{1}{(7k+1)(7k-6)} = \frac{n}{7n+1}$$

for every positive integer $n \geq 1$.

Answer: Let:

$$L(n) = \sum_{k=1}^{n} \frac{1}{(7k+1)(7k-6)}$$

$$R(n) = \frac{n}{7n+1}$$

The claim is that $L(n) = R(n)$ for all $n \geq 1$.

The base case occurs for $n = 1$:

$$L(1) = \sum_{k=1}^{1} \frac{1}{(7k+1)(7k-6)} = \frac{1}{(7 \cdot 1 + 1)(7 \cdot 1 - 6)} = \frac{1}{8 \cdot 1} = \frac{1}{8}$$

$$R(1) = \frac{1}{7 \cdot 1 + 1} = \frac{1}{8}$$

Since $L(1) = R(1)$, the base case is verified.

For the inductive step, assume that:

$$L(n) = R(n)$$

for some $n \geq 1$. We need to prove that:

$$L(n+1) = R(n+1)$$

Indeed:

$$L(n+1) =$$

$$\triangle \qquad = \sum_{k=1}^{n+1} \frac{1}{(7k+1)(7k-6)}$$

$$\clubsuit \qquad = \left(\sum_{k=1}^{n} \frac{1}{(7k+1)(7k-6)} \right) + \frac{1}{(7(n+1)+1)(7(n+1)-6)}$$

$$\triangle \qquad = L(n) + \frac{1}{(7(n+1)+1)(7(n+1)-6)}$$

$$\bullet \qquad = R(n) + \frac{1}{(7(n+1)+1)(7(n+1)-6)}$$

$$\triangledown \qquad = \frac{n}{7n+1} + \frac{1}{(7(n+1)+1)(7(n+1)-6)} = \frac{n}{7n+1} + \frac{1}{(7n+8)(7n+1)}$$

$$= \frac{1}{(7n+1)} \cdot \left(n + \frac{1}{(7n+8)} \right) = \frac{1}{(7n+1)} \cdot \frac{7n^2+8n+1}{(7n+8)}$$

$$= \frac{1}{(7n+1)} \cdot \frac{(7n+1)(n+1)}{(7n+8)} = \frac{n+1}{7n+8} = \frac{n+1}{7(n+1)+1}$$

$$\triangledown \qquad = R(n+1)$$

The equalities $\triangle$ follow from the definition of L; the equalities $\triangledown$ follow from the definition of R; the equality $\clubsuit$ follows from the definition of $\sum$; the inequality $\bullet$ follows from the inductive hypothesis; while the other equalities are obtained by arithmetic transformations.

Problem 159   Use mathematical induction to prove that:

$$\prod_{k=2}^{n} \left(1 + \frac{1}{k} - \frac{2}{k^2}\right) = \frac{(n+1)(n+2)}{6n}$$

for every integer $n \geq 2$.

Answer: Let:

$$L(n) = \prod_{k=2}^{n} \left(1 + \frac{1}{k} - \frac{2}{k^2}\right)$$

$$R(n) = \frac{(n+1)(n+2)}{6n}$$

The claim is that $L(n) = R(n)$ for every integer $n \geq 2$.

The base case occurs for $n = 2$:

$$L(2) = \prod_{k=2}^{2} \left(1 + \frac{1}{k} - \frac{2}{k^2}\right) = \left(1 + \frac{1}{2} - \frac{2}{2^2}\right) = \left(1 + \frac{1}{2} - \frac{1}{2}\right) = 1$$

$$R(2) = \frac{(2+1)(2+2)}{6 \cdot 2} = 1$$

Since $L(2) = R(2)$, the base case is verified.

Inductively, assume that:

$$L(n) = R(n)$$

for some $n \geq 2$. We need to prove that:

$$L(n+1) = R(n+1)$$

Indeed:

$$
\begin{aligned}
L(n+1) \quad &= \\
\triangle \qquad &= \prod_{k=2}^{n+1} \left(1 + \frac{1}{k} - \frac{2}{k^2}\right) \\
\clubsuit \qquad &= \left(\prod_{k=2}^{n} \left(1 + \frac{1}{k} - \frac{2}{k^2}\right)\right) \cdot \left(1 + \frac{1}{n+1} - \frac{2}{(n+1)^2}\right) \\
\triangle \qquad &= L(n) \cdot \left(1 + \frac{1}{n+1} - \frac{2}{(n+1)^2}\right) \\
\bullet \qquad &= R(n) \cdot \left(1 + \frac{1}{n+1} - \frac{2}{(n+1)^2}\right) \\
\triangledown \qquad &= \frac{(n+1)(n+2)}{6n} \cdot \left(1 + \frac{1}{n+1} - \frac{2}{(n+1)^2}\right) \\
&= \frac{(n+1)(n+2)}{6n} \cdot \frac{(n+1)^2 + (n+1) - 2}{(n+1)^2} = \frac{(n+2)}{6n} \cdot \frac{n^2 + 2n + 1 + n + 1 - 2}{(n+1)} \\
&= \frac{(n+2)}{6n} \cdot \frac{(n^2 + 3n)}{(n+1)} = \frac{(n+2)}{6n} \cdot \frac{n(n+3)}{(n+1)} = \frac{(n+2)}{6} \cdot \frac{(n+3)}{(n+1)} \\
\triangledown \qquad &= R(n+1)
\end{aligned}
$$

The equalities $\triangle$ follow from the definition of L; the equalities $\triangledown$ follow from the definition of R; the equality $\clubsuit$ follows from the definition of $\prod$; the equality $\bullet$ follows from the inductive hypothesis; while the other equalities are obtained by arithmetic transformations.

Problem 160   Use mathematical induction to prove that:

$$\sum_{k=1}^{n} \frac{1}{(2k-1)(2k+3)} = \frac{n(4n+5)}{3(2n+1)(2n+3)}$$

for every positive integer $n \geq 1$.

Answer: Let:

$$L(n) = \sum_{k=1}^{n} \frac{1}{(2k-1)(2k+3)}$$

$$R(n) = \frac{n(4n+5)}{3(2n+1)(2n+3)}$$

The claim is that $L(n) = R(n)$ for all $n \geq 1$.

The base case occurs for $n = 1$:

$$L(1) = \sum_{k=1}^{1} \frac{1}{(2k-1)(2k+3)} = \frac{1}{(2 \cdot 1 - 1)(2 \cdot 1 + 3)} = \frac{1}{5}$$

$$R(1) = \frac{1 \cdot (4 \cdot 1 + 5)}{3(2 \cdot 1 + 1)(2 \cdot 1 + 3)} = \frac{9}{3 \cdot 3 \cdot 5} = \frac{1}{5}$$

Since $L(1) = R(1)$, the base case is verified.

For the inductive step, assume that:

$$L(n) = R(n)$$

for some $n \geq 1$. We need to prove that:

$$L(n+1) = R(n+1)$$

Indeed:

$$
\begin{aligned}
L(n+1) \quad &= \\
\triangle \quad &= \sum_{k=1}^{n+1} \frac{1}{(2k-1)(2k+3)} \\
\clubsuit \quad &= \left(\sum_{k=1}^{n} \frac{1}{(2k-1)(2k+3)} \right) + \frac{1}{(2(n+1)-1)(2(n+1)+3)} \\
\triangle \quad &= L(n) + \frac{1}{(2(n+1)-1)(2(n+1)+3)} \\
\bullet \quad &= R(n) + \frac{1}{(2(n+1)-1)(2(n+1)+3)} \\
\triangledown \quad &= \frac{n(4n+5)}{3(2n+1)(2n+3)} + \frac{1}{(2(n+1)-1)(2(n+1)+3)} \\
&= \frac{n(4n+5)}{3(2n+1)(2n+3)} + \frac{1}{(2n+1)(2n+5)} \\
&= \frac{1}{(2n+1)} \cdot \left(\frac{n(4n+5)}{3(2n+3)} + \frac{1}{(2n+5)} \right) = \frac{1}{(2n+1)} \cdot \left(\frac{n(4n+5)(2n+5) + 3(2n+3)}{3(2n+3)(2n+5)} \right) \\
&= \frac{1}{(2n+1)} \cdot \frac{(8n^3 + 30n^2 + 25n) + (6n+9)}{3(2n+3)(2n+5)} = \frac{8n^3 + 30n^2 + 31n + 9}{(2n+1) \cdot 3(2n+3)(2n+5)}
\end{aligned}
$$

$$R(n+1) =$$

$$\triangledown \qquad = \frac{(n+1)(4(n+1)+5)}{3(2(n+1)+1)(2(n+1)+3)} = \frac{(n+1)(4n+9)}{3(2n+3)(2n+5)} = \frac{(2n+1)}{(2n+1)} \cdot \frac{(4n^2+13n+9)}{3(2n+3)(2n+5)}$$

$$= \frac{8n^3 + 26n^2 + 18n + 4n^2 + 13n + 9}{(2n+1) \cdot 3(2n+3)(2n+5)} = \frac{8n^3 + 30n^2 + 31n + 9}{(2n+1) \cdot 3(2n+3)(2n+5)}$$

The equalities $\triangle$ follow from the definition of L; the equalities $\triangledown$ follow from the definition of R; the equality $\clubsuit$ follows from the definition of $\sum$; the inequality $\bullet$ follows from the inductive hypothesis; while the other equalities are obtained by arithmetic transformations. To complete the proof, observe that $L(n+1) = R(n+1)$.

Problem 161   Use mathematical induction to prove that:

$$\sum_{k=0}^{2n-1} (k-1)(3k-8) = 4n(2n^2 + 7(1-n))$$

for every positive integer $n \geq 1$.

Answer: Let:

$$L(n) = \sum_{k=0}^{2n-1} (k-1)(3k-8)$$
$$R(n) = 4n(2n^2 + 7(1-n))$$

The claim is that $L(n) = R(n)$ for all $n \geq 1$.

The base case occurs for $n = 1$:

$$L(1) = \sum_{k=0}^{2 \cdot 1 - 1} (k-1)(3k-8) = \sum_{k=0}^{1} (k-1)(3k-8) = (0-1)(3 \cdot 0 - 8) + (1-1)(3 \cdot 1 - 8)$$

$$= (-1)(-8) + 0 = 8$$

$$R(1) = 4 \cdot 1 \cdot (2 \cdot 1^2 + 7(1-1)) = 4 \cdot 2 = 8$$

Since $L(1) = R(1)$, the base case is verified.

For the inductive step, assume that:

$$L(n) = R(n)$$

for some $n \geq 1$. We need to prove that:

$$L(n+1) = R(n+1)$$

Indeed:

$$R(n+1) =$$

$$\triangledown \qquad = 4(n+1)\left(2(n+1)^2 + 7(1 - (n+1))\right) = 4(n+1)\left((2n^2 + 4n + 2) - 7n\right)$$

$$= 4(n+1)(2n^2 - 3n + 2) = 4(2n^3 - 3n^2 + 2n + 2n^2 - 3n + 2) = 4(2n^3 - n^2 - n + 2)$$

$$= 8n^3 - 4n^2 - 4n + 8$$

$$L(n+1) \quad =$$

$$\triangle \qquad = \sum_{k=0}^{2(n+1)-1} (k-1)(3k-8) = \sum_{k=0}^{2n+1} (k-1)(3k-8)$$

$$\clubsuit \qquad = \left(\sum_{k=0}^{2n-1} (k-1)(3k-8) \right) + (2n-1)(3(2n)-8) + ((2n+1)-1)(3(2n+1)-8)$$

$$\triangle \qquad = L(n) + (2n-1)(3(2n)-8) + ((2n+1)-1)(3(2n+1)-8)$$

$$\bullet \qquad = R(n) + (2n-1)(3(2n)-8) + ((2n+1)-1)(3(2n+1)-8)$$

$$\triangledown \qquad = \left(4n(2n^2 + 7(1-n)) \right) + (2n-1)(3(2n)-8) + ((2n+1)-1)(3(2n+1)-8)$$

$$= \left(8n^3 + 28n - 28n^2 \right) + (2n-1)(6n-8) + 2n(6n-5)$$

$$= (8n^3 - 28n^2 + 28n) + (12n^2 - 22n + 8) + (12n^2 - 10n)$$

$$= 8n^3 - 4n^2 - 4n + 8$$

The equalities $\triangle$ follow from the definition of L; the equalities $\triangledown$ follow from the definition of R; the equality $\clubsuit$ follows from the definition of $\sum$; the equality $\bullet$ follows from the inductive hypothesis; while the other equalities are obtained by arithmetic transformations. To complete the proof, observe that $L(n+1) = R(n+1)$.

Problem 162 Use mathematical induction to prove that:

$$\sum_{k=1}^{n} \frac{1}{(5k-1)(5k+4)} = \frac{n}{4(5n+4)}$$

for every positive integer $n \geq 1$.

Answer: Let:

$$L(n) = \sum_{k=1}^{n} \frac{1}{(5k-1)(5k+4)}$$

$$R(n) = \frac{n}{4(5n+4)}$$

The claim is that $L(n) = R(n)$ for all $n \geq 1$.

The base case occurs for $n = 1$:

$$L(1) \quad = \sum_{k=1}^{1} \frac{1}{(5k-1)(5k+4)} = \frac{1}{(5 \cdot 1 - 1)(5 \cdot 1 + 4)} = \frac{1}{4 \cdot 9} = \frac{1}{36}$$

$$R(1) \quad = \frac{1}{4(5 \cdot 1 + 4)} = \frac{1}{4 \cdot 9} = \frac{1}{36}$$

Since $L(1) = R(1)$, the base case is verified.

For the inductive step, assume that:

$$L(n) = R(n)$$

for some $n \geq 1$. We need to prove that:

$$L(n+1) = R(n+1)$$

Indeed:

$$L(n+1) \quad =$$

$$\triangle \qquad = \sum_{k=1}^{n+1} \frac{1}{(5k-1)(5k+4)}$$

$$\clubsuit \qquad = \left(\sum_{k=1}^{n} \frac{1}{(5k-1)(5k+4)} \right) + \frac{1}{(5(n+1)-1)(5(n+1)+4)}$$

$$\triangle \qquad = L(n) + \frac{1}{(5(n+1)-1)(5(n+1)+4)}$$

$$\bullet \qquad = R(n) + \frac{1}{(5(n+1)-1)(5(n+1)+4)}$$

$$\triangledown \qquad = \frac{n}{4(5n+4)} + \frac{1}{(5(n+1)-1)(5(n+1)+4)} = \frac{n}{4(5n+4)} + \frac{1}{(5n+4)(5n+9)}$$

$$= \frac{5n^2+9n+4}{4(5n+4)(5n+9)}$$

$$R(n+1) \quad =$$

$$\triangledown \qquad = \frac{n+1}{4(5(n+1)+4)} = \frac{n+1}{4(5n+9)} = \frac{(n+1)(5n+4)}{4(5n+9)(5n+4)} = \frac{5n^2+9n+4}{4(5n+4)(5n+9)}$$

The equalities $\triangle$ follow from the definition of L; the equalities $\triangledown$ follow from the definition of R; the equality $\clubsuit$ follows from the definition of $\sum$; the inequality $\bullet$ follows from the inductive hypothesis; while the other equalities are obtained by arithmetic transformations. To complete the proof, observe that $L(n+1) = R(n+1)$.

Problem 163   Use mathematical induction to prove that:

$$\sum_{k=2n}^{4n} (k+1) = (3n+1)(2n+1)$$

for every natural number n.

Answer: Let:

$$L(n) = \sum_{k=2n}^{4n} (k+1)$$

$$R(n) = (3n+1)(2n+1)$$

The claim is that $L(n) = R(n)$ for all $n \geq 0$.

The base case occurs for $n = 0$:

$$L(0) = \sum_{k=2\cdot 0}^{4\cdot 0} (k+1) = \sum_{k=0}^{0}(k+1) = (0+1) = 1$$

$$R(0) = (3 \cdot 0 + 1)(2 \cdot 0 + 1) = 1$$

Since $L(0) = R(0)$, the base case is verified.

For the inductive step, assume that:

$$L(n) = R(n)$$

for some $n \geq 0$.

We need to prove that:

$$L(n + 1) = R(n + 1)$$

Indeed:

$L(n + 1) \quad =$

$\triangle \qquad = \qquad \displaystyle\sum_{k=2(n+1)}^{4(n+1)} (k + 1) = \sum_{k=2n+2}^{4n+4} (k + 1)$

$\clubsuit \qquad = \qquad \left(\displaystyle\sum_{k=2n}^{4n+4} (k + 1) \right) - (((2n + 1) + 1) + (2n + 1))$

$\clubsuit \qquad = \qquad \displaystyle\sum_{k=2n}^{4n} (k + 1)$

$\qquad \qquad \quad + \quad ((4n + 1) + 1) + ((4n + 2) + 1) + ((4n + 3) + 1) + ((4n + 4) + 1)$

$\qquad \qquad \quad - \quad (((2n + 1) + 1) + (2n + 1))$

$\triangle \qquad = \qquad L(n)$

$\qquad \qquad \quad + \quad ((4n + 1) + 1) + ((4n + 2) + 1) + ((4n + 3) + 1) + ((4n + 4) + 1)$

$\qquad \qquad \quad - \quad (((2n + 1) + 1) + (2n + 1))$

$\qquad \qquad = \qquad L(n) + (4n + 2) + (4n + 3) + (4n + 4) + (4n + 5) - (2n + 2 + 2n + 1)$

$\qquad \qquad = \qquad L(n) + 12n + 11$

$\bullet \qquad = \qquad R(n) + 12n + 11$

$\triangledown \qquad = \qquad (3n + 1)(2n + 1) + 12n + 11 = (6n^2 + 5n + 1) + 12n + 11$

$\qquad \qquad = \qquad 6n^2 + 17n + 12$

$R(n + 1) \quad =$

$\triangledown \qquad = \quad (3(n + 1) + 1)(2(n + 1) + 1) = (3n + 4)(2n + 3) = 6n^2 + 17n + 12$

The equalities $\triangle$ follow from the definition of L; the equalities $\triangledown$ follow from the definition of R; the equalities $\clubsuit$ follow from the definition of $\sum$; the equality $\bullet$ follows from the inductive hypothesis; while the other equalities are obtained by arithmetic transformations. To complete the proof, observe that $L(n + 1) = R(n + 1)$.

Problem 164   Use mathematical induction to prove that:

$$\sum_{k=n}^{3n-1} (2k + 1) = 8n^2$$

for every positive integer n.

Answer: Let:

$$L(n) = \sum_{k=n}^{3n-1} (2k + 1)$$

$$R(n) = 8n^2$$

The claim is that $L(n) = R(n)$ for all $n \geq 1$.

The base case occurs for $n = 1$:

$$L(1) = \sum_{k=1}^{3 \cdot 1 - 1} (2k + 1) = \sum_{k=1}^{2} (2k + 1) = (2 \cdot 1 + 1) + (2 \cdot 2 + 1) = 3 + 5 = 8$$
$$R(1) = 8 \cdot 1^2 = 8$$

Since $L(1) = R(1)$, the base case is verified.

For the inductive step, assume that:

$$L(n) = R(n)$$

for some $n \geq 1$. We need to prove that:

$$L(n + 1) = R(n + 1)$$

Indeed:

$$L(n + 1) \quad =$$

$$\triangle \qquad = \sum_{k=n+1}^{3(n+1)-1} (2k + 1) = \sum_{k=n+1}^{3n+2} (2k + 1)$$

$$\clubsuit \qquad = \left(\sum_{k=n}^{3n+2} (2k + 1) \right) - (2n + 1)$$

$$\clubsuit \qquad = \left(\left(\sum_{k=n}^{3n-1} (2k + 1) \right) + (2(3n) + 1) + (2(3n + 1) + 1) + (2(3n + 2) + 1) \right) - (2n + 1)$$

$$\triangle \qquad = L(n) + (2(3n) + 1) + (2(3n + 1) + 1) + (2(3n + 2) + 1) - (2n + 1)$$

$$\qquad = L(n) + (6n + 1) + (6n + 3) + (6n + 5) - (2n + 1)$$

$$\qquad = L(n) + 16n + 8$$

$$\bullet \qquad = R(n) + 16n + 8$$

$$\triangledown \qquad = 8n^2 + 16n + 8 = 8(n^2 + 2n + 1) = 8(n + 1)^2$$

$$\triangledown \qquad = R(n + 1)$$

The equalities $\triangle$ follow from the definition of L; the equalities $\triangledown$ follow from the definition of R; the equalities $\clubsuit$ follow from the definition of $\sum$; the equality $\bullet$ follows from the inductive hypothesis; while the other equalities are obtained by arithmetic transformations.

Problem 165   Use mathematical induction to prove that:

$$\prod_{k=1}^{3n} \left(1 + \frac{2}{k} \right) = 1 + \left(\frac{9}{2} \right) \cdot n(n + 1)$$

for every integer $n \geq 1$.

Answer: Let:

$$L(n) = \prod_{k=1}^{3n} \left(1 + \frac{2}{k} \right)$$
$$R(n) = 1 + \left(\frac{9}{2} \right) \cdot n(n + 1)$$

The claim is that $L(n) = R(n)$ for every integer $n \geq 1$.

The base case occurs for $n = 1$:

$$L(1) = \prod_{k=1}^{3}\left(1 + \frac{2}{k}\right) = \left(1 + \frac{2}{1}\right) \cdot \left(1 + \frac{2}{2}\right) \cdot \left(1 + \frac{2}{3}\right) = 3 \cdot 2 \cdot \frac{5}{3} = 10$$

$$R(1) = 1 + \left(\frac{9}{2}\right) \cdot 1 \cdot (1 + 1) = 1 + 9 = 10$$

Since $L(1) = R(1)$, the base case is verified.

Inductively, assume that:

$$L(n) = R(n)$$

for some $n \geq 1$. We need to prove that:

$$L(n + 1) = R(n + 1)$$

Indeed:

$L(n + 1) \quad =$

$\triangle \qquad = \displaystyle\prod_{k=1}^{3(n+1)}\left(1 + \frac{2}{k}\right) = \prod_{k=1}^{3n+3}\left(1 + \frac{2}{k}\right)$

$\clubsuit \qquad = \displaystyle\prod_{k=1}^{3n}\left(1 + \frac{2}{k}\right) \cdot \left(1 + \frac{2}{3n + 1}\right) \cdot \left(1 + \frac{2}{3n + 2}\right) \cdot \left(1 + \frac{2}{3n + 3}\right)$

$\triangle \qquad = L(n) \cdot \left(1 + \frac{2}{3n + 1}\right) \cdot \left(1 + \frac{2}{3n + 2}\right) \cdot \left(1 + \frac{2}{3n + 3}\right)$

$\bullet \qquad = R(n) \cdot \left(1 + \frac{2}{3n + 1}\right) \cdot \left(1 + \frac{2}{3n + 2}\right) \cdot \left(1 + \frac{2}{3n + 3}\right)$

$\triangledown \qquad = \left(1 + \left(\frac{9}{2}\right) \cdot n(n + 1)\right) \cdot \left(1 + \frac{2}{3n + 1}\right) \cdot \left(1 + \frac{2}{3n + 2}\right) \cdot \left(1 + \frac{2}{3n + 3}\right)$

$\qquad = \left(\frac{2 + 9n(n + 1)}{2}\right) \cdot \left(\frac{3n + 1 + 2}{3n + 1}\right) \cdot \left(\frac{3n + 2 + 2}{3n + 2}\right) \cdot \left(\frac{3n + 3 + 2}{3n + 3}\right)$

$\qquad = \left(\frac{9n^2 + 9n + 2}{2}\right) \cdot \left(\frac{3n + 3}{3n + 1}\right) \cdot \left(\frac{3n + 4}{3n + 2}\right) \cdot \left(\frac{3n + 5}{3n + 3}\right)$

$\qquad = \left(\frac{3n(3n + 2) + (3n + 2)}{2}\right) \cdot \left(\frac{3n + 3}{3n + 3}\right) \cdot \frac{(3n + 4)(3n + 5)}{(3n + 1)(3n + 2)}$

$\qquad = \frac{(3n + 2)(3n + 1)}{2} \cdot \frac{(3n + 4)(3n + 5)}{(3n + 1)(3n + 2)} = \frac{(3n + 4)(3n + 5)}{2} = \frac{2 + (9n^2 + 27n + 18)}{2}$

$\qquad = 1 + \left(\frac{9}{2}\right) \cdot (n^2 + 3n + 2) = 1 + \left(\frac{9}{2}\right) \cdot (n + 1)(n + 2)$

$\triangledown \qquad = R(n + 1)$

The equalities $\triangle$ follow from the definition of L; the equalities $\triangledown$ follow from the definition of R; the equality $\clubsuit$ follows from the definition of $\prod$; the equality $\bullet$ follows from the inductive hypothesis; while the other equalities are obtained by arithmetic transformations.

2.4   Recursive Functions

Problem 166   (a) Write a recursive definition of the function $f : N \to N$ whose closed form is:

$$f_n = 2^n, \text{ for } n \geq 0$$

using only the addition operator. (Multiplication and exponentiation are forbidden.)
Answer:

$$f_0 = 1$$
$$f_n = f_{n-1} + f_{n-1}, \text{ for } n \geq 1$$

(b) Use mathematical induction to prove your answer given in part (a). That is, prove that the recursive definition that you have proposed in part (a) indeed defines the function f_n.

Answer: The base case occurs for $n = 0$, and is verified by an observation that:

$$f_0 = 1 = 2^0$$

which follows from the base case of the recursive definition of f. For the inductive step, assume that:

$$f_n = 2^n$$

for some $n \geq 0$. We need to prove that:

$$f_{n+1} = 2^{n+1}$$

Indeed:

$$
\begin{aligned}
f_{n+1} \quad &= \\
\bullet \quad &= f_n + f_n \\
* \quad &= 2^n + 2^n = 2 \cdot 2^n = 2^{n+1}
\end{aligned}
$$

The equality $\bullet$ is obtained from the recursive definition of the sequence f; the equality $*$ is obtained from the inductive hypothesis; and the other equalities are obtained by arithmetic transformations.

Problem 167   For each of the following functions, write a complete recursive definition of the function, using only the addition operator. (Multiplication and exponentiation are forbidden.) Explain your answer.

(a) $f(n) = 3n^2$
Answer: To relate $f(n + 1)$ to $f(n)$, observe:

$$f(n + 1) = 3(n + 1)^2 = 3(n^2 + 2n + 1) = 3n^2 + 6n + 3 = f(n) + 6n + 3$$

Hence, the required definition is:

$$f(0) = 0$$
$$f(n + 1) = f(n) + n + n + n + n + n + n + 3$$

(b) $g(n) = 3 \cdot 5^n$
Answer: To relate $g(n + 1)$ to $g(n)$, observe:

$$g(n + 1) = 3 \cdot 5^{n+1} = 3 \cdot 5^n \cdot 5 = g(n) \cdot 5$$

Hence, the required definition is:

$$g(0) = 3$$
$$g(n + 1) = g(n) + g(n) + g(n) + g(n) + g(n)$$

Problem 168   For each of the following functions, write a complete recursive definition of the function, using only the addition and subtraction operators.

(a) $f(n) = 2^n + 1$

Answer: Observe that:

$$2^{n+1} + 1 = 2 \cdot 2^n + 1 = 2^n + 2^n + 1 + (1 - 1) = (2^n + 1) + (2^n + 1) - 1$$

whence the recurrence:

$$f(0) = 2$$
$$f(n) = f(n-1) + f(n-1) - 1, \quad \text{for } n \geq 1$$

(b) $g(n) = 3n + 5$

Answer: Observe that:

$$3(n + 1) + 5 = 3n + 3 + 5 = (3n + 5) + 3$$

whence the recurrence:

$$f(0) = 5$$
$$f(n) = f(n-1) + 3, \quad \text{for } n \geq 1$$

Problem 169   For each of the following functions, write a complete recursive definition of the function, using only the addition and subtraction operators.

(a) $f(n) = 4 \cdot 3^n + 3n + 2$

Answer: Observe that:

$$f(n + 1) \quad = 4 \cdot 3^{n+1} + 3(n + 1) + 2 = 3 \cdot 4 \cdot 3^n + 3n + 5 = 3 \cdot 4 \cdot 3^n + (9n - 6n) + (6 - 1)$$
$$= 3\left(4 \cdot 3^n + 3n + 2\right) - (6n + 1) = 3f(n) - 6n - 1$$

whence the recursive definition:

$$f(0) = 6$$
$$f(n + 1) = f(n) + f(n) + f(n) - n - n - n - n - n - n - 1$$

(b) $g(n) = 2 \cdot n^2 + 7n - 3$

Answer: Observe that:

$$g(n + 1) \quad = 2(n + 1)^2 + 7(n + 1) - 3 = 2(n^2 + 2n + 1) + (7n + 7) - 3 = 2n^2 + 11n + 6$$
$$= 2n^2 + (7 + 4)n + (9 - 3) = \left(2n^2 + 7n - 3\right) + 4n + 9 = g(n) + 4n + 9$$

whence the definition:

$$g(0) = -3$$
$$g(n + 1) = g(n) + n + n + n + n + 9$$

Problem 170   **(a)** Write a recursive definition of the function $f : N^+ \rightarrow N^+$ whose closed form is:

$$f_n = n!, \text{ for } n \geq 1$$

using only the addition operator.

Answer:

$$f_1 = 1$$

$$f_n = \sum_{k=1}^{n} f_{n-1}, \text{ for } n \geq 1$$

(b) Use mathematical induction to prove your answer given in part **(a)**. That is, prove that the recursive definition that you have proposed in part **(a)** indeed defines the function $n!$.

Answer: The base case occurs for $n = 1$, and is verified by an observation that:

$$f_1 = 1$$

which follows from the base case of the recursive definition of f. For the inductive step, assume that:

$$f_n = n!$$

for some $n \geq 1$. We need to prove that:

$$f_{n+1} = (n+1)!$$

Indeed:

$$
\begin{aligned}
f_{n+1} \quad &= \\
\bullet \quad &= \sum_{k=1}^{n+1} f_n \\
\clubsuit \quad &= f_n \left(\sum_{k=1}^{n+1} 1 \right) = f_n \cdot (n+1) \\
* \quad &= n! \cdot (n+1) \\
\spadesuit \quad &= (n+1)!
\end{aligned}
$$

The equality $\bullet$ is obtained from the recursive definition of the sequence f; the equality $\clubsuit$ follows from the definition of $\sum$; the equality $*$ is obtained from the inductive hypothesis; the equality $\spadesuit$ follows from the definition of the factorial function; and the other equalities are obtained by arithmetic transformations.

Problem 171   Sequence a is defined recursively as follows.

$$
\begin{aligned}
a_0 &= -4 \\
a_1 &= 0 \\
a_n &= 4a_{n-1} - 3a_{n-2} - 4 \quad \text{for } n \geq 2
\end{aligned}
$$

Use mathematical induction to prove that:

$$a_n = 3^n + 2n - 5$$

for every natural number n.

Answer: The base case occurs for $n = 0, 1$ and follows from the base case of the recursive definition of the sequence a. Indeed:

$$
\begin{aligned}
a_0 &= -4 = 3^0 + 2 \cdot 0 - 5 \\
a_1 &= 0 = 3^1 + 2 \cdot 1 - 5
\end{aligned}
$$

For the inductive step, assume that:

$$a_k = 3^k + 2k - 5$$

for every k such that $0 \leq k \leq n$, for some $n \geq 1$.

We have to prove that:

$$a_{n+1} = 3^{n+1} + 2(n+1) - 5$$

Indeed:

$a_{n+1} =$

• $= 4a_{(n+1)-1} - 3a_{(n+1)-2} - 4 = 4a_n - 3a_{n-1} - 4$

∗ $= 4(3^n + 2n - 5) - 3(3^{n-1} + 2(n-1) - 5) - 4$

$= (4 \cdot 3^n + 8n - 20) - (3^n + 6(n-1) - 15) - 4$

$= (4 \cdot 3^n - 3^n) + (8n - 6n) + (-20 + 6 + 15 - 4) = 3 \cdot 3^n + 2n - 3 = 3^{n+1} + 2(n+1) - 5$

The equality • follows from the recursive definition of a_{n+1} for $n + 1 \geq 2$; the equality ∗ is obtained by two applications of the inductive hypothesis: to a_n and to a_{n-1}; and the other equalities are obtained by arithmetic transformations.

Problem 172   Sequence a is defined recursively as follows.

$$a_0 = 1$$
$$a_1 = 4$$
$$a_n = 2a_{n-1} - a_{n-2} + 18 \quad \text{for } n \geq 2$$

Use mathematical induction to prove that:

$$a_n = (3n - 1)^2$$

for every natural number n.

Answer: The base case occurs for $n = 0, 1$ and follows from the base case of the recursive definition of the sequence a. Indeed:

$$a_0 = 1 = (-1)^2 = (3 \cdot 0 - 1)^2$$
$$a_1 = 4 = 2^2 = (3 \cdot 1 - 1)^2$$

For the inductive step, assume that:

$$a_k = (3k - 1)^2$$

for every k such that $0 \leq k \leq n$, for some $n \geq 1$. We have to prove that:

$$a_{n+1} = (3(n + 1) - 1)^2$$

Indeed:

$a_{n+1} =$

• $= 2a_{(n+1)-1} - a_{(n+1)-2} + 18 = 2a_n - a_{n-1} + 18$

∗ $= 2(3n - 1)^2 - (3(n-1) - 1)^2 + 18 = 2(9n^2 - 6n + 1) - (3n - 4)^2 + 18$

$= (18n^2 - 12n + 2) - (9n^2 - 24n + 16) + 18 = 9n^2 + 12n + 4 = (3n + 2)^2$

$= (3(n + 1) - 1)^2$

The equality • follows from the recursive definition of a_{n+1} for $n + 1 \geq 2$; the equality ∗ is obtained by two applications of the inductive hypothesis: to a_n and to a_{n-1}; and the other equalities are obtained by arithmetic transformations.

Problem 173   Sequence a is defined recursively as follows.

$$a_0 = -1$$
$$a_1 = 2$$
$$a_n = 2a_{n-1} - a_{n-2} \quad \text{for } n \geq 2$$

Use mathematical induction to prove that:

$$a_n = 3n - 1$$

for every natural number n.

Answer: The base case occurs for $n = 0, 1$ and follows from the base case of the recursive definition of the sequence a. Indeed:

$$a_0 = -1 = 3 \cdot 0 - 1$$
$$a_1 = 2 = 3 \cdot 1 - 1$$

For the inductive step, assume that:

$$a_k = 3k - 1$$

for every k such that $0 \leq k \leq n$, for some $n \geq 1$. We have to prove that:

$$a_{n+1} = 3(n + 1) - 1$$

Indeed:

$$a_{n+1} =$$
$$\bullet \qquad = 2a_{(n+1)-1} - a_{(n+1)-2} = 2a_n - a_{n-1}$$
$$* \qquad = 2(3n - 1) - (3(n - 1) - 1) = 6n - 2 - 3n + 4 = 3n + 2 = 3(n + 1) - 1$$

The equality $\bullet$ follows from the recursive definition of a_{n+1} for $n + 1 \geq 2$; the equality $*$ is obtained by two applications of the inductive hypothesis: to a_n and to a_{n-1}; and the other equalities are obtained by arithmetic transformations.

Problem 174   Sequence a is defined recursively as follows.

$$a_0 = 9$$
$$a_1 = 25$$
$$a_n = 2a_{n-1} - a_{n-2} + 8 \quad \text{for } n \geq 2$$

Use mathematical induction to prove that:

$$a_n = (2n + 3)^2$$

for every natural number n.

Answer: The base case occurs for $n = 0, 1$ and follows from the base case of the recursive definition of the sequence a. Indeed:

$$a_0 = 9 = (2 \cdot 0 + 3)^2$$
$$a_1 = 25 = (2 \cdot 1 + 3)^2$$

For the inductive step, assume that:

$$a_k = (2k + 3)^2$$

for every k such that $0 \leq k \leq n$, for some $n \geq 1$.

We have to prove that:

$$a_{n+1} = (2(n+1) + 3)^2$$

Indeed:

$$
\begin{aligned}
a_{n+1} &= \\
\bullet \quad &= 2a_{(n+1)-1} - a_{(n+1)-2} + 8 = 2a_n - a_{n-1} + 8 \\
* \quad &= 2(2n+3)^2 - (2(n-1)+3)^2 + 8 = 2(4n^2 + 12n + 9) - (2n+1)^2 + 8 \\
&= (8n^2 + 24n + 18) - (4n^2 + 4n + 1) + 8 = 4n^2 + 20n + 25 = (2n+5)^2 \\
&= (2(n+1) + 3)^2
\end{aligned}
$$

The equality $\bullet$ follows from the recursive definition of a_{n+1} for $n+1 \geq 2$; the equality $*$ is obtained by two applications of the inductive hypothesis: to a_n and to a_{n-1}; and the other equalities are obtained by arithmetic transformations.

Problem 175   Sequence a is defined recursively as follows.

$$
\begin{aligned}
a_0 &= 0 \\
a_1 &= 10 \\
a_2 &= 40 \\
a_n &= 2a_{n-1} + \frac{a_{n-1}^2}{2a_{n-3}} \left(\frac{1}{n-1} - \frac{2}{(n-1)^2} \right) \quad \text{for } n \geq 3
\end{aligned}
$$

Use mathematical induction to prove that:

$$a_n = 5n \cdot 2^n$$

for every natural number n.

Answer: The base case occurs for $n = 0, 1, 2$ and follows from the base case of the recursive definition of the sequence a. Indeed:

$$
\begin{aligned}
a_0 &= 0 = 5 \cdot 0 \cdot 2^0 \\
a_1 &= 10 = 5 \cdot 1 \cdot 2^1 \\
a_2 &= 40 = 5 \cdot 2 \cdot 2^2
\end{aligned}
$$

For the inductive step, assume that:

$$a_k = 5k \cdot 2^k$$

for every k such that $0 \leq k \leq n$, for some $n \geq 2$. We have to prove that:

$$a_{n+1} = 5(n+1) \cdot 2^{n+1}$$

Indeed:

$$
\begin{aligned}
a_{n+1} &= \\
\bullet \quad &= 2a_{(n+1)-1} + \frac{a_{(n+1)-1}^2}{2a_{(n+1)-3}} \left(\frac{1}{(n+1)-1} - \frac{2}{((n+1)-1)^2} \right) = 2a_n + \frac{a_n^2}{2a_{n-2}} \left(\frac{1}{n} - \frac{2}{n^2} \right) \\
* \quad &= 2(5n \cdot 2^n) + \frac{(5 \cdot 2^n)^2}{2 \cdot 5(n-2) \cdot 2^{n-2}} \cdot \frac{(n-2)}{n^2} = 5n \cdot 2^{n+1} + \frac{25n^2 \cdot 2^{2n}}{10(n-2) \cdot 2^{n-2}} \cdot \frac{(n-2)}{n^2} \\
&= 5n \cdot 2^{n+1} + \frac{25 \cdot 2^{2n-(n-2)}}{10} = 5n \cdot 2^{n+1} + \frac{25 \cdot 2^{n+2}}{10} = 5n \cdot 2^{n+1} + \frac{25 \cdot 2 \cdot 2^{n+1}}{10} \\
&= 5n \cdot 2^{n+1} + 5 \cdot 2^{n+1} = 5 \cdot 2^{n+1}(n+1) = 5(n+1) \cdot 2^{n+1}
\end{aligned}
$$

The equality $\bullet$ follows from the recursive definition of a_{n+1} for $n+1 \geq 3$; the equality $*$ is obtained by two applications of the inductive hypothesis: to a_n and to a_{n-2}; and the other equalities are obtained by arithmetic transformations.

Problem 176   Sequence a is defined recursively as follows.

$$a_0 = -1$$
$$a_1 = 4$$
$$a_n = 6a_{n-1} - 5a_{n-2} - 4 \quad \text{for } n \geq 2$$

Use mathematical induction to prove that:

$$a_n = 5^n + n - 2$$

for every natural number n.

Answer: The base case occurs for $n = 0, 1$ and follows from the base case of the recursive definition of the sequence a. Indeed:

$$a_0 = -1 = 5^0 + 0 - 2$$
$$a_1 = 4 = 5^1 + 1 - 2$$

For the inductive step, assume that:

$$a_k = 5^k + k - 2$$

for every k such that $0 \leq k \leq n$, for some $n \geq 1$. We have to prove that:

$$a_{n+1} = 5^{n+1} + (n + 1) - 2$$

Indeed:

$$a_{n+1} =$$
$$\bullet \quad = 6a_{(n+1)-1} - 5a_{(n+1)-2} - 4 = 6a_n - 5a_{n-1} - 4$$
$$* \quad = 6(5^n + n - 2) - 5(5^{n-1} + (n - 1) - 2) - 4$$
$$= (6 \cdot 5^n + 6n - 12) - (5^n + 5(n - 1) - 10) - 4$$
$$= (6 - 1)5^n + n - 1 = 5 \cdot 5^n + n - 1 = 5^{n+1} + (n + 1) - 2$$

The equality $\bullet$ follows from the recursive definition of a_{n+1} for $n + 1 \geq 2$; the equality $*$ is obtained by two applications of the inductive hypothesis: to a_n and to a_{n-1}; and the other equalities are obtained by arithmetic transformations.

Problem 177   Sequence a is defined recursively as follows.

$$a_0 = 49$$
$$a_1 = 64$$
$$a_2 = 81$$
$$a_n = 3a_{n-2} - 2a_{n-3} + 6 \quad \text{for } n \geq 3$$

Use mathematical induction to prove that:

$$a_n = (n + 7)^2$$

for every natural number n.

Answer: The base case occurs for $n = 0, 1, 2$ and follows from the base case of the recursive definition of the sequence a. Indeed:

$$a_0 = 49 = (0 + 7)^2$$
$$a_1 = 64 = (1 + 7)^2$$
$$a_2 = 81 = (2 + 7)^2$$

For the inductive step, assume that:

$$a_k = (k + 7)^2$$

for every k such that $0 \le k \le n$, for some $n \ge 2$. We have to prove that:

$$a_{n+1} = ((n + 1) + 7)^2$$

Indeed:

$a_{n+1} \quad =$

$\bullet \qquad = 3a_{(n+1)-2} - 2a_{(n+1)-3} + 6 = 3a_{n-1} - 2a_{n-2} + 6$

$* \qquad = 3((n - 1) + 7)^2 - 2((n - 2) + 7)^2 + 6 = 3(n + 6)^2 - 2(n + 5)^2 + 6$

$\qquad = 3(n^2 + 12n + 36) - 2(n^2 + 10n + 25) + 6 = n^2 + 16n + 64 = (n + 8)^2 = ((n + 1) + 7)^2$

The equality $\bullet$ follows from the recursive definition of a_{n+1} for $n + 1 \ge 3$; the equality $*$ is obtained by two applications of the inductive hypothesis: to a_{n-1} and to a_{n-2}; and the other equalities are obtained by arithmetic transformations.

Problem 178 Sequence a is defined recursively as follows.

$$a_0 = 12$$
$$a_1 = 15$$
$$a_n = 16a_{n-2} - 165 \quad \text{for } n \ge 2$$

Use mathematical induction to prove that:

$$a_n = 4^n + 11$$

for every natural number n.

Answer: The base case occurs for $n = 0, 1$ and follows from the base case of the recursive definition of the sequence a. Indeed:

$$a_0 = 12 = 4^0 + 11$$
$$a_1 = 15 = 4^1 + 11$$

For the inductive step, assume that:

$$a_k = 4^k + 11$$

for every k such that $0 \le k \le n$, for some $n \ge 1$.
We have to prove that:

$$a_{n+1} = 4^{n+1} + 11$$

Indeed:

$a_{n+1} \quad =$

$\bullet \qquad = 16a_{(n+1)-2} - 165 = 16a_{n-1} - 165$

$* \qquad = 16(4^{n-1} + 11) - 165 = 16 \cdot 4^{n-1} + 176 - 165 = 4^2 \cdot 4^{n-1} + 11 = 4^{n+1} + 11$

The equality $\bullet$ follows from the recursive definition of a_{n+1} for $n + 1 \ge 2$; the equality $*$ is obtained by an application of the inductive hypothesis to a_{n-1}; and the other equalities are obtained by arithmetic transformations.

Problem 179   Sequence a is defined recursively as follows.

$$a_0 = 0$$
$$a_1 = 6$$
$$a_n = \frac{a_{n-1}^2}{a_{n-2}}\left(1 - \frac{1}{(n-1)^2}\right) \quad \text{for } n \geq 2$$

Use mathematical induction to prove that:

$$a_n = 2n \cdot 3^n$$

for every natural number n.

Answer: The base case occurs for $n = 0, 1$ and follows from the base case of the recursive definition of the sequence a. Indeed:

$$a_0 = 0 = 2 \cdot 0 \cdot 3^0$$
$$a_1 = 6 = 2 \cdot 1 \cdot 3^1$$

For the inductive step, assume that:

$$a_k = 2k \cdot 3^k$$

for every k such that $0 \leq k \leq n$, for some $n \geq 1$. We have to prove that:

$$a_{n+1} = 2(n+1) \cdot 3^{n+1}$$

Indeed:

$$a_{n+1} =$$

$$\bullet \qquad = \frac{a_{(n+1)-1}^2}{a_{(n+1)-2}}\left(1 - \frac{1}{((n+1)-1)^2}\right) = \frac{a_n^2}{a_{n-1}}\left(1 - \frac{1}{n^2}\right)$$

$$* \qquad = \frac{(2n \cdot 3^n)^2}{2(n-1) \cdot 3^{n-1}}\left(1 - \frac{1}{n^2}\right) = \frac{4n^2 \cdot 3^{2n}}{2(n-1) \cdot 3^{n-1}} \cdot \frac{n^2 - 1}{n^2} = \frac{4n^2 \cdot 3^{2n-(n-1)}}{2(n-1)} \cdot \frac{(n-1)(n+1)}{n^2}$$

$$= \frac{4 \cdot 3^{n+1}}{2} \cdot (n+1) = 2(n+1) \cdot 3^{n+1}$$

The equality $\bullet$ follows from the recursive definition of a_{n+1} for $n+1 \geq 2$; the equality $*$ is obtained by two applications of the inductive hypothesis: to a_n and to a_{n-1}; and the other equalities are obtained by arithmetic transformations.

Problem 180   Sequence a is defined recursively as follows.

$$a_0 = 6$$
$$a_1 = 6$$
$$a_2 = 5$$
$$a_n = a_{n-1} + 4(a_{n-2} - a_{n-3}) + 6n - 19 \quad \text{for } n \geq 3$$

Use mathematical induction to prove that:

$$a_n = 2^n - n^2 + 5$$

for every natural number n.

Answer: The base case occurs for $n = 0, 1, 2$ and follows from the base case of the recursive definition of the sequence a. Indeed:

$$a_0 = 6 = 2^0 - 0^2 + 5$$
$$a_1 = 6 = 2^1 - 1^2 + 5$$
$$a_2 = 5 = 2^2 - 2^2 + 5$$

For the inductive step, assume that:

$$a_k = 2^k - k^2 + 5$$

for every k such that $0 \le k \le n$, for some $n \ge 2$. We have to prove that:

$$a_{n+1} = 2^{n+1} - (n+1)^2 + 5$$

Indeed:

$a_{n+1} =$

$\bullet \quad = a_{(n+1)-1} + 4(a_{(n+1)-2} - a_{(n+1)-3}) + 6(n+1) - 19 = a_n + 4(a_{n-1} - a_{n-2}) + 6n - 13$

$* \quad = (2^n - n^2 + 5) + 4\left((2^{n-1} - (n-1)^2 + 5) - (2^{n-2} - (n-2)^2 + 5)\right) + 6n - 13$

$\quad = (2^n - n^2 + 5) + 4\left((2^{n-1} - n^2 + 2n + 4) - (2^{n-2} - n^2 + 4n + 1)\right) + 6n - 13$

$\quad = (2^n - n^2 + 5) + 4\left((2 - 1) \cdot 2^{n-2} - 2n + 3\right) + 6n - 13 = 2^n + 4 \cdot 2^{n-2} - n^2 - 2n + 4$

$\quad = 2^n + 2^n - (n^2 + 2n + 1) + 5 = 2^{n+1} - (n+1)^2 + 5$

The equality $\bullet$ follows from the recursive definition of a_{n+1} for $n+1 \ge 3$; the equality $*$ is obtained by three applications of the inductive hypothesis: to a_n, to a_{n-1} and to a_{n-2}; and the other equalities are obtained by arithmetic transformations.

Problem 181 Sequence b is defined recursively as follows.

$$b_1 = \frac{65}{8}$$
$$b_2 = 65$$
$$b_n = 4b_{n-1} + 36b_{n-2} \quad \text{for } n \ge 2$$

(a) Use mathematical induction to prove that:

$$b_n > 8^n$$

for every positive integer n.

Answer: The base case occurs for $n = 1, 2$ and follows from the base case of the recursive definition of the sequence b. Indeed:

$$b_1 = \frac{65}{8} > \frac{64}{8} = (8)^1$$
$$b_2 = 65 => 64 = 8^2$$

For the inductive step, assume that:

$$b_k > 8^k$$

for every k such that $1 \le k \le n$, for some $n \ge 2$. We have to prove that:

$$b_{n+1} > 8^{n+1}$$

Indeed:

$$b_{n+1} \;=$$

$$\bullet \qquad = 4b_{(n+1)-1} + 36b_{(n+1)-2} = 4b_n + 36b_{n-1}$$

$$* \qquad > 4 \cdot 8^n + 36 \cdot 8^{n-1} = 8^{n-1} \cdot (4 \cdot 8 + 36) = 8^{n-1} \cdot 68$$

$$> 8^{n-1} \cdot 64 = 8^{n-1} \cdot 8^2 = 8^{n+1}$$

The equality $\bullet$ follows from the recursive definition of b_{n+1} for $n+1 \geq 2$; the inequality $*$ is obtained by two applications of the inductive hypothesis: to b_n and to b_{n-1}; and the other (in)equalities are obtained by arithmetic transformations.

(b) Use mathematical induction to prove that:

$$b_n < \left(\frac{26}{3}\right)^n$$

for every positive integer n.

Answer: The base case occurs for $n = 1, 2$ and follows from the base case of the recursive definition of the sequence b. Indeed:

$$b_1 = \frac{65}{8} = \frac{195}{24} < \frac{208}{24} = \left(\frac{26}{3}\right)^1$$

$$b_2 = 65 = \frac{65 \cdot 9}{9} = \frac{585}{9} < \frac{676}{9} = \left(\frac{26}{3}\right)^2$$

For the inductive step, assume that:

$$b_k < \left(\frac{26}{3}\right)^k$$

for every k such that $1 \leq k \leq n$, for some $n \geq 2$. We have to prove that:

$$b_{n+1} < \left(\frac{26}{3}\right)^{n+1}$$

Indeed:

$$b_{n+1} \;=$$

$$\bullet \qquad = 4b_{(n+1)-1} + 36b_{(n+1)-2} = 4b_n + 36b_{n-1}$$

$$* \qquad < 4 \cdot \left(\frac{26}{3}\right)^n + 36 \cdot \left(\frac{26}{3}\right)^{n-1} = \left(\frac{26}{3}\right)^{n-1} \cdot \left(4 \cdot \frac{26}{3} + 36\right)$$

$$= \left(\frac{26}{3}\right)^{n-1} \cdot \left(\frac{4 \cdot 26 + 108}{3}\right) = \left(\frac{26}{3}\right)^{n-1} \cdot \frac{212}{3} = \left(\frac{26}{3}\right)^{n-1} \cdot \frac{636}{9}$$

$$< \left(\frac{26}{3}\right)^{n-1} \cdot \frac{676}{9} = \left(\frac{26}{3}\right)^{n-1} \cdot \left(\frac{26}{3}\right)^2 = \left(\frac{26}{3}\right)^{n+1}$$

The equality $\bullet$ follows from the recursive definition of b_{n+1} for $n+1 \geq 2$; the inequality $*$ is obtained by two applications of the inductive hypothesis: to b_n and to b_{n-1}; and the other (in)equalities are obtained by arithmetic transformations.

Problem 182   Sequence b is defined recursively as follows.

$$b_1 = \frac{17}{4}$$
$$b_2 = 20$$
$$b_n = 2b_{n-1} + 10b_{n-2} \quad \text{for } n \geq 2$$

(a) Use mathematical induction to prove that:

$$b_n > 4^n$$

for every positive integer n.

Answer: The base case occurs for $n = 1, 2$ and follows from the base case of the recursive definition of the sequence b. Indeed:

$$b_1 = \frac{17}{4} > \frac{16}{4} = 4^1$$
$$b_2 = 20 > 16 = 4^2$$

For the inductive step, assume that:

$$b_k > 4^k$$

for every k such that $1 \leq k \leq n$, for some $n \geq 2$.
We have to prove that:

$$b_{n+1} > 4^{n+1}$$

Indeed:

b_{n+1} =
- $= 2b_{(n+1)-1} + 10b_{(n+1)-2} = 2b_n + 10b_{n-1}$
* $> 2 \cdot 4^n + 10 \cdot 4^{n-1} = 4^{n-1}(2 \cdot 4 + 10) = 4^{n-1} \cdot 18 > 4^{n-1} \cdot 16 = 4^{n-1} \cdot 4^2 = 4^{n+1}$

The equality • follows from the recursive definition of b_{n+1} for $n+1 \geq 2$; the inequality * is obtained by two applications of the inductive hypothesis: to b_n and to b_{n-1}; and the other (in)equalities are obtained by arithmetic transformations.

(b) Use mathematical induction to prove that:

$$b_n < \left(\frac{9}{2}\right)^n$$

for every positive integer n.

Answer: The base case occurs for $n = 1, 2$ and follows from the base case of the recursive definition of the sequence b. Indeed:

$$b_1 = \frac{17}{4} < \frac{18}{4} = \left(\frac{9}{2}\right)^1$$
$$b_2 = 20 = \frac{80}{4} < \frac{81}{4} = \left(\frac{9}{2}\right)^2$$

For the inductive step, assume that:

$$b_k < \left(\frac{9}{2}\right)^k$$

for every k such that $1 \leq k \leq n$, for some $n \geq 2$.

We have to prove that:

$$b_{n+1} < \left(\frac{9}{2}\right)^{n+1}$$

Indeed:

$b_{n+1} =$

• $\quad = 2b_{(n+1)-1} + 10b_{(n+1)-2} = 2b_n + 10b_{n-1}$

* $\quad < 2 \cdot \left(\frac{9}{2}\right)^n + 10 \cdot \left(\frac{9}{2}\right)^{n-1} = \left(\frac{9}{2}\right)^{n-1} \cdot \left(2 \cdot \frac{9}{2} + 10\right) = \left(\frac{9}{2}\right)^{n-1} \cdot 19 = \left(\frac{9}{2}\right)^{n-1} \cdot \frac{76}{4}$

$\quad < \left(\frac{9}{2}\right)^{n-1} \cdot \frac{81}{4} = \left(\frac{9}{2}\right)^{n-1} \cdot \left(\frac{9}{2}\right)^2 = \left(\frac{9}{2}\right)^{n+1}$

The equality • follows from the recursive definition of b_{n+1} for $n+1 \geq 2$; the inequality * is obtained by two applications of the inductive hypothesis: to b_n and to b_{n-1}; and the other (in)equalities are obtained by arithmetic transformations.

Problem 183   Sequence b is defined recursively as follows.

$$b_1 = \frac{41}{8}$$
$$b_2 = 27$$
$$b_n = 2b_{n-1} + 16b_{n-2} \quad \text{for } n \geq 2$$

(a) Use mathematical induction to prove that:

$$b_n > 5^n$$

for every positive integer n.

Answer: The base case occurs for $n = 1, 2$ and follows from the base case of the recursive definition of the sequence b. Indeed:

$$b_1 = \frac{41}{8} > \frac{40}{8} = 5^1$$
$$b_2 = 27 > 25 = 5^2$$

For the inductive step, assume that:

$$b_k > 5^k$$

for every k such that $1 \leq k \leq n$, for some $n \geq 2$. We have to prove that:

$$b_{n+1} > 5^{n+1}$$

Indeed:

$b_{n+1} =$

• $\quad = 2b_{(n+1)-1} + 16b_{(n+1)-2} = 2b_n + 16b_{n-1}$

* $\quad > 2 \cdot 5^n + 16 \cdot 5^{n-1} = 5^{n-1}(2 \cdot 5 + 16) = 5^{n-1} \cdot 26 > 5^{n-1} \cdot 25 = 5^{n-1} \cdot 5^2 = 5^{n+1}$

The equality • follows from the recursive definition of b_{n+1} for $n+1 \geq 2$; the inequality * is obtained by two applications of the inductive hypothesis: to b_n and to b_{n-1}; and the other (in)equalities are obtained by arithmetic transformations.

(b) Use mathematical induction to prove that:

$$b_n < \left(\frac{21}{4}\right)^n$$

for every positive integer n.

Answer: The base case occurs for $n = 1, 2$ and follows from the base case of the recursive definition of the sequence b. Indeed:

$$b_1 = \frac{41}{8} < \frac{42}{8} = \left(\frac{21}{4}\right)^1$$

$$b_2 = 27 = \frac{432}{16} < \frac{441}{16} = \left(\frac{21}{4}\right)^2$$

For the inductive step, assume that:

$$b_k < \left(\frac{21}{4}\right)^k$$

for every k such that $1 \le k \le n$, for some $n \ge 2$. We have to prove that:

$$b_{n+1} < \left(\frac{21}{4}\right)^{n+1}$$

Indeed:

$$b_{n+1} =$$

$$\bullet \qquad = 2b_{(n+1)-1} + 16b_{(n+1)-2} = 2b_n + 16b_{n-1}$$

$$* \qquad < 2 \cdot \left(\frac{21}{4}\right)^n + 16 \cdot \left(\frac{21}{4}\right)^{n-1} = \left(\frac{21}{4}\right)^{n-1} \cdot \left(2 \cdot \frac{21}{4} + 16\right) = \left(\frac{21}{4}\right)^{n-1} \cdot \left(\frac{2 \cdot 21 + 64}{4}\right)$$

$$= \left(\frac{21}{4}\right)^{n-1} \cdot \frac{106}{4} = \left(\frac{21}{4}\right)^{n-1} \cdot \frac{424}{16}$$

$$< \left(\frac{21}{4}\right)^{n-1} \cdot \frac{441}{16} = \left(\frac{21}{4}\right)^{n-1} \cdot \left(\frac{21}{4}\right)^2 = \left(\frac{21}{4}\right)^{n+1}$$

The equality $\bullet$ follows from the recursive definition of b_{n+1} for $n+1 \ge 2$; the inequality $*$ is obtained by two applications of the inductive hypothesis: to b_n and to b_{n-1}; and the other (in)equalities are obtained by arithmetic transformations.

Problem 184 Sequence b is defined recursively as follows.

$$b_1 = \frac{65}{24}$$
$$b_2 = \frac{15}{2}$$
$$b_n = 2b_{n-1} + 2b_{n-2} \quad \text{for } n \ge 2$$

(a) Use mathematical induction to prove that:

$$b_n > \left(\frac{8}{3}\right)^n$$

for every positive integer n.

Answer: The base case occurs for $n = 1, 2$ and follows from the base case of the recursive definition of the sequence b. Indeed:

$$b_1 = \frac{65}{24} > \frac{64}{24} = \left(\frac{8}{3}\right)^1$$

$$b_2 = \frac{15}{2} = \frac{135}{18} > \frac{128}{18} = \frac{64}{9} = \left(\frac{8}{3}\right)^2$$

For the inductive step, assume that:

$$b_k > \left(\frac{8}{3}\right)^k$$

for every k such that $1 \le k \le n$, for some $n \ge 2$. We have to prove that:

$$b_{n+1} > \left(\frac{8}{3}\right)^{n+1}$$

Indeed:

$$
\begin{aligned}
b_{n+1} &= \\
\bullet \quad &= 2b_{(n+1)-1} + 2b_{(n+1)-2} = 2b_n + 2b_{n-1} \\
* \quad &> 2 \cdot \left(\frac{8}{3}\right)^n + 2 \cdot \left(\frac{8}{3}\right)^{n-1} = \left(\frac{8}{3}\right)^{n-1} \cdot \left(2 \cdot \frac{8}{3} + 2\right) \\
&= \left(\frac{8}{3}\right)^{n-1} \cdot \left(\frac{2 \cdot 8 + 6}{3}\right) = \left(\frac{8}{3}\right)^{n-1} \cdot \frac{22}{3} = \left(\frac{8}{3}\right)^{n-1} \cdot \frac{66}{9} \\
&> \left(\frac{8}{3}\right)^{n-1} \cdot \frac{64}{9} = \left(\frac{8}{3}\right)^{n-1} \cdot \left(\frac{8}{3}\right)^2 = \left(\frac{8}{3}\right)^{n+1}
\end{aligned}
$$

The equality $\bullet$ follows from the recursive definition of b_{n+1} for $n+1 \ge 2$; the inequality $*$ is obtained by two applications of the inductive hypothesis: to b_n and to b_{n-1}; and the other (in)equalities are obtained by arithmetic transformations.

(b) Use mathematical induction to prove that:

$$b_n < \left(\frac{11}{4}\right)^n$$

for every positive integer n.

Answer: The base case occurs for $n = 1, 2$ and follows from the base case of the recursive definition of the sequence b. Indeed:

$$b_1 = \frac{65}{24} < \frac{66}{24} = \left(\frac{11}{4}\right)^1$$

$$b_2 = \frac{15}{2} = \frac{120}{16} < \frac{121}{16} = \left(\frac{11}{4}\right)^2$$

For the inductive step, assume that:

$$b_k < \left(\frac{11}{4}\right)^k$$

for every k such that $1 \le k \le n$, for some $n \ge 2$. We have to prove that:

$$b_{n+1} < \left(\frac{11}{4}\right)^{n+1}$$

Indeed:

$$b_{n+1} =$$

$$\bullet \quad = 2b_{(n+1)-1} + 2b_{(n+1)-2} = 2b_n + 2b_{n-1}$$

$$* \quad < 2 \cdot \left(\frac{11}{4}\right)^n + 2 \cdot \left(\frac{11}{4}\right)^{n-1} = \left(\frac{11}{4}\right)^{n-1} \cdot \left(2 \cdot \frac{11}{4} + 2\right) = \left(\frac{11}{4}\right)^{n-1} \cdot \left(\frac{2 \cdot 11 + 8}{4}\right)$$

$$= \left(\frac{11}{4}\right)^{n-1} \cdot \frac{30}{4} = \left(\frac{11}{4}\right)^{n-1} \cdot \frac{120}{16}$$

$$< \left(\frac{11}{4}\right)^{n-1} \cdot \frac{121}{16} = \left(\frac{11}{4}\right)^{n-1} \cdot \left(\frac{11}{4}\right)^2 = \left(\frac{11}{4}\right)^{n+1}$$

The equality $\bullet$ follows from the recursive definition of b_{n+1} for $n+1 \geq 2$; the inequality $*$ is obtained by two applications of the inductive hypothesis: to b_n and to b_{n-1}; and the other (in)equalities are obtained by arithmetic transformations.

Problem 185   Sequence b is defined recursively as follows.

$$b_1 = \frac{19}{5}$$
$$b_2 = 15$$
$$b_3 = 57$$
$$b_n = b_{n-1} + 8b_{n-2} + 6b_{n-3} \quad \text{for } n \geq 3$$

(a) Use mathematical induction to prove that:

$$b_n > \left(\frac{18}{5}\right)^n$$

for every positive integer n.

Answer: The base case occurs for $n = 1, 2, 3$ and follows from the base case of the recursive definition of the sequence b. Indeed:

$$b_1 = \frac{19}{5} > \left(\frac{18}{5}\right)^1$$

$$b_2 = 15 = \frac{375}{25} > \frac{324}{25} = \left(\frac{18}{5}\right)^2$$

$$b_2 = 57 = \frac{19 \cdot 3 \cdot 125}{125} = \frac{19 \cdot 375}{125} > \frac{18 \cdot 324}{125} = \frac{18 \cdot 18^2}{5^3} = \left(\frac{18}{5}\right)^3$$

For the inductive step, assume that:

$$b_k > \left(\frac{18}{5}\right)^k$$

for every k such that $1 \leq k \leq n$, for some $n \geq 3$. We have to prove that:

$$b_{n+1} > \left(\frac{18}{5}\right)^{n+1}$$

Indeed:

b_{n+1} =

• $\quad = b_{(n+1)-1} + 8b_{(n+1)-2} + 6b_{(n+1)-3} = b_n + 8b_{n-1} + 6b_{n-2}$

$*\quad > \left(\dfrac{18}{5}\right)^n + 8 \cdot \left(\dfrac{18}{5}\right)^{n-1} + 6 \cdot \left(\dfrac{18}{5}\right)^{n-2} = \left(\dfrac{18}{5}\right)^{n-2} \cdot \left(\left(\dfrac{18}{5}\right)^2 + 8 \cdot \dfrac{18}{5} + 6\right)$

$\quad = \left(\dfrac{18}{5}\right)^{n-2} \cdot \left(\dfrac{18^2 + 40 \cdot 18 + 6 \cdot 25}{25}\right) = \left(\dfrac{18}{5}\right)^{n-2} \cdot \dfrac{324 + 720 + 150}{25}$

$\quad = \left(\dfrac{18}{5}\right)^{n-2} \cdot \dfrac{1194}{25} = \left(\dfrac{18}{5}\right)^{n-2} \cdot \dfrac{5970}{125}$

$\quad > \left(\dfrac{18}{5}\right)^{n-2} \cdot \dfrac{5832}{125} = \left(\dfrac{18}{5}\right)^{n-2} \cdot \left(\dfrac{18}{5}\right)^3 = \left(\dfrac{18}{5}\right)^{n+1}$

The equality • follows from the recursive definition of b_{n+1} for $n + 1 \geq 3$; the inequality $*$ is obtained by three applications of the inductive hypothesis: to b_n, to b_{n-1}, and to b_{n-2}; and the other (in)equalities are obtained by arithmetic transformations.

(b) Use mathematical induction to prove that:

$$b_n < 4^n$$

for every positive integer n.

Answer: The base case occurs for $n = 1, 2, 3$ and follows from the base case of the recursive definition of the sequence b. Indeed:

$$b_1 = \frac{19}{5} < \frac{20}{5} = 4^1$$
$$b_2 = 15 < 16 = 4^2$$
$$b_3 = 57 < 64 = 4^3$$

For the inductive step, assume that:

$$b_k < 4^k$$

for every k such that $1 \leq k \leq n$, for some $n \geq 3$. We have to prove that:

$$b_{n+1} < 4^{n+1}$$

Indeed:

b_{n+1} =

• $\quad = b_{(n+1)-1} + 8b_{(n+1)-2} + 6b_{(n+1)-3} = b_n + 8b_{n-1} + 6b_{n-2}$

$*\quad < 4^n + 8 \cdot 4^{n-1} + 6 \cdot 4^{n-2} = 4^{n-2} \cdot \left(4^2 + 8 \cdot 4 + 6\right) = 4^{n-2} \cdot 54$

$\quad < 4^{n-2} \cdot 64 = 4^{n-2} \cdot 4^3 = 4^{n+1}$

The equality • follows from the recursive definition of b_{n+1} for $n + 1 \geq 3$; the inequality $*$ is obtained by three applications of the inductive hypothesis: to b_n, to b_{n-1}, and to b_{n-2}; and the other (in)equalities are obtained by arithmetic transformations.

Problem 186   Sequence d is defined recursively as follows.

$$d_0 = 4$$
$$d_1 = 2$$
$$d_n = d_{n-1} + d_{n-2} \quad \text{for } n \geq 2$$

Use mathematical induction to prove that:

$$\sum_{k=0}^{2n}(-1)^k d_k = d_{2n-1} + 6$$

for every positive integer n.

Answer: Let:

$$L(n) = \sum_{k=0}^{2n}(-1)^k d_k$$

$$R(n) = d_{2n-1} + 6$$

The claim is that $L(n) = R(n)$ for all $n \geq 1$.

The base case occurs for $n = 1$:

$$L(1) \;=\; \sum_{k=0}^{2 \cdot 1}(-1)^k d_k = \sum_{k=0}^{2}(-1)^k d_k = (-1)^0 d_0 + (-1)^1 d_1 + (-1)^2 d_2 = d_0 - d_1 + d_2$$

$$= d_0 - d_1 + (d_0 + d_1) = 2d_0 = 2 \cdot 4 = 8$$

$$R(1) \;=\; d_{2 \cdot 1 - 1} + 6 = d_1 + 6 = 2 + 6 = 8$$

where the calculation of d_2 follows from the recursive definition of d_n for $n = 2$. Since $L(1) = R(1)$, the base case is verified.

For the inductive step, assume that:

$$L(n) = R(n)$$

for some $n \geq 1$. We have to prove that:

$$L(n + 1) = R(n + 1)$$

Indeed:

$$L(n + 1) =$$

$$\triangle \qquad = \sum_{k=0}^{2(n+1)}(-1)^k d_k = \sum_{k=0}^{2n+2}(-1)^k d_k$$

$$\clubsuit \qquad = \sum_{k=0}^{2n}(-1)^k d_k + \left((-1)^{2n+1} d_{2n+1}\right) + \left((-1)^{2n+2} d_{2n+2}\right)$$

$$\triangle \qquad = L(n) + \left((-1)^{2n+1} d_{2n+1}\right) + \left((-1)^{2n+2} d_{2n+2}\right)$$

$$= L(n) + \left(\left((-1)^2\right)^n \cdot (-1) \cdot d_{2n+1}\right) + \left(\left((-1)^2\right)^{n+1} \cdot d_{2n+2}\right)$$

$$= L(n) + (1^n \cdot (-1) \cdot d_{2n+1}) + \left(1^{n+1} \cdot d_{2n+2}\right) = L(n) - d_{2n+1} + d_{2n+2}$$

$$* \qquad = R(n) - d_{2n+1} + d_{2n+2}$$

$$\triangledown \qquad = d_{2n-1} + 6 - d_{2n+1} + d_{2n+2} = d_{2n-1} + 6 + (d_{2n+2} - d_{2n+1})$$

$$\bullet \qquad = d_{2n-1} + 6 + d_{2n} = (d_{2n} + d_{2n-1}) + 6$$

$$\bullet \qquad = d_{2n+1} + 6 = d_{2(n+1)-1} + 6$$

$$\triangledown \qquad = R(n + 1)$$

The equalities $\triangle$ follow from the definition of L; the equalities $\triangledown$ follow from the definition of R; the equality $\clubsuit$ follows from the definition of $\sum$; the equality $*$ is obtained by an application of the inductive hypothesis; the equalities • are obtained by application of the recursive rule of the definition of d to d_{2n+2} and to d_{2n+1}, justified by the fact that $2n + 2 \geq 2$ and $2n + 1 \geq 2$ whenever $n \geq 1$; while the other equalities are obtained by arithmetic transformations.

Problem 187   Sequence d is defined recursively as follows.

$$d_0 = 3$$
$$d_1 = 4$$
$$d_n = d_{n-1} + d_{n-2} \quad \text{for } n \geq 2$$

Use mathematical induction to prove that:

$$d_{n+1}d_{n-1} - d_n^2 = 5 \cdot (-1)^{n-1}$$

for every positive integer n.

Answer: Let:

$$L(n) = d_{n+1}d_{n-1} - d_n^2$$
$$R(n) = 5 \cdot (-1)^{n-1}$$

The claim is that $L(n) = R(n)$ for all $n \geq 1$.

The base case occurs for $n = 1$:

$$L(1) = d_{1+1}d_{1-1} - d_1^2 = d_2 d_0 - d_1^2 = (d_1 + d_0)d_0 - d_1^2 = (4 + 3) \cdot 3 - 4^2 = 7 \cdot 3 - 16 = 5$$
$$R(1) = 5 \cdot (-1)^{1-1} = 5 \cdot (-1)^0 = 5$$

where the calculation of d_2 follows from the recursive definition of d_n for $n = 2$. Since $L(1) = R(1)$, the base case is verified.

For the inductive step, assume that:

$$L(n) = R(n)$$

for some $n \geq 1$. We have to prove that:

$$L(n + 1) = R(n + 1)$$

Indeed:

$L(n + 1) =$

$\triangle$ $= d_{(n+1)+1}d_{(n+1)-1} - d_{n+1}^2 = d_{n+2}d_n - d_{n+1}^2$

• $= (d_{n+1} + d_n)d_n - d_{n+1}^2 = d_{n+1}d_n + d_n^2 - d_{n+1}^2 = d_{n+1}(d_n - d_{n+1}) + d_n^2$

       $= -\left(d_{n+1}(d_{n+1} - d_n) - d_n^2\right)$

• $= -\left(d_{n+1}d_{n-1} - d_n^2\right)$

$\triangle$ $= -L(n)$

$*$ $= -R(n)$

$\triangledown$ $= -\left(5 \cdot (-1)^{n-1}\right) = (-1) \cdot 5 \cdot (-1)^{n-1} = 5 \cdot (-1)^n = 5 \cdot (-1)^{(n+1)-1}$

$\triangledown$ $= R(n + 1)$

The equalities $\triangle$ follow from the definition of L; the equalities $\triangledown$ follow from the definition of R; the equality $*$ is obtained by an application of the inductive hypothesis; the equalities • are obtained by application of the recursive rule of the definition of d to d_{n+2} and d_{n+1} , justified by the fact that $n + 2 \geq 2$ and $n + 1 \geq 2$ whenever $n \geq 1$; while the other equalities are obtained by arithmetic transformations.

Problem 188   Sequence d is defined recursively as follows.

$$d_0 = 11$$
$$d_1 = 5$$
$$d_n = d_{n-1} + d_{n-2} \quad \text{for } n \geq 2$$

Use mathematical induction to prove that:

$$\sum_{k=0}^{n-1} d_{2k+1} = d_{2n} - 11$$

for every positive integer n.

Answer: Let:

$$L(n) = \sum_{k=0}^{n-1} d_{2k+1}$$

$$R(n) = d_{2n} - 11$$

The claim is that $L(n) = R(n)$ for all $n \geq 1$.

The base case occurs for $n = 1$:

$$L(1) = \sum_{k=0}^{1-1} d_{2k+1} = \sum_{k=0}^{0} d_{2k+1} = d_{2 \cdot 0 + 1} = d_1 = 5$$
$$R(1) = d_{2 \cdot 1} - 11 = d_2 - 11 = (d_1 + d_0) - 11 = (11 + 5) - 11 = 5$$

where the calculation of d_2 follows from the recursive definition of d_n for $n = 2$. Since $L(1) = R(1)$, the base case is verified.

For the inductive step, assume that:

$$L(n) = R(n)$$

for some $n \geq 1$. We have to prove that:

$$L(n+1) = R(n+1)$$

Indeed:

$$L(n+1) =$$

$\triangle$	$= \displaystyle\sum_{k=0}^{(n+1)-1} d_{2k+1} = \sum_{k=0}^{n} d_{2k+1}$
$\clubsuit$	$= \displaystyle\sum_{k=0}^{n-1} d_{2k+1} + d_{2n+1}$
$\triangle$	$= L(n) + d_{2n+1}$
$*$	$= R(n) + d_{2n+1}$
$\triangledown$	$= (d_{2n} - 11) + d_{2n+1} = (d_{2n} + d_{2n+1}) - 11$
$\bullet$	$= d_{2n+2} - 11 = d_{2(n+1)} - 11$
$\triangledown$	$= R(n+1)$

The equalities $\triangle$ follow from the definition of L; the equalities $\triangledown$ follow from the definition of R; the equality $\clubsuit$ follows from the definition of $\sum$; the equality $*$ is obtained by an application of the inductive hypothesis; the equality $\bullet$ is obtained by an application of the recursive rule of the definition of d to d_{2n+2}, justified by the fact that $2n + 2 \geq 2$; while the other equalities are obtained by arithmetic transformations.

Problem 189   Sequence d is defined recursively as follows.

$$d_0 = 2$$
$$d_1 = 4$$
$$d_n = d_{n-1} + d_{n-2} \quad \text{for } n \geq 2$$

Use mathematical induction to prove that:

$$\sum_{k=0}^{n} \frac{2^k d_k}{3^{k+1}} = \frac{2^{n+1} d_{n+4}}{3^{n+1}} - 10$$

for every natural number n.

Answer: Let:

$$L(n) = \sum_{k=0}^{n} \frac{2^k d_k}{3^{k+1}}$$

$$R(n) = \frac{2^{n+1} d_{n+4}}{3^{n+1}} - 10$$

The claim is that $L(n) = R(n)$ for all $n \geq 0$.

The base case occurs for $n = 0$:

$$L(0) = \sum_{k=0}^{0} \frac{2^k d_k}{3^{k+1}} = \frac{2^0 d_0}{3^{0+1}} = \frac{d_0}{3} = \frac{2}{3}$$

$$R(0) = \frac{2^{0+1} d_{0+4}}{3^{0+1}} - 10 = \frac{2 d_4}{3} - 10$$

To calculate $R(0)$, we employ the recursive rule of the definition of d_n for $2 \leq n \leq 4$ to calculate d_4:

$$d_2 = d_1 + d_0 = 4 + 2 = 6$$
$$d_3 = d_2 + d_1 = 6 + 4 = 10$$
$$d_4 = d_3 + d_2 = 10 + 6 = 16$$

whence:

$$R(0) = \frac{2 d_4}{3} - 10 = \frac{2 \cdot 16}{3} - 10 = \frac{2 \cdot 16 - 3 \cdot 10}{3} = \frac{32 - 30}{3} = \frac{2}{3}$$

Since $L(0) = R(0)$, the base case is verified.

For the inductive step, assume that:

$$L(n) = R(n)$$

for some $n \geq 0$.

We have to prove that:

$$L(n + 1) = R(n + 1)$$

Indeed:

$$L(n+1) =$$

$\triangle$
$$= \sum_{k=0}^{n+1} \frac{2^k d_k}{3^{k+1}}$$

$\clubsuit$
$$= \left(\sum_{k=0}^{n} \frac{2^k d_k}{3^{k+1}} \right) + \frac{2^{n+1} d_{n+1}}{3^{(n+1)+1}}$$

$\triangle$
$$= L(n) + \frac{2^{n+1} d_{n+1}}{3^{(n+1)+1}}$$

$*$
$$= R(n) + \frac{2^{n+1} d_{n+1}}{3^{(n+1)+1}}$$

$\triangledown$
$$= \frac{2^{n+1} d_{n+4}}{3^{n+1}} - 10 + \frac{2^{n+1} d_{n+1}}{3^{(n+1)+1}}$$

$$= -10 + \frac{2^{n+1}}{3^{n+2}} \left(3 d_{n+4} + d_{n+1} \right)$$

$$= -10 + \frac{2^{n+1}}{3^{n+2}} \left(2 d_{n+4} + d_{n+4} + d_{n+1} \right)$$

$\bullet$
$$= -10 + \frac{2^{n+1}}{3^{n+2}} \left(2 d_{n+4} + (d_{n+3} + d_{n+2}) + d_{n+1} \right)$$

$$= -10 + \frac{2^{n+1}}{3^{n+2}} \left(2 d_{n+4} + d_{n+3} + (d_{n+2} + d_{n+1}) \right)$$

$\bullet$
$$= -10 + \frac{2^{n+1}}{3^{n+2}} \left(2 d_{n+4} + d_{n+3} + d_{n+3} \right)$$

$$= -10 + \frac{2^{n+1}}{3^{n+2}} \left(2(d_{n+4} + d_{n+3}) \right)$$

$\bullet$
$$= -10 + \frac{2^{n+1}}{3^{n+2}} \left(2 d_{n+5} \right) = -10 + \frac{2^{n+2}}{3^{n+2}} \cdot d_{n+5} = \frac{2^{(n+1)+1} d_{(n+1)+4}}{3^{(n+1)+1}} - 10$$

$\triangledown$
$$= R(n+1)$$

The equalities $\triangle$ follow from the definition of L; the equalities $\triangledown$ follow from the definition of R; the equality $\clubsuit$ follows from the definition of $\sum$; the equality $*$ is obtained by an application of the inductive hypothesis; the equalities $\bullet$ are obtained by application of the recursive rule of the definition of d to $d_{n+\ell}$ for $3 \leq \ell \leq 5$; while the other equalities are obtained by arithmetic transformations.

Problem 190 Sequence d is defined recursively as follows.

$$d_0 = 9$$
$$d_1 = 6$$
$$d_n = d_{n-1} + d_{n-2} \quad \text{for } n \geq 2$$

Use mathematical induction to prove that:

$$\sum_{k=0}^{n} d_k^2 = d_n d_{n+1} + 27$$

for every positive integer n.

Answer: Let:

$$L(n) = \sum_{k=0}^{n} d_k^2$$
$$R(n) = d_n d_{n+1} + 27$$

The claim is that $L(n) = R(n)$ for all $n \geq 1$.

The base case occurs for $n = 1$:

$$L(1) \quad = \sum_{k=0}^{1} d_k^2 = d_0^2 + d_1^2 = 9^2 + 6^2 = 117$$

$$R(1) \quad = d_1 d_{1+1} + 27 = d_1 d_2 + 27 = d_1 (d_1 + d_0) + 27 = 6(6 + 9) + 27 = 117$$

where the calculation of d_2 follows from the recursive definition of d_n for $n = 2$. Since $L(1) = R(1)$, the base case is verified.

For the inductive step, assume that:

$$L(n) = R(n)$$

for some $n \geq 1$. We have to prove that:

$$L(n + 1) = R(n + 1)$$

Indeed:

$$L(n + 1) =$$

$$\triangle \qquad = \sum_{k=0}^{n+1} d_k^2$$

$$\clubsuit \qquad = \sum_{k=0}^{n} d_k^2 + d_{n+1}^2$$

$$\triangle \qquad = L(n) + d_{n+1}^2$$

$$* \qquad = R(n) + d_{n+1}^2$$

$$\triangledown \qquad = d_n d_{n+1} + 27 + d_{n+1}^2 = d_{n+1}(d_n + d_{n+1}) + 27$$

$$\bullet \qquad = d_{n+1} d_{n+2} + 27 = d_{n+1} d_{(n+1)+1} + 27$$

$$\triangledown \qquad = R(n + 1)$$

The equalities $\triangle$ follow from the definition of L; the equalities $\triangledown$ follow from the definition of R; the equality $\clubsuit$ follows from the definition of $\sum$; the equality $*$ is obtained by an application of the inductive hypothesis; the equality $\bullet$ is obtained by an application of the recursive rule of the definition of d to d_{n+2}, justified by the fact that $n + 2 \geq 2$; while the other equalities are obtained by arithmetic transformations.

Problem 191   Sequence d is defined recursively as follows.

$$d_0 = 5$$
$$d_1 = 3$$
$$d_n = d_{n-1} + d_{n-2} \quad \text{for } n \geq 2$$

Use mathematical induction to prove that:

$$\sum_{k=0}^{2n-1} d_k d_{k+1} = d_{2n}^2 - 25$$

for every positive integer n.

Answer: Let:

$$L(n) = \sum_{k=0}^{2n-1} d_k d_{k+1}$$

$$R(n) = d_{2n}^2 - 25$$

The claim is that $L(n) = R(n)$ for all $n \geq 1$.

The base case occurs for $n = 1$:

$$L(1) = \sum_{k=0}^{2 \cdot 1 - 1} d_k d_{k+1} = \sum_{k=0}^{1} d_k d_{k+1} = d_0 d_1 + d_1 d_2 = d_0 d_1 + d_1(d_0 + d_1) = 5 \cdot 3 + 3(5 + 3) = 39$$

$$R(1) = d_{2 \cdot 1}^2 - 25 = d_2^2 - 25 = (d_0 + d_1)^2 - 25 = (5 + 3)^2 - 25 = 39$$

where the calculation of d_2 follows from the recursive definition of d_n for $n = 2$. Since $L(1) = R(1)$, the base case is verified.

For the inductive step, assume that:

$$L(n) = R(n)$$

for some $n \geq 1$. We have to prove that:

$$L(n + 1) = R(n + 1)$$

Indeed:

$L(n + 1) =$

$\triangle \qquad = \displaystyle\sum_{k=0}^{2(n+1)-1} d_k d_{k+1} = \sum_{k=0}^{2n+1} d_k d_{k+1}$

$\clubsuit \qquad = \displaystyle\sum_{k=0}^{2n-1} d_k d_{k+1} + d_{2n} d_{2n+1} + d_{2n+1} d_{2n+2}$

$\triangle \qquad = L(n) + d_{2n} d_{2n+1} + d_{2n+1} d_{2n+2}$

$* \qquad = R(n) + d_{2n} d_{2n+1} + d_{2n+1} d_{2n+2}$

$\triangledown \qquad = d_{2n}^2 - 25 + d_{2n} d_{2n+1} + d_{2n+1} d_{2n+2} = d_{2n}(d_{2n} + d_{2n+1}) + d_{2n+1} d_{2n+2} - 25$

$\bullet \qquad = d_{2n} d_{2n+2} + d_{2n+1} d_{2n+2} - 25 = d_{2n+2}(d_{2n} + d_{2n+1}) - 25$

$\bullet \qquad = d_{2n+2} d_{2n+2} - 25 = d_{2n+2}^2 - 25 = d_{2(n+1)}^2 - 25$

$\triangledown \qquad = R(n + 1)$

The equalities $\triangle$ follow from the definition of L; the equalities $\triangledown$ follow from the definition of R; the equality $\clubsuit$ follows from the definition of $\sum$; the equality $*$ is obtained by an application of the inductive hypothesis; the equalities $\bullet$ are obtained by application of the recursive rule of the definition of d to d_{2n+2}, justified by the fact that $2n + 2 \geq 2$; while the other equalities are obtained by arithmetic transformations.

Problem 192   Sequence d is defined recursively as follows.

$$d_0 = 3$$
$$d_1 = 2$$
$$d_n = d_{n-1} + d_{n-2} \quad \text{for } n \geq 2$$

Use mathematical induction to prove that:

$$\sum_{k=0}^{n} \frac{3^k d_k}{5^{k+1}} = 12 - \frac{3^{n+1} d_{n+5}}{5^{n+1}}$$

for every natural number n.

Answer: Let:

$$L(n) = \sum_{k=0}^{n} \frac{3^k d_k}{5^{k+1}}$$

$$R(n) = 12 - \frac{3^{n+1} d_{n+5}}{5^{n+1}}$$

The claim is that $L(n) = R(n)$ for all $n \geq 0$.

The base case occurs for $n = 0$:

$$L(0) \quad = \sum_{k=0}^{0} \frac{3^k d_k}{5^{k+1}} = \frac{3^0 d_0}{5^{0+1}} = \frac{d_0}{5} = \frac{3}{5}$$

$$R(0) \quad = 12 - \frac{3^{0+1} d_{0+5}}{5^{0+1}} = 12 - \frac{3 d_5}{5}$$

To calculate $R(0)$, we employ the recursive rule of the definition of d_n for $2 \leq n \leq 5$ to calculate d_5:

$$d_2 = d_1 + d_0 = 2 + 3 = 5$$
$$d_3 = d_2 + d_1 = 5 + 2 = 7$$
$$d_4 = d_3 + d_2 = 7 + 5 = 12$$
$$d_5 = d_4 + d_3 = 12 + 7 = 19$$

whence:

$$R(0) = 12 - \frac{3 d_5}{5} = 12 - \frac{3 \cdot 19}{5} = \frac{12 \cdot 5 - 3 \cdot 19}{5} = \frac{60 - 57}{5} = \frac{3}{5}$$

Since $L(0) = R(0)$, the base case is verified.

For the inductive step, assume that:

$$L(n) = R(n)$$

for some $n \geq 0$.

We have to prove that:

$$L(n + 1) = R(n + 1)$$

Indeed:

$$L(n+1) =$$

$$\triangle \qquad = \sum_{k=0}^{n+1} \frac{3^k d_k}{5^{k+1}}$$

$$\clubsuit \qquad = \left(\sum_{k=0}^{n} \frac{3^k d_k}{5^{k+1}} \right) + \frac{3^{n+1} d_{n+1}}{5^{(n+1)+1}}$$

$$\triangle \qquad = L(n) + \frac{3^{n+1} d_{n+1}}{5^{(n+1)+1}}$$

$$* \qquad = R(n) + \frac{3^{n+1} d_{n+1}}{5^{(n+1)+1}}$$

$$\triangledown \qquad = 12 - \frac{3^{n+1} d_{n+5}}{5^{n+1}} + \frac{3^{n+1} d_{n+1}}{5^{(n+1)+1}}$$

$$= 12 - \frac{3^{n+1}}{5^{n+2}} \left(5 d_{n+5} - d_{n+1} \right)$$

$$= 12 - \frac{3^{n+1}}{5^{n+2}} \left(3 d_{n+5} + 2 d_{n+5} - d_{n+1} \right)$$

$$\bullet \qquad = 12 - \frac{3^{n+1}}{5^{n+2}} \left(3 d_{n+5} + 2(d_{n+4} + d_{n+3}) - d_{n+1} \right)$$

$$\bullet \qquad = 12 - \frac{3^{n+1}}{5^{n+2}} \left(3 d_{n+5} + 2 d_{n+4} + (d_{n+3} + d_{n+3}) - d_{n+1} \right)$$

$$= 12 - \frac{3^{n+1}}{5^{n+2}} \left(3 d_{n+5} + 2 d_{n+4} + d_{n+3} + (d_{n+3} - d_{n+1}) \right)$$

$$\bullet \qquad = 12 - \frac{3^{n+1}}{5^{n+2}} \left(3 d_{n+5} + 2 d_{n+4} + (d_{n+3} + d_{n+2}) \right)$$

$$\bullet \qquad = 12 - \frac{3^{n+1}}{5^{n+2}} \left(3 d_{n+5} + 2 d_{n+4} + d_{n+4} \right)$$

$$= 12 - \frac{3^{n+1}}{5^{n+2}} \left(3(d_{n+5} + d_{n+4}) \right)$$

$$\bullet \qquad = 12 - \frac{3^{n+1}}{5^{n+2}} \left(3 d_{n+6} \right)$$

$$= 12 - \frac{3^{n+2}}{5^{n+2}} \cdot d_{n+6}$$

$$= 12 - \frac{3^{(n+1)+1} d_{(n+1)+5}}{5^{(n+1)+1}}$$

$$\triangledown \qquad = R(n+1)$$

The equalities $\triangle$ follow from the definition of L; the equalities $\triangledown$ follow from the definition of R; the equality $\clubsuit$ follows from the definition of $\sum$; the equality $*$ is obtained by an application of the inductive hypothesis; the equalities $\bullet$ are obtained by application of the recursive rule of the definition of d to $d_{n+\ell}$ for $3 \le \ell \le 6$; while the other equalities are obtained by arithmetic transformations.

Problem 193   Sequence d is defined recursively as follows.

$$d_0 = 3$$
$$d_1 = 4$$
$$d_n = d_{n-1} + d_{n-2} \quad \text{for } n \geq 2$$

Use mathematical induction to prove that:

$$\sum_{k=0}^{n} \frac{5^k d_k}{8^{k+1}} = \frac{5^{n+1} d_{n+6}}{8^{n+1}} - 29$$

for every natural number n.

Answer: Let:

$$L(n) = \sum_{k=0}^{n} \frac{5^k d_k}{8^{k+1}}$$
$$R(n) = \frac{5^{n+1} d_{n+6}}{8^{n+1}} - 29$$

The claim is that $L(n) = R(n)$ for all $n \geq 0$.

The base case occurs for $n = 0$:

$$L(0) = \sum_{k=0}^{0} \frac{5^k d_k}{8^{k+1}} = \frac{5^0 d_0}{8^{0+1}} = \frac{d_0}{8} = \frac{3}{8}$$
$$R(0) = \frac{5^{0+1} d_{0+6}}{8^{0+1}} - 29 = \frac{5 d_6}{8} - 29$$

To calculate $R(0)$, we employ the recursive rule of the definition of d_n for $2 \leq n \leq 6$ to calculate d_6:

$$d_2 = d_1 + d_0 = 4 + 3 = 7$$
$$d_3 = d_2 + d_1 = 7 + 4 = 11$$
$$d_4 = d_3 + d_2 = 11 + 7 = 18$$
$$d_5 = d_4 + d_3 = 18 + 11 = 29$$
$$d_6 = d_5 + d_4 = 29 + 18 = 47$$

whence:

$$R(0) = \frac{5 d_6}{8} - 29 = \frac{5 \cdot 47}{8} - 29 = \frac{5 \cdot 47 - 8 \cdot 29}{8} = \frac{235 - 232}{8} = \frac{3}{8}$$

Since $L(0) = R(0)$, the base case is verified.

For the inductive step, assume that:

$$L(n) = R(n)$$

for some $n \geq 0$. We have to prove that:

$$L(n+1) = R(n+1)$$

Indeed:

$$L(n+1) =$$

$\triangle$
$$= \sum_{k=0}^{n+1} \frac{5^k d_k}{8^{k+1}}$$

$\clubsuit$
$$= \left(\sum_{k=0}^{n} \frac{5^k d_k}{8^{k+1}} \right) + \frac{5^{n+1} d_{n+1}}{8^{(n+1)+1}}$$

$\triangle$
$$= L(n) + \frac{5^{n+1} d_{n+1}}{8^{(n+1)+1}}$$

$*$
$$= R(n) + \frac{5^{n+1} d_{n+1}}{8^{(n+1)+1}}$$

$\triangledown$
$$= \frac{5^{n+1} d_{n+6}}{8^{n+1}} - 29 + \frac{5^{n+1} d_{n+1}}{8^{(n+1)+1}}$$

$$= -29 + \frac{5^{n+1}}{8^{n+2}} (8 d_{n+6} + d_{n+1})$$

$$= -29 + \frac{5^{n+1}}{8^{n+2}} (5 d_{n+6} + 3 d_{n+6} + d_{n+1})$$

$\bullet$
$$= -29 + \frac{5^{n+1}}{8^{n+2}} (5 d_{n+6} + 3(d_{n+5} + d_{n+4}) + d_{n+1})$$

$$= -29 + \frac{5^{n+1}}{8^{n+2}} (5 d_{n+6} + 3 d_{n+5} + 2 d_{n+4} + d_{n+4} + d_{n+1})$$

$\bullet$
$$= -29 + \frac{5^{n+1}}{8^{n+2}} (5 d_{n+6} + 3 d_{n+5} + 2 d_{n+4} + (d_{n+3} + d_{n+2}) + d_{n+1})$$

$$= -29 + \frac{5^{n+1}}{8^{n+2}} (5 d_{n+6} + 3 d_{n+5} + 2 d_{n+4} + d_{n+3} + (d_{n+2} + d_{n+1}))$$

$\bullet$
$$= -29 + \frac{5^{n+1}}{8^{n+2}} (5 d_{n+6} + 3 d_{n+5} + 2 d_{n+4} + d_{n+3} + d_{n+3})$$

$$= -29 + \frac{5^{n+1}}{8^{n+2}} (5 d_{n+6} + 3 d_{n+5} + 2(d_{n+4} + d_{n+3}))$$

$\bullet$
$$= -29 + \frac{5^{n+1}}{8^{n+2}} (5 d_{n+6} + 3 d_{n+5} + 2 d_{n+5})$$

$\bullet$
$$= -29 + \frac{5^{n+1}}{8^{n+2}} (5 d_{n+6} + 5 d_{n+5})$$

$$= -29 + \frac{5^{n+1}}{8^{n+2}} (5(d_{n+6} + d_{n+5}))$$

$\bullet$
$$= -29 + \frac{5^{n+1}}{8^{n+2}} (5 d_{n+7})$$

$$= -29 + \frac{5^{n+2}}{8^{n+2}} \cdot d_{n+7}$$

$$= \frac{5^{(n+1)+1} d_{(n+1)+6}}{8^{(n+1)+1}} - 29$$

$\triangledown$
$$= R(n+1)$$

The equalities $\triangle$ follow from the definition of L; the equalities $\triangledown$ follow from the definition of R; the equality $\clubsuit$ follows from the definition of $\sum$; the equality $*$ is obtained by an application of the inductive hypothesis; the equalities $\bullet$ are obtained by application of the recursive rule of the definition of d to $d_{n+\ell}$ for $3 \leq \ell \leq 7$; while the other equalities are obtained by arithmetic transformations.

Problem 194   Sequence d is defined recursively as follows.

$$d_0 = -18$$
$$d_1 = 41$$
$$d_n = d_{n-1} + d_{n-2} \quad \text{for } n \geq 2$$

Use mathematical induction to prove that:

$$\sum_{k=0}^{2n+1} d_{2k+1} = d_{4(n+1)} + 18$$

for every natural number n.

Answer: Let:

$$L(n) = \sum_{k=0}^{2n+1} d_{2k+1}$$
$$R(n) = d_{4(n+1)} + 18$$

The claim is that $L(n) = R(n)$ for all $n \geq 0$.

To verify the base case, calculate the values of d_2, d_3, d_4, according to the recursive rule of the definition of d_n, for $n \geq 2$.

$$d_2 = d_1 + d_0 = 41 - 18 = 23$$
$$d_3 = d_2 + d_1 = 23 + 41 = 64$$
$$d_4 = d_3 + d_2 = 64 + 23 = 87$$

The base case occurs for $n = 0$.

$$L(0) = \sum_{k=0}^{2 \cdot 0 + 1} d_{2k+1} = \sum_{k=0}^{1} d_{2k+1} = d_1 + d_3 = 41 + 64 = 105$$
$$R(0) = d_{4(0+1)} + 18 = d_4 + 18 = 87 + 18 = 105$$

Since $L(0) = R(0)$, the base case is verified.

For the inductive step, assume that:

$$L(n) = R(n)$$

for some $n \geq 0$. We have to prove that:

$$L(n+1) = R(n+1)$$

Indeed:

$L(n+1) =$

$\triangle \qquad = \displaystyle\sum_{k=0}^{2(n+1)+1} d_{2k+1} = \sum_{k=0}^{2n+3} d_{2k+1}$

$\clubsuit \qquad = \displaystyle\sum_{k=0}^{2n+1} d_{2k+1} + d_{2(2n+2)+1} + d_{2(2n+3)+1} = \sum_{k=0}^{2n+1} d_{2k+1} + d_{4n+5} + d_{4n+7}$

$\triangle \qquad = L(n) + d_{4n+5} + d_{4n+7}$

$* \qquad = R(n) + d_{4n+5} + d_{4n+7}$

$\triangledown \qquad = \left(d_{4(n+1)} + 18 \right) + d_{4n+5} + d_{4n+7} = d_{4n+4} + d_{4n+5} + d_{4n+7} + 18$

$\bullet \qquad = d_{4n+6} + d_{4n+7} + 18$

$\bullet \qquad = d_{4n+8} + 18 = d_{4(n+2)} + 18 = d_{4((n+1)+1)} + 18$

$\triangledown \qquad = R(n+1)$

The equalities $\triangle$ follow from the definition of L; the equalities $\triangledown$ follow from the definition of R; the equality $\clubsuit$ follows from the definition of $\sum$; the equality $*$ is obtained by an application of the inductive hypothesis; the equalities $\bullet$ are obtained by applications of the recursive rule of the definition of d to d_{4n+6} and d_{4n+8} , justified by the fact that $4n + 6 \geq 2$; while the other equalities are obtained by arithmetic transformations.

Problem 195 Sequence d is defined recursively as follows.

$$d_0 = 3$$
$$d_1 = 3$$
$$d_n = d_{n-1} + d_{n-2} \quad \text{for } n \geq 2$$

Use mathematical induction to prove that:

$$\sum_{k=0}^{n} \frac{8^k d_k}{13^{k+1}} = 39 - \frac{8^{n+1} d_{n+7}}{13^{n+1}}$$

for every natural number n.

Answer: Let:

$$L(n) = \sum_{k=0}^{n} \frac{8^k d_k}{13^{k+1}}$$

$$R(n) = 39 - \frac{8^{n+1} d_{n+7}}{13^{n+1}}$$

The claim is that $L(n) = R(n)$ for all $n \geq 0$.

The base case occurs for $n = 0$:

$$L(0) \quad = \sum_{k=0}^{0} \frac{8^k d_k}{13^{k+1}} = \frac{8^0 d_0}{13^{0+1}} = \frac{d_0}{13} = \frac{3}{13}$$

$$R(0) \quad = 39 - \frac{8^{0+1} d_{0+7}}{13^{0+1}} = 39 - \frac{8 d_7}{13}$$

To calculate $R(0)$, we employ the recursive rule of the definition of d_n for $2 \leq n \leq 7$ to calculate d_7:

$$d_2 = d_1 + d_0 = 3 + 3 = 6$$
$$d_3 = d_2 + d_1 = 6 + 3 = 9$$
$$d_4 = d_3 + d_2 = 9 + 6 = 15$$
$$d_5 = d_4 + d_3 = 15 + 9 = 24$$
$$d_6 = d_5 + d_4 = 24 + 15 = 39$$
$$d_7 = d_6 + d_5 = 39 + 24 = 63$$

whence:

$$R(0) = 39 - \frac{8 d_7}{13} = 39 - \frac{8 \cdot 63}{13} = \frac{39 \cdot 13 - 8 \cdot 63}{13} = \frac{507 - 504}{13} = \frac{3}{13}$$

Since $L(0) = R(0)$, the base case is verified.

For the inductive step, assume that:

$$L(n) = R(n)$$

for some $n \geq 0$. We have to prove that:

$$L(n + 1) = R(n + 1)$$

Indeed:

$$L(n+1) =$$

$\triangle \qquad \displaystyle = \sum_{k=0}^{n+1} \frac{8^k d_k}{13^{k+1}}$

$\clubsuit \qquad \displaystyle = \left(\sum_{k=0}^{n} \frac{8^k d_k}{13^{k+1}} \right) + \frac{8^{n+1} d_{n+1}}{13^{(n+1)+1}}$

$\triangle \qquad \displaystyle = L(n) + \frac{8^{n+1} d_{n+1}}{13^{(n+1)+1}}$

$* \qquad \displaystyle = R(n) + \frac{8^{n+1} d_{n+1}}{13^{(n+1)+1}}$

$\triangledown \qquad \displaystyle = 39 - \frac{8^{n+1} d_{n+7}}{13^{n+1}} + \frac{8^{n+1} d_{n+1}}{13^{(n+1)+1}}$

$\qquad \displaystyle = 39 - \frac{8^{n+1}}{13^{n+2}} \left(13 d_{n+7} - d_{n+1} \right)$

$\qquad \displaystyle = 39 - \frac{8^{n+1}}{13^{n+2}} \left(8 d_{n+7} + 5 d_{n+7} - d_{n+1} \right)$

$\bullet \qquad \displaystyle = 39 - \frac{8^{n+1}}{13^{n+2}} \left(8 d_{n+7} + 5(d_{n+6} + d_{n+5}) - d_{n+1} \right)$

$\qquad \displaystyle = 39 - \frac{8^{n+1}}{13^{n+2}} \left(8 d_{n+7} + 5 d_{n+6} + 3 d_{n+5} + 2 d_{n+5} - d_{n+1} \right)$

$\bullet \qquad \displaystyle = 39 - \frac{8^{n+1}}{13^{n+2}} \left(8 d_{n+7} + 5 d_{n+6} + 3 d_{n+5} + 2(d_{n+4} + d_{n+3}) - d_{n+1} \right)$

$\qquad \displaystyle = 39 - \frac{8^{n+1}}{13^{n+2}} \left(8 d_{n+7} + 5 d_{n+6} + 3 d_{n+5} + 2 d_{n+4} + d_{n+3} + d_{n+3} - d_{n+1} \right)$

$\bullet \qquad \displaystyle = 39 - \frac{8^{n+1}}{13^{n+2}} \left(8 d_{n+7} + 5 d_{n+6} + 3 d_{n+5} + 2 d_{n+4} + d_{n+3} + (d_{n+2} + d_{n+1}) - d_{n+1} \right)$

$\qquad \displaystyle = 39 - \frac{8^{n+1}}{13^{n+2}} \left(8 d_{n+7} + 5 d_{n+6} + 3 d_{n+5} + 2 d_{n+4} + (d_{n+3} + d_{n+2}) \right)$

$\bullet \qquad \displaystyle = 39 - \frac{8^{n+1}}{13^{n+2}} \left(8 d_{n+7} + 5 d_{n+6} + 3 d_{n+5} + 2 d_{n+4} + d_{n+4} \right)$

$\qquad \displaystyle = 39 - \frac{8^{n+1}}{13^{n+2}} \left(8 d_{n+7} + 5 d_{n+6} + 3(d_{n+5} + d_{n+4}) \right)$

$\bullet \qquad \displaystyle = 39 - \frac{8^{n+1}}{13^{n+2}} \left(8 d_{n+7} + 5 d_{n+6} + 3 d_{n+6} \right)$

$\qquad \displaystyle = 39 - \frac{8^{n+1}}{13^{n+2}} \left(8(d_{n+7} + d_{n+6}) \right)$

$\bullet \qquad \displaystyle = 39 - \frac{8^{n+1}}{13^{n+2}} \left(8 d_{n+8} \right)$

$\qquad \displaystyle = 39 - \frac{8^{n+2}}{13^{n+2}} \cdot d_{n+8} = 39 - \frac{8^{(n+1)+1} d_{(n+1)+7}}{13^{(n+1)+1}}$

$\triangledown \qquad = R(n+1)$

The equalities $\triangle$ follow from the definition of L; the equalities $\triangledown$ follow from the definition of R; the equality $\clubsuit$ follows from the definition of $\sum$; the equality $*$ is obtained by an application of the inductive hypothesis; the equalities $\bullet$ are obtained by application of the recursive rule of the definition of d to $d_{n+\ell}$ for $3 \leq \ell \leq 8$; while the other equalities are obtained by arithmetic transformations.

Chapter 3

Counting

3.1 Elementary Counting

Problem 196 Sets A, B, and C are defined as follows:

$$A = \{1, 2, 3\}$$
$$B = \{a, b, c, d\}$$
$$C = \{P, Q, R\}$$

(a) How many functions are there from B to A?

Answer:

$$3^4 = 81$$

(b) How many injections are there from B to C?

Answer: Zero.

(c) How many injections are there from A to A?

Answer:

$$3! = 6$$

(d) How many length-6 strings are there whose letters belong to the set B?

Answer:

$$4^6 = 4096$$

(e) How many length-9 strings are there whose first 7 letters belong to the set C, while the last 2 letters belong to the set B?

Answer:

$$3^7 \cdot 4^2 = 34{,}992$$

(f) If 30 distinct strings of equal length are required for identifier assignment, what is the smallest possible length of the identifier string, if its elements belong to the set A?

Answer: The smallest possible length is 4, since:

$$3^3 = 27 < 30 < 81 = 3^4$$

(g) If 30 distinct strings of equal length are required for identifier assignment, what is the smallest possible length of the identifier string, if its elements belong to the set B?

Answer: The smallest possible length is 3, since:

$$4^2 = 16 < 30 < 64 = 4^3$$

Problem 197

(a) How many subsets are there in an 8-element set?

Answer:

$$2^8 = 256$$

(b) How many 3-element subsets are there in an 8-element set?

Answer:

$$\binom{8}{3} = 56$$

(c) How many elements should a set have in order to have at least 30 different subsets that have exactly 4 elements?

Answer: The set should have at least 7 elements, since:

$$\binom{6}{4} = 15 < 30 < 35 = \binom{7}{4}$$

(d) How many functions are there from an 8-element set to a 3-element set?

Answer:

$$3^8 = 6561$$

(e) How many injective functions are there from an 8-element set to a 3-element set?

Answer: Zero, because the codomain has fewer elements than the domain.

(f) How many length-5 strings are there that consist of letters $\{a, b, c\}$ and do not begin with c?

Answer:

$$2 \cdot 3^4 = 162$$

(g) How many length-5 strings are there that consist of letters $\{a, b, c\}$ and begin and end with the same letter?

Answer:

$$3 \cdot 3^3 = 81$$

(h) How many length-5 strings are there that consist of letters $\{a, b, c\}$ and begin and end with different letters?

Answer:

$$3 \cdot 2 \cdot 3^3 = 162$$

(i) How many elements are there in a set X, if there are more functions from $\{0, 1\}$ to X than subsets of X?

Answer: The set X has 3 elements, since:

$$2^3 = 8 < 9 = 3^2$$

Problem 198   Sets A and B are defined as follows:

$$A = \{1, 2, 3, 4, 5, 6\}$$
$$B = \{a, b, c, d\}$$

(a) How many elements are there in the set $A \times B \times B$?

Answer:
$$6 \cdot 4 \cdot 4 = 96$$

(b) How many functions are there from A to B?

Answer:
$$4^6 = 4096$$

(c) How many injections are there from A to B?

Answer: Zero.

(d) How many injections are there from B to A?

Answer:
$$6 \cdot 5 \cdot 4 \cdot 3 = 360$$

(e) How many permutations of the set A are there?

Answer:
$$6! = 720$$

(f) How many elements does a set X have, if X has fewer than 1000 permutations?

Answer: X should have no more than 7 elements, since:

$$6! = 720 < 1000 < 5040 = 7!$$

(g) How many subsets are there in the set A?

Answer:
$$2^6 = 64$$

(h) How many subsets are there in the set A that have 3 elements?

Answer:
$$\binom{6}{3} = 20$$

(i) How many strings of lenght 6 are there that consist of the symbols that belong to the set B?

Answer:
$$4^6 = 4096$$

(j) How many strings of lenght 6 are there that consist of the symbols that belong to the set $A \cup B$?

Answer:
$$10^6 = 1,000,000$$

(k) How many strings of lenght 6 are there that consist of 3 symbols that belong to the set A, followed by 3 symbols that belong to the set B?

Answer:
$$6^3 \cdot 4^3 = 13,824$$

(l) If identifiers are constructed as strings of fixed length, which begin with 2 letters that belong to the set B and end with a certain number of digits that belong to the set A, then what is the minimum number of digits that suffices for obtaining 200 distinct identifiers?

Answer: Two digits are necessary and sufficient, since:
$$4^2 \cdot 6^1 = 96 < 200 < 576 = 4^2 \cdot 6^2$$

Problem 199   Sets A and B are defined as follows:
$$A = \{1, 2, 3, 4\}$$
$$B = \{a, b, c, d, e, f, g\}$$

(a) How many elements are there in the set $A \times A \times B$?

Answer:
$$4 \cdot 4 \cdot 7 = 112$$

(b) How many subsets are there in the set A?

Answer:
$$2^4 = 16$$

(c) How many subsets are there in the set B that have exactly 5 elements?

Answer:
$$\binom{7}{5} = 21$$

(d) How many subsets are there in the set B that have no more than 3 elements?

Answer:
$$\binom{7}{0} + \binom{7}{1} + \binom{7}{2} + \binom{7}{3} = 1 + 7 + 21 + 35 = 64$$

(e) How many strings of length 7 are there that consist of the symbols that belong to the set $A \cup B$?

Answer:
$$11^7 = 19,487,171$$

(f) How many strings of length 7 are there that consist of 4 symbols that belong to the set A, followed by 3 symbols that belong to the set B?

Answer:
$$4^4 \cdot 7^3 = 87,808$$

(g) How many strings of length 7 are there that consist of symbols that belong to the set B, if no symbol is allowed to appear more than once?

Answer:
$$7 \cdot 6 \cdot 5 \cdot 4 \cdot 3 \cdot 2 \cdot 1 = 5040$$

Problem 200   Sets A, B, and C are defined as follows:

$$A = \{1, 2, 3\}$$
$$B = \{a, b, c, d\}$$
$$C = \{P, Q, R\}$$

(a) How many functions are there from A to B?

Answer:
$$4^3 = 64$$

(b) How many injections are there from A to B?

Answer:
$$4 \cdot 3 \cdot 2 = 24$$

(c) How many injections are there from $\mathcal{P}(A)$ to $\mathcal{P}(B)$?

Answer:
$$16 \cdot 15 \cdot 14 \cdot 13 \cdot 12 \cdot 11 \cdot 10 \cdot 9 = 518,918,400$$

(d) How many surjections are there from A to B?

Answer: Zero.

(e) How many bijections are there from B to C?

Answer: Zero.

(f) How many bijections are there from A to C?

Answer:
$$3! = 6$$

(g) How many elements are there in a set Y, if A can be mapped to Y in at least 100 different ways?

Answer: The set Y has at least 5 elements, since:

$$4^3 = 64 < 100 < 125 = 5^3$$

(h) How many elements are there in a set X, if A can be mapped injectively into X in at least 100 different ways?

Answer: The set X has at least 6 elements, since:

$$5 \cdot 4 \cdot 3 = 60 < 100 < 120 = 6 \cdot 5 \cdot 4$$

(i) How many elements are there in a set Z, if $\mathcal{P}(A)$ can be mapped injectively into Z?

Answer: The set Z has at least 8 elements, since:

$$2^3 = 8$$

Problem 201   Sets A, B, and C are defined as follows:

$$A = \{1, 2, 3\}$$
$$B = \{2, 3, 4, 5, 6\}$$
$$C = \{3, 4\}$$

(a) How many functions are there from A to B?

   Answer:

$$5^3 = 125$$

(b) How many injections are there from A to B?

   Answer:

$$5 \cdot 4 \cdot 3 = 60$$

(c) How many surjections are there from B to C?

   Answer:
$$2^5 - 2 = 32 - 2 = 30$$

(Observe that there can be only two functions to a 2-element set that are not surjective: one whose entire image is one element and the other whose entire image is the other element, whence the count.)

(d) How many elements does $B \times C$ have?

   Answer:

$$5 \cdot 2 = 10$$

(e) How many subsets does B have?

   Answer:

$$2^5 = 32$$

(f) How many subsets does $B \times B$ have?

   Answer:

$$2^{25} = 33,554,432$$

(g) How many injections are there from B to $A \times C$?

   Answer:

$$6 \cdot 5 \cdot 4 \cdot 3 \cdot 2 = 720$$

(h) How big should n be, if the set C can be mapped injectively into A^n in at least $10,000$ different ways?

   Answer: $n \geq 5$, since:

$$3^4 \cdot (3^4 - 1) = 81 \cdot 80 = 6,480 < 10,000 < 40,000 < 243 \cdot 242 = 3^5 \cdot (3^5 - 1)$$

Problem 202   Sets A and B are defined as follows:

$$A = \{1, 2, 3, 4, 5\}$$
$$B = \{y \mid (\exists x \in A)(y = x \vee y = 2x)\}$$

(All variables assume values from the set of natural numbers $N = \{0, 1, \ldots\}$.)

(a) List all elements of B, or explain why it is impossible.

Answer:

$$1, 2, 3, 4, 5, 6, 8, 10$$

(b) How many subsets does the set A have?

Answer:

$$2^5 = 32$$

(c) How many 3-element subsets does the set A have?

Answer:

$$\binom{5}{3} = 10$$

(d) How many elements does a set X have, if X can be mapped to B in at least 400 different ways?

Answer: The set X should have at least 3 elements, since:

$$8^2 = 64 < 400 < 512 = 8^3$$

(e) How many elements does a set Y have, if Y can be mapped into B injectively in at least 400 different ways?

Answer: The set Y should have at least 4 elements, since:

$$8 \cdot 7 \cdot 6 = 336 < 400 < 1680 = 8 \cdot 7 \cdot 6 \cdot 5$$

Problem 203   Sets A, B, and C are defined as follows:

$$A = \{5, 6\}$$
$$B = \{a, b, c, d, e\}$$
$$C = \{y \mid (\exists x \in N)(\exists z \in A)(2y + x = z)\}$$

(All variables assume values from the set of natural numbers $N = \{0, 1, \ldots\}$.)

(a) List all elements of C, or explain why it is impossible.

Answer:

$$0, 1, 2, 3$$

(b) How many functions are there from B to A?

Answer:

$$2^5 = 32$$

(c) How many injections are there from B to C?

Answer: Zero.

(d) How many injections are there from C to C?

    Answer:

$$4! = 24$$

(e) How many length-5 strings are there whose letters belong to the set B?

    Answer:

$$5^5 = 3125$$

(f) How many elements are there in a set X, if there are more functions from X to C than permutations of X?

    Answer: The set X has no more than 8 elements since:

$$4^8 = 65,536 > 40,320 = 8!$$
$$4^9 = 262,144 < 362,880 = 9!$$

Problem 204   Sets A, B, C are defined as follows:

$$A = \{1, 2, 3\}$$
$$B = \{14, 15, 16, 17, 18\}$$
$$C = \{z \mid (\exists x \in A)(\exists y \in B)(z = y - x)\}$$

(All variables assume values from the set of natural numbers $N = \{0, 1, \ldots\}$.)

(a) List all elements of C, or explain why this is impossible.

    Answer:

$$11, 12, 13, 14, 15, 16, 17$$

(b) How many subsets are there in the set C?

    Answer:

$$2^7 = 128$$

(c) How many subsets that do not contain number 14 are there in the set C?

    Answer:

$$2^6 = 64$$

(d) How many subsets that contain number 14 are there in the set C?

    Answer:

$$2^6 = 64$$

(e) How many subsets that contain exactly 4 elements are there in the set C?

    Answer:

$$\binom{7}{4} = 35$$

(f) How many functions are there from B to C?

Answer:

$$7^5 = 16,807$$

(g) How many injective functions are there from B to C?

Answer:

$$7 \cdot 6 \cdot 5 \cdot 4 \cdot 3 = 2520$$

(h) How many injective functions are there from C to B?

Answer: Zero.

(i) How many surjective functions are there from A to C?

Answer: Zero.

(j) How many injective functions are there from C to C?

Answer:

$$7! = 5040$$

(k) How many strings of length 8 are there that consist of elements of the set A?

Answer:

$$3^8 = 6561$$

(l) How many elements are there in the set $A \times B \times C$?

Answer:

$$3 \cdot 5 \cdot 7 = 105$$

(m) What should be the subset size in order for C to have at least 40 subsets of exactly that size?

Answer: The set C cannot have 40 subsets of any given size. The largest number of equal-sized subsets of C occurs when the size is equal to 3 or to 4, and the number of such subsets is:

$$\binom{7}{4} = \binom{7}{3} = 35$$

(n) What should be the subset size in order for C to have at least 20 subsets of exactly that size?

Answer: The subset size should be at least equal to 2, but not greater than 5, since:

$$\binom{7}{0} = \binom{7}{7} = 1$$
$$\binom{7}{1} = \binom{7}{6} = 7$$
$$\binom{7}{2} = \binom{7}{5} = 21$$
$$\binom{7}{3} = \binom{7}{4} = 35$$

Problem 205   Sets A and B are defined as follows:

$$A = \{1, 2, 3, 4, 5\}$$
$$B = \{x \mid (\exists y \in A)(x = y \lor x = 2^y)\}$$

(All variables assume values from the set of natural numbers $N = \{0, 1, \ldots\}$.)

(a) List all elements of B, or explain why it is impossible.

    Answer:

$$B = \{1, 2, 3, 4, 5, 8, 16, 32\}$$

(b) How many subsets does the set B have?

    Answer:

$$2^8 = 256$$

(c) How many elements should a set X have in order to have at least 1000 subsets?

    Answer: X should have at least 10 elements, since:

$$2^9 = 512 < 1000 < 1024 = 2^{10}$$

(d) How many 3-element subsets does the set A have?

    Answer:

$$\binom{5}{3} = 10$$

(e) How many elements should a set X have in order to have at least 1000 3-element subsets?

    Answer: X should have at least 20 elements, since:

$$\binom{19}{3} = 969 < 1000$$
$$\binom{20}{3} = 1140 > 1000$$

(f) How many strings of length 4 are there, if their symbols belong to the set B?

    Answer:

$$8^4 = 4096$$

(g) How many distinct sets of length-4 strings are there, if the symbols of these strings belong to the set B?

    Answer:

$$2^{4096}$$

(h) How many subsets of the set B contain numbers 4, 8, and 16?

    Answer:

$$2^{8-3} = 32$$

(i) How many subsets of the set B do not contain the set $\{4, 8, 16\}$ as a subset?

    Answer:

$$2^8 - 2^5 = 224$$

Problem 206   Let sets A, B, and C be defined as follows:

$$A = \{a, b, c\}$$
$$B = \{1, 2, 3, 4, 5\}$$
$$C = \left\{ x \mid (\exists y \in B) \left(x = y^2 + 1 \vee x = y^2 - 1 \right) \right\}$$

(All variables assume values from the the set of natural numbers $N = \{0, 1, \ldots\}$.)

(a) How many functions are there from the set A to the set B?

Answer:

$$5^3 = 125$$

(b) How many injections are there from the set A to the set B?

Answer:

$$5 \cdot 4 \cdot 3 = 60$$

(c) How many strings of length 4 are there that consist of symbols from the set B?

Answer:

$$5^4 = 625$$

(d) How many subsets does the set B have?

Answer:

$$2^5 = 32$$

(e) How many 3-element subsets does the set B have?

Answer:

$$\binom{5}{3} = 10$$

(f) How many strings of length 8 are there such that the first 4 symbols belong to the set A, while the last 4 symbols belong to the set B?

Answer:

$$3^4 \cdot 5^4 = 50,625$$

(g) How many strings of length 8 are there such that their symbols belong to the set $A \cup B$?

Answer:

$$8^8 = 16,777,216$$

(h) List all elements of the set C, or explain why it is impossible.

Answer:

$$C = \{0, 2, 3, 5, 8, 10, 15, 17, 24, 26\}$$

(i) How many subsets does the set C have?

Answer:

$$2^{10} = 1024$$

(j) Does the set C have more than 200 equal-sized subsets of any given size?

Answer: Set C contains 252 subsets of size equal to 5, since:

$$\binom{10}{5} = 252 > 200$$

(k) How many functions from the set A to the set C are not injective?

Answer:
$$10^3 - 10 \cdot 9 \cdot 8 = 280$$

Problem 207   A course identification code is made up of four (upper-case) letters followed by three decimal digits. The four letters form the department code, while the three digits form the code of the course subject area. The English alphabet has 26 letters.

(a) How many different four-letter department codes are there?

Answer:
$$26^4 = 456,976$$

(b) How many different three-digit subject codes are there?

Answer:
$$10^3 = 1000$$

(c) How many different three-digit subject codes are there that begin with 1?

Answer:
$$10^2 = 100$$

(d) How many different three-digit subject codes are there that do not begin with 1?

Answer:

$$(10 - 1) \cdot 10^2 = 900$$

(e) How many different four-letter department codes are there, if no letter can be repeated?

Answer:

$$26 \cdot 25 \cdot 24 \cdot 23 = 358,800$$

(f) How many different course codes are there?

Answer:

$$26^4 \cdot 10^3 = 456,976,000$$

Problem 208 The Student Computing Facility is giving user names and passwords to students. The user name is a string of three (upper-case) letters, while the password is a string of four decimal digits. The English alphabet has 26 letters.

(a) How many different three-letter user names are there?

Answer:
$$26^3 = 17,576$$

(b) How many different four-digit passwords are there?

Answer:
$$10^4 = 10,000$$

(c) How many different three-letter user names are there that begin with X?

$$26^2 = 676$$

(d) How many different three-letter user names are there that begin with X or Z?

Answer:

$$2 \cdot 26^2 = 1352$$

(e) How many different three-letter user names are there that do not begin with X?

Answer:

$$(26 - 1) \cdot 26^2 = 16,900$$

(f) How many different four-digit passwords are there, if no digit can be repeated?

Answer:

$$10 \cdot 9 \cdot 8 \cdot 7 = 5040$$

(g) How many different three-letter user names are there that contain exactly two different letters, one of which appears twice?

Answer:
$$26 \cdot 25 \cdot \binom{3}{1} = 1,950$$

(h) How many different three-letter user names are there if the first letter is equal to the last?

Answer:
$$26 \cdot 26 = 676$$

Problem 209   Dangerous Professor has prepared a total of 12 questions for her first midterm exam. Four of these 12 questions are in the subject area of Logic, five are in Induction, and three are in Combinatorics.

(a) How many ways are there to select the first question, if this question may belong to any of the three subject areas?

Answer:

$$4 + 5 + 3 = 12$$

(b) How many ways are there to select the first three questions, if the first question has to be in Logic, the second question has to be in Induction, and the third question has to be in Combinatorics?

Answer:

$$4 \cdot 5 \cdot 3 = 60$$

(c) The Professor has decided to eliminate two questions in the subject area of Induction. How many ways are there to select them?

Answer:

$$\binom{5}{2} = 10$$

(d) How many ways are there to select question number 1 and question number 2, if both of them have to be in the subject area of Logic?

Answer:

$$4 \cdot 3 = 12$$

(e) If all the Logic questions have to appear in a separate examination booklet, how many ways are there to number them from 1 to 4 (without repeating numbers?)

Answer:

$$4! = 24$$

(f) How many ways are there to number all the questions from 1 to 12, (without repeating numbers), if the three questions in the area of Combinatorics have to receive consecutive numbers?

Answer:

$$(4 + 5 + 1)! \cdot 3! = 21,772,800$$

(g) If a total of 10 questions are actually administered, and each is graded as a **pass** or fail, how many different score lists can a problem set have? (A score list is a table containing question numbers with corresponding grades.)

Answer:

$$2^{10} = 1024$$

Problem 210   Dangerous Professor has prepared a total of 11 questions for her second midterm exam. Five of these 11 questions are easy, two are hard, and four are moderately difficult.

(a) How many ways are there to select the first question, if it can be any of the available 11 questions?

Answer:

$$11$$

(b) How many ways are there to select question number 1 and question number 2, if the Professor wishes both of them to be moderately difficult?

Answer:

$$4 \cdot 3 = 12$$

(c) How many ways are there to number the questions from 1 to 11 (without repeating numbers), if all the easy questions precede all the moderately difficult questions, which in turn precede all the hard questions?

Answer:

$$5! \cdot 4! \cdot 2! = 5760$$

(d) How many ways are there to number the questions from 1 to 11 (without repeating numbers), if none of the first three questions may be hard?

Answer:

$$9 \cdot 8 \cdot 7 \cdot 8! = 20,321,280$$

(e) How many ways are there to number the questions from 1 to 11 (without repeating numbers), if none of the last three questions may be easy?

Answer:

$$6 \cdot 5 \cdot 4 \cdot 8! = 4,838,400$$

(f) The Professor is wondering how to order the moderately difficult questions in the problem set. How many arrangements of these four questions are possible?

Answer:

$$4! = 24$$

(g) A colleague of the Professor forgot to set his exam on time, and is asking the Professor to "lend" him three questions. How many ways are there to select these three questions, if one of them has to be easy, one hard, and one moderately difficult?

Answer:

$$5 \cdot 2 \cdot 4 = 40$$

(h) The colleague from the previous part is returning the favor by proofreading the 11 questions, which the Professor has arranged into a problem set, where they appear numbered from 1 to 11. He makes a check-list of question numbers, with a remark next to each number: ok if the questions is correct or not ok if the question requires revision. How many different check-lists can the colleague generate?

Answer:

$$2^{11} = 2048$$

Problem 211   A company is getting ready to introduce 12 new products into the market.

(a) The company is selecting 5 products out of these 12 products for an aggressive advertising campaign. How many distinct sets of 5 products can be formed out of the 12 candidates?

Answer:
$$\binom{12}{5} = 792$$

(b) The janitor is requested to bring 12 glass boxes for the 12 products into the exhibit hall. All boxes are identical, except for their color, which may be red, blue, or green. There are two hundred boxes of each color available. How many distinct sets of 12 boxes can the janitor prepare?

Answer:
$$\binom{12 + 3 - 1}{12} = 91$$

(c) The company is assigning production priorities to the 12 products. There are four such priorities: 1, 2, 3, and 4. How many distinct priority lists can the company come up with? the other list.

Answer:
$$4^{12} = 16,777,216$$

(d) The janitor is now requested to prepare 12 labels for the 12 products, and to put on each label the name of the product. All empty labels are identical, except for their color, which may be blue, yellow, or white. There are two hundred labels of each color available. How many distinct label arrangements can the janitor prepare?

Answer:
$$3^{12} = 531,441$$

(e) The company is issuing a sales rank list of the 12 products, which assigns to each product a unique placement between 1 and 12, depending on the income from the sales of that product. How many distinct sales rank lists can occur?

Answer:
$$12! = 479,001,600$$

(f) The company is giving a total of 12 awards to the 12 designers of the 12 products. There are three $1000-awards, four $500-awards, and five $200-awards. How many distinct award lists can the company come up with?

Answer:
$$\binom{12}{3} \cdot \binom{9}{4} \cdot \binom{5}{5} = 27,720$$

Problem 212   A department Awards Committee is preparing a celebration for 10 distinguished students.

(a) The janitor is requested to bring 10 chairs for the award winners into the hall. All chairs are identical, except for their color, which may be red, blue, or green. There are two hundred chairs of each color available. How many distinct sets of 10 chairs can the janitor prepare?

Answer:

$$\binom{10 + 3 - 1}{10} = 66$$

(b) The janitor is now requested to prepare 10 tables where the 10 award winners will sit, and to put on each table the name of the student who sits there. All tables are identical, except for their color, which may be blue, yellow, or white. There are two hundred tables of each color available. How many distinct table arrangements can the janitor prepare?

Answer:

$$3^{10} = 59,049$$

(c) The committee is selecting 4 students out of the awarded ten for a trip to New York. How many distinct sets of 4 travelers can be formed out of the 10 winners?

Answer:

$$\binom{10}{4} = 210$$

(d) The committee is assigning final grades for the honors projects of the 10 awarded students. There are three possible final grades: A^{++}, A^{+}, and A. How many distinct grade sheets can the committee come up with? (A grade sheet is an alphabetically ordered list of student names, with the grade written next to each name.)

Answer:

$$3^{10} = 59,049$$

(e) How many ways are there to select a sequence of 4 winners out of the 10 to give short presentations about their honors projects?

Answer:

$$10 \cdot 9 \cdot 8 \cdot 7 = 5040$$

(f) The committee is giving a total of 10 gifts to the 10 distinguished students: two \$1000-gifts, three \$500-gifts, and five \$200-gifts. How many distinct gift decisions can the committee come up with?

Answer:

$$\binom{10}{2} \cdot \binom{8}{3} \cdot \binom{5}{5} = 2520$$

(g) Finally, the committee is issuing a rank list of the 10 distinguished students, which assigns to each student a unique number between 1 and 10. How many distinct rank lists can the committee come up with?

Answer:

$$10! = 3,628,800$$

Problem 213   Twelve students of the Dangerous Professor are taking a midterm exam.

(a) How many ways are there to send each of the three proctors to one corner of the exam room?

Answer:
$$4^3 = 64$$

(b) How many ways are there to assign each of the 12 students to one of the three proctors?

Answer:
$$3^{12} = 531,441$$

(c) How many ways are there for each of the 3 proctors to select one of the 12 students for intensive observation, if no student is observed intensively by more than one proctor?

Answer:
$$12 \cdot 11 \cdot 10 = 1320$$

(d) If the class has 12 students, and there are only two grades: pass and fail, how many ways are there for the Professor to pass 8 students and fail 4?

Answer:
$$\binom{12}{8} = \binom{12}{4} = 495$$

(e) If the class has 12 students, and there are only two grades: pass and fail, how many ways are there for the Professor to pass 4 students and fail 8?

Answer:
$$\binom{12}{8} = \binom{12}{4} = 495$$

(f) If the class has 12 students, and there are only two grades: pass and fail, how many ways are there for the Professor to pass at least 10 students?

Answer:
$$\binom{12}{10} + \binom{12}{11} + \binom{12}{12} = 79$$

(g) The class has 12 students, and there are only two grades: pass and fail. How many possible grade sheets are there? (A grade sheet is an alphabetic list of student names with their assigned grades.)

Answer:
$$2^{12} = 4096$$

(h) The class has 12 students, and there are only two grades: pass and fail. We know that student MARY CLEVER passed. How many possible grade sheets are there? (A grade sheet is an alphabetic list of student names with their assigned grades.)

Answer:
$$2^{11} = 2048$$

(i) If the class has 12 students, and all of them are sitting in one line, how many ways are there to arrange them in that line?

Answer:
$$12! = 479,001,600$$

Problem 214   Dangerous Professor has prepared twenty different short questions for the combinatorics part of her test.

(a) If 8 different questions are to be given on the test, how many different collections of questions for the test can the Professor make out of the 20 questions that she has prepared?

Answer:
$$\binom{20}{8} = 125,970$$

(b) The test page contains 8 places for 8 different questions, numbered 1 through 8. How many different test pages can the Professor generate, using the 20 questions that she has?

Answer:
$$20 \cdot 19 \cdot 18 \cdot 17 \cdot 16 \cdot 15 \cdot 14 \cdot 13 = 5,079,110,400$$

(c) The test page contains 8 places for questions, numbered 1 through 8. Once the Professor has decided which 8 questions to give, how many ways are there to write them on the page?

Answer:
$$8! = 40,320$$

(d) Each of the 8 questions on the page is graded by 1 or 0. The grades are recorded in a table printed on the bottom of the page, each grade underneath its question number. How many different grade tables are there?

Answer:
$$2^8 = 256$$

(e) Each of the 8 questions on the page is graded by 1 or 0. How many ways are there to earn exactly 5 points?

Answer:
$$\binom{8}{5} = 56$$

(f) Each of the 8 questions on the page is graded by 1 or 0. How many ways are there to earn at most 3 points?

Answer:
$$\binom{8}{0} + \binom{8}{1} + \binom{8}{2} + \binom{8}{3} = 93$$

(g) Each of the 8 questions on the page is graded by 1 or 0. How many ways are there to earn an odd number of points?

Answer:
$$\binom{8}{1} + \binom{8}{3} + \binom{8}{5} + \binom{8}{7} = 128$$

(h) The Professor has 3 days to grade the 8 questions. The Professor may grade questions in any order, but she always finishes the grading of an individual question on the same day on which she begins grading it. How many ways are there to assign each of the 8 questions to one of the 3 days?

Answer:
$$3^8 = 6561$$

Problem 215   Dangerous Professor has selected 20 different problems, as practice exercises for her 12 students.

(a) If the Professor wishes to split the 20 problems into two (disjoint) groups, each having 10 problems, how many ways are there to do it?

Answer:

$$\binom{20}{10} = 184,756$$

(b) If the Professor wishes to give exactly one problem to each of the 12 students, so that no two students receive the same problem, how many ways are there to do it?

Answer:

$$20 \cdot 19 \cdot 18 \cdot 17 \cdot 16 \cdot 15 \cdot 14 \cdot 13 \cdot 12 \cdot 11 \cdot 10 \cdot 9 = 60,339,831,552,000$$

(c) If the Professor wishes to select 3 problems to be omitted from the problem set, how many ways are there to do it?

Answer:

$$\binom{20}{3} = 1140$$

(d) If an industrious student is asking for some problems to work on (as many or as few as the Professor would give him, as long as he has something to do), how many ways are there to select some of the 20 problems for him?

Answer:

$$2^{20} - 1 = 1,048,575$$

(e) If the first 10 of the 20 available problems are easy and the remaining 10 are hard, how many ways are there to select some of the 20 problems for an industrious student, if the selection has to contain at least one hard problem and at least one easy problem?

Answer:

$$(2^{10} - 1) \cdot (2^{10} - 1) = 1,046,529$$

(f) If the Professor has selected 4 problems for her presentation, how many ways are there to assign numbers 1–4 to the problems (if no number can be repeated?)

Answer:

$$4! = 24$$

(g) If the Professor has (yet) to select 4 problems for her presentation (out of all the 20), how many ways are there to produce a numbered list of 4 problems?

Answer:

$$20 \cdot 19 \cdot 18 \cdot 17 = 116,280$$

(h) If the Professor has 3 lectures left, and she wishes to present all of the 20 problems, how many ways are there to assign the 20 problems to the 3 lectures? (Since the problems are very short, there is no limit on the number of problems she may present in one lecture—it is also all right not to present any problems in some lecture(s). However, all the 20 problems have to appear eventually.)

Answer:

$$3^{20} = 3,486,784,401$$

Problem 216 Dangerous Professor is giving a final exam.

(a) There are 8 questions in the exam problem set. How many ways are there to number them, using numbers 1 through 8, if no number is repeated?

Answer:

$$8! = 40,320$$

(b) The Professor has prepared 10 questions for the exam, but she needs only 8. How many ways are there to select the 8 questions to be inflicted on the students?

Answer:

$$\binom{10}{8} = 45$$

(c) The Professor has decided to give an award (a copy of her workbook with 300 recursion exercises) to three students in her class. How many ways are there to select the three recipients of the award, if her class has 11 students?

Answer:

$$\binom{11}{3} = 165$$

(d) The Professor is giving out three prizes: one grand prize (a velvet file folder with a stylized text of the Induction Axiom), one first prize (a pin holder decorated by the text of the Master Theorem), and one consolation prize (a sheet with the Greek alphabet). How many ways are there to distribute the prizes to some of her 11 students, if nobody can get more than one prize?

Answer:

$$11 \cdot 10 \cdot 9 = 990$$

(e) The Professor had decided to give only 3 grades this semester: A, C, and F. Her class has 11 students. How many different grade sheets can she generate? (A grade sheet is an alphabetically ordered list of student names with grades written next to the names.)

Answer:

$$3^{11} = 177,147$$

(f) The Professor had decided to give only 3 grades this semester: A, C, and F. Also, she has decided that the number of students who are getting an F has to be exactly 5. Her class has 11 students. How many different grade sheets can she generate? (A grade sheet is an alphabetically ordered list of student names with grades written next to the names.)

Answer:

$$\binom{11}{5} \cdot 2^{11-5} = 29,568$$

(g) There are 11 students in the class of Dangerous Professor. Her students have decided that those who pass her course will assemble for a special study session on advanced counting techniques. How many different study groups are possible?

Answer:

$$2^{11} = 2048$$

Problem 217   Student Henry Prudent is attending 7 courses this semester. A report card is a list of courses attended by a student, ordered by course number, with a final grade listed next to each course name. There are 12 possible grades altogether (F through $A+$.)

(a) How many ways are there for Henry's report card to be filled out this semester, if Henry already knows that no two of his grades are equal?

Answer:
$$12 \cdot 11 \cdot 10 \cdot 9 \cdot 8 \cdot 7 \cdot 6 = 3,991,680$$

(b) How many ways are there for Henry's report card to be filled out this semester, if Henry already knows that he is getting an A in exactly two courses: Discrete Structures and Swimming on Dry?

Answer:
$$11^5 = 161,051$$

(c) How many ways are there for Henry's report card to be filled out this semester, if Henry already knows that he is getting an A in at least two courses: Discrete Structures and Swimming on Dry?

Answer:
$$12^5 = 248,832$$

(d) How many ways are there for Henry's report card to be filled out this semester, if Henry already knows that he is getting an A in exactly two courses, but he does not know which courses they are?

Answer:
$$\binom{7}{2} \cdot 11^5 = 3,382,071$$

(e) How many ways are there for Henry's report card to be filled out this semester, if Henry already knows that at most one of his grades is $A+$, but he does not know in which course he may be getting it?

Answer:
$$11^7 + \binom{7}{1} \cdot 11^6 = 31,888,098$$

(f) How many ways are there for Henry's report card to be filled out this semester, if Henry already knows that he is getting an A in exactly one course and a C in exactly one course, but he does not know which courses they are?

Answer:
$$\binom{7}{2} \cdot 2! \cdot 10^5 = 4,200,000$$

(g) How many ways are there for Henry's report card to be filled out this semester, if Henry already knows that he is getting an A in exactly one course, an $A+$ in exactly one course, and a C in exactly one course, but he does not know which courses they are?

Answer:
$$\binom{7}{3} \cdot 3! \cdot 9^4 = 1,377,810$$

Problem 218 There are 20 faculty members in the Department of Computer Science, 11 of which are tenured.

(a) How many ways are there to form the Committee for Personnel and Budget, which consists of 4 faculty members, if the faculty members without tenure cannot serve on this committee?

Answer:
$$\binom{11}{4} = 330$$

(b) How many ways are there to form the Curriculum Committee, which consists of 6 faculty members, if at least 4 members have to be tenured?

Answer:
$$\binom{11}{4} \cdot \binom{9}{2} + \binom{11}{5} \cdot \binom{9}{1} + \binom{11}{6} \cdot \binom{9}{0} = 16,500$$

(c) How many ways are there to elect three faculty members to serve as the Department Chair, the Assistant Chair for Academic Affairs, and the Assistant Chair for Administrative Affairs, if the Chair and the Assistant Chair for Academic Affairs have to be tenured?

Answer:
$$11 \cdot 10 \cdot (11 - 2 + 9) = 1980$$

(d) On a Department outing, how many ways are there to assign the 20 faculty members and the Secretary to three vans, numbered 1, 2, and 3, if each van has exactly 7 seats?

Answer:
$$\binom{21}{7} \cdot \binom{14}{7} \cdot \binom{7}{7} = 399,072,960$$

(e) On a visit to an industrial sponsor, how many ways are there for the corporation representative to assemble a gift package of 20 T-shirts for the Department members, if shirts come in three different colors (red, blue, and white) but are otherwise identical?

Answer:
$$\binom{20 + 3 - 1}{20} = 231$$

(f) On a visit to an industrial sponsor, the Department has received a gift package with 20 T-shirts, 3 of which are red, 4 of which are blue, and 13 of which are white. The shirts are identical, except for the color. How many ways are there to distribute these 20 shirts to the 20 faculty members?

Answer:
$$\binom{20}{3} \cdot \binom{17}{4} \cdot \binom{13}{13} = 2,713,200$$

Problem 219 A major software manufacturer is arranging a job fair with the intention of hiring 6 recent graduates. No candidate can receive more than one offer. In response to the company's invitation, 136 candidates have appeared at the fair.

(a) How many ways are there to extend job offers to 6 of the 136 candidates?

Answer:
$$\binom{136}{6} = 7,858,539,612$$

(b) How many ways are there to extend job offers to 6 of the 136 candidates, if we already know that MARY CLEVER is getting an offer?

Answer:

$$\binom{135}{5} = 346,700,277$$

(c) How many ways are there to extend job offers to 6 of the 136 candidates, if we already know that ANNE EXPENSIVE is not getting an offer?

Answer:

$$\binom{135}{6} = 7,511,839,335$$

(d) How many ways are there to extend job offers to 6 of the 136 candidates, if we already know that GEORGE SMART and JENNY WISE are getting offers?

Answer:

$$\binom{134}{4} = 12,840,751$$

(e) How many ways are there to extend job offers to 6 of the 136 candidates, if we already know that at least one of GEORGE SMART and JENNY WISE gets an offer?

Answer:

$$\binom{134}{4} + 2 \cdot \binom{134}{5} = 680,559,803$$

(f) How many ways are there to distribute the 136 resumes to 3 interviewers?

Answer:

$$3^{136}$$

(g) How many ways are there for 3 interviewers to select 3 resumes (one resume for each interviewer) from the pile of 136 resumes for the first interview round?

Answer:

$$136 \cdot 135 \cdot 134 = 2,460,240$$

Problem 220   A major software manufacturer is arranging a job fair with the intention of hiring 6 recent graduates. The 6 jobs are different, and numbered 1 through 6. No candidate can receive more than one offer. In response to the company's invitation, 136 candidates have appeared at the fair.

(a) How many ways are there to extend the 6 offers to 6 of the 136 candidates?

Answer:
$$136 \cdot 135 \cdot 134 \cdot 133 \cdot 132 \cdot 131 = 5,658,148,520,640$$

(b) How many ways are there to extend the 6 offers to 6 of the 136 candidates, if we already know that MARY CLEVER is getting an offer, but we do not know which?

Answer:
$$6 \cdot 135 \cdot 134 \cdot 133 \cdot 132 \cdot 131 = 249,624,199,440$$

(c) How many ways are there to extend the 6 offers to 6 of the 136 candidates, if we already know that MARY CLEVER is getting an offer for the job number 2?

Answer:

$$135 \cdot 134 \cdot 133 \cdot 132 \cdot 131 = 41,604,033,240$$

(d) How many ways are there to extend the 6 offers to 6 of the 136 candidates, if we already know that ANNE EXPENSIVE is not getting any offers?

Answer:

$$135 \cdot 134 \cdot 133 \cdot 132 \cdot 131 \cdot 130 = 5,408,524,321,200$$

(e) How many ways are there to extend the 6 offers to 6 of the 136 candidates, if we already know that ANNE EXPENSIVE is not getting the offer for the job number 1?

Answer:

$$135 \cdot 135 \cdot 134 \cdot 133 \cdot 132 \cdot 131 = 5,616,544,487,400$$

(f) How many ways are there to extend the 6 offers to 6 of the 136 candidates, if we already know that GEORGE SMART and JENNY WISE are getting offers, but we do not know for which jobs?

Answer:

$$\binom{6}{2} \cdot 2 \cdot 134 \cdot 133 \cdot 132 \cdot 131 = 9,245,340,720$$

(g) How many ways are there to extend the 6 offers to 6 of the 136 candidates, if we already know that at least one of GEORGE SMART and JENNY WISE gets an offer, but we do not know for which job(s)?

Answer:

$$\binom{6}{1} \cdot 2 \cdot 134 \cdot 133 \cdot 132 \cdot 131 \cdot 130 + \binom{6}{2} \cdot 2 \cdot 134 \cdot 133 \cdot 132 \cdot 131 = 12,943,477,008$$

(h) How many ways are there to distribute the 136 resumes to 6 interviewers, each one hiring for one job, if no candidate is considered for more than one job?

Answer:

$$6^{136}$$

Problem 221 A major software manufacturer is arranging a job fair with the intention of hiring 6 recent graduates. The 6 jobs are distinguished by rank. There is only one job of the first rank; there are two jobs of the second rank, and there are three jobs of the third rank. No candidate can receive more than one offer. In response to the company's invitation, 136 candidates have appeared at the fair.

(a) How many ways are there to extend the 6 offers to 6 of the 136 candidates?

Answer:

$$\binom{136}{1} \cdot \binom{135}{2} \cdot \binom{133}{3} = 471,512,376,720$$

(b) How many ways are there to extend the 6 offers to 6 of the 136 candidates, if we already know that MARY CLEVER is getting an offer, but we do not know of which rank?

Answer:

$$\binom{135}{2} \cdot \binom{133}{3} + \binom{135}{1} \cdot \binom{134}{1} \cdot \binom{133}{3} + \binom{135}{1} \cdot \binom{134}{2} \binom{132}{2}$$
$$= 20,802,016,620$$

(c) How many ways are there to extend the 6 offers to 6 of the 136 candidates, if we already know that MARY CLEVER is getting an offer for one of the jobs ranked 2?

Answer:

$$\binom{135}{1} \cdot \binom{134}{1} \cdot \binom{133}{3} = 6,934,005,540$$

(d) How many ways are there to extend the 6 offers to 6 of the 136 candidates, if we already know that ANNE EXPENSIVE is not getting any offers?

Answer:

$$\binom{135}{1} \cdot \binom{134}{2} \cdot \binom{132}{3} = 450,710,360,100$$

(e) How many ways are there to extend the 6 offers to 6 of the 136 candidates, if we already know that ANNE EXPENSIVE is not getting an offer for a job of the second rank?

Answer:

$$\binom{136}{1} \cdot \binom{135}{2} \cdot \binom{133}{3} - \binom{135}{1} \cdot \binom{134}{1} \cdot \binom{133}{3} = 464,578,371,180$$

Problem 222   Consider all strings whose letters belong to the set:

$$A = \{a, b, c, d, e\}$$

(a) How many strings of length 6 are there that begin with a vowel and end with a vowel?

Answer:

$$2 \cdot 5^4 \cdot 2 = 2500$$

(b) How many strings of length 6 are there that contain exactly one a?

Answer:

$$6 \cdot 4^5 = 6144$$

(c) How many strings of length 6 are there that contain at least one a?

Answer:

$$5^6 - 4^6 = 11,529$$

(d) How many strings of length 6 are there that contain at most one a?

Answer:

$$4^6 + 6 \cdot 4^5 = 10,240$$

(e) How many strings of length 6 are there that contain exactly one vowel?

Answer:

$$2 \cdot 6 \cdot 3^5 = 2916$$

(f) How many strings of length 6 are there that contain exactly two consonants?

Answer:

$$\binom{6}{2} \cdot 3^2 \cdot 2^4 = 2160$$

(g) How many strings of length 6 are there that contain two or three consonants?

Answer:

$$\binom{6}{2} \cdot 3^2 \cdot 2^4 + \binom{6}{3} \cdot 3^3 \cdot 2^3 = 6480$$

(h) How many strings of length 6 are there that contain at most one consonant?

Answer:

$$2^6 + 6 \cdot 3 \cdot 2^5 = 640$$

Problem 223   Consider the letters that appear in the words:

DISCRETE STRUCTURES

excluding the space.

(a) How many distinct strings of length 5 can be formed of these letters, if there are no restrictions on the number of occurrences of any letter?

Answer:

$$8^5 = 32,768$$

(b) How many distinct strings of length 5 can be formed of these letters, if each letter can occur at most once?

Answer:

$$8 \cdot 7 \cdot 6 \cdot 5 \cdot 4 = 6720$$

(c) How many distinct strings can be formed of these letters, if each letter must occur exactly once?

Answer:

$$8! = 40,320$$

(d) How many distinct strings can be formed of these letters, if each letter must occur exactly as many times as it occurs in the original string DISCRETE STRUCTURES?

Answer:

$$\frac{18!}{2! \cdot 3! \cdot 3! \cdot 3! \cdot 3! \cdot 2!} = 19,760,412,672,000$$

(e) If the original string DISCRETE STRUCTURES is written using a deck of cards, so that one letter is written on the face of one card, while each card has a unique identifier printed on its back, how many distinct arrangements can be formed of these cards, if each card has to be used?

Answer:

$$18! = 6,402,373,705,728,000$$

Problem 224   Set A has 5 elements.

(a) How many subsets are there in A?

Answer:

$$2^5 = 32$$

(b) How many elements are there in $A \times A$?

Answer:

$$5 \cdot 5 = 25$$

(c) How many binary relations are there on A?

> **Answer:** A binary relation on A is a subset of $A \times A$, whence the count:
>
> $$2^{5^2} = 2^{25} = 33,554,432$$

(d) How many binary relations on A are functions?

> **Answer:**
>
> $$5^5 = 3125$$

(e) How many binary relations on A are injective functions?

> **Answer:**
>
> $$5! = 120$$

(f) How many binary relations on A are surjective functions?

> **Answer:**
>
> $$5! = 120$$

(g) How many elements are there in $A \times A \times A$?

> **Answer:**
>
> $$5 \cdot 5 \cdot 5 = 125$$

(h) How many ternary relations are there on A?

> **Answer:** A ternary relation on A is a subset of $A \times A \times A$, whence the count:
>
> $$2^{5^3} = 2^{125}$$

3.2   Inclusion-Exclusion Theorem

Problem 225   Sets A and B are subsets of a finite set U.

(a) Order the following sets by inclusion:

$$A$$
$$A \cup B$$
$$A \cap B$$
$$U$$
$$\emptyset$$

Answer:

$$\emptyset \subseteq A \cap B \subseteq A \subseteq A \cup B \subseteq U$$

(b) Order the following sets by inclusion:

$$(A \cup C) \setminus (B \setminus C)$$
$$((A \cup C) \setminus B) \cup (A \cap B \cap C)$$
$$(A \setminus B) \setminus C$$
$$A \cup C$$
$$A \setminus (B \setminus C)$$
$$A \setminus B$$

Answer:

$$(A \setminus B) \setminus C \subseteq A \setminus B \subseteq A \setminus (B \setminus C) \subseteq ((A \cup C) \setminus B) \cup (A \cap B \cap C) \subseteq (A \cup C) \setminus (B \setminus C) \subseteq A \cup C$$

Problem 226   At a school, the students study mathematics, physics, chemistry, biology, and astronomy. All students of physics and all students of astronomy are also students of mathematics. All students of biology also study chemistry. 5 students study all five subjects. 3 students study all subjects except physics. 4 students study all subjects except astronomy. 11 students study only mathematics. 12 students study only chemistry. 9 students of astronomy do not study chemistry. 10 students of physics do not study chemistry. There are 2 students that do not study chemistry, but study both astronomy and physics. There are 6 students that do not study biology, but study astronomy, physics, and chemistry. There are 9 students studying both chemistry and astronomy that study neither biology nor physics. There are 10 students studying both chemistry and physics that study neither biology nor astronomy.

(a) How many students study (simultaneously) biology, chemistry, mathematics, and at least one of physics and astronomy?

Answer: 12

(b) How many students study only one subject?

Answer: 23

(c) How many students study at least one of physics and astronomy?

Answer: 54

(d) How many students study at least one of physics and astronomy, but do not study chemistry?

Answer: 17

(e) How many students study both astronomy and physics?

Answer: 13

(f) How many students study astronomy?

Answer: 32

(g) How many students study physics?

Answer: 35

(h) How many students study chemistry and physics?

Answer: 25

(i) How many students study chemistry and astronomy?

Answer: 23

(j) How many students study mathematics but do not study chemistry?

Answer: 28

For an explanation of the answers, see the Venn diagram of the five sets on Figure 3.1.

Problem 227   There are 67 students in the class of Dangerous Professor. 38 of these students have studied for the third midterm. 40 students in the class regularly come to lectures. Of all students, 44 got A^+ on the third midterm. 30 students in the class bought a new book.

Exactly 19 of the students who bought a new book got A^+ on the midterm; only 12 of these 19 have studied for the midterm; and only 9 of these 12 regularly come to lectures.

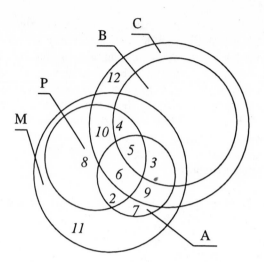

Figure 3.1: The Venn diagram of the five sets

Among the students who studied for the midterm, 25 got A^+ on the midterm; 23 of the students who studied for the midterm also come to lectures regularly, but only 16 of these 23 got A^+ on the midterm. 18 of the students who studied for the midterm bought a new book.

There are 17 students that regularly come to lectures and have bought a new book; only 11 of these 17 got A^+ on the midterm. However, among the 40 students that regularly come to lectures, exactly 26 students got A^+ on the midterm.

How many of the students who bought a new book come to lectures regularly and have studied for the midterm? Explain your answer carefully.

Answer: Define sets A, B, C, D as follows:

  A is the set of students who got A^+ on the third midterm;
  B is the set of students who bought a new book;
  C is the set of students who come to lectures regularly;
  D is the set of students who studied for the third midterm.

The following are sizes of all intersections and the union of these sets:

$$|A \cup B \cup C \cup D| = 67$$
$$|A| = 44, \, |B| = 30, \, |C| = 40, \, |D| = 38$$
$$|AB| = 19, \, |AC| = 26, \, |AD| = 25, \, |BC| = 17, \, |BD| = 18, \, |CD| = 23$$
$$|ABC| = 11, \, |ABD| = 12, \, |ACD| = 16, \, |BCD| = x$$
$$|ABCD| = 9$$

The Inclusion-Exclusion Theorem implies:

$$
\begin{aligned}
|A \cup B \cup C \cup D| \quad = \quad & |A| + |B| + |C| + |D| \\
- \quad & (|AB| + |AC| + |AD| + |BC| + |BD| + |CD|) \\
+ \quad & |ABC| + |ABD| + |ACD| + |BCD| \\
- \quad & |ABCD|
\end{aligned}
$$

which yields:

$$67 = \begin{aligned} & 44 + 30 + 40 + 38 \\ - \ & (19 + 26 + 25 + 17 + 18 + 23) \\ + \ & 11 + 12 + 16 + x \\ - \ & 9 \end{aligned}$$

and:

$$67 = 152 - 128 + 39 + x - 9$$

whence:

$$x = 67 - 54 = 13$$

Recall that $x = |BCD|$, meaning that 13 of the students who bought a new book come to lectures regularly and have studied for the midterm.

Problem 228   A School of Computer Science has 135 professors. Among these professors, 68 are distinguished researchers in the Theory of Computer Science, 82 are career programmers, 88 teach extremely good courses, and 81 give hard exams.

Among the 68 theorists, 35 are also career programmers, while 42 teach extremely good courses. There are exactly 21 theorists who are career programmers and extremely good teachers. Among the 35 theorists who are career programmers, only 18 give hard exams. Only 19 of those theorists that teach extremely good courses are found to give hard exams.

There are 49 career programmers who teach extremely good courses, and only 25 of them give hard exams. Exactly 10 of these 25 are theorists.

Among the professors who give hard exams, 49 are career programmers and 48 teach extremely good courses.

How many theorists give hard exams?

Answer:

Answer: Define sets T, P, C, E as follows:

 T is the set of professors who are distinguished theorists;
 P is the set of professors who are career programmers;
 C is the set of professors who teach extremely good courses;
 E is the set of professors who give hard exams.

The following are sizes of all intersections and the union of these sets:

$$|T \cup P \cup C \cup E| = 135$$
$$|T| = 68, \ |P| = 82, \ |C| = 88, \ |E| = 81$$
$$|TP| = 35, \ |TC| = 42, \ |TE| = x, \ |PC| = 49, \ |PE| = 49, \ |CE| = 48$$
$$|TPC| = 21, \ |TPE| = 18, \ |TCE| = 19, \ |PCE| = 25$$
$$|TPCE| = 10$$

The Inclusion-Exclusion Theorem implies:

$$|T \cup P \cup C \cup E| = \begin{aligned} & |T| + |P| + |C| + |E| \\ - \ & (|TP| + |TC| + |TE| + |PC| + |PE| + |CE|) \\ + \ & |TPC| + |TPE| + |TCE| + |PCE| \\ - \ & |TPCE| \end{aligned}$$

which yields:

$$
\begin{aligned}
135 \;=\;\;& 68 + 82 + 88 + 81 \\
-\;\;& (35 + 42 + x + 49 + 49 + 48) \\
+\;\;& 21 + 18 + 19 + 25 \\
-\;\;& 10
\end{aligned}
$$

and:

$$135 = 319 - (223 + x) + 83 - 10$$

whence:

$$x = 319 - 135 - 223 + 83 - 10 = 34$$

Recall that $x = |TE|$, meaning that there are 34 theorists that give hard exams.

Problem 229   Set A has 6 elements.

(a) How many elements should a set X have, if there are at least 5000 functions from the set A to the set X?

Answer: The set X should have at least 5 elements. To see this, observe that the requirement is that:

$$|X|^{|A|} \geq 5000$$

However:

$$4^{|A|} = 4^6 = 4096 < 5000$$
$$5^{|A|} = 5^6 = 15625 > 5000$$

(b) How many elements should a set Y have, if there are at least 5000 injective functions from the set A to the set Y?

Answer: The set Y should have at least 7 elements. To see this, observe that the requirement is that:

$$(|Y|)(|Y| - 1)\ldots(|Y| - |A| + 1) \geq 5000$$

or, equivalently, given that $|A| = 6$:

$$(|Y|)(|Y| - 1)(|Y| - 2)(|Y| - 3)(|Y| - 4)(|Y| - 5) \geq 5000$$

However:

$$6 \cdot 5 \cdot 4 \cdot 3 \cdot 2 \cdot 1 = 720 < 5000$$
$$7 \cdot 6 \cdot 5 \cdot 4 \cdot 3 \cdot 2 = 5040 > 5000$$

(c) How many elements should a set Z have, if there are at least 5000 surjective functions from the set A to the set Z?

Answer: The set A cannot be mapped surjectively to any set in 5000 different ways. To see this, let $s(a, z)$ be the number of surjections from an a-element set to a z-elements set, and recall that:

$$s(a, z) = z^a - \sum_{k=1}^{z-1}(-1)^{k-1} \binom{z}{k} (z - k)^a$$

whenever $z \leq a$, and $s(a, z) = 0$ if $z > a$. The requirement is that:

$$s(6, z) \geq 5000$$

since $|A| = a = 6$. All nontrivial values of $s(6, z)$ occur for $1 \leq z \leq 6$, and they are as follows.

z	$s(6, z)$
1	1
2	62
3	540
4	1548
5	1790
6	720

Evidently, all of these values are less than 5000.

(d) Find a concise closed form for the following expression:

$$\Psi(n) = \sum_{k=1}^{n-1} (-1)^{k-1} \binom{n}{k} (n-k)^n$$

and prove your answer.

Answer:

$$\Psi(n) = n^n - n!$$

To prove this, first recall that the number of injections from an n-element set to itself is equal to $n!$. Next, observe that the number of surjections from an n-element set to itself is equal to $n^n - \Psi(n)$, which is verified by a substitution $z = a = n$ into the expression stated in the answer to the part (c). Finally, these two counts are equal, since every function between two sets of equal cardinality is injective if and only if it is surjective. Hence:

$$n! = n^n - \Psi(n)$$

whence the claim.

Problem 230   Dangerous Professor is giving an exam.

(a) How many ways are there to assign five proctors to three exam rooms, so that each exam is proctored by at least one person?

Answer:

$$3^5 - \binom{3}{1} \cdot (3-1)^5 + \binom{3}{2} \cdot (3-2)^5 = 150$$

(b) Five lecturers have shown up for the beginning of the exam, which is taking place in three rooms simultaneously. How many ways are there to assign one lecturer to be in charge of the exam in each of the three rooms?

Answer:

$$5 \cdot 4 \cdot 3 = 60$$

(c) In the middle of the exam, five additional proctors have arrived, as reinforcements. How many ways are there to dispatch them all to the three exam rooms?

Answer:

$$3^5 = 243$$

(d) In the middle of the exam, five additional proctors have arrived, as reinforcements. How many ways are there to dispatch them to the three exam rooms, if it is permissible for any of them to stay behind rather than go to a classroom?

Answer:

$$4^5 = 1024$$

(e) In the middle of the exam, five additional proctors have arrived, as reinforcements. How many ways are there to dispatch them to the three exam rooms, if it is permissible for any of them to stay behind rather than go to a classroom, as long as at least one of them goes?

Answer:

$$4^5 - 1 = 1023$$

(f) In the middle of the exam, five additional proctors have arrived, as reinforcements. How many ways are there to dispatch them to the three exam rooms, if it is permissible for one of them to stay behind rather than go to a classroom?

Answer:

$$3^5 + 5 \cdot 3^4 = 648$$

Problem 231   Dangerous Professor is grading an exam. The Professor has 4 days to grade 5 questions. The Professor may grade questions in any order, but she always finishes the grading of an individual question on the same day on which she begins grading it.

(a) How many ways are there to assign each of the 5 questions to one of the 4 days?

Answer:

$$4^5 = 1024$$

(b) How many ways are there to assign each of the 5 questions to one of the 4 days, if she has decided to do some grading on each of the 4 days?

Answer:

$$4^5 - \binom{4}{1} \cdot (4-1)^5 + \binom{4}{2} \cdot (4-2)^5 - \binom{4}{3} \cdot (4-3)^5 = 240$$

(c) Assume that the Professor always finishes one question before proceeding to the next. How many ways are there to schedule the grading of 5 questions over 4 days, if the Professor is sensitive to the order in which individual questions are graded on the same day?

Answer:

$$5! \cdot \binom{5+4-1}{5} = 6720$$

Problem 232   Dangerous Professor has selected 6 different problems, as practice exercises for her 4 students.

(a) If the Professor wishes to give at least one problem to each student to work on, how many ways are there to distribute all of the 6 problems to the 4 students, if no two students are to receive the same problem?

Answer:

$$4^6 - \binom{4}{1} \cdot (4-1)^6 + \binom{4}{2} \cdot (4-2)^6 - \binom{4}{3} \cdot (4-3)^6 = 1560$$

(b) How many ways are there to distribute all of the 6 problems to the 4 students, if no two students are to receive the same problem, but it is permissible for a student not to work on any problems?

Answer:

$$4^6 = 4096$$

(c) How many ways are there to distribute the 6 problems to the 4 students, if no two students are to receive the same problem, it is permissible for a student not to work on any problems, and it it permissible for any number of problems to remain unassigned?

Answer:

$$5^6 = 15,625$$

(d) How many ways are there to distribute the 6 problems to the 4 students, if it is permissible for more than one student to work on the same problem, it is permissible for a student not to work on any problems, and it it permissible for any number of problems to remain unassigned?

Answer:

$$\left(2^6\right)^4 = 2^{24} = 16,777,216$$

Problem 233 Dangerous Professor is giving a final exam. Her class has 7 students. A grade sheet is an alphabetically ordered list of student names with a grade written next to each name.

(a) The Professor had decided to give only 3 grades this semester: A, C, and F. How many different grade sheets can she generate?

Answer:

$$3^7 = 2187$$

(b) The Professor had decided to give only 3 grades this semester: A, C, and F. Also, she has decided that every grade has to be assigned to somebody. How many different grade sheets can she generate?

Answer:

$$3^7 - \binom{3}{1} \cdot (3-1)^7 + \binom{3}{2} \cdot (3-2)^7 = 1806$$

Problem 234 A company is hiring 4 recent graduates. No candidate can receive more than one offer. 7 candidates are considered.

(a) How many ways are there to distribute the 7 resumes to 4 interviewers, if no interviewer is to remain idle and only one copy of any resume is available?

Answer:

$$4^7 - \binom{4}{1} \cdot (4-1)^7 + \binom{4}{2}(4-2)^7 - \binom{4}{3}(4-3)^7 = 8400$$

(b) If each interviewer is entitled to a copy of each candidate's resume, how many ways are there to distribute copies of the 7 resumes to 4 interviewers, if no interviewer is to remain idle, but the company has no obligation to interview any individual candidate?

Answer:

$$(2^7 - 1)^4 = 260,144,641$$

Problem 235   There are 20 faculty members in the Department of Computer Science.

(a) During a site visit from a funding agency, how many ways are there to assign the 20 faculty members to 3 (simultaneous) poster sessions, if at least one member has to be present in each session, and every member has to attend some session?

Answer:

$$3^{20} - \binom{3}{1} \cdot (3-1)^{20} + \binom{3}{2} \cdot (3-2)^{20} = 3,483,638,676$$

(b) During a site visit from a funding agency, how many ways are there to find among the 20 faculty members one member for each of the three simultaneous sessions, to serve as the host?

Answer:

$$20 \cdot 19 \cdot 18 = 6840$$

Problem 236   The School Principal is planning a reception in honor of 9 professors of Computer Science, who just got tenured.

(a) The Principal wishes to honor each professor with a monetary prize. The customary prize levels are 10, 12, 14, 16, and 18 dollars. How many ways are there to for the Principal to award each of the nine professors, at some level?

Answer:

$$5^9 = 1,953,125$$

(b) To entertain the audience, the Principal has bought five different toys, for the professors to play with. How many ways are there to give away the five toys to the nine professors, if toys are not shared, and nobody can play with more than one toy?

Answer:

$$9 \cdot 8 \cdot 7 \cdot 6 \cdot 5 = 15,120$$

(c) To entertain the Principal, the professors have decided to acquire five identical personal fireworks to activate during the party. How many ways are there for the nine professors to decide which five of them will activate a firework session?

Answer:

$$\binom{9}{5} = 126$$

(d) The five fireworks of the previous part turned out to come in five different colors, and the professors have preferences. How many ways are there for the nine professors to distribute among themselves the five different fireworks, if no professor can get more than one?

Answer:

$$9 \cdot 8 \cdot 7 \cdot 6 \cdot 5 = 15,120$$

(e) To add to the humorous aspect of the event, the professors have decided to wear colored hats. The local department store reports large supplies of hats in five different colors. How many ways are there for the nine professors to select one hat each?

Answer:

$$5^9 = 1,953,125$$

(f) To emphasize team spirit, the professors have decided to appear in identical T-shirts, and they ordered nine from the local store. However, they forgot to specify the color, and the store attendant decided that any shirt may appear in any one from the five shirt colors (available in unlimited supplies.) The professors are wondering how many different T-shirt packages may arrive.

Answer:
$$\binom{9+5-1}{9} = 715$$

(g) The nine shirts ordered in the previous part turned out to be of three different colors only: four blue shirts, three yellow shirts, and two green shirts. How many ways are there for the nine professors to dress?

Answer:
$$\binom{9}{4} \cdot \binom{9-4}{3} \cdot \binom{9-(4+3)}{2} = 1260$$

(h) Right before the beginning of the party, two of the nine shirts received in the previous part were discovered to have large moth holes, disqualifying them for the evening—two of the nine professors will have to appear in their original clothes. If one blue shirt and one green shirt are damaged, how many ways are there for the nine professors to dress?

Answer:
$$\binom{9}{3} \cdot \binom{9-3}{3} \cdot \binom{9-(3+3)}{1} = 5040$$

(i) The Principal, in principle, recognizes five principal ways for a professor of Computer Science to contribute to the School: teaching, scientific research, engineering expertise, administrative service, and book writing. The Principal wishes to single out the most prominent of the five aspects in which each professor has contributed to the School, and to give special thanks to each of them for that aspect of their contribution. How many possible ways are there for the Principal to thank each of the nine professors for one of the five possible points of merit?

Answer:
$$5^9 = 1,953,125$$

(j) How many possible ways are there for the Principal to thank each of the nine professors for one of the five possible aspects of merit (as is described in the previous part), if the Principal desires that each of the five aspects has at least one representative among the nine professors?

Answer:
$$5^9 - \binom{5}{1}(5-1)^9 + \binom{5}{2}(5-2)^9 - \binom{5}{3}(5-3)^9 + \binom{5}{4}(5-4)^9 = 834,120$$

(k) The five toys, which the Principal purchased in the part (b), turned out to be all identical, once the shipment arrived. Moreover, they are rather dull, and the Principal wishes only that the professors take them home, each taking as many or as few as desired, as long as all the five are taken. How many ways are there for the nine professors to collect the five identical toys?

Answer:
$$\binom{5+9-1}{5} = 1287$$

(l) The professors agree to collect the toys from the previous part, but none of them will take home more than two toys. How many ways are there for the nine professors to collect the five identical toys in this case?

Answer:

$$\binom{9}{5} + \binom{9}{3} \cdot \binom{9-3}{1} + \binom{9}{1} \cdot \binom{9-1}{2} = 882$$

(m) The principal has brought many flowers of five different types, and is decorating each professor with two different flowers. How many ways are there for the Principal to choose two different flowers for each of the nine professors, from the five available flower types?

Answer:

$$\binom{5}{2}^9 = 1,000,000,000$$

(n) The school photographer has made many copies of five different snapshots featuring the nine professors. At the end of the event, the professors are invited to take as many copies of each picture as they wish. The professors decline duplicates, but politely agree not to leave empty-handed. How many ways are there for the nine professors to take different copies of some of the five pictures, if each professor takes at least one copy?

Answer:

$$\left(2^5 - 1\right)^9 = 26,439,622,160,671$$

(o) Five graduate students need thesis supervisors and committee members. The Principal is asking the nine professors to serve. Each professor may serve on any one of the five thesis committees, and in fact may also decline to do so (at this time), as long as each of the five students acquires at least one committee member. How many ways are there to accomplish this?

Answer: First, calculate s_j, for $5 \le j \le 9$, as follows.

$$s_9 = 5^9 - \binom{5}{1}(5-1)^9 + \binom{5}{2}(5-2)^9 - \binom{5}{3}(5-3)^9 + \binom{5}{4}(5-4)^9 = 834,120$$

$$s_8 = 5^8 - \binom{5}{1}(5-1)^8 + \binom{5}{2}(5-2)^8 - \binom{5}{3}(5-3)^8 + \binom{5}{4}(5-4)^8 = 126,000$$

$$s_7 = 5^7 - \binom{5}{1}(5-1)^7 + \binom{5}{2}(5-2)^7 - \binom{5}{3}(5-3)^7 + \binom{5}{4}(5-4)^7 = 16,800$$

$$s_6 = 5^6 - \binom{5}{1}(5-1)^6 + \binom{5}{2}(5-2)^6 - \binom{5}{3}(5-3)^6 + \binom{5}{4}(5-4)^6 = 1800$$

$$s_5 = 5^5 - \binom{5}{1}(5-1)^5 + \binom{5}{2}(5-2)^5 - \binom{5}{3}(5-3)^5 + \binom{5}{4}(5-4)^5 = 120$$

(Note that $s_5 = 5!$, as follows from the argument given in the part (d) of Problem 229.)

Finally, the required count is equal to:

$$\binom{9}{9} \cdot s_9 + \binom{9}{8} \cdot s_8 + \binom{9}{7} \cdot s_7 + \binom{9}{6} \cdot s_6 + \binom{9}{5} \cdot s_5 = 2,739,240$$

3.3 Linear Recurrences

Problem 237 (a) Find a closed form for the sequence a_n, defined by the following recurrence relation, and prove your answer.

$$a_0 = 2$$
$$a_1 = -8$$
$$a_n = -8a_{n-1} - 15a_{n-2} \quad \text{for } n \geq 2$$

Answer: This is a linear homogeneous recurrence relation of second order with constant coefficients, whose characteristic equation is:

$$x^2 + 8x + 15 = 0$$

or, equivalently:

$$(x + 5)(x + 3) = 0$$

which yields two distinct roots $x_1 = -5$, $x_2 = -3$. The solutions are of the form:

$$a_n = \alpha x_1^n + \beta x_2^n = \alpha \cdot (-5)^n + \beta \cdot (-3)^n$$

where α and β are calculated from the initial conditions:

$$a_0 = 2 = \alpha + \beta$$
$$a_1 = -8 = -5\alpha - 3\beta$$

Hence: $\alpha = 1$, $\beta = 1$. The required closed form is:

$$a_n = (-5)^n + (-3)^n$$

(b) Find a closed form for the sequence b_n, defined by the following recurrence relation, and prove your answer.

$$b_0 = 2$$
$$b_1 = -16$$
$$b_n = -8b_{n-1} - 16b_{n-2} \quad \text{for } n \geq 2$$

Answer: This is a linear homogeneous recurrence relation of second order with constant coefficients, whose characteristic equation is:

$$x^2 + 8x + 16 = 0$$

or, equivalently:

$$(x + 4)(x + 4) = 0$$

which yields a single root: $x_0 = -4$. The solutions are of the form:

$$b_n = \alpha x_0^n + \beta n x_0^n = \alpha \cdot (-4)^n + \beta \cdot n \cdot (-4)^n = (-4)^n(\alpha + \beta n)$$

where α and β are calculated from the initial conditions:

$$b_0 = \alpha = 2$$
$$b_1 = -4(\alpha + \beta) = -16$$

Hence: $\alpha = 2$, $\beta = 2$. The required closed form is:

$$b_n = 2(1 + n)(-4)^n$$

Problem 238   Sequence a_n is defined recursively as follows.

$$a_0 = -\frac{43}{4}$$
$$a_1 = -\frac{21}{2}$$
$$a_n = 3a_{n-1} - 2a_{n-2} \quad \text{for } n \geq 2$$

(a) Find a closed form (non-recursive definition) of the sequence a_n and prove your answer.

Answer: This is a linear homogeneous recurrence relation of second order with constant coefficients, whose characteristic equation is:

$$x^2 - 3x + 2 = 0$$

or, equivalently:

$$(x - 2)(x - 1) = 0$$

which yields two distinct roots $x_1 = 2$, $x_2 = 1$. The solutions are of the form:

$$a_n = \alpha x_1^n + \beta x_2^n = \alpha \cdot 2^n + \beta \cdot 1^n = \alpha \cdot 2^n + \beta$$

where α and β are calculated from the initial conditions:

$$a_0 = \alpha + \beta = -\frac{43}{4}$$
$$a_1 = 2\alpha + \beta = -\frac{21}{2}$$

Hence:

$$\alpha = -\frac{21}{2} + \frac{43}{4} = \frac{1}{4}$$
$$\beta = -\frac{43}{4} - \frac{1}{4} = -11$$

These values of α and β correspond to the closed form:

$$a_n = \frac{1}{4} \cdot 2^n - 11 = 2^{-2} \cdot 2^n - 11$$

or, equivalently:

$$a_n = 2^{n-2} - 11$$

(b) Calculate a_{12} exactly and prove your answer.

Answer: By substitution into the closed form obtained in the answer given in part **(a)**:

$$a_{12} = 2^{12-2} - 11 = 2^{10} - 11 = 1024 - 11 = 1013$$

Problem 239   Sequence b_n is defined recursively as follows.

$$b_0 = \frac{1}{4}$$
$$b_1 = \frac{26}{48}$$
$$b_n = 4b_{n-1} - 4b_{n-2} \quad \text{for } n \geq 2$$

(a) Find a closed form (non-recursive definition) of the sequence b_n and prove your answer.

Answer: This is a linear homogeneous recurrence relation of second order with constant coefficients, whose characteristic equation is:

$$x^2 - 4x + 4 = 0$$

or, equivalently:

$$(x - 2)^2 = 0$$

which yields a single root $x_0 = 2$. The solutions are of the form:

$$b_n = \alpha x_0^n + \beta n x_0^n = \alpha \cdot 2^n + \beta \cdot n \cdot 2^n = (\alpha + \beta n) \cdot 2^n$$

where α and β are calculated from the initial conditions:

$$b_0 = \alpha = \frac{1}{4}$$

$$b_1 = 2(\alpha + \beta) = \frac{26}{48}$$

Hence:

$$\alpha = \frac{1}{4}$$

$$\beta = \frac{13}{48} - \frac{1}{4} = \frac{1}{48}$$

These values of α and β correspond to the closed form:

$$b_n = \left(\frac{1}{4} + \frac{n}{48}\right) \cdot 2^n = \left(1 + \frac{n}{12}\right) \cdot \left(\frac{1}{4}\right) \cdot 2^n$$

or, equivalently:

$$b_n = \left(1 + \frac{n}{12}\right) \cdot 2^{n-2}$$

(b) Calculate b_{12} exactly and prove your answer.

Answer: By substitution into the closed form obtained in the answer given in part **(a)**:

$$b_{12} = \left(1 + \frac{12}{12}\right) \cdot 2^{12-2} = 2 \cdot 2^{10} = 2^{11} = 2048$$

Problem 240   Sequences a_n, b_n, c_n are defined by the same recursive rule and they assume identical values when $n = 0$. However, their values for $n = 1$ differ slightly—by no more than 1:

$a_0 = 1$	$b_0 = 1$	$c_0 = 1$
$a_1 = 3$	$b_1 = 4$	$c_1 = 3 + \frac{1}{4}$
$a_n = 7a_{n-1} - 12a_{n-2}$	$b_n = 7b_{n-1} - 12b_{n-2}$	$c_n = 7c_{n-1} - 12c_{n-2}$ for $n \geq 2$

(a) Find a closed form for the sequence a_n and prove your answer.

Answer: This is a linear homogeneous recurrence relation of second order with constant coefficients, whose characteristic equation is:

$$x^2 - 7x + 12 = 0$$

or, equivalently:

$$(x - 3)(x - 4) = 0$$

which yields two distinct roots $x_1 = 3$, $x_2 = 4$. The solutions are of the form:

$$a_n = \alpha x_1^n + \beta x_2^n = \alpha \cdot 3^n + \beta \cdot 4^n$$

where α and β are calculated from the initial conditions:

$$a_0 = \alpha + \beta = 1$$
$$a_1 = 3\alpha + 4\beta = 3$$

Hence: $\alpha = 1$, $\beta = 0$. These values of α and β correspond to the closed form:

$$a_n = 3^n$$

(b)   Find a closed form for the sequence b_n and prove your answer.

Answer: We conclude from the analysis given in the answer to part **(a)**:

$$b_n = \alpha \cdot 3^n + \beta \cdot 4^n$$

where α and β are calculated from the initial conditions:

$$b_0 = \alpha + \beta = 1$$
$$b_1 = 3\alpha + 4\beta = 4$$

Hence: $\alpha = 0$, $\beta = 1$. These values of α and β correspond to the closed form:

$$b_n = 4^n$$

(c)   Find a closed form for the sequence c_n and prove your answer.

Answer: We conclude from the analysis given in the answer to part **(a)**:

$$c_n = \alpha \cdot 3^n + \beta \cdot 4^n$$

where α and β are calculated from the initial conditions:

$$c_0 = \alpha + \beta = 1$$
$$c_1 = 3\alpha + 4\beta = \frac{13}{4}$$

Hence: $\alpha = \dfrac{3}{4}$, $\beta = \dfrac{1}{4}$. These values of α and β correspond to the closed form:

$$c_n = \frac{3}{4} \cdot 3^n + \frac{1}{4} \cdot 4^n$$

or, equivalently:

$$c_n = \frac{3^{n+1}}{4} + 4^{n-1}$$

(d)   Calculate and list the difference between c_n and a_n for $n = 0, 1, 2, 5, 10$.

Answer: We conclude from the answers given in parts **(a)** and **(c)**:

$$c_n - a_n = \left(\frac{3^{n+1}}{4} + 4^{n-1} \right) - 3^n = 4^{n-1} - \frac{3^n}{4}$$

whence the listing:

n	$c_n - a_n$
0	0
1	0.25
2	1.75
5	195.25
10	247381.75

Problem 241   Sequences a_n, b_n, c_n are defined by the same recursive rule and they assume identical values when $n = 1$. However, their values for $n = 0$ differ slightly—by no more than 0.25:

$$a_0 = \frac{1}{2} \qquad b_0 = \frac{1}{4} \qquad c_0 = \frac{3}{8}$$
$$a_1 = 1 \qquad b_1 = 1 \qquad c_1 = 1$$
$$a_n = 6a_{n-1} - 8a_{n-2} \quad b_n = 6b_{n-1} - 8b_{n-2} \quad c_n = 6c_{n-1} - 8c_{n-2} \quad \text{for } n \geq 2$$

(a) Find a closed form for the sequence a_n and prove your answer.

Answer: This is a linear homogeneous recurrence relation of second order with constant coefficients, whose characteristic equation is:

$$x^2 - 6x + 8 = 0$$

or, equivalently:

$$(x - 2)(x - 4) = 0$$

which yields two distinct roots $x_1 = 2$, $x_2 = 4$. The solutions are of the form:

$$a_n = \alpha x_1^n + \beta x_2^n = \alpha \cdot 2^n + \beta \cdot 4^n$$

where α and β are calculated from the initial conditions:

$$a_0 = \alpha + \beta = \frac{1}{2}$$
$$a_1 = 2\alpha + 4\beta = 1$$

Hence: $\alpha = \frac{1}{2}$, $\beta = 0$. These values of α and β correspond to the closed form:

$$a_n = 2^{n-1}$$

(b) Find a closed form for the sequence b_n and prove your answer.

Answer: We conclude from the analysis given in the answer to part **(a)**:

$$b_n = \alpha \cdot 2^n + \beta \cdot 4^n$$

where α and β are calculated from the initial conditions:

$$b_0 = \alpha + \beta = \frac{1}{4}$$
$$b_1 = 2\alpha + 4\beta = 1$$

Hence: $\alpha = 0$, $\beta = \frac{1}{4}$. These values of α and β correspond to the closed form:

$$b_n = 4^{n-1}$$

(c) Find a closed form for the sequence c_n and prove your answer.

Answer: We conclude from the analysis given in the answer to part **(a)**:

$$c_n = \alpha \cdot 2^n + \beta \cdot 4^n$$

where α and β are calculated from the initial conditions:

$$c_0 = \alpha + \beta = \frac{3}{8}$$
$$c_1 = 2\alpha + 4\beta = 1$$

Hence: $\alpha = \frac{1}{4}$, $\beta = \frac{1}{8}$. These values of α and β correspond to the closed form:

$$c_n = \frac{1}{4} \cdot 2^n + \frac{1}{8} \cdot 4^n$$

or, equivalently:

$$c_n = 2^{n-2} + 2 \cdot 4^{n-2}$$

(d) Express the quotient $\dfrac{b_n}{a_n}$ as a function of n.

Answer: We conclude from the answers given in parts **(a)** and **(b)**:

$$\frac{b_n}{a_n} = \frac{4^{n-1}}{2^{n-1}} = \left(\frac{4}{2}\right)^{n-1} = 2^{n-1} = a_n$$

Problem 242   Sequences a_n, b_n are defined by the same recursive rule and different initial conditions:

$$a_0 = 1 \qquad\qquad b_0 = 0$$
$$a_1 = 1 \qquad\qquad b_1 = 4$$
$$a_n = 2a_{n-1} + a_{n-2} \quad b_n = 2b_{n-1} + b_{n-2} \quad \text{for } n \geq 2$$

(a) Find a closed form for the sequence a_n and prove your answer.

Answer: This is a linear homogeneous recurrence relation of second order with constant coefficients, whose characteristic equation is:

$$x^2 - 2x - 1 = 0$$

The roots of the characteristic equation are calculated as follows.

$$x_{1,2} = \frac{2 \pm \sqrt{(-2)^2 - 4(-1)}}{2} = \frac{2 \pm \sqrt{8}}{2} = \frac{2 \pm \sqrt{4} \cdot \sqrt{2}}{2} = 1 \pm \sqrt{2}$$

The solutions are of the form:

$$a_n = \alpha x_1^n + \beta x_2^n = \alpha \left(1 + \sqrt{2}\right)^n + \beta \left(1 - \sqrt{2}\right)^n$$

where α and β are calculated from the initial conditions:

$$a_0 = \alpha + \beta = 1$$
$$a_1 = \alpha \left(1 + \sqrt{2}\right) + \beta \left(1 - \sqrt{2}\right) = \alpha + \beta + \sqrt{2}(\alpha - \beta) = 1$$

yielding:

$$\alpha + \beta = 1$$
$$\alpha - \beta = 0$$

Hence: $\alpha = \beta = \dfrac{1}{2}$. These values of α and β correspond to the closed form:

$$a_n = \frac{1}{2}\left(\left(1 + \sqrt{2}\right)^n + \left(1 - \sqrt{2}\right)^n\right)$$

(b) Find a closed form for the sequence b_n and prove your answer.

Answer: We conclude from the analysis given in the answer to part **(a)**:

$$b_n = \alpha\left(1 + \sqrt{2}\right)^n + \beta\left(1 - \sqrt{2}\right)^n$$

where α and β are calculated from the initial conditions:

$$b_0 = \alpha + \beta = 0$$
$$b_1 = \alpha\left(1 + \sqrt{2}\right) + \beta\left(1 - \sqrt{2}\right) = \alpha + \beta + \sqrt{2}(\alpha - \beta) = 4$$

yielding:

$$\alpha + \beta = 0$$
$$\alpha - \beta = \frac{4}{\sqrt{2}} = 2 \cdot \sqrt{2}$$

whence: $\alpha = \sqrt{2}$, $\beta = -\sqrt{2}$. These values of α and β correspond to the closed form:

$$b_n = \sqrt{2}\left(\left(1 + \sqrt{2}\right)^n - \left(1 - \sqrt{2}\right)^n\right)$$

Problem 243 Sequences a_n, b_n are defined by the same recursive rule and different initial conditions:

$$a_0 = \frac{3}{5} \qquad\qquad b_0 = 0$$
$$a_1 = 3 \qquad\qquad b_1 = 2$$
$$a_n = 10a_{n-1} - 18a_{n-2} \quad b_n = 10b_{n-1} - 18b_{n-2} \quad \text{for } n \geq 2$$

(a) Find a closed form for the sequence a_n and prove your answer.

Answer: This is a linear homogeneous recurrence relation of second order with constant coefficients, whose characteristic equation is:

$$x^2 - 10x + 18 = 0$$

The roots of the characteristic equation are calculated as follows.

$$x_{1,2} = \frac{10 \pm \sqrt{(-10)^2 - 4 \cdot 18}}{2} = \frac{10 \pm \sqrt{28}}{2} = \frac{10 \pm \sqrt{4} \cdot \sqrt{7}}{2} = 5 \pm \sqrt{7}$$

The solutions are of the form:

$$a_n = \alpha x_1^n + \beta x_2^n = \alpha\left(5 + \sqrt{7}\right)^n + \beta\left(5 - \sqrt{7}\right)^n$$

where α and β are calculated from the initial conditions:

$$a_0 = \alpha + \beta = \frac{3}{5}$$
$$a_1 = \alpha\left(5 + \sqrt{7}\right) + \beta\left(5 - \sqrt{7}\right) = 5(\alpha + \beta) + \sqrt{7}(\alpha - \beta) = 3$$

yielding:

$$\alpha + \beta = \frac{3}{5}$$
$$\alpha - \beta = 0$$

Hence: $\alpha = \beta = \frac{3}{10}$. These values of α and β correspond to the closed form:

$$a_n = \frac{3}{10}\left(\left(5 + \sqrt{7}\right)^n + \left(5 - \sqrt{7}\right)^n\right)$$

(b) Find a closed form for the sequence b_n and prove your answer.

Answer: We conclude from the analysis given in the answer to part **(a)**:

$$b_n = \alpha\left(5 + \sqrt{7}\right)^n + \beta\left(5 - \sqrt{7}\right)^n$$

where α and β are calculated from the initial conditions:

$$b_0 = \alpha + \beta = 0$$
$$b_1 = \alpha\left(5 + \sqrt{7}\right) + \beta\left(5 - \sqrt{7}\right) = 5(\alpha + \beta) + \sqrt{7}(\alpha - \beta) = 2$$

yielding:

$$\alpha + \beta = 0$$
$$\alpha - \beta = \frac{2}{\sqrt{7}}$$

whence: $\alpha = \frac{1}{\sqrt{7}}$, $\beta = -\frac{1}{\sqrt{7}}$. These values of α and β correspond to the closed form:

$$b_n = \frac{1}{\sqrt{7}}\left(\left(5 + \sqrt{7}\right)^n - \left(5 - \sqrt{7}\right)^n\right)$$

Problem 244   Sequences a_n, b_n are defined by the same recursive rule and different initial conditions:

$$a_0 = 2 \qquad\qquad b_0 = 2$$
$$a_1 = 26 \qquad\qquad b_1 = 24$$
$$a_n = 4a_{n-1} + 7a_{n-2} \quad b_n = 4b_{n-1} + 7b_{n-2} \quad \text{for } n \geq 2$$

(a) Find a closed form for the sequence a_n and prove your answer.

Answer: This is a linear homogeneous recurrence relation of second order with constant coefficients, whose characteristic equation is:

$$x^2 - 4x - 7 = 0$$

The roots of the characteristic equation are calculated as follows.

$$x_{1,2} = \frac{4 \pm \sqrt{(-4)^2 - 4(-7)}}{2} = \frac{4 \pm \sqrt{44}}{2} = \frac{4 \pm \sqrt{4} \cdot \sqrt{11}}{2} = 2 \pm \sqrt{11}$$

The solutions are of the form:

$$a_n = \alpha x_1^n + \beta x_2^n = \alpha \left(2 + \sqrt{11}\right)^n + \beta \left(2 - \sqrt{11}\right)^n$$

where α and β are calculated from the initial conditions:

$$a_0 = \alpha + \beta = 2$$
$$a_1 = \alpha \left(2 + \sqrt{11}\right) + \beta \left(2 - \sqrt{11}\right) = 2(\alpha + \beta) + \sqrt{11}(\alpha - \beta) = 26$$

yielding:

$$\alpha + \beta = 2$$
$$\alpha - \beta = \frac{22}{\sqrt{11}} = \frac{2 \cdot 11}{\sqrt{11}} = 2 \cdot \sqrt{11}$$

Hence: $\alpha = 1 + \sqrt{11}$, $\beta = 1 - \sqrt{11}$. These values of α and β correspond to the closed form:

$$a_n = \left(1 + \sqrt{11}\right) \left(2 + \sqrt{11}\right)^n + \left(1 - \sqrt{11}\right) \left(2 - \sqrt{11}\right)^n$$

(b) Find a closed form for the sequence b_n and prove your answer.

Answer: We conclude from the analysis given in the answer to part **(a)**:

$$b_n = \alpha \left(2 + \sqrt{11}\right)^n + \beta \left(2 - \sqrt{11}\right)^n$$

where α and β are calculated from the initial conditions:

$$b_0 = \alpha + \beta = 2$$
$$b_1 = \alpha \left(2 + \sqrt{11}\right) + \beta \left(2 - \sqrt{11}\right) = 2(\alpha + \beta) + \sqrt{11}(\alpha - \beta) = 24$$

yielding:

$$\alpha + \beta = 2$$
$$\alpha - \beta = \frac{20}{\sqrt{11}}$$

whence: $\alpha = 1 + \dfrac{10}{\sqrt{11}}$, $\beta = 1 - \dfrac{10}{\sqrt{11}}$. These values of α and β correspond to the closed form:

$$b_n = \left(1 + \frac{10}{\sqrt{11}}\right) \left(2 + \sqrt{11}\right)^n + \left(1 - \frac{10}{\sqrt{11}}\right) \left(2 - \sqrt{11}\right)^n$$

Problem 245   Sequence a_n is defined by the following recursive rule:

$$a_n = 9a_{n-1} - 18a_{n-2} \quad \text{for } n \geq 2$$

In each of the following cases, select the initial conditions so that the sequence a_n has the stated property, and prove your answer. If this is impossible, prove it.

(a) $a_n = 3^{n-2}$ for $n \geq 0$

    Answer: Sequence a_n is defined by a linear homogeneous recurrence relation of second order with constant coefficients, whose characteristic equation is:

$$x^2 - 9x + 18 = 0$$

or, equivalently:

$$(x - 3)(x - 6) = 0$$

which yields two distinct roots $x_1 = 3$, $x_2 = 6$. The solutions are of the form:

$$a_n = \alpha x_1^n + \beta x_2^n = \alpha \cdot 3^n + \beta \cdot 6^n$$

where α and β are selected so as to honor the required closed form for a_n:

$$\alpha \cdot 3^n + \beta \cdot 6^n = 3^{n-2}$$

Hence: $\alpha = \dfrac{1}{3^2} = \dfrac{1}{9}$ and $\beta = 0$.

The initial conditions are obtained by evaluating the closed form for $n = 0, 1$:

$$a_0 = 3^{0-2} = \frac{1}{9}$$
$$a_1 = 3^{1-2} = \frac{1}{3}$$

(b) $a_n = 6^{n+1}$ for $n \geq 0$

Answer: We conclude from the analysis given in the answer to part **(a)**: $a_n = \alpha \cdot 3^n + \beta \cdot 6^n$, where α and β are selected so as to honor the required closed form for a_n:

$$\alpha \cdot 3^n + \beta \cdot 6^n = 6^{n+1}$$

Hence: $\alpha = 0$ and $\beta = 6$.

The initial conditions are obtained by evaluating the closed form for $n = 0, 1$:

$$a_0 = 6^{0+1} = 6$$
$$a_1 = 6^{1+1} = 36$$

(c) $a_n = 4^n$ for $n \geq 0$

Answer: This is impossible. To prove this, assume the opposite, and calculate the necessary initial conditions from the required closed form of a_n:

$$a_0 = 4^0 = 1$$
$$a_1 = 4^1 = 4$$

The recursive rule of the definition of a_n yields:

$$a_2 = 9a_{2-1} - 18a_{2-2} = 9a_1 - 18a_0 = 9 \cdot 4 - 18 \cdot 1 = 18$$

However, the required closed form mandates:

$$a_2 = 4^2 = 16$$

Finally, $18 \neq 16$, whence the contradiction.

(d) $a_n = 4n$ for $n \geq 0$

> **Answer:** This is impossible. To prove this, assume the opposite, and calculate the necessary initial conditions from the required closed form of a_n:
>
> $$a_0 = 4 \cdot 0 = 0$$
> $$a_1 = 4 \cdot 1 = 4$$
>
> The recursive rule of the definition of a_n yields:
>
> $$a_2 = 9a_1 - 18a_0 = 9 \cdot 4 - 18 \cdot 0 = 36$$
>
> However, the required closed form mandates:
>
> $$a_2 = 4 \cdot 2 = 8$$
>
> Finally, $36 \neq 8$, whence the contradiction.

(e) $a_n = 0$ for $n \geq 0$

> **Answer:** We conclude from the analysis given in the answer to part **(a)**: $a_n = \alpha \cdot 3^n + \beta \cdot 6^n$, where α and β are selected so as to honor the required closed form for a_n:
>
> $$\alpha \cdot 3^n + \beta \cdot 6^n = 0$$
>
> Hence: $\alpha = 0$ and $\beta = 0$.
>
> The initial conditions are obtained by evaluating the closed form for $n = 0, 1$:
>
> $$a_0 = 0$$
> $$a_1 = 0$$

(f) $a_n = t$ for $n \geq 0$, where $t > 0$ is a real constant of your choice.

> **Answer:** This is impossible. To prove this, assume the opposite, and calculate the necessary initial conditions from the required closed form of a_n:
>
> $$a_0 = t$$
> $$a_1 = t$$
>
> The recursive rule of the definition of a_n yields:
>
> $$a_2 = 9a_1 - 18a_0 = 9t - 18t = -9t$$
>
> However, the required closed form mandates:
>
> $$a_2 = t$$
>
> Finally, $-9t \neq t$ (since $t > 0$), whence the contradiction.

(g) $a_n = 2 \cdot 3^{n+1} - 7 \cdot 6^n$ for $n \geq 0$

> **Answer:** We conclude from the analysis given in the answer to part **(a)**: $a_n = \alpha \cdot 3^n + \beta \cdot 6^n$, where α and β are selected so as to honor the required closed form for a_n:
>
> $$\alpha \cdot 3^n + \beta \cdot 6^n = 2 \cdot 3^{n+1} - 7 \cdot 6^n$$
>
> Hence: $\alpha = 2 \cdot 3$ and $\beta = -7$.
>
> The initial conditions are obtained by evaluating the closed form for $n = 0, 1$:
>
> $$a_0 = 2 \cdot 3^{0+1} - 7 \cdot 6^0 = 6 - 7 = -1$$
> $$a_1 = 2 \cdot 3^{1+1} - 7 \cdot 6^1 = 18 - 42 = -24$$

(h) $a_n^2 = 5 \cdot 9^n$ for $n \geq 0$

   Answer: We conclude from the analysis given in the answer to part **(a)**: $a_n = \alpha \cdot 3^n + \beta \cdot 6^n$, where α and β are selected so as to honor the required closed form for a_n:

$$(\alpha \cdot 3^n + \beta \cdot 6^n)^2 = 5 \cdot 9^n$$

   or, equivalently:

$$\alpha^2 \cdot 9^n + \beta^2 \cdot 36^n + 2\alpha\beta \cdot 18^n = 5 \cdot 9^n$$

   Hence: $\alpha^2 = 5$ and $\beta = 0$, or $\alpha = \sqrt{5}$ and $\beta = 0$. These values of α and β correspond to the closed form:

$$a_n = \sqrt{5} \cdot 3^n$$

   The initial conditions are obtained by evaluating the closed form for $n = 0, 1$:

$$a_0 = \sqrt{5} \cdot 3^0 = \sqrt{5}$$
$$a_1 = \sqrt{5} \cdot 3^1 = 3\sqrt{5}$$

Problem 246   Sequence a_n is defined by the following recursive rule:

$$a_n = 10a_{n-1} - 25a_{n-2}   \text{ for } n \geq 2$$

In each of the following cases, select the initial conditions so that the sequence a_n has the stated property, and prove your answer. If this is impossible, prove it.

(a) $a_n = 4 \cdot 5^n$ for $n \geq 0$

   Answer: Sequence a_n is defined by a linear homogeneous recurrence relation of second order with constant coefficients, whose characteristic equation is:

$$x^2 - 10x + 25 = 0$$

   or, equivalently:

$$(x - 5)^2 = 0$$

   which has a single root $x_0 = 5$. The solutions are of the form:

$$a_n = \alpha x_0^n + \beta n \cdot x_0^n = \alpha \cdot 5^n + \beta n \cdot 5^n$$

   where α and β are selected so as to honor the required closed form for a_n:

$$\alpha \cdot 5^n + \beta n \cdot 5^n = 4 \cdot 5^n$$

   Hence: $\alpha = 4$ and $\beta = 0$.

   The initial conditions are obtained by evaluating the closed form for $n = 0, 1$:

$$a_0 = 4 \cdot 5^0 = 4$$
$$a_1 = 4 \cdot 5^1 = 20$$

(b) $a_n = (7 - 11n) \cdot 5^{n-3}$ for $n \geq 0$

Answer: We conclude from the analysis given in the answer to part **(a)**: $a_n = \alpha \cdot 5^n + \beta n \cdot 5^n$, where α and β are selected so as to honor the required closed form for a_n:

$$\alpha \cdot 5^n + \beta n \cdot 5^n = 7 \cdot 5^{n-3} - 11n \cdot 5^{n-3}$$

Hence: $\alpha = \dfrac{7}{5^3}$ and $\beta = -\dfrac{11}{5^3}$.

The initial conditions are obtained by evaluating the closed form for $n = 0, 1$:

$$a_0 = (7 - 11 \cdot 0) \cdot 5^{0-3} = \frac{7}{125}$$

$$a_1 = (7 - 11 \cdot 1) \cdot 5^{1-3} = -\frac{4}{25}$$

(c) $a_n = (-3) \cdot 5^{n+1}$ for $n \geq 0$

Answer: We conclude from the analysis given in the answer to part **(a)**: $a_n = \alpha \cdot 5^n + \beta n \cdot 5^n$, where α and β are selected so as to honor the required closed form for a_n:

$$\alpha \cdot 5^n + \beta n \cdot 5^n = (-3) \cdot 5^{n+1}$$

Hence: $\alpha = -15$ and $\beta = 0$.

The initial conditions are obtained by evaluating the closed form for $n = 0, 1$:

$$a_0 = (-3) \cdot 5^{0+1} = -15$$

$$a_1 = (-3) \cdot 5^{1+1} = -75$$

(d) $a_n = (-3n) \cdot 5^{n+1}$ for $n \geq 0$

Answer: We conclude from the analysis given in the answer to part **(a)**: $a_n = \alpha \cdot 5^n + \beta n \cdot 5^n$, where α and β are selected so as to honor the required closed form for a_n:

$$\alpha \cdot 5^n + \beta n \cdot 5^n = (-3n) \cdot 5^{n+1}$$

Hence: $\alpha = 0$ and $\beta = -15$.

The initial conditions are obtained by evaluating the closed form for $n = 0, 1$:

$$a_0 = (-3 \cdot 0) \cdot 5^{0+1} = 0$$

$$a_1 = (-3 \cdot 1) \cdot 5^{1+1} = -75$$

(e) $a_n = (-3n^2) \cdot 5^{n+1}$ for $n \geq 0$

Answer: This is impossible. To prove this, assume the opposite, and calculate the necessary initial conditions from the required closed form of a_n:

$$a_0 = (-3 \cdot 0^2) \cdot 5^{0+1} = 0$$

$$a_1 = (-3 \cdot 1^2) \cdot 5^{1+1} = -75$$

The recursive rule of the definition of a_n yields:

$$a_2 = 10a_{2-1} - 25a_{2-2} = 10a_1 - 25a_0 = 10 \cdot (-75) - 25 \cdot 0 = -750$$

However, the required closed form mandates:

$$a_2 = (-3 \cdot 2^2) \cdot 5^{2+1} = -1500$$

Finally, $-1500 \neq -750$, whence the contradiction.

(f) $a_n = t$ for $n \geq 0$, where $t > 0$ is a real constant of your choice.

> **Answer:** This is impossible. To prove this, assume the opposite, and calculate the necessary initial conditions from the required closed form of a_n:
>
> $$a_0 = t$$
> $$a_1 = t$$
>
> The recursive rule of the definition of a_n yields:
>
> $$a_2 = 10a_1 - 25a_0 = 10t - 25t = -15t$$
>
> However, the required closed form mandates:
>
> $$a_2 = t$$
>
> Finally, $-15t \neq t$ (since $t > 0$), whence the contradiction.

(g) $a_n = tn$ for $n \geq 0$, where $t > 0$ is a real constant of your choice.

> **Answer:** This is impossible. To prove this, assume the opposite, and calculate the necessary initial conditions from the required closed form of a_n:
>
> $$a_0 = t \cdot 0 = 0$$
> $$a_1 = t \cdot 1 = t$$
>
> The recursive rule of the definition of a_n yields:
>
> $$a_2 = 10a_1 - 25a_0 = 10t - 25 \cdot 0 = 10t$$
>
> However, the required closed form mandates:
>
> $$a_2 = 2t$$
>
> Finally, $10t \neq 2t$ (since $t > 0$), whence the contradiction.

(h) $a_n = tn^2$ for $n \geq 0$, where $t > 0$ is a real constant of your choice.

> **Answer:** This is impossible. To prove this, assume the opposite, and calculate the necessary initial conditions from the required closed form of a_n:
>
> $$a_0 = t \cdot 0^2 = 0$$
> $$a_1 = t \cdot 1^2 = t$$
>
> The recursive rule of the definition of a_n yields:
>
> $$a_2 = 10a_1 - 25a_0 = 10t - 25 \cdot 0 = 10t$$
>
> However, the required closed form mandates:
>
> $$a_2 = t \cdot 2^2 = 4t$$
>
> Finally, $10t \neq 4t$ (since $t > 0$), whence the contradiction.

Problem 247   Sequence a_n is defined by the following recursive rule:

$$a_n = 2a_{n-1} - a_{n-2} \quad \text{for } n \geq 2$$

In each of the following cases, select the initial conditions so that the sequence a_n has the stated property, and prove your answer. If this is impossible, prove it.

(a) $a_n = 2^n$ for $n \geq 0$

>   **Answer:** To prove that this is impossible, observe that sequence a_n is defined by a linear homogeneous recurrence relation of second order with constant coefficients, whose characteristic equation is:
>
> $$x^2 - 2x + 1 = 0$$
>
>   or, equivalently:
>
> $$(x - 1)^2 = 0$$
>
>   which has a single root $x_0 = 1$. The solutions are of the form:
>
> $$a_n = \alpha x_0^n + \beta n x_0^n = \alpha \cdot 1^n + \beta n \cdot 1^n = \alpha + \beta n$$
>
>   where α and β depend on the initial conditions. If it was possible for for the sequence a_n to assume the required closed form, the initial conditions would be:
>
> $$a_0 = 2^0 = 1$$
> $$a_1 = 2^1 = 2$$
>
>   The recursive rule of the definition of a_n yields:
>
> $$a_2 = 2a_1 - a_0 = 2 \cdot 2 - 1 = 3$$
>
>   However, the required closed form mandates:
>
> $$a_2 = 2^2 = 4$$
>
>   Finally, $4 \neq 3$, whence the contradiction.

(b) $a_n = 7 - 11n$ for $n \geq 0$

>   **Answer:** We conclude from the analysis given in the answer to part **(a)**: $a_n = \alpha + \beta n$, where α and β are selected so as to honor the required closed form for a_n:
>
> $$\alpha + \beta n = 7 - 11n$$

Hence: $\alpha = 7$ and $\beta = -11$.

The initial conditions are obtained by evaluating the closed form for $n = 0, 1$:

$$a_0 = 7 - 11 \cdot 0 = 7$$
$$a_1 = 7 - 11 \cdot 1 = -4$$

(c) $a_n = -3$ for $n \geq 0$

>   **Answer:** We conclude from the analysis given in the answer to part **(a)**: $a_n = \alpha + \beta n$, where α and β are selected so as to honor the required closed form for a_n:
>
> $$\alpha + \beta n = -3$$

Hence: $\alpha = -3$ and $\beta = 0$.

The initial conditions are obtained by evaluating the closed form for $n = 0, 1$:

$$a_0 = -3$$
$$a_1 = -3$$

(d) $a_n = -3n$ for $n \geq 0$

> **Answer:** We conclude from the analysis given in the answer to part **(a)**: $a_n = \alpha + \beta n$, where α and β are selected so as to honor the required closed form for a_n:
>
> $$\alpha + \beta n = -3n$$
>
> Hence: $\alpha = 0$ and $\beta = -3$.
>
> The initial conditions are obtained by evaluating the closed form for $n = 0, 1$:
>
> $$a_0 = -3 \cdot 0 = 0$$
> $$a_1 = -3 \cdot 1 = -3$$

(e) $a_n = -3n^2$ for $n \geq 0$

> **Answer:** This is impossible. To prove this, assume the opposite, and calculate the necessary initial conditions from the required closed form of a_n:
>
> $$a_0 = -3 \cdot 0^2 = 0$$
> $$a_1 = -3 \cdot 1^2 = -3$$
>
> The recursive rule of the definition of a_n yields:
>
> $$a_2 = 2a_1 - a_0 = 2 \cdot (-3) - 0 = -6$$
>
> However, the required closed form mandates:
>
> $$a_2 = -3 \cdot 2^2 = -12$$
>
> Finally, $-12 \neq -6$, whence the contradiction.

(f) $a_n = 5$ for $n \geq 0$

> **Answer:** We conclude from the analysis given in the answer to part **(a)**: $a_n = \alpha + \beta n$, where α and β are selected so as to honor the required closed form for a_n:
>
> $$\alpha + \beta n = 5$$
>
> Hence: $\alpha = 5$ and $\beta = 0$.
>
> The initial conditions are obtained by evaluating the closed form for $n = 0, 1$:
>
> $$a_0 = 5$$
> $$a_1 = 5$$

Problem 248   Sequences a_n, b_n, c_n, d_n are defined by their closed forms. For each of these sequences, construct a recursive definition which employs only the four elementary arithmetic operators, and prove that your construction is correct.

(a) $a_n = \dfrac{2^{n+2}}{5} + \dfrac{7^n}{5}$

Answer: Sequence a_n satisfies the closed form of a solution to a linear homogeneous recurrence relation of second order with constant coefficients:

$$a_n = \alpha \cdot 2^n + \beta \cdot 7^n$$

where $\alpha = \dfrac{2^2}{5}$ and $\beta = \dfrac{1}{5}$. The characteristic equation of this recurrence has two distinct roots: $x_1 = 2$ and $x_2 = 7$, which corresponds to the characteristic equation:

$$(x - 2)(x - 7) = 0$$

or, equivalently:

$$x^2 - 9x + 14 = 0$$

The coefficients of the characteristic equation are equal to the coefficients of the recurrence. The initial conditions are obtained by evaluating the closed form for $n = 0, 1$:

$$a_0 = \frac{2^{0+2}}{5} + \frac{7^0}{5} = \frac{4}{5} + \frac{1}{5} = 1$$
$$a_1 = \frac{2^{1+2}}{5} + \frac{7^1}{5} = \frac{8}{5} + \frac{7}{5} = 3$$

whence the complete recursive definition of the sequence a_n:

$$a_0 = 1$$
$$a_1 = 3$$
$$a_n = 9a_{n-1} - 14a_{n-2}, \quad \text{for } n \geq 2$$

(b) $b_n = 8^n - 5n \cdot 8^{n-1}$

Answer: Sequence b_n satisfies the closed form of a solution to a linear homogeneous recurrence relation of second order with constant coefficients:

$$b_n = \alpha \cdot 8^n + \beta n \cdot 8^n$$

where $\alpha = 1$ and $\beta = -\dfrac{5}{8}$. The characteristic equation of this recurrence has a single root: $x_0 = 8$, which corresponds to the characteristic equation:

$$(x - 8)^2 = 0$$

or, equivalently:

$$x^2 - 16x + 64 = 0$$

The coefficients of the characteristic equation are equal to the coefficients of the recurrence. The initial conditions are obtained by evaluating the closed form for $n = 0, 1$:

$$b_0 = 8^0 - 5 \cdot 0 \cdot 8^{0-1} = 1$$
$$b_1 = 8^1 - 5 \cdot 1 \cdot 8^{1-1} = 3$$

whence the complete recursive definition of the sequence b_n:

$$b_0 = 1$$
$$b_1 = 3$$
$$b_n = 16b_{n-1} - 64b_{n-2}, \quad \text{for } n \geq 2$$

(c) $c_n = (-1)^n + 4n \cdot (-1)^{n+1}$

> **Answer:** Sequence c_n satisfies the closed form of a solution to a linear homogeneous recurrence relation of second order with constant coefficients:
>
> $$c_n = \alpha \cdot (-1)^n + \beta n \cdot (-1)^n$$
>
> where $\alpha = 1$ and $\beta = -4$. The characteristic equation of this recurrence has a single root: $x_0 = -1$, which corresponds to the characteristic equation:
>
> $$(x+1)^2 = 0$$
>
> or, equivalently:
>
> $$x^2 + 2x + 1 = 0$$
>
> The coefficients of the characteristic equation are equal to the coefficients of the recurrence. The initial conditions are obtained by evaluating the closed form for $n = 0, 1$:
>
> $$c_0 = (-1)^0 + 4 \cdot 0 \cdot (-1)^{0+1} = 1$$
> $$c_1 = (-1)^1 + 4 \cdot 1 \cdot (-1)^{1+1} = 3$$
>
> whence the complete recursive definition of the sequence c_n:
>
> $$c_0 = 1$$
> $$c_1 = 3$$
> $$c_n = -2c_{n-1} - c_{n-2}, \quad \text{for } n \geq 2$$

(d) $d_n = \frac{1}{2}\left(1 + \frac{1}{\sqrt{3}}\right)\left(2 + \sqrt{3}\right)^n + \frac{1}{2}\left(1 - \frac{1}{\sqrt{3}}\right)\left(2 - \sqrt{3}\right)^n$

> **Answer:** Sequence d_n satisfies the closed form of a solution to a linear homogeneous recurrence relation of second order with constant coefficients:
>
> $$d_n = \alpha \left(2 + \sqrt{3}\right)^n + \beta \left(2 - \sqrt{3}\right)^n$$
>
> where $\alpha = \frac{1}{2}\left(1 + \frac{1}{\sqrt{3}}\right)$ and $\beta = \frac{1}{2}\left(1 - \frac{1}{\sqrt{3}}\right)$. The characteristic equation of this recurrence has two distinct roots: $x_1 = 2 + \sqrt{3}$ and $x_2 = 2 - \sqrt{3}$, which corresponds to the characteristic equation:
>
> $$\left(x - 2 - \sqrt{3}\right)\left(x - 2 + \sqrt{3}\right) = 0$$
>
> or, equivalently:
>
> $$x^2 - 4x + 1 = 0$$
>
> The coefficients of the characteristic equation are equal to the coefficients of the recurrence. The initial conditions are obtained by evaluating the closed form for $n = 0, 1$:
>
> $$d_0 = \frac{1}{2}\left(1 + \frac{1}{\sqrt{3}}\right)\left(2 + \sqrt{3}\right)^0 + \frac{1}{2}\left(1 - \frac{1}{\sqrt{3}}\right)\left(2 - \sqrt{3}\right)^0 = \frac{1}{2}\left(1 + \frac{1}{\sqrt{3}} + 1 - \frac{1}{\sqrt{3}}\right) = 1$$
>
> $$d_1 = \frac{1}{2}\left(1 + \frac{1}{\sqrt{3}}\right)\left(2 + \sqrt{3}\right)^1 + \frac{1}{2}\left(1 - \frac{1}{\sqrt{3}}\right)\left(2 - \sqrt{3}\right)^1$$
>
> $$= \frac{1}{2}\left(2 + 1 + \sqrt{3} + \frac{2}{\sqrt{3}}\right) + \frac{1}{2}\left(2 + 1 - \sqrt{3} - \frac{2}{\sqrt{3}}\right) = \frac{3}{2} + \frac{3}{2} = 3$$

whence the complete recursive definition of the sequence d_n:

$$a_0 = 1$$
$$a_1 = 3$$
$$a_n = 4a_{n-1} - a_{n-2}, \quad \text{for } n \geq 2$$

Note that all of the four sequences a_n, b_n, c_n, d_n have identical initial conditions (but different recursive rules.)

Problem 249 Dangerous Professor gives a test to her students at the beginning of every lecture. She selects a certain number of questions from her Green Booklet for every test, and often repeats questions already administered in previous tests. However, she keeps changing constants and other minor details, so that no two copies of the same question are exactly identical.

The Professor composes her tests in the following way. The first test consists of 5 questions, selected from the Green Booklet. The second test consists of 13 questions, selected from the Green Booklet. The test number n, for $n \geq 3$, consists of all the questions that appeared in the test number $n - 1$, together with two copies of each question that appeared in the test number $n - 2$.

(a) Let $q(n)$ be the number of questions that appear in the test number n. Write the recurrence relation for $q(n)$.

Answer:

$$q_1 = 5$$
$$q_2 = 13$$
$$q_n = q_{n-1} + 2q_{n-2}, \text{ for } n \geq 3$$

(b) Find the closed form (non-recursive definition) of the function $q(n)$. Explain your answer.

Answer: This is a linear homogeneous recurrence relation of second order with constant coefficients. The characteristic equation is:

$$r^2 - r - 2 = 0$$

the roots of which are:

$$r_1 = -1 \text{ and } r_2 = 2$$

The solutions are of the form:

$$q_n = \alpha \cdot (-1)^n + \beta \cdot 2^n$$

The initial conditions mandate:

$$q_1 = \alpha \cdot (-1)^1 + \beta \cdot 2^1 = -\alpha + 2\beta = 5$$
$$q_2 = \alpha \cdot (-1)^2 + \beta \cdot 2^2 = \alpha + 4\beta = 13$$

whence:

$$\beta = 3 \text{ and } \alpha = 1$$

The closed form is:

$$q_n = (-1)^n + 3 \cdot 2^n$$

(c) Calculate the number of questions that appear in the test number 9.

Answer: By substitution into the closed form obtained in the answer to part **(b)**:

$$q_9 = (-1)^9 + 3 \cdot 2^9 = -1 + 3 \cdot 512 = 1535$$

Problem 250 A vertical wall has to be coated (in its entirety, from the bottom to the top) with a certain number of horizontal rows of decorative bricks. The bricks come in three colors: blue, white, and red. All bricks in any individual row are of the same color. The height of each red brick is two inches; the height of each blue brick is three inches; and the height of each white brick is four inches.

(a) Write a recurrence relation with the initial conditions for the number of ways the wall of height equal to n inches can be overlaid with bricks of these three types.

Answer: Assume that the bricks are laid out row by row, starting from the bottom of the wall. Let $f(n)$ be the number of ways the wall can be coated if there are n inches left to the top of the wall. Observe that n is the difference between the height of the wall and the height of the top row of bricks (if any). The job is finished when n becomes equal to zero.

For a sufficiently large n, the workers are free to choose any one from the three available brick heights for the next row. Depending on the brick height, the next row of bricks reduces n by two, three, or four inches. If the height of the next row of bricks is k inches, then the workers are left with $f(n-k)$ ways to coat the remaining $n-k$ inches of the wall, whence the recurrence:

$$f(n) = f(n-2) + f(n-3) + f(n-4) \ \text{for} \ n \geq 4$$

For the initial conditions, it is sufficient to observe that:

$$f(0) = 1$$
$$f(-n) = 0, \ \text{for all} \ n \geq 1$$

To verify this, recall that the value of $f(-n)$ would correspond to the impossible situation of laying bricks above the wall (that is, coating a wall of negative height), which indeed has no way of happening. Straightforwardly, the condition in which $n = 0$ indicates that the entire wall has been overlaid, leaving the workers with a single option to do nothing. If these observations are employed in successive applications of the recursive rule for $f(n)$ to the initial positive values of n, precisely such that $0 \leq n \leq 3$, a set of actual initial conditions is obtained, where the arguments correspond to plausible (non-negative) values of the height of the wall.

$$
\begin{aligned}
f(0) =&\ 1 \\
f(1) =&\ f(1-2) + f(1-3) + f(1-4) = f(-1) + f(-2) + f(-3) = 0+0+0 = 0 \\
f(2) =&\ f(2-2) + f(2-3) + f(2-4) = f(0) + f(-1) + f(-2) = 1+0+0 = 1 \\
f(3) =&\ f(3-2) + f(3-3) + f(3-4) = f(1) + f(0) + f(-1) = 0+1+0 = 1
\end{aligned}
$$

(b) Calculate the number of ways to coat a wall of height equal to 12 inches.

Answer:

$$
\begin{aligned}
f(4) =&\ f(4-2) + f(4-3) + f(4-4) = f(2) + f(1) + f(0) = 1+0+1 = 2 \\
f(5) =&\ f(5-2) + f(5-3) + f(5-4) = f(3) + f(2) + f(1) = 1+1+0 = 2 \\
f(6) =&\ f(6-2) + f(6-3) + f(6-4) = f(4) + f(3) + f(2) = 2+1+1 = 4 \\
f(7) =&\ f(7-2) + f(7-3) + f(7-4) = f(5) + f(4) + f(3) = 2+2+1 = 5 \\
f(8) =&\ f(8-2) + f(8-3) + f(8-4) = f(6) + f(5) + f(4) = 4+2+2 = 8 \\
f(9) =&\ f(9-2) + f(9-3) + f(9-4) = f(7) + f(6) + f(5) = 5+4+2 = 11 \\
f(10) =&\ f(10-2) + f(10-3) + f(10-4) = f(8) + f(7) + f(6) = 8+5+4 = 17 \\
f(11) =&\ f(11-2) + f(11-3) + f(11-4) = f(9) + f(8) + f(7) = 11+8+5 = 24 \\
f(12) =&\ f(12-2) + f(12-3) + f(12-4) = f(10) + f(9) + f(8) = 17+11+8 = 36
\end{aligned}
$$

Problem 251   Dangerous Professor is giving an oral exam to her class. Each student, when it is his turn, answers a sequence of questions. As soon as the student gives his answer to an individual question, the Professor grades it by giving it a certain number of points, which may be one of the following: $\{0, 1, 2, 4, 8\}$. If the student receives a zero as the grade for any question, the exam is over and the student fails. To pass the exam, the student has to collect n points, and the questioning continues until this happens or the student runs away, whichever comes first.

(a) Write a recurrence relation with the initial conditions for the number of ways a student can collect the necessary n points and pass the exam.

Answer: Let $f(n)$ be the number of ways the student can pass if he is n points away from the requirement. Observe that n is the difference between the number of points required to pass the exam and the number of points earned so far. The student answers until n for the first time becomes equal to zero or to a negative number.

At any stage of the test, the next question may reduce n by one, two, four or eight points. (Zero is not an option, since we are analyzing the case of a student who is passing.) If the student receives k points for his next answer, then he is left with $f(n - k)$ ways to earn the required $n - k$ points, whence the recurrence:

$$f(n) = f(n-1) + f(n-2) + f(n-4) + f(n-8) \text{ for } n \geq 8$$

To obtain the initial conditions, note that:

$$f(-n) = 1, \text{ for all } n \geq 0$$

since a student who has met or exceeded his requirement is left with a single option to do nothing. If this observation is employed in successive applications of the recursive rule for $f(n)$ to the initial positive values of n, precisely such that $0 \leq n \leq 7$, a set of actual initial conditions is obtained, where the arguments correspond to realistic (positive) values of the passing score n.

$$
\begin{aligned}
f(0) =&\ 1 \\
f(1) =&\ f(0) + f(-1) + f(-3) + f(-7) = & 1 + 1 + 1 + 1 =&\ 4 \\
f(2) =&\ f(1) + f(0) + f(-2) + f(-6) = & 4 + 1 + 1 + 1 =&\ 7 \\
f(3) =&\ f(2) + f(1) + f(-1) + f(-5) = & 7 + 4 + 1 + 1 =&\ 13 \\
f(4) =&\ f(3) + f(2) + f(0) + f(-4) = & 13 + 7 + 1 + 1 =&\ 22 \\
f(5) =&\ f(4) + f(3) + f(1) + f(-3) = & 22 + 13 + 4 + 1 =&\ 40 \\
f(6) =&\ f(5) + f(4) + f(2) + f(-2) = & 40 + 22 + 7 + 1 =&\ 70 \\
f(7) =&\ f(6) + f(5) + f(3) + f(-1) = & 70 + 40 + 13 + 1 =&\ 124
\end{aligned}
$$

(b) Calculate the number of ways a student can collect 9 points.

Answer:

$$f(8) = f(7) + f(6) + f(4) + f(0) = 124 + 70 + 22 + 1 = 217$$
$$f(9) = f(8) + f(7) + f(5) + f(1) = 217 + 124 + 40 + 4 = 385$$

Problem 252   Stamps are dispensed by a vending machine according to the following protocol. First, the customer deposits any amount of money in any acceptable denominations. Once the customer indicates that the intended amount is deposited, the machine counts the money, and acknowledges the amount as the customer credit. Next, the customer selects stamps, one by one, as long as the credit lasts. The available face values of stamps are: 1 cent, 2 cents, 5 cents, and 10 cents.

At any time, the customer may request the transaction to be concluded and the remaining credit refunded. However, the machine complies with this request only if the remaining credit does not exceed 6 cents. (If the credit is 7 cents or more, the customer has to make another stamp selection.)

(a) Write a recurrence relation with the initial conditions for the number of possible selection histories that can occur with a deposit of n cents.

Answer: Let $f(n)$ be the number of possible selection histories with the remaining credit of n cents. The transaction continues until n is reduced to zero or to a value less than 7, when the customer may issue a legal refund request.

For a sufficiently large value of n, the selection of the next stamp may reduce n by one, two, five or ten cents. If a stamp with the face value of k cents is selected, then the customer is left with $f(n - k)$ ways to complete the transaction, whence the recurrence:

$$f(n) = f(n - 1) + f(n - 2) + f(n - 5) + f(n - 10) \text{ for } n \geq 10$$

To obtain the initial conditions, first observe that:

$$f(0) = 1$$
$$f(-n) = 0, \text{ for all } n \geq 1$$

To verify this, note that the value of $f(-n)$ would correspond to the impossible situation of a customer having taken money from the vending machine (negative credit), which indeed has no way of happening. Straightforwardly, the condition in which $n = 0$ indicates that the customer's credit has been exhausted, which leaves the conclusion of the transaction as a single option to continue.

Next, recall that the transaction protocol in those cases when $n \leq 6$ differs from that specified by the recursive rule in that the customer has a an option to conclude the sale (in addition to the stamp selection options, accounted for by the recursive rule.) Hence, the initial conditions are calculated as follows.

$$
\begin{aligned}
f(0) &= 1 \\
f(1) &= 1 + f(0) + f(-1) + f(-4) + f(-9) = & 1 + 1 + 0 + 0 + 0 = & 2 \\
f(2) &= 1 + f(1) + f(0) + f(-3) + f(-8) = & 1 + 2 + 1 + 0 + 0 = & 4 \\
f(3) &= 1 + f(2) + f(1) + f(-2) + f(-7) = & 1 + 4 + 2 + 0 + 0 = & 7 \\
f(4) &= 1 + f(3) + f(2) + f(-1) + f(-6) = & 1 + 7 + 4 + 0 + 0 = & 12 \\
f(5) &= 1 + f(4) + f(3) + f(0) + f(-5) = & 1 + 12 + 7 + 1 + 0 = & 21 \\
f(6) &= 1 + f(5) + f(4) + f(1) + f(-4) = & 1 + 21 + 12 + 2 + 0 = & 36 \\
f(7) &= f(6) + f(5) + f(2) + f(-3) = & 36 + 21 + 4 + 0 = & 61 \\
f(8) &= f(7) + f(6) + f(3) + f(-2) = & 61 + 36 + 7 + 0 = & 104 \\
f(9) &= f(8) + f(7) + f(4) + f(-1) = & 104 + 61 + 12 + 0 = & 177
\end{aligned}
$$

(b) Calculate the number of possible selection histories with a deposit of 11 cents.

Answer:

$$f(10) = f(9) + f(8) + f(5) + f(0) = 177 + 104 + 21 + 1 = 303$$
$$f(11) = f(10) + f(9) + f(6) + f(1) = 303 + 177 + 36 + 2 = 518$$

Problem 253   Breakfast packages are sold by a vending machine according to the following protocol. First, the customer deposits any amount of money in any acceptable denominations of whole dollars. Once the customer indicates that the intended amount is deposited, the machine counts the money, and acknowledges the amount as the customer credit. Next, the customer selects food packages, one by one, as long as the credit lasts. The available menu comprises: two types of chocolate bars at 2 dollars each; three types of beverages at 1 dollar each; four types of sandwiches at 4 dollars each; and a large salad at 5 dollars. The customer cannot request a refund of the remaining credit amount, but has to continue making selections as long as the credit lasts.

(a) Write a recurrence relation with the initial conditions for the number of possible selection histories that can occur with a deposit of n dollars.

Answer: Let $f(n)$ be the number of possible selection histories with the remaining credit of n dollars. The transaction continues until n is reduced to zero.

For a sufficiently large value of n, the purchase of the next food item may reduce n by one, two, four or five dollars, depending on the choice of the item. If an item with a price of k dollars is selected, then the customer is left with $f(n - k)$ ways to complete the transaction. By taking into account all possible items and their respective costs, we arrive at the recurrence:

$$f(n) = 3f(n - 1) + 2f(n - 2) + 4f(n - 4) + f(n - 5) \text{ for } n \geq 5$$

To obtain the initial conditions, first observe that:

$$f(0) = 1$$
$$f(-n) = 0, \text{ for all } n \geq 1$$

To verify this, note that the value of $f(-n)$ would correspond to the impossible situation of a customer having taken money from the vending machine (negative credit), which indeed has no way of happening. Straightforwardly, the condition in which $n = 0$ indicates that the customer's credit has been exhausted, which leaves the conclusion of the transaction as the single option to continue. The initial conditions are calculated from the recursive rule as follows.

$$
\begin{aligned}
f(0) &= 1 \\
f(1) &= 3f(0) + 2f(-1) + 4f(-3) + f(-4) = & 3 \cdot 1 + 2 \cdot 0 + 4 \cdot 0 + 0 = & 3 \\
f(2) &= 3f(1) + 2f(0) + 4f(-2) + f(-3) = & 3 \cdot 3 + 2 \cdot 1 + 4 \cdot 0 + 0 = & 11 \\
f(3) &= 3f(2) + 2f(1) + 4f(-1) + f(-2) = & 3 \cdot 11 + 2 \cdot 3 + 4 \cdot 0 + 0 = & 39 \\
f(4) &= 3f(3) + 2f(2) + 4f(0) + f(-1) = & 3 \cdot 39 + 2 \cdot 11 + 4 \cdot 1 + 0 = & 143
\end{aligned}
$$

(b) Calculate the number of possible transaction histories with a deposit of 5 dollars.

Answer:

$$f(5) = 3f(4) + 2f(3) + 4f(1) + f(0) = 3 \cdot 143 + 2 \cdot 39 + 4 \cdot 3 + 1 = 520$$

Problem 254 A friendly vending machine in a summer camp for children is selling toys according to the following protocol. First, the customer deposits any amount of money in any acceptable denominations of whole dollars. Once the customer indicates that the intended amount is deposited, the machine counts the money, and acknowledges the amount as the customer credit. Next, the customer selects toys, one by one, as long as the credit lasts. The available assortment of toy models comprises: five models at 3 dollars; three models at 4 dollars; and two models at 5 dollars.

If and when the customer requests that the transaction be concluded and the remaining credit refunded, the machine complies as long as if the remaining credit does not exceed 3 dollars. (If the credit is 4 dollars or more, the customer has to make another toy selection.) However, the machine will grant the last selection request even if the price of the selected toy exceeds the remaining credit by no more than 2 dollars.

(a) Write a recurrence relation with the initial conditions for the number of possible selection histories that can occur with a deposit of n dollars.

Answer: Let $f(n)$ be the number of possible selection histories with the remaining credit of n dollars. The transaction continues until n assumes one of the following three values: $0, -1, -2$.

For a sufficiently large value of n, the selection of the next toy may reduce n by three, four, or five dollars, depending on the choice of the toy. If a toy with a price of k dollars is selected, then the customer is left with $f(n-k)$ ways to complete the transaction. By taking into account all possible toy models and their respective costs, we arrive at the recurrence:

$$f(n) = 5f(n-3) + 3f(n-4) + 2f(n-5) \text{ for } n \geq 5$$

To obtain the initial conditions, first observe that:

$$f(0) = f(-1) = f(-2) = 1$$
$$f(-n) = 0, \text{ for all } n > 2$$

To verify this, note that the value of $f(-n)$ corresponds to the a situation in which the customer has overdrawn credit from the vending machine (negative credit). This is impossible for any $n > 2$, but permissible for $n \leq 2$, in which case it leaves the customer with a single option to conclude the transaction, in very much the same way as it happens in the case when $n = 0$ (fully consumed credit.)

Next, recall that in those cases when $n \leq 3$ the transaction protocol differs from that specified by the recursive rule, in that the customer has a an option to conclude the sale (in addition to the toy selection options, accounted for by the recursive rule.) Hence, the initial conditions are calculated as follows.

$$
\begin{aligned}
f(0) &= 1 \\
f(1) &= 1 + 5f(-2) + 3f(-3) + 2f(-4) = 1+5+0+0 = 6 \\
f(2) &= 1 + 5f(-1) + 3f(-2) + 2f(-3) = 1+5+3+0 = 9 \\
f(3) &= 1 + 5f(0) + 3f(-1) + 2f(-2) = 1+5+3+2 = 11 \\
f(4) &= 5f(1) + 3f(0) + 2f(-1) = 30+3+2 = 35
\end{aligned}
$$

(b) Calculate the number of possible selection histories with a deposit of 6 dollars.
Answer:

$$f(5) = 5f(2) + 3f(1) + 2f(0) = 45 + 18 + 2 = 65$$
$$f(6) = 5f(3) + 3f(2) + 2f(1) = 55 + 27 + 12 = 94$$

Problem 255   Let x_n be the number of strings of length n over the alphabet $\{a, b, c\}$ that contain an odd number of b's.

(a) Write a recurrence relation with the initial conditions for x_n.

Answer: Consider sequences x_n and y_n, described as follows.

x_n :   the number of strings of length n over $\{a, b, c\}$ that contain an odd number of b's,

y_n :   the number of strings of length n over $\{a, b, c\}$ that contain an even number of b's.

There are exactly x_n strings of length $n \geq 1$ that contain an odd number of b's. Every such string, say σ, can be represented by exactly one of the following three forms:

$\sigma = b\alpha$

where α is a string of length $n-1$ that contains an even number of b's; there are exactly y_{n-1} such strings;

$\sigma = a\beta$

where β is a string of length $n-1$ that contains an odd number of b's; there are exactly x_{n-1} such strings;

$\sigma = c\beta$

where β is a string of length $n-1$ that contains an odd number of b's; there are exactly x_{n-1} such strings;

To verify this, observe that append a letter b to a string creates a new string, whose parity of the number of b's is the opposite of the original. In contrast, appending any of the letters a, c creates a new string, whose number of b's (and its parity) is equal to the original. Hence, x_n is calculated by summing over these three cases:

$$x_n = y_{n-1} + 2x_{n-1}$$

On the other hand, there are exactly y_n strings of length $n \geq 1$ that contain an even number of b's. Every such string, say σ, can be represented by exactly one of the following three forms.

$\sigma = b\beta$

where β is a string of length $n-1$ that contains an odd number of b's; there are exactly x_{n-1} such strings;

$\sigma = a\alpha$

where α is a string of length $n-1$ that contains an even number of b's; there are exactly y_{n-1} such strings;

$\sigma = c\alpha$

where α is a string of length $n-1$ that contains an even number of b's; there are exactly y_{n-1} such strings.

Hence, y_n is calculated by summing over these three cases:

$$y_n = x_{n-1} + 2y_{n-1}$$

Sequences x_n and y_n, for $n \geq 1$, are defined by the following system of two recurrences.

$$x_n = y_{n-1} + 2x_{n-1} \tag{3.1}$$
$$y_n = x_{n-1} + 2y_{n-1} \tag{3.2}$$

Recurrence 3.1 is rewritten as follows.

$$x_{n+1} = y_n + 2x_n$$

or, equivalently:

$$y_n = x_{n+1} - 2x_n \tag{3.3}$$

After a multiplication by 2, recurrence 3.1 becomes:

$$2x_n = 2y_{n-1} + 4x_{n-1}$$

or, equivalently:

$$2y_{n-1} = 2x_n - 4x_{n-1} \tag{3.4}$$

By a substitution with the right-hand sides of 3.3 and 3.4 into 3.2, we obtain:

$$x_{n+1} - 2x_n = x_{n-1} + (2x_n - 4x_{n-1})$$

or, equivalently:

$$x_{n+1} = 4x_n - 3x_{n-1} \quad \text{for } n \geq 1 \tag{3.5}$$

For the initial conditions, observe that the empty string contains an even number, precisely zero, of occurrences of the letter b. Hence, there are no strings of length equal to 0 that contain an odd number of b's. Among the three strings of length equal to 1, exactly one string, namely b, has an odd number of occurrences of the letter b—one occurence, precisely. These observations and the recurrence 3.5 lead to the complete recurrence for x_n:

$$x_0 = 0$$
$$x_1 = 1$$
$$x_n = 4x_{n-1} - 3x_{n-2} \quad \text{for } n \geq 2$$

(b) Find a closed form for x_n.

This is a linear homogeneous recurrence relation of second order with constant coefficients. The characteristic equation is:

$$r^2 - 4r + 3 = 0$$

the roots of which are: $r_1 = 1$ and $r_2 = 3$. The solutions are of the form:

$$x_n = \alpha \cdot 1^n + \beta \cdot 3^n = \alpha + \beta \cdot 3^n$$

The initial conditions mandate:

$$x_0 = \alpha + \beta \cdot 3^0 = \alpha + \beta = 0$$
$$x_1 = \alpha + \beta \cdot 3^1 = \alpha + 3\beta = 1$$

whence: $\beta = \dfrac{1}{2}$ and $\alpha = -\dfrac{1}{2}$. The closed form is:

$$x_n = \frac{3^n - 1}{2}$$

(c) Calculate the number of strings of length equal to 6 over the alphabet $\{a, b, c\}$ that contain an odd number of b's.

Answer: By a substitution into the closed form, constructed in the answer given in part **(b)**, we obtain:

$$x_6 = \frac{3^6 - 1}{2} = \frac{729 - 1}{2} = 364$$

Problem 256   Let x_n be the number of strings of length n over the alphabet $\{a, b, c\}$ in which the sum of the number of a's and the number of c's is odd.

(a) Write a recurrence relation with the initial conditions for x_n.

Answer: For convenience, let any string in which the sum of the number of a's and the number of c's is odd be termed a good string. Consider sequences x_n and y_n, described as follows.

x_n :   the number of good strings of length n over $\{a, b, c\}$,

y_n :   the number of strings of length n over $\{a, b, c\}$ that are not good.

Each of the x_n good strings (whenever $n \geq 1$) can be generated by exactly one of the following three steps.

append a letter b to any of the x_{n-1} good strings of length $n - 1$;

append a letter a to any of the y_{n-1} strings of length $n - 1$ that are not good;

append a letter c to any of the y_{n-1} strings of length $n - 1$ that are not good.

Hence, x_n is calculated by summing over these three cases:

$$x_n = 2y_{n-1} + x_{n-1}$$

On the other hand, each of the y_n strings of length $n \geq 1$ that are not good can be generated by exactly one of the following three steps.

append a letter b to any of the y_{n-1} strings of length $n - 1$ that are not good;

append a letter a to any of the x_{n-1} good strings of length $n - 1$;

append a letter c to any of the x_{n-1} good strings of length $n - 1$.

Hence, y_n is calculated by summing over these three cases:

$$y_n = 2x_{n-1} + y_{n-1}$$

Sequences x_n and y_n, for $n \geq 1$, are defined by the following system of two recurrences.

$$x_n = 2y_{n-1} + x_{n-1} \tag{3.6}$$
$$y_n = 2x_{n-1} + y_{n-1} \tag{3.7}$$

Recurrence 3.6 is equivalent to:

$$2y_{n-1} = x_n - x_{n-1} \tag{3.8}$$

which is rewritten as follows.

$$2y_n = x_{n+1} - x_n \tag{3.9}$$

After a multiplication by 2, recurrence 3.7 becomes:

$$2y_n = 4x_{n-1} + 2y_{n-1}$$

or, equivalently:

$$2y_n - 2y_{n-1} = 4x_{n-1} \tag{3.10}$$

By a substitution with the right-hand sides of 3.8 and 3.9 into 3.10, we obtain:

$$(x_{n+1} - x_n) - (x_n - x_{n-1}) = 4x_{n-1}$$

or, equivalently:

$$x_{n+1} = 2x_n + 3x_{n-1} \quad \text{for } n \geq 1 \tag{3.11}$$

For the initial conditions, observe that the total number of occurrences of a and c in the empty string is equal to zero, and thereby even. Hence, there are no good strings of length equal to 0.

Among the three strings of length equal to 1, exactly two string, namely a and c, are good. These observations and the recurrence 3.11 lead to the complete recurrence for x_n:

$$x_0 = 0$$
$$x_1 = 2$$
$$x_n = 2x_{n-1} + 3x_{n-2} \quad \text{for } n \geq 2$$

(b) Find a closed form for x_n.

Answer: This is a linear homogeneous recurrence relation of second order with constant coefficients. The characteristic recurrence is:

$$r^2 - 2r - 3 = 0$$

the roots of which are: $r_1 = -1$ and $r_2 = 3$. The solutions are of the form:

$$x_n = \alpha \cdot (-1)^n + \beta \cdot 3^n$$

The initial conditions mandate:

$$x_0 = \alpha \cdot (-1)^0 + \beta \cdot 3^0 = \alpha + \beta = 0$$
$$x_1 = \alpha \cdot (-1)^1 + \beta \cdot 3^1 = -\alpha + 3\beta = 2$$

whence: $\beta = \dfrac{1}{2}$ and $\alpha = -\dfrac{1}{2}$. The closed form is:

$$x_n = \frac{3^n + (-1)^{n+1}}{2}$$

(c) Calculate the number of strings of length equal to 7 over the alphabet $\{a, b, c\}$ that contain an odd number of b's.

Answer: By a substitution into the closed form, constructed in the answer given in part **(b)**, we obtain:

$$x_7 = \frac{3^7 + (-1)^{7+1}}{2} = \frac{2187 + 1}{2} = 1094$$

Problem 257   Let x_n be the number of strings of length n over the alphabet $\{a, b\}$ such that the difference between the number of a's and the number of b's is by one greater than an integer multiple of 3.

(a) Write a recurrence relation with the initial conditions for x_n.

Answer: For an arbitrary string ξ over the alphabet $\{a, b\}$, let $\Delta(\xi)$ be the difference between the number of a's and the number of b's in the string ξ. Define three sequences of sets of strings X_n, Y_n, Z_n as follows.

$$\begin{aligned}
X_n &= \{\xi \mid |\xi| = n \wedge (\exists k)(\Delta(\xi) = 3k + 1)\} \\
Y_n &= \{\xi \mid |\xi| = n \wedge (\exists k)(\Delta(\xi) = 3k + 2)\} \\
Z_n &= \{\xi \mid |\xi| = n \wedge (\exists k)(\Delta(\xi) = 3k)\}
\end{aligned}$$

where k is an integer. Observe that any string ξ of length n belongs to exactly one of the three sets X_n, Y_n, Z_n, according to the remainder of $\Delta(\xi)$ after the division by 3. Define three sequences of numbers x_n, y_n, z_n as the cardinalities of the corresponding sets:

$$x_n = |X_n|, \quad y_n = |Y_n|, \quad z_n = |Z_n|$$

To construct the recurrence rules for x_n, y_n, z_n, observe that, for an arbitrary string ξ:

$$\Delta(\xi a) = \Delta(\xi) + 1$$
$$\Delta(\xi b) = \Delta(\xi) - 1$$

In other words:

$$\xi \in X_{n-1} \implies (\xi a \in Y_n \land \xi b \in Z_n)$$
$$\xi \in Y_{n-1} \implies (\xi a \in Z_n \land \xi b \in X_n)$$
$$\xi \in Z_{n-1} \implies (\xi a \in X_n \land \xi b \in Y_n)$$

Consider all possible ways to create a string of length n, by appending a character to a string of length $n - 1$, in each of the three sets. The possibilities are as follows.

Each of the x_n members of X_n is created by exactly one of the following two steps:

append a letter b to any of the y_{n-1} members of Y_{n-1};
append a letter a to any of the z_{n-1} members of Z_{n-1}.

Each of the y_n members of Y_n is created by exactly one of the following two steps:

append a letter a to any of the x_{n-1} members of X_{n-1};
append a letter b to any of the z_{n-1} members of Z_{n-1}.

Each of the z_n members of Z_n is created by exactly one of the following two steps:

append a letter b to any of the x_{n-1} members of X_{n-1};
append a letter a to any of the y_{n-1} members of Y_{n-1}.

By summing over all possible ways to create all possible strings, we calculate cardinalities of the three sets, as sequences x_n, y_n, z_n, for $n \geq 1$, defined by the following system of three recurrences.

$$x_n = z_{n-1} + y_{n-1} \tag{3.12}$$
$$y_n = x_{n-1} + z_{n-1} \tag{3.13}$$
$$z_n = y_{n-1} + x_{n-1} \tag{3.14}$$

Recurrence 3.12 is equivalent to:

$$y_{n-1} = x_n - z_{n-1} \tag{3.15}$$

which is rewritten as follows.

$$y_n = x_{n+1} - z_n \tag{3.16}$$

By a substitution with the right-hand sides of 3.15 and 3.16 into 3.13 and 3.14, we obtain the following system of two recurrences.

$$x_{n+1} - z_n = x_{n-1} + z_{n-1}$$
$$z_n = x_n - z_{n-1} + x_{n-1}$$

This system is equivalent to:

$$z_n + z_{n-1} = x_{n+1} - x_{n-1} \tag{3.17}$$
$$z_n + z_{n-1} = x_n + x_{n-1} \tag{3.18}$$

By equating the right-hand sides of 3.17 and 3.18, we obtain:

$$x_{n+1} - x_{n-1} = x_n + x_{n-1}$$

or, equivalently:

$$x_{n+1} = x_n + 2x_{n-1} \quad \text{for } n \geq 1 \tag{3.19}$$

For the initial conditions, we calculate $\Delta(\xi)$ for all strings ξ such that $0 \leq |\xi| \leq 1$.

ξ	$\Delta(\xi)$
empty string	$0 - 0 = 3 \cdot 0 + 0$
a	$1 - 0 = 3 \cdot 0 + 1$
b	$0 - 1 = 3 \cdot (-1) + 2$

We conclude that $X_0 = \emptyset$ and $X_1 = \{a\}$. This observation and 3.19 lead to the recurrence for x_n:

$$x_0 = 0$$
$$x_1 = 1$$
$$x_n = x_{n-1} + 2x_{n-2} \quad \text{for } n \geq 2$$

(b) Find a closed form for x_n.

Answer: This is a linear homogeneous recurrence relation of second order with constant coefficients. The characteristic recurrence is:

$$r^2 - r - 2 = 0$$

the roots of which are: $r_1 = -1$ and $r_2 = 2$. The solutions are of the form:

$$x_n = \alpha \cdot (-1)^n + \beta \cdot 2^n$$

The initial conditions mandate:

$$x_0 = \alpha \cdot (-1)^0 + \beta \cdot 2^0 = \alpha + \beta = 0$$
$$x_1 = \alpha \cdot (-1)^1 + \beta \cdot 2^1 = -\alpha + 2\beta = 1$$

whence: $\beta = \dfrac{1}{3}$ and $\alpha = -\dfrac{1}{3}$. The closed form is:

$$x_n = \frac{2^n + (-1)^{n+1}}{3}$$

(c) Calculate the number of strings of length equal to 12 over the alphabet $\{a, b\}$ such that the difference between the number of a's and the number of b's is by one greater than an integer multiple of 3.

Answer: By a substitution into the closed form, constructed in the answer given in part **(b)**, we obtain:

$$x_{12} = \frac{2^{12} + (-1)^{13}}{3} = \frac{4095}{3} = 1365$$

Problem 258   Let x_n be the number of strings of length n over the alphabet $\{a, b, c\}$ that contain bc as a substring.

(a) Write a recurrence relation with the initial conditions for x_n.

Answer: Define three sequences of sets of strings X_n, Y_n, Z_n over the alphabet $\{a, b, c\}$ as follows.

X_n : the set of strings of length n that contain substring bc,

Y_n : the set of strings of length n that do not contain substring bc but end with b,

Z_n : the set of strings of length n that do not contain substring bc and do not end with b.

Observe that any string ξ of length n belongs to exactly one of the three sets X_n, Y_n, Z_n. Define three sequences of numbers x_n, y_n, z_n as the cardinalities of the corresponding sets:

$$x_n = |X_n|, \quad y_n = |Y_n|, \quad z_n = |Z_n|$$

To construct the recurrence rules for x_n, y_n, z_n, observe that, for an arbitrary string ξ:

$$\xi \in X_{n-1} \implies (\xi a \in X_n \wedge \xi b \in X_n \wedge \xi c \in X_n)$$
$$\xi \in Y_{n-1} \implies (\xi a \in Z_n \wedge \xi b \in Y_n \wedge \xi c \in X_n)$$
$$\xi \in Z_{n-1} \implies (\xi a \in Z_n \wedge \xi b \in Y_n \wedge \xi c \in Z_n)$$

Consider all possible ways to create a string of length n, by appending a character to a string of length $n-1$, in each of the three sets. The possibilities are as follows.

Each of the x_n members of X_n is created by exactly one of the following four steps:

append a letter a to any of the x_{n-1} members of X_{n-1};
append a letter b to any of the x_{n-1} members of X_{n-1};
append a letter c to any of the x_{n-1} members of X_{n-1};
append a letter c to any of the y_{n-1} members of Y_{n-1}.

Each of the y_n members of Y_n is created by exactly one of the following two steps:

append a letter b to any of the y_{n-1} members of Y_{n-1};
append a letter b to any of the z_{n-1} members of Z_{n-1}.

Each of the z_n members of Z_n is created by exactly one of the following three steps:

append a letter a to any of the z_{n-1} members of Z_{n-1};
append a letter c to any of the z_{n-1} members of Z_{n-1};
append a letter a to any of the y_{n-1} members of Y_{n-1}.

By summing over all possible ways to create all possible strings, we calculate cardinalities of the three sets, as sequences x_n, y_n, z_n, for $n \geq 1$, defined by the following system of three recurrences.

$$x_n = 3x_{n-1} + y_{n-1} \tag{3.20}$$
$$y_n = y_{n-1} + z_{n-1} \tag{3.21}$$
$$z_n = 2z_{n-1} + y_{n-1} \tag{3.22}$$

Recurrence 3.21, after a multiplication by 2, is equivalent to:

$$2y_n - 2y_{n-1} = 2z_{n-1} \tag{3.23}$$

Furthermore, recurrence 3.21 is rewritten as follows.

$$y_{n+1} - y_n = z_n \tag{3.24}$$

By a substitution with the left-hand sides of 3.23 and 3.24 into 3.22, we obtain:

$$y_{n+1} - y_n = (2y_n - 2y_{n-1}) + y_{n-1}$$

or, equivalently:

$$y_{n+1} - 3y_n + y_{n-1} = 0 \tag{3.25}$$

Recurrence 3.20, after successive rewriting, is equivalent to all of the following three recurrences.

$$y_{n-1} = x_n - 3x_{n-1} \tag{3.26}$$
$$y_n = x_{n+1} - 3x_n \tag{3.27}$$
$$y_{n+1} = x_{n+2} - 3x_{n+1} \tag{3.28}$$

By a substitution with the right-hand side of each of 3.26, 3.27, and 3.28 into 3.25, we obtain:

$$(x_{n+2} - 3x_{n+1}) - 3(x_{n+1} - 3x_n) + (x_n - 3x_{n-1}) = 0$$

or, equivalently:

$$x_{n+2} = 6x_{n+1} - 10x_n + 3x_{n-1} \tag{3.29}$$

For the initial conditions, observe that $X_0 = X_1 = \emptyset$, $X_2 = \{bc\}$. This observation and 3.29 lead to the complete recurrence for x_n:

$$x_0 = 0$$
$$x_1 = 0$$
$$x_2 = 1$$
$$x_n = 6x_{n-1} - 10x_{n-2} + 3x_{n-3} \quad \text{for } n \geq 3$$

(b) Calculate the number of strings of length equal to 6 over the alphabet $\{a, b, c\}$ that contain bc as a substring.

Answer: The recurrence developed in the answer given in part **(a)** yields:

$$x_3 = 6x_2 - 10x_1 + 3x_0 = 6 \cdot 1 - 10 \cdot 0 + 3 \cdot 0 = 6$$
$$x_4 = 6x_3 - 10x_2 + 3x_1 = 6 \cdot 6 - 10 \cdot 1 + 3 \cdot 0 = 26$$
$$x_5 = 6x_4 - 10x_3 + 3x_2 = 6 \cdot 26 - 10 \cdot 6 + 3 \cdot 1 = 99$$
$$x_6 = 6x_5 - 10x_4 + 3x_3 = 6 \cdot 99 - 10 \cdot 26 + 3 \cdot 6 = 352$$

Problem 259   Let x_n be the number of strings of length n over the alphabet $\{a, b\}$ that contain aba as a substring.

(a) Write a recurrence relation with the initial conditions for x_n.

Answer: Define four sequences of sets of strings X_n, Y_n, Z_n, W_n over the alphabet $\{a, b\}$ as follows.

X_n : the set of length-n strings that contain substring aba,

Y_n : the set of length-n strings that do not contain substring aba but end with ab,

Z_n : the set of length-n strings that do not contain substring aba but end with a,

W_n : the set of length-n strings that do not contain substring aba and do not end with a or ab.

Observe that any string ξ of length n belongs to exactly one of the four sets X_n, Y_n, Z_n, W_n. Define four sequences of numbers x_n, y_n, z_n, w_n as the cardinalities of the corresponding sets:

$$x_n = |X_n|, \quad y_n = |Y_n|, \quad z_n = |Z_n|, \quad w_n = |W_n|$$

To construct the recurrence rules for x_n, y_n, z_n, w_n, observe that, for an arbitrary string ξ:

$$\xi \in X_{n-1} \Longrightarrow (\xi a \in X_n \wedge \xi b \in X_n)$$
$$\xi \in Y_{n-1} \Longrightarrow (\xi a \in X_n \wedge \xi b \in W_n)$$
$$\xi \in Z_{n-1} \Longrightarrow (\xi a \in Z_n \wedge \xi b \in Y_n)$$
$$\xi \in W_{n-1} \Longrightarrow (\xi a \in Z_n \wedge \xi b \in W_n)$$

Consider all possible ways to create a string of length n, by appending a character to a string of length $n-1$, in each of the four sets. By summing over all possible ways to do this, we calculate cardinalities of these sets, as sequences x_n, y_n, z_n, w_n, for $n \geq 1$, defined by the following system of four recurrences.

$$x_n = 2x_{n-1} + y_{n-1} \tag{3.30}$$
$$y_n = z_{n-1} \tag{3.31}$$
$$z_n = z_{n-1} + w_{n-1} \tag{3.32}$$
$$w_n = w_{n-1} + y_{n-1} \tag{3.33}$$

Recurrence 3.31 is rewritten as:

$$y_{n+1} = z_n \tag{3.34}$$

By a substitution with the left-hand sides of 3.31 and 3.34 into 3.32, we obtain:

$$y_{n+1} = y_n + w_{n-1}$$

or, equivalently:

$$w_{n-1} = y_{n+1} - y_n \tag{3.35}$$

which is rewritten as:

$$w_n = y_{n+2} - y_{n+1} \tag{3.36}$$

By a substitution with the right-hand sides of 3.35 and 3.36 into 3.33, we obtain:

$$y_{n+2} - y_{n+1} = (y_{n+1} - y_n) + y_{n-1}$$

or, equivalently:

$$y_{n+2} - 2y_{n+1} + y_n - y_{n-1} = 0 \tag{3.37}$$

Recurrence 3.30, after successive rewriting, is equivalent to all of the following four recurrences.

$$y_{n-1} = x_n - 2x_{n-1} \tag{3.38}$$
$$y_n = x_{n+1} - 2x_n \tag{3.39}$$
$$y_{n+1} = x_{n+2} - 2x_{n+1} \tag{3.40}$$
$$y_{n+2} = x_{n+3} - 2x_{n+2} \tag{3.41}$$

By a substitution with the right-hand side of each of 3.38, 3.39, 3.40, and 3.41 into 3.37, we obtain:

$$(x_{n+3} - 2x_{n+2}) - 2(x_{n+2} - 2x_{n+1}) + (x_{n+1} - 2x_n) - (x_n - 2x_{n-1}) = 0$$

or, equivalently:

$$x_{n+3} = 4x_{n+2} - 5x_{n+1} + 3x_n - 2x_{n-1} \qquad (3.42)$$

For the initial conditions, observe that $X_0 = X_1 = X_2 = \emptyset$, $X_3 = \{aba\}$. This observation and **3.42** lead to the complete recurrence for x_n:

$$x_0 = 0$$
$$x_1 = 0$$
$$x_2 = 0$$
$$x_3 = 1$$
$$x_n = 4x_{n-1} - 5x_{n-2} + 3x_{n-3} - 2x_{n-4} \quad \text{for } n \geq 4$$

(b) Calculate the number of strings of length equal to 7 over the alphabet $\{a, b\}$ that contain *aba* as a substring.

Answer: The recurrence developed in the answer given in part **(a)** yields:

$$x_4 = 4x_3 - 5x_2 + 3x_1 - 2x_0 = 4 \cdot 1 - 5 \cdot 0 + 3 \cdot 0 - 2 \cdot 0 = 4$$
$$x_5 = 4x_4 - 5x_3 + 3x_2 - 2x_1 = 4 \cdot 4 - 5 \cdot 1 + 3 \cdot 0 - 2 \cdot 0 = 11$$
$$x_6 = 4x_5 - 5x_4 + 3x_3 - 2x_2 = 4 \cdot 11 - 5 \cdot 4 + 3 \cdot 1 - 2 \cdot 0 = 27$$
$$x_7 = 4x_6 - 5x_5 + 3x_4 - 2x_3 = 4 \cdot 27 - 5 \cdot 11 + 3 \cdot 4 - 2 \cdot 1 = 63$$

Chapter 4

Algorithm Analysis

4.1 Asymptotic Order of Functions

Problem 260 Sets A, B, C, D, E are defined as follows.

$$A = \{0, 1, 2, 3, 4\}$$
$$B = \{x \mid x \in A \land n^x = \omega\left(n^2\right)\}$$
$$C = \{z \mid 3 \leq z \leq 12\}$$
$$D = \{x \mid (\exists y \in C)\left(n^{3x} = \Theta\left(n^y\right)\right)\}$$
$$E = \{x \mid (\exists y \in C)\left(n^y = \Omega\left(n^{2x}\right)\right)\}$$

All variables assume values from the set of natural numbers $N = \{0, 1, \ldots\}$.

(a) List all elements of B, or explain why it is impossible.

   Answer: Observe that:
$$B = \{x \mid x \in A \land x > 2\}$$

   whence the listing:
$$3, 4$$

(b) List all elements of D, or explain why it is impossible.

   Answer: Observe that:
$$D = \{x \mid (\exists y \in C)(3x = y)\}$$

   whence the listing:
$$1, 2, 3, 4$$

(c) List all elements of E, or explain why it is impossible.

   Answer: Observe that:
$$E = \{x \mid (\exists y \in C)(y \geq 2x)\}$$

   whence the listing:
$$0, 1, 2, 3, 4, 5, 6$$

Problem 261   Sets A, B, C, D, E are defined as follows.

$$A = \{0, 1, 2, 3, 4, 5, 6\}$$
$$B = \left\{x \mid x \in A \wedge n^x = \omega\left(n^2\right)\right\}$$
$$C = \left\{y \mid (\forall x \in B)\left(n^y = O\left(n^x\right)\right)\right\}$$
$$D = \left\{x \mid (\exists y \in C)\left(n^x = \Theta\left(n^{y+1}\right)\right)\right\}$$
$$E = \left\{x \mid (\exists y \in B)\left(n^{y+1} = \Omega\left(n^x\right)\right)\right\}$$

All variables assume values from the set of natural numbers $N = \{0, 1, \ldots\}$.

(a) List all elements of B, or explain why it is impossible.

   Answer: Observe that the definitions imply:

$$B = \{x \mid x \in A \wedge x > 2\}$$

and the elements of B are:

$$3, 4, 5, 6$$

(b) List all elements of C, or explain why it is impossible.

   Answer: Observe that the definitions imply:

$$C = \{y \mid (\forall x \in B)(y \le x)\}$$

and the elements of C are:

$$0, 1, 2, 3$$

(c) List all elements of D, or explain why it is impossible.

   Answer: Observe that the definitions imply:

$$D = \{x \mid (\exists y \in C)(x = y + 1)\}$$

and the elements of D are:

$$1, 2, 3, 4$$

(d) List all elements of E, or explain why it is impossible.

   Answer: Observe that the definitions imply:

$$E = \{x \mid (\exists y \in B)(y + 1 \ge x))\}$$

and the elements of E are:

$$0, 1, 2, 3, 4, 5, 6, 7$$

Problem 262   Sets A, B, C, D, E are defined as follows.

$$A = \{0, 1, 2, 3, 4, 5, 6, 7, 8\}$$
$$B = \left\{x \mid x \in A \wedge n^{2x} = o\left(2^n\right)\right\}$$
$$C = \left\{y \mid (\exists x \in A)\left((2y)^n = \Theta\left(x^n\right)\right)\right\}$$
$$D = \left\{p \mid (\forall y \in A)\left(\lg^p n = \Omega\left(n^y\right)\right)\right\}$$
$$E = \left\{q \mid q \in A \wedge n^q = \omega\left(\lg n\right)\right\}$$

All variables assume values from the set of natural numbers $N = \{0, 1, \ldots\}$.

(a) List all elements of B, or explain why it is impossible.

Answer: Observe that the definitions imply:

$$B = \{x \mid x \in A\}$$

and the elements of B are:

$$0, 1, 2, 3, 4, 5, 6, 7, 8$$

(b) List all elements of C, or explain why it is impossible.

Answer: Observe that the definitions imply:

$$C = \{y \mid (\exists x \in A)\,(2y = x)\}$$

and the elements of C are:

$$0, 1, 2, 3, 4$$

(c) List all elements of D, or explain why it is impossible.

Answer: Observe that the definitions imply:

$$D = \left\{p \mid \lg^p n = \Omega\left(n^8\right)\right\}$$

There does not exist a nonempty list of elements of D, since D is empty.

(d) List all elements of E, or explain why it is impossible.

Answer: Observe that the definitions imply:

$$E = \{q \mid q \in A \wedge q > 0\}$$

and the elements of E are:

$$1, 2, 3, 4, 5, 6, 7, 8$$

Problem 263 Sets A, B, C are defined as follows.

$$A = \{0, 1, 2, 3, 4, 5, 6, 7, 8\}$$
$$B = \left\{x \mid (\exists y \in A)\left((\sqrt{n})^y = \Theta\left(n^x\right)\right)\right\}$$
$$C = \left\{z \mid (z \in A \wedge n^z = o\left(\lg n\right))\right\}$$

All variables assume values from the set of natural numbers $N = \{0, 1, \ldots\}$.

(a) List all elements of B, or explain why it is impossible.

Answer: Observe that the definitions imply:

$$B = \{x \mid (\exists y \in A)\,((y/2) = x)\}$$

and the elements of B are:

$$0, 1, 2, 3, 4$$

(b) List all elements of C, or explain why it is impossible.

Answer: Observe that the definitions imply:

$$C = \{z \mid (z \in A \wedge z = 0)\}$$

and the single element of C is:

$$0$$

Problem 264   Sets, A, B, C, D are defined as follows.

$$A = \{0,1,2,3,4,5\}$$
$$B = \{8,9,10\}$$
$$C = \left\{ k \mid (\exists x \in A)(\exists y \in B)\left(n^{y+x} = \Theta\left((n^k)^3\right)\right) \right\}$$
$$D = \left\{ \ell \mid (\forall x \in A)(\exists y \in B)\left(n^{y-x} = \omega\left(n^\ell\right)\right) \right\}$$

All variables assume values from the set of natural numbers $N = \{0,1,\ldots\}$.

(a) List all elements of C, or explain why it is impossible.

   Answer: Observe that:

$$C = \{ k \mid (\exists x \in A)(\exists y \in B)(y + x = 3k)\}$$

   whence the listing:
$$3,4,5$$

(b) List all elements of D, or explain why it is impossible.

   Answer: Observe that:

$$D = \{ \ell \mid (\forall x \in A)(\exists y \in B)(y - x > \ell)\}$$

   whence the listing:
$$0,1,2,3,4$$

Problem 265   Sets, A, B, C, D are defined as follows.

$$A = \{0,1,2,3,4,5\}$$
$$B = \{8,9,10\}$$
$$C = \left\{ t \mid (\exists x \in A)(\exists y \in B)\left(n^{y-x} = \omega\left((n^t)^2\right)\right) \right\}$$
$$D = \left\{ t \mid (\forall x \in A)(\exists y \in B)\left(\sqrt{n^{y-x}} = \Theta\left(n^t\right)\right) \right\}$$

All variables assume values from the set of natural numbers $N = \{0,1,\ldots\}$.

(a) List all elements of C, or explain why it is impossible.

   Answer: Observe that:

$$C = \{ t \mid (\exists x \in A)(\exists y \in B)(y - x > 2t)\}$$

   whence the listing:
$$0,1,2,3,4$$

(b) List all elements of D, or explain why it is impossible.

   Answer: Observe that:

$$D = \{ t \mid (\forall x \in A)(\exists y \in B)(y - x = 2t)\}$$

which means that $D = \emptyset$, and there are no elements to list.

Problem 266   In each of the following formulae, write in the empty box one of the following letters: o, ω, Θ (little-o, little-ω, Θ) so as to obtain a correct claim about the asymptotic order of the two functions.

Answer:

$\sqrt{n} \cdot \lg n = \boxed{} \left(n^{2/3}\right)$ $\qquad$ $\sqrt{n} \cdot \lg n = \boxed{o} \left(n^{2/3}\right)$

$\sqrt{n} \cdot \lg n = \boxed{} \left(n^{1/3} \cdot \lg n\right)$ $\qquad$ $\sqrt{n} \cdot \lg n = \boxed{\omega} \left(n^{1/3} \cdot \lg n\right)$

$\sqrt{n} \cdot \lg n = \boxed{} \left(n^{1/2} \cdot \lg\lg n\right)$ $\qquad$ $\sqrt{n} \cdot \lg n = \boxed{\omega} \left(n^{1/2} \cdot \lg\lg n\right)$

$\sqrt{n} \cdot \lg n = \boxed{} \left(n^{1/2} + \lg\lg n\right)$ $\qquad$ $\sqrt{n} \cdot \lg n = \boxed{\omega} \left(n^{1/2} + \lg\lg n\right)$

$\sqrt{n} \cdot \lg n = \boxed{} \left(n/\left(\lg^2 n\right)\right)$ $\qquad$ $\sqrt{n} \cdot \lg n = \boxed{o} \left(n/\left(\lg^2 n\right)\right)$

$\lg\left(n^{3/2}\right) = \boxed{} \left(\log_{10} n\right)$ $\qquad$ $\lg\left(n^{3/2}\right) = \boxed{\Theta} \left(\log_{10} n\right)$

$\lg\left(n^{3/2}\right) = \boxed{} \left(\log_{10}\left(n^2\right)\right)$ $\qquad$ $\lg\left(n^{3/2}\right) = \boxed{\Theta} \left(\log_{10}\left(n^2\right)\right)$

$\lg\left(n^{3/2}\right) = \boxed{} \left(\sqrt{n}\right)$ $\qquad$ $\lg\left(n^{3/2}\right) = \boxed{o} \left(\sqrt{n}\right)$

$\lg\left(n^{3/2}\right) = \boxed{} \left(n^{1/10}\right)$ $\qquad$ $\lg\left(n^{3/2}\right) = \boxed{o} \left(n^{1/10}\right)$

$\lg\left(n^{3/2}\right) = \boxed{} \left(\left(\log_{10} n\right)^2\right)$ $\qquad$ $\lg\left(n^{3/2}\right) = \boxed{o} \left(\left(\log_{10} n\right)^2\right)$

$\lg\left(n^{3/2}\right) = \boxed{} \left((3/2)^n\right)$ $\qquad$ $\lg\left(n^{3/2}\right) = \boxed{o} \left((3/2)^n\right)$

$(7/2)^n = \boxed{} \left(n^7\right)$ $\qquad$ $(7/2)^n = \boxed{\omega} \left(n^7\right)$

$(7/2)^n = \boxed{} \left(n^3\right)$ $\qquad$ $(7/2)^n = \boxed{\omega} \left(n^3\right)$

$(7/2)^n = \boxed{} \left(7^n\right)$ $\qquad$ $(7/2)^n = \boxed{o} \left(7^n\right)$

$(7/2)^n = \boxed{} \left(3^n\right)$ $\qquad$ $(7/2)^n = \boxed{\omega} \left(3^n\right)$

$(7/3)^n = \boxed{} \left(3^n\right)$ $\qquad$ $(7/3)^n = \boxed{o} \left(3^n\right)$

$(7/2)^n = \boxed{} \left((\lg n)^{7/2}\right)$ $\qquad$ $(7/2)^n = \boxed{\omega} \left((\lg n)^{7/2}\right)$

$(7/2)^n = \boxed{} \left(\lg\left(n^{7/2}\right)\right)$ $\qquad$ $(7/2)^n = \boxed{\omega} \left(\lg\left(n^{7/2}\right)\right)$

$(3/2)^n = \boxed{} \left(n^{3/2}\right)$ $\qquad$ $(3/2)^n = \boxed{\omega} \left(n^{3/2}\right)$

$(3/2)^n = \boxed{} \left(n^3\right)$ $\qquad$ $(3/2)^n = \boxed{\omega} \left(n^3\right)$

$2^n = \boxed{} \left(3^n\right)$ $\qquad$ $2^n = \boxed{o} \left(3^n\right)$

$2^{n+1} = \boxed{} \left(2^n\right)$ $\qquad$ $2^{n+1} = \boxed{\Theta} \left(2^n\right)$

$5n^2 + 11 = \boxed{} \left(n^3 + 4n^2\right)$ $\qquad$ $5n^2 + 11 = \boxed{o} \left(n^3 + 4n^2\right)$

$5n^2 + 11 = \boxed{} \left(6n^2 + n + 1\right)$ $\qquad$ $5n^2 + 11 = \boxed{\Theta} \left(6n^2 + n + 1\right)$

$5n^2 + 11 = \boxed{} \left((5n + 11)^2\right)$ $\qquad$ $5n^2 + 11 = \boxed{\Theta} \left((5n + 11)^2\right)$

$n^2 + 3^n = \boxed{} \left(n^3 + 2^n\right)$ $\qquad$ $n^2 + 3^n = \boxed{\omega} \left(n^3 + 2^n\right)$

$2^n \cdot n^3 = \boxed{} \left(3^n \cdot n^2\right)$ $\qquad$ $2^n \cdot n^3 = \boxed{o} \left(3^n \cdot n^2\right)$

$n! = \boxed{} \left(16^n\right)$ $\qquad$ $n! = \boxed{\omega} \left(16^n\right)$

Problem 267 In each of the following formulae, write in the empty box one of the following letters: o, ω, Θ (little-o, little-ω, Θ) so as to obtain a correct claim about the asymptotic order of the two functions.

Answer:

$\log_{10}\left(n^4\right) = \boxed{}\left(\log_3\left(n\cdot\sqrt{n^3}\right)\right)$ $\log_{10}\left(n^4\right) = \boxed{\Theta}\left(\log_3\left(n\cdot\sqrt{n^3}\right)\right)$

$\lg\left(n^3\right) = \boxed{}\left(\lg\left(n^2\right)\cdot\lg n\right)$ $\lg\left(n^3\right) = \boxed{o}\left(\lg\left(n^2\right)\cdot\lg n\right)$

$\lg\left(n^2\right) = \boxed{}\left(\lg\left(n^3\right)\right)$ $\lg\left(n^2\right) = \boxed{\Theta}\left(\lg\left(n^3\right)\right)$

$\lg^3 n = \boxed{}\left(\lg\left(n^6\right)\right)$ $\lg^3 n = \boxed{\omega}\left(\lg\left(n^6\right)\right)$

$(\lg\lg n)^3 = \boxed{}\left(\lg\left(n^6\right)\right)$ $(\lg\lg n)^3 = \boxed{o}\left(\lg\left(n^6\right)\right)$

$n^2\cdot\lg n = \boxed{}\left(n\cdot\lg^2 n\right)$ $n^2\cdot\lg n = \boxed{\omega}\left(n\cdot\lg^2 n\right)$

$n^2\cdot\lg n = \boxed{}\left((n\cdot\lg n)^2\right)$ $n^2\cdot\lg n = \boxed{o}\left((n\cdot\lg n)^2\right)$

$n^2\cdot\lg n = \boxed{}\left(n^2 + \lg n\right)$ $n^2\cdot\lg n = \boxed{\omega}\left(n^2 + \lg n\right)$

$n^2\cdot\lg n = \boxed{}\left(n^2 + \lg^2 n\right)$ $n^2\cdot\lg n = \boxed{\omega}\left(n^2 + \lg^2 n\right)$

$n^2\cdot\lg n = \boxed{}\left(n^2\cdot\lg^2 n\right)$ $n^2\cdot\lg n = \boxed{o}\left(n^2\cdot\lg^2 n\right)$

$n\cdot\lg n = \boxed{}\left(n^2/\left(\lg^2 n\right)\right)$ $n\cdot\lg n = \boxed{o}\left(n^2/\left(\lg^2 n\right)\right)$

$n\cdot\lg n = \boxed{}\left(n\cdot\lg\lg n\right)$ $n\cdot\lg n = \boxed{\omega}\left(n\cdot\lg\lg n\right)$

$n\cdot\lg n = \boxed{}\left(n^{3/2}\right)$ $n\cdot\lg n = \boxed{o}\left(n^{3/2}\right)$

$n\cdot\lg n = \boxed{}\left(n\right)$ $n\cdot\lg n = \boxed{\omega}\left(n\right)$

$n\cdot\lg n = \boxed{}\left(\left(n\cdot\lg\lg\left(n^2\right)\right)^2\right)$ $n\cdot\lg n = \boxed{o}\left(\left(n\cdot\lg\lg\left(n^2\right)\right)^2\right)$

$n\cdot\lg n = \boxed{}\left(n^2\right)$ $n\cdot\lg n = \boxed{o}\left(n^2\right)$

$n^2 + 3 = \boxed{}\left(n^3 + 2\right)$ $n^2 + 3 = \boxed{o}\left(n^3 + 2\right)$

$n^2 + 3 = \boxed{}\left((n+2)^2\right)$ $n^2 + 3 = \boxed{\Theta}\left((n+2)^2\right)$

$n^2 + 3 = \boxed{}\left((n+2)^2 + n\right)$ $n^2 + 3 = \boxed{\Theta}\left((n+2)^2 + n\right)$

$n^2 + 3 = \boxed{}\left(n^2\cdot\lg\lg n\right)$ $n^2 + 3 = \boxed{o}\left(n^2\cdot\lg\lg n\right)$

$n^2 + 3 = \boxed{}\left(n^2 + \lg\lg n\right)$ $n^2 + 3 = \boxed{\Theta}\left(n^2 + \lg\lg n\right)$

$n^2 + 3 = \boxed{}\left((3/2)^n\right)$ $n^2 + 3 = \boxed{o}\left((3/2)^n\right)$

$(3/2)^n = \boxed{}\left(2^n\right)$ $(3/2)^n = \boxed{o}\left(2^n\right)$

$(3/2)^n = \boxed{}\left((4/3)^n\right)$ $(3/2)^n = \boxed{\omega}\left((4/3)^n\right)$

$3^n + n^2 = \boxed{}\left(2^n + n^3\right)$ $3^n + n^2 = \boxed{\omega}\left(2^n + n^3\right)$

$3^n + n^2 = \boxed{}\left(3^n + n^3\right)$ $3^n + n^2 = \boxed{\Theta}\left(3^n + n^3\right)$

$3^n + n^2 = \boxed{}\left(3^n\cdot\lg n\right)$ $3^n + n^2 = \boxed{o}\left(3^n\cdot\lg n\right)$

Problem 268   In each of the following formulae, write in the empty box one of the following letters: o, ω, Θ (little-o, little-ω, Θ) so as to obtain a correct claim about the asymptotic order of the two functions.

Answer:

$\log_4\left(\sqrt{n}\right) = \boxed{}\ \left(\log_2\left(n^2 \cdot \sqrt{n}\right)\right) \qquad \log_4\left(\sqrt{n}\right) = \boxed{\Theta}\ \left(\log_2\left(n^2 \cdot \sqrt{n}\right)\right)$

$\lg\left(n^2\right) = \boxed{}\ \left(\lg\left(n \cdot \lg n\right)\right) \qquad \lg\left(n^2\right) = \boxed{\Theta}\ \left(\lg\left(n \cdot \lg n\right)\right)$

$n^2 = \boxed{}\ \left(n \cdot \lg n\right) \qquad n^2 = \boxed{\omega}\ \left(n \cdot \lg n\right)$

$\lg n = \boxed{}\ \left(\lg\left(n+1\right)\right) \qquad \lg n = \boxed{\Theta}\ \left(\lg\left(n+1\right)\right)$

$\lg n = \boxed{}\ \left(\lg\left(2n\right)\right) \qquad \lg n = \boxed{\Theta}\ \left(\lg\left(2n\right)\right)$

$\lg n = \boxed{}\ \left(\lg\left(n^2\right)\right) \qquad \lg n = \boxed{\Theta}\ \left(\lg\left(n^2\right)\right)$

$\lg n = \boxed{}\ \left(\left(\lg n\right)^2\right) \qquad \lg n = \boxed{o}\ \left(\left(\lg n\right)^2\right)$

$\lg n = \boxed{}\ \left(\lg\lg n\right) \qquad \lg n = \boxed{\omega}\ \left(\lg\lg n\right)$

$\lg n = \boxed{}\ \left(\lg\lg\left(n^2\right)\right) \qquad \lg n = \boxed{\omega}\ \left(\lg\lg\left(n^2\right)\right)$

$\lg n = \boxed{}\ \left(\left(\lg\lg n\right)^2\right) \qquad \lg n = \boxed{\omega}\ \left(\left(\lg\lg n\right)^2\right)$

$\lg n \cdot \lg n = \boxed{}\ \left(\left(\lg\lg n\right)^2\right) \qquad \lg n \cdot \lg n = \boxed{\omega}\ \left(\left(\lg\lg n\right)^2\right)$

$\lg n \cdot \lg n = \boxed{}\ \left(\lg\lg\left(n^2\right)\right) \qquad \lg n \cdot \lg n = \boxed{\omega}\ \left(\lg\lg\left(n^2\right)\right)$

$\lg n \cdot \lg n = \boxed{}\ \left(\left(\lg n\right)^2\right) \qquad \lg n \cdot \lg n = \boxed{\Theta}\ \left(\left(\lg n\right)^2\right)$

$\lg n \cdot \lg n = \boxed{}\ \left(\lg\left(n^2\right)\right) \qquad \lg n \cdot \lg n = \boxed{\omega}\ \left(\lg\left(n^2\right)\right)$

$\lg n \cdot \lg n = \boxed{}\ \left(\lg\left(2n\right)\right) \qquad \lg n \cdot \lg n = \boxed{\omega}\ \left(\lg\left(2n\right)\right)$

$\lg n \cdot \lg n = \boxed{}\ \left(\sqrt{\lg n}\right) \qquad \lg n \cdot \lg n = \boxed{\omega}\ \left(\sqrt{\lg n}\right)$

$\lg\left(n \cdot \lg n\right) = \boxed{}\ \left(\left(\lg\lg n\right)^2\right) \qquad \lg\left(n \cdot \lg n\right) = \boxed{\omega}\ \left(\left(\lg\lg n\right)^2\right)$

$\lg\left(n \cdot \lg n\right) = \boxed{}\ \left(\lg\lg\left(n^2\right)\right) \qquad \lg\left(n \cdot \lg n\right) = \boxed{\omega}\ \left(\lg\lg\left(n^2\right)\right)$

$\lg\left(n \cdot \lg n\right) = \boxed{}\ \left(\left(\lg n\right)^2\right) \qquad \lg\left(n \cdot \lg n\right) = \boxed{o}\ \left(\left(\lg n\right)^2\right)$

$\lg\left(n \cdot \lg n\right) = \boxed{}\ \left(\lg\left(n^2\right)\right) \qquad \lg\left(n \cdot \lg n\right) = \boxed{\Theta}\ \left(\lg\left(n^2\right)\right)$

$\lg\left(n \cdot \lg n\right) = \boxed{}\ \left(\lg\left(2n\right)\right) \qquad \lg\left(n \cdot \lg n\right) = \boxed{\Theta}\ \left(\lg\left(2n\right)\right)$

$\lg\left(n \cdot \lg n\right) = \boxed{}\ \left(\sqrt{\lg n}\right) \qquad \lg\left(n \cdot \lg n\right) = \boxed{\omega}\ \left(\sqrt{\lg n}\right)$

$3n^2 + 5 = \boxed{}\ \left(2n^3 + 6\right) \qquad 3n^2 + 5 = \boxed{o}\ \left(2n^3 + 6\right)$

$n^2 \cdot \lg n = \boxed{}\ \left(n^3 / \left(\lg n\right)\right) \qquad n^2 \cdot \lg n = \boxed{o}\ \left(n^3 / \left(\lg n\right)\right)$

$n^3 / \left(\lg n\right) = \boxed{}\ \left(n^3 / \left(\lg\lg n\right)\right) \qquad n^3 / \left(\lg n\right) = \boxed{o}\ \left(n^3 / \left(\lg\lg n\right)\right)$

$2^n \cdot n^5 = \boxed{}\ \left(3^n \cdot n^2\right) \qquad 2^n \cdot n^5 = \boxed{o}\ \left(3^n \cdot n^2\right)$

$n^5 = \boxed{}\ \left(\left(n+17\right)^5\right) \qquad n^5 = \boxed{\Theta}\ \left(\left(n+17\right)^5\right)$

Problem 269   In each of the following formulae, write in the empty box one of the following letters: o, ω, Θ (little-o, little-ω, Θ) so as to obtain a correct claim about the asymptotic order of the two functions.

Answer:

$\log_{10}\left(\sqrt{n^5}\right) = \boxed{} \ (\log_4(n\cdot\sqrt{n}))$ $\log_{10}\left(\sqrt{n^5}\right) = \boxed{\Theta} \ (\log_4(n\cdot\sqrt{n}))$

$\sqrt{n^5} = \boxed{} \ (n\cdot\sqrt{n})$ $\sqrt{n^5} = \boxed{\omega} \ (n\cdot\sqrt{n})$

$\lg\left(n^2\right) = \boxed{} \ (\lg\left(n^2\cdot\lg n\right))$ $\lg\left(n^2\right) = \boxed{\Theta} \ (\lg\left(n^2\cdot\lg n\right))$

$n\cdot\lg n = \boxed{} \ ((n+1)\cdot\lg(n+1))$ $n\cdot\lg n = \boxed{\Theta} \ ((n+1)\cdot\lg(n+1))$

$n\cdot\lg n = \boxed{} \ ((2n)\cdot\lg(2n))$ $n\cdot\lg n = \boxed{\Theta} \ ((2n)\cdot\lg(2n))$

$n\cdot\lg n = \boxed{} \ \left(n^2\cdot\lg\left(n^2\right)\right)$ $n\cdot\lg n = \boxed{o} \ \left(n^2\cdot\lg\left(n^2\right)\right)$

$n\cdot\lg n = \boxed{} \ \left(n^2/\lg\left(n^2\right)\right)$ $n\cdot\lg n = \boxed{o} \ \left(n^2/\lg\left(n^2\right)\right)$

$n\cdot\lg n = \boxed{} \ \left(n\cdot\lg^2 n\right)$ $n\cdot\lg n = \boxed{o} \ \left(n\cdot\lg^2 n\right)$

$n\cdot\lg n = \boxed{} \ \left(n\cdot\lg\left(n^2\right)\right)$ $n\cdot\lg n = \boxed{\Theta} \ \left(n\cdot\lg\left(n^2\right)\right)$

$n\cdot\lg n = \boxed{} \ (n\cdot\lg\lg n)$ $n\cdot\lg n = \boxed{\omega} \ (n\cdot\lg\lg n)$

$n\cdot\lg n = \boxed{} \ \left(n+(\sqrt{n})^3\right)$ $n\cdot\lg n = \boxed{o} \ \left(n+(\sqrt{n})^3\right)$

$n^2 = \boxed{} \ \left(n^2\cdot\lg n\right)$ $n^2 = \boxed{o} \ \left(n^2\cdot\lg n\right)$

$n^2 = \boxed{} \ \left(n^2/\lg n\right)$ $n^2 = \boxed{\omega} \ \left(n^2/\lg n\right)$

$n^2 = \boxed{} \ \left(n^2+\lg n\right)$ $n^2 = \boxed{\Theta} \ \left(n^2+\lg n\right)$

$n^2 = \boxed{} \ \left(n^2+\lg^3 n\right)$ $n^2 = \boxed{\Theta} \ \left(n^2+\lg^3 n\right)$

$3^n = \boxed{} \ \left(n^3+n^4\right)$ $3^n = \boxed{\omega} \ \left(n^3+n^4\right)$

$3^n = \boxed{} \ \left((10/3)^n\right)$ $3^n = \boxed{o} \ \left((10/3)^n\right)$

$3^n = \boxed{} \ \left(n^{10}\right)$ $3^n = \boxed{\omega} \ \left(n^{10}\right)$

$3^n = \boxed{} \ \left(3^{n-1}\right)$ $3^n = \boxed{\Theta} \ \left(3^{n-1}\right)$

$3^n = \boxed{} \ \left(4^{n-4}\right)$ $3^n = \boxed{o} \ \left(4^{n-4}\right)$

$3^n = \boxed{} \ (3^n\cdot\lg\lg n)$ $3^n = \boxed{o} \ (3^n\cdot\lg\lg n)$

$3^n = \boxed{} \ \left(3^n+\lg^3 n\right)$ $3^n = \boxed{\Theta} \ \left(3^n+\lg^3 n\right)$

$3^n = \boxed{} \ (3n!)$ $3^n = \boxed{o} \ (3n!)$

$3^n = \boxed{} \ (n^n/3)$ $3^n = \boxed{o} \ (n^n/3)$

$\lg^2\left(n^5\right) = \boxed{} \ \left(\lg^5\left(n^2\right)\right)$ $\lg^2\left(n^5\right) = \boxed{o} \ \left(\lg^5\left(n^2\right)\right)$

$\lg^2\left(n^5\right) = \boxed{} \ \left(\lg^5\left(n^5\right)\right)$ $\lg^2\left(n^5\right) = \boxed{o} \ \left(\lg^5\left(n^5\right)\right)$

$\lg^2\left(n^5\right) = \boxed{} \ \left(\lg\lg\left(n^5\right)\right)$ $\lg^2\left(n^5\right) = \boxed{\omega} \ \left(\lg\lg\left(n^5\right)\right)$

Problem 270   In each of the following formulae, write in the empty box one of the following letters: o, ω, Θ (little-o, little-ω, Θ) so as to obtain a correct claim about the asymptotic order of the two functions.

Answer:

$$\log_2 n^8 = \boxed{} \ (\log_8 n^2) \qquad \log_2 n^8 = \boxed{\Theta} \ (\log_8 n^2)$$

$$\log_2\left((n^{10})\right)^2 = \boxed{} \ (\log_{10} n^{10}) \qquad \log_2\left((n^{10})\right)^2 = \boxed{\Theta} \ (\log_{10} n^{10})$$

$$\lg\lg\left(n^2\right) = \boxed{} \ (\lg\lg n)^2 \qquad \lg\lg\left(n^2\right) = \boxed{o} \ (\lg\lg n)^2$$

$$\lg\lg\left(3n\right) = \boxed{} \ (\lg\lg n) \qquad \lg\lg\left(3n\right) = \boxed{\Theta} \ (\lg\lg n)$$

$$\lg^2 n = \boxed{} \ (\lg\lg\left(2^n\right)) \qquad \lg^2 n = \boxed{\omega} \ (\lg\lg\left(2^n\right))$$

$$\lg^3 n = \boxed{} \ (\lg\left(3^n\right)) \qquad \lg^3 n = \boxed{o} \ (\lg\left(3^n\right))$$

$$n^3 \cdot \lg n = \boxed{} \ (n^{3.5}) \qquad n^3 \cdot \lg n = \boxed{o} \ (n^{3.5})$$

$$n^2 + 17n = \boxed{} \ (n^3 + 1) \qquad n^2 + 17n = \boxed{o} \ (n^3 + 1)$$

$$\sqrt{n^3} = \boxed{} \ (n + \sqrt{n}) \qquad \sqrt{n^3} = \boxed{\omega} \ (n + \sqrt{n})$$

$$n/\lg n = \boxed{} \ ((n+1)/\lg(n+1)) \qquad n/\lg n = \boxed{\Theta} \ ((n+1)/\lg(n+1))$$

$$n/\lg^2 n = \boxed{} \ (n/(\lg\lg n)) \qquad n/\lg^2 n = \boxed{o} \ (n/(\lg\lg n))$$

$$n^2 + 3 = \boxed{} \ (n^3 + 3n^2) \qquad n^2 + 3 = \boxed{o} \ (n^3 + 3n^2)$$

$$5^n = \boxed{} \ (17^n) \qquad 5^n = \boxed{o} \ (17^n)$$

$$n/(\log\log n) = \boxed{} \ (n/\lg n) \qquad n/(\log\log n) = \boxed{\omega} \ (n/\lg n)$$

$$n^2\left(\log^2 n\right) = \boxed{} \ (n^{2.1}) \qquad n^2\left(\log^2 n\right) = \boxed{o} \ (n^{2.1})$$

$$n^2\left(\sqrt{n}\right) = \boxed{} \ (n^{2.1}) \qquad n^2\left(\sqrt{n}\right) = \boxed{\omega} \ (n^{2.1})$$

$$5n^2 + 3^n = \boxed{} \ (3n^2 + 5^n) \qquad 5n^2 + 3^n = \boxed{o} \ (3n^2 + 5^n)$$

$$5n^2 + 2^n = \boxed{} \ (5n^3 + (5/2)^n) \qquad 5n^2 + 2^n = \boxed{o} \ (5n^3 + (5/2)^n)$$

$$n/(\lg n) = \boxed{} \ (\sqrt{n} \cdot \lg n^3) \qquad n/(\lg n) = \boxed{\omega} \ (\sqrt{n} \cdot \lg n^3)$$

$$n^2/(\lg^2 n) = \boxed{} \ (n \cdot \lg n) \qquad n^2/(\lg^2 n) = \boxed{\omega} \ (n \cdot \lg n)$$

$$3^n \cdot \lg n = \boxed{} \ (3^n \cdot \lg^2 n) \qquad 3^n \cdot \lg n = \boxed{o} \ (3^n \cdot \lg^2 n)$$

$$5^n = \boxed{} \ (3^n \cdot 2^n) \qquad 5^n = \boxed{o} \ (3^n \cdot 2^n)$$

$$5^n = \boxed{} \ (3^n + 2^n) \qquad 5^n = \boxed{\omega} \ (3^n + 2^n)$$

$$6^n = \boxed{} \ (3^{n+1} \cdot 2^{n+1}) \qquad 6^n = \boxed{\Theta} \ (3^{n+1} \cdot 2^{n+1})$$

$$2^n = \boxed{} \ (\log_2 n) \qquad 2^n = \boxed{\omega} \ (\log_2 n)$$

$$n! = \boxed{} \ (n^n) \qquad n! = \boxed{o} \ (n^n)$$

$$\lg(n!) = \boxed{} \ (\lg\left(n^n\right)) \qquad \lg(n!) = \boxed{\Theta} \ (\lg\left(n^n\right))$$

Problem 271   In each of the following formulae, write in the empty box one of the following letters: o, ω, Θ (little-o, little-ω, Θ) so as to obtain a correct claim about the asymptotic order of the two functions.

Answer:

$$5\log n = \boxed{} \ (\sqrt{n}) \qquad 5\log n = \boxed{o} \ (\sqrt{n})$$

$$5\log\log n = \boxed{} \ (1) \qquad 5\log\log n = \boxed{\omega} \ (1)$$

$$5\log\log n = \boxed{} \ (\log^2 n) \qquad 5\log\log n = \boxed{o} \ (\log^2 n)$$

$$5\log^2 n = \boxed{} \ (2n^2) \qquad 5\log^2 n = \boxed{o} \ (2n^2)$$

$$5\log^2 n = \boxed{} \ (4n^{(1/2)}) \qquad 5\log^2 n = \boxed{o} \ (4n^{(1/2)})$$

$$5\log_2 n = \boxed{} \ (10\log_{10} n) \qquad 5\log_2 n = \boxed{\Theta} \ (10\log_{10} n)$$

$$5\log\log n = \boxed{} \ (2\log n) \qquad 5\log\log n = \boxed{o} \ (2\log n)$$

$$5n^2 = \boxed{} \ (n \cdot \sqrt{n}) \qquad 5n^2 = \boxed{\omega} \ (n \cdot \sqrt{n})$$

$$5n^2 \cdot \log n = \boxed{} \ (n^2) \qquad 5n^2 \cdot \log n = \boxed{\omega} \ (n^2)$$

$$15n^2 + n^3 = \boxed{} \ \left(n^2 (n+1)^2\right) \qquad 15n^2 + n^3 = \boxed{o} \ \left(n^2 (n+1)^2\right)$$

$$5n/(\log n) = \boxed{} \ (\sqrt{n}) \qquad 5n/(\log n) = \boxed{\omega} \ (\sqrt{n})$$

$$5n/(\log n) = \boxed{} \ (n/(\log\log n)) \qquad 5n/(\log n) = \boxed{o} \ (n/(\log\log n))$$

$$5n^4 = \boxed{} \ (4^n) \qquad 5n^4 = \boxed{o} \ (4^n)$$

$$5n^4 = \boxed{} \ (2^n) \qquad 5n^4 = \boxed{o} \ (2^n)$$

$$2^n = \boxed{} \ (3^n) \qquad 2^n = \boxed{o} \ (3^n)$$

$$5^n = \boxed{} \ (3^n) \qquad 5^n = \boxed{\omega} \ (3^n)$$

$$3^n + n^3 = \boxed{} \ (n^4) \qquad 3^n + n^3 = \boxed{\omega} \ (n^4)$$

$$3^n + n^5 = \boxed{} \ (n^4) \qquad 3^n + n^5 = \boxed{\omega} \ (n^4)$$

$$n \cdot 2^n = \boxed{} \ (4^n) \qquad n \cdot 2^n = \boxed{o} \ (4^n)$$

$$n \cdot 3^n = \boxed{} \ (3^n) \qquad n \cdot 3^n = \boxed{\omega} \ (3^n)$$

$$\lg(n \cdot 3^n) = \boxed{} \ (\lg(3^n)) \qquad \lg(n \cdot 3^n) = \boxed{\Theta} \ (\lg(3^n))$$

$$\sqrt{n^3} = \boxed{} \ (5n) \qquad \sqrt{n^3} = \boxed{\omega} \ (5n)$$

$$n^2 \cdot \log(n^2) = \boxed{} \ (n^2 \cdot \log^2 n) \qquad n^2 \cdot \log(n^2) = \boxed{o} \ (n^2 \cdot \log^2 n)$$

$$15n^5 + n^3 = \boxed{} \ (n^4 \cdot \sqrt{n}) \qquad 15n^5 + n^3 = \boxed{\omega} \ (n^4 \cdot \sqrt{n})$$

$$\sqrt{n} = \boxed{} \ (n^{(5/6)}) \qquad \sqrt{n} = \boxed{o} \ (n^{(5/6)})$$

$$n/(\log n) = \boxed{} \ (\sqrt{n}) \qquad n/(\log n) = \boxed{\omega} \ (\sqrt{n})$$

$$n^2 = \boxed{} \ (2^n) \qquad n^2 = \boxed{o} \ (2^n)$$

$$n^3 = \boxed{} \ (2^n) \qquad n^3 = \boxed{o} \ (2^n)$$

Problem 272   In each of the following formulae, write in the empty box one of the following letters: o, ω, Θ (little-o, little-ω, Θ) so as to obtain a correct claim about the asymptotic order of the two functions.

Answer:

$$\log\left(n^5\right) = \boxed{}\ \left(\sqrt{n^5}\right) \qquad \log\left(n^5\right) = \boxed{o}\ \left(\sqrt{n^5}\right)$$

$$\log\log n = \boxed{}\ \left(\log^2 n\right) \qquad \log\log n = \boxed{o}\ \left(\log^2 n\right)$$

$$\log\log n = \boxed{}\ \left(\log\log n^3\right) \qquad \log\log n = \boxed{\Theta}\ \left(\log\log n^3\right)$$

$$\log^2 n = \boxed{}\ (2^n) \qquad \log^2 n = \boxed{o}\ (2^n)$$

$$\log^3 n = \boxed{}\ (3^n) \qquad \log^3 n = \boxed{o}\ (3^n)$$

$$\log^4 n = \boxed{}\ \left(n^{(1/8)}\right) \qquad \log^4 n = \boxed{o}\ \left(n^{(1/8)}\right)$$

$$\log^4 n = \boxed{}\ \left(n^{(1/4)}\right) \qquad \log^4 n = \boxed{o}\ \left(n^{(1/4)}\right)$$

$$\log_2 n = \boxed{}\ (\log_{10} n) \qquad \log_2 n = \boxed{\Theta}\ (\log_{10} n)$$

$$\log\left(n^5\right) = \boxed{}\ \left(\sqrt{\log n}\right) \qquad \log\left(n^5\right) = \boxed{\omega}\ \left(\sqrt{\log n}\right)$$

$$\log_{10}\left(n^5\right) = \boxed{}\ \left(\log_2\left(n^{10}\right)\right) \qquad \log_{10}\left(n^5\right) = \boxed{\Theta}\ \left(\log_2\left(n^{10}\right)\right)$$

$$\sqrt{n} = \boxed{}\ \left(n^{(1/4)}\right) \qquad \sqrt{n} = \boxed{\omega}\ \left(n^{(1/4)}\right)$$

$$n^2 \cdot \log n = \boxed{}\ \left(n^3/\log n\right) \qquad n^2 \cdot \log n = \boxed{o}\ \left(n^3/\log n\right)$$

$$n^5 + n^3 = \boxed{}\ \left(15n^4 \cdot \log n\right) \qquad n^5 + n^3 = \boxed{\omega}\ \left(15n^4 \cdot \log n\right)$$

$$n/(\log n) = \boxed{}\ (\sqrt{n}) \qquad n/(\log n) = \boxed{\omega}\ (\sqrt{n})$$

$$n^3 = \boxed{}\ (2^n) \qquad n^3 = \boxed{o}\ (2^n)$$

$$n^2 = \boxed{}\ (2^n) \qquad n^2 = \boxed{o}\ (2^n)$$

$$2^n + n^5 = \boxed{}\ \left(n^4\right) \qquad 2^n + n^5 = \boxed{\omega}\ \left(n^4\right)$$

$$4^n + n^3 = \boxed{}\ \left(2^n + n^4\right) \qquad 4^n + n^3 = \boxed{\omega}\ \left(2^n + n^4\right)$$

$$3^n + n^5 = \boxed{}\ \left(n^6\right) \qquad 3^n + n^5 = \boxed{\omega}\ \left(n^6\right)$$

$$5 \cdot 2^n = \boxed{}\ (2^n) \qquad 5 \cdot 2^n = \boxed{\Theta}\ (2^n)$$

$$5 \cdot 2^n = \boxed{}\ (3^n) \qquad 5 \cdot 2^n = \boxed{o}\ (3^n)$$

$$\lg\left(5 \cdot 2^n\right) = \boxed{}\ (\lg\left(3^n\right)) \qquad \lg\left(5 \cdot 2^n\right) = \boxed{\Theta}\ (\lg\left(3^n\right))$$

$$n! = \boxed{}\ (2^n) \qquad n! = \boxed{\omega}\ (2^n)$$

$$\lg(n!) = \boxed{}\ (\lg\left(2^n\right)) \qquad \lg(n!) = \boxed{\omega}\ (\lg\left(2^n\right))$$

$$\lg(n!) = \boxed{}\ \left(\lg n \cdot \lg\left(2^n\right)\right) \qquad \lg(n!) = \boxed{\Theta}\ \left(\lg n \cdot \lg\left(2^n\right)\right)$$

$$\lg(n^n) = \boxed{}\ (\lg\left(2^n\right)) \qquad \lg(n^n) = \boxed{\omega}\ (\lg\left(2^n\right))$$

$$\lg(n^n) = \boxed{}\ \left(\lg n \cdot \lg\left(2^n\right)\right) \qquad \lg(n^n) = \boxed{\Theta}\ \left(\lg n \cdot \lg\left(2^n\right)\right)$$

$$\lg(2^n) = \boxed{}\ (n) \qquad \lg(2^n) = \boxed{\Theta}\ (n)$$

Problem 273   In each of the following formulae, write in the empty box one of the following letters: o, ω, Θ (little-o, little-ω, Θ) so as to obtain a correct claim about the asymptotic order of the two functions.

Answer:

$$\log\log n = \boxed{} \left(\log\log\left(n^3\right)\right) \qquad \log\log n = \boxed{\Theta} \left(\log\log\left(n^3\right)\right)$$

$$\log\log\left(n^2\right) = \boxed{} \left(\log\log n\right) \qquad \log\log\left(n^2\right) = \boxed{\Theta} \left(\log\log n\right)$$

$$\log^3 n = \boxed{} \left(\log\left(n^3\right)\right) \qquad \log^3 n = \boxed{\omega} \left(\log\left(n^3\right)\right)$$

$$\log\left(n^3\right) = \boxed{} \left(\log\left(n^2\right)\right) \qquad \log\left(n^3\right) = \boxed{\Theta} \left(\log\left(n^2\right)\right)$$

$$\log^2 n = \boxed{} \left(\log^3 n\right) \qquad \log^2 n = \boxed{o} \left(\log^3 n\right)$$

$$\log^2 n = \boxed{} \left(\log^3 n\right) \qquad \log^2 n = \boxed{o} \left(\log^3 n\right)$$

$$\log^2\left(n\right) = \boxed{} \left(n^{0.2}\right) \qquad \log^2\left(n\right) = \boxed{o} \left(n^{0.2}\right)$$

$$\log^4 n = \boxed{} \left(\log\left(n^4\right)\right) \qquad \log^4 n = \boxed{\omega} \left(\log\left(n^4\right)\right)$$

$$\log\left(n^4\right) = \boxed{} \left(\log\left(n^2\right)\right) \qquad \log\left(n^4\right) = \boxed{\Theta} \left(\log\left(n^2\right)\right)$$

$$n\log\log n = \boxed{} \left(n + \log n\right) \qquad n\log\log n = \boxed{\omega} \left(n + \log n\right)$$

$$n^3\log\log n = \boxed{} \left(n^3 + \log n\right) \qquad n^3\log\log n = \boxed{\omega} \left(n^3 + \log n\right)$$

$$n^2 + n^3 = \boxed{} \left(5n^2 + n\right) \qquad n^2 + n^3 = \boxed{\omega} \left(5n^2 + n\right)$$

$$4n^3 = \boxed{} \left(n^3/\left(\log n\right)\right) \qquad 4n^3 = \boxed{\omega} \left(n^3/\left(\log n\right)\right)$$

$$n^2 + 3n^3 = \boxed{} \left(n^2 \cdot \log n\right) \qquad n^2 + 3n^3 = \boxed{\omega} \left(n^2 \cdot \log n\right)$$

$$3n^2 + n^3 = \boxed{} \left(n^3/\left(\log n\right)\right) \qquad 3n^2 + n^3 = \boxed{\omega} \left(n^3/\left(\log n\right)\right)$$

$$\sqrt{n} = \boxed{} \left(n^{(3/2)}\right) \qquad \sqrt{n} = \boxed{o} \left(n^{(3/2)}\right)$$

$$\sqrt{n} = \boxed{} \left(\log n\right) \qquad \sqrt{n} = \boxed{\omega} \left(\log n\right)$$

$$2^n + n^2 = \boxed{} \left(n^3\right) \qquad 2^n + n^2 = \boxed{\omega} \left(n^3\right)$$

$$2^n + n^5 = \boxed{} \left(n^6\right) \qquad 2^n + n^5 = \boxed{\omega} \left(n^6\right)$$

$$2^n + 3^n = \boxed{} \left(2^n \cdot n^4\right) \qquad 2^n + 3^n = \boxed{\omega} \left(2^n \cdot n^4\right)$$

$$2^n + 3^n = \boxed{} \left(4^n\right) \qquad 2^n + 3^n = \boxed{o} \left(4^n\right)$$

$$2^n + 4^n = \boxed{} \left(3^n \cdot n\right) \qquad 2^n + 4^n = \boxed{\omega} \left(3^n \cdot n\right)$$

$$2^n + 3^n = \boxed{} \left(n!\right) \qquad 2^n + 3^n = \boxed{o} \left(n!\right)$$

$$5n^2 + 2^n = \boxed{} \left(n^3 \cdot (1/2)^n\right) \qquad 5n^2 + 2^n = \boxed{\omega} \left(n^3 \cdot (1/2)^n\right)$$

$$5n^2 + 2^n = \boxed{} \left(n^3 \cdot (5/2)^n\right) \qquad 5n^2 + 2^n = \boxed{o} \left(n^3 \cdot (5/2)^n\right)$$

$$8^n + n^3 = \boxed{} \left(4^n \cdot 2^n + n^2\right) \qquad 8^n + n^3 = \boxed{\Theta} \left(4^n \cdot 2^n + n^2\right)$$

$$2^n + n^3 = \boxed{} \left(n^3\right) \qquad 2^n + n^3 = \boxed{\omega} \left(n^3\right)$$

$$3^n = \boxed{} \left(3^{n+1}\right) \qquad 3^n = \boxed{\Theta} \left(3^{n+1}\right)$$

Problem 274 In each of the following formulae, write in the empty box one of the following letters: o, ω, Θ (little-o, little-ω, Θ) so as to obtain a correct claim about the asymptotic order of the two functions.

Answer:

$\log\log\left(n^2\right) = \boxed{} \left(\log\log\log n\right)$ $\log\log\left(n^2\right) = \boxed{\omega} \left(\log\log\log n\right)$

$\log^2 n = \boxed{} \left(\log\left(n^2\right)\right)$ $\log^2 n = \boxed{\omega} \left(\log\left(n^2\right)\right)$

$\log^2\left(n^2\right) = \boxed{} \left(\log^2\left(n^3\right)\right)$ $\log^2\left(n^2\right) = \boxed{\Theta} \left(\log^2\left(n^3\right)\right)$

$\log\left(n^4\right) = \boxed{} \left(\log\left(2^n\right)\right)$ $\log\left(n^4\right) = \boxed{o} \left(\log\left(2^n\right)\right)$

$\log n = \boxed{} \left(\log\left(\sqrt{n}\right)\right)$ $\log n = \boxed{\Theta} \left(\log\left(\sqrt{n}\right)\right)$

$\log\left(n^5\right) = \boxed{} \left(5 + \log\left(n^2 \cdot \sqrt{n}\right)\right)$ $\log\left(n^5\right) = \boxed{\Theta} \left(5 + \log\left(n^2 \cdot \sqrt{n}\right)\right)$

$\log\log\log n = \boxed{} \left(\log^3 n\right)$ $\log\log\log n = \boxed{o} \left(\log^3 n\right)$

$\log^3 n = \boxed{} \left(n^{(1/3)}\right)$ $\log^3 n = \boxed{o} \left(n^{(1/3)}\right)$

$n \cdot \sqrt{n} = \boxed{} \left(n \cdot \log n\right)$ $n \cdot \sqrt{n} = \boxed{\omega} \left(n \cdot \log n\right)$

$n + n^2 + \log n = \boxed{} \left(5n^2/\left(\log n\right)\right)$ $n + n^2 + \log n = \boxed{\omega} \left(5n^2/\left(\log n\right)\right)$

$n \cdot \log n + n^2 = \boxed{} \left(n \cdot \log^2 n\right)$ $n \cdot \log n + n^2 = \boxed{\omega} \left(n \cdot \log^2 n\right)$

$n^2 \log\log n = \boxed{} \left(n^2 + \log^2 n\right)$ $n^2 \log\log n = \boxed{\omega} \left(n^2 + \log^2 n\right)$

$n^2 = \boxed{} \left(\log\left(2^n\right)\right)$ $n^2 = \boxed{\omega} \left(\log\left(2^n\right)\right)$

$n^2 + n^3 = \boxed{} \left(5n^2 + n\right)$ $n^2 + n^3 = \boxed{\omega} \left(5n^2 + n\right)$

$5n^4 + 4n^3 = \boxed{} \left(n^3 \cdot \log n\right)$ $5n^4 + 4n^3 = \boxed{\omega} \left(n^3 \cdot \log n\right)$

$\sqrt{n} = \boxed{} \left(n^{(2/3)}\right)$ $\sqrt{n} = \boxed{o} \left(n^{(2/3)}\right)$

$\sqrt{n} = \boxed{} \left(n^{(1/4)}\right)$ $\sqrt{n} = \boxed{\omega} \left(n^{(1/4)}\right)$

$n/\left(\log n\right) = \boxed{} \left(n + \log\log n\right)$ $n/\left(\log n\right) = \boxed{o} \left(n + \log\log n\right)$

$n + n^2 = \boxed{} \left(n^2 \cdot \log n\right)$ $n + n^2 = \boxed{o} \left(n^2 \cdot \log n\right)$

$n^3 + n^2 = \boxed{} \left(n^3/\left(\log n\right)\right)$ $n^3 + n^2 = \boxed{\omega} \left(n^3/\left(\log n\right)\right)$

$2^n \cdot \log n = \boxed{} \left(3^n\right)$ $2^n \cdot \log n = \boxed{o} \left(3^n\right)$

$2^n + 3^n = \boxed{} \left(2^n \cdot \log n\right)$ $2^n + 3^n = \boxed{\omega} \left(2^n \cdot \log n\right)$

$2^n + 3^n = \boxed{} \left(n^4\right)$ $2^n + 3^n = \boxed{\omega} \left(n^4\right)$

$2^n + n^3 = \boxed{} \left(n^4\right)$ $2^n + n^3 = \boxed{\omega} \left(n^4\right)$

$n \cdot \log^2 n = \boxed{} \left(n^2\right)$ $n \cdot \log^2 n = \boxed{o} \left(n^2\right)$

$n \cdot \log n = \boxed{} \left(\lg\left(n!\right)\right)$ $n \cdot \log n = \boxed{\Theta} \left(\lg\left(n!\right)\right)$

$n \cdot \log n = \boxed{} \left(\lg\left(n^n\right)\right)$ $n \cdot \log n = \boxed{\Theta} \left(\lg\left(n^n\right)\right)$

$n \cdot \log n = \boxed{} \left(\lg\left(8^n\right)\right)$ $n \cdot \log n = \boxed{\omega} \left(\lg\left(8^n\right)\right)$

Problem 275   In each of the following formulae, write in the empty box one of the following letters: o, ω, Θ (little-o, little-ω, Θ) so as to obtain a correct claim about the asymptotic order of the two functions.

$$n^2 \cdot \log^2 n = \boxed{} \left(n^2 \cdot \log\log n\right) \qquad n^2 \cdot \log^2 n = \boxed{\omega} \left(n^2 \cdot \log\log n\right)$$

$$15n^5 + n^3 = \boxed{} \left(n^4 \cdot n^{1.5}\right) \qquad 15n^5 + n^3 = \boxed{o} \left(n^4 \cdot n^{1.5}\right)$$

$$n \cdot \log n = \boxed{} \left(\sqrt{n^3}\right) \qquad n \cdot \log n = \boxed{o} \left(\sqrt{n^3}\right)$$

$$\sqrt{n} = \boxed{} \left(n^{(1/3)}\right) \qquad \sqrt{n} = \boxed{\omega} \left(n^{(1/3)}\right)$$

$$n^2 / \left(\log^2 n\right) = \boxed{} \left(n \cdot \sqrt{n}\right) \qquad n^2 / \left(\log^2 n\right) = \boxed{\omega} \left(n \cdot \sqrt{n}\right)$$

$$n^3 \cdot \log n = \boxed{} \left(\sqrt{n^7}\right) \qquad n^3 \cdot \log n = \boxed{o} \left(\sqrt{n^7}\right)$$

$$15n^2 + n^3 = \boxed{} \left(n^2 \left(n + \lg n\right)\right) \qquad 15n^2 + n^3 = \boxed{\Theta} \left(n^2 \left(n + \lg n\right)\right)$$

$$\sqrt{n^7} = \boxed{} \left(n \left(n + 1\right)\right) \qquad \sqrt{n^7} = \boxed{\omega} \left(n \left(n + 1\right)\right)$$

$$n / \left(\log n\right) = \boxed{} \left(\sqrt{n} \cdot \lg^2 n\right) \qquad n / \left(\log n\right) = \boxed{\omega} \left(\sqrt{n} \cdot \lg^2 n\right)$$

$$n / \left(\log n\right) = \boxed{} \left(n / \left(\log^2 n\right)\right) \qquad n / \left(\log n\right) = \boxed{\omega} \left(n / \left(\log^2 n\right)\right)$$

$$n^3 = \boxed{} \left(2^n \cdot n^2\right) \qquad n^3 = \boxed{o} \left(2^n \cdot n^2\right)$$

$$\log\log \sqrt{n} = \boxed{} \left(\sqrt{\log n}\right) \qquad \log\log \sqrt{n} = \boxed{o} \left(\sqrt{\log n}\right)$$

$$\log\log n^3 = \boxed{} \left(\log n \cdot \log n^3\right) \qquad \log\log n^3 = \boxed{o} \left(\log n \cdot \log n^3\right)$$

$$\log^2 n = \boxed{} \left(\log\log n^2\right) \qquad \log^2 n = \boxed{\omega} \left(\log\log n^2\right)$$

$$\log^3 n = \boxed{} \left((1/3)^n\right) \qquad \log^3 n = \boxed{\omega} \left((1/3)^n\right)$$

$$n \cdot \log\log n = \boxed{} \left(n^2 / \lg n\right) \qquad n \cdot \log\log n = \boxed{o} \left(n^2 / \lg n\right)$$

$$n^2 / \left(\log^2 n\right) = \boxed{} \left(n^{(1.9)}\right) \qquad n^2 / \left(\log^2 n\right) = \boxed{\omega} \left(n^{(1.9)}\right)$$

$$n^2 / \left(\sqrt{n}\right) = \boxed{} \left(n^{(1.1)}\right) \qquad n^2 / \left(\sqrt{n}\right) = \boxed{\omega} \left(n^{(1.1)}\right)$$

$$n / \left(\log n\right) = \boxed{} \left(\sqrt{n} \cdot \lg n^3\right) \qquad n / \left(\log n\right) = \boxed{\omega} \left(\sqrt{n} \cdot \lg n^3\right)$$

$$n / \left(\log n\right) = \boxed{} \left(n / \left(\log \lg n\right)\right) \qquad n / \left(\log n\right) = \boxed{o} \left(n / \left(\log \lg n\right)\right)$$

$$5 \cdot 2^n = \boxed{} \left(4^n\right) \qquad 5 \cdot 2^n = \boxed{o} \left(4^n\right)$$

$$\left(5 \cdot 2^n\right)^2 = \boxed{} \left(4^n\right) \qquad \left(5 \cdot 2^n\right)^2 = \boxed{\Theta} \left(4^n\right)$$

$$3^n = \boxed{} \left(2^n\right) \qquad 3^n = \boxed{\omega} \left(2^n\right)$$

$$3^n = \boxed{} \left((\sqrt{2})^n\right) \qquad 3^n = \boxed{\omega} \left((\sqrt{2})^n\right)$$

$$3^n = \boxed{} \left(\left(\left(\sqrt{2}\right)^2\right)^n\right) \qquad 3^n = \boxed{\omega} \left(\left(\left(\sqrt{2}\right)^2\right)^n\right)$$

$$3^n = \boxed{} \left(\left(\left(\sqrt{2}\right)^3\right)^n\right) \qquad 3^n = \boxed{\omega} \left(\left(\left(\sqrt{2}\right)^3\right)^n\right)$$

$$3^n = \boxed{} \left(\left(\left(\sqrt{2}\right)^4\right)^n\right) \qquad 3^n = \boxed{o} \left(\left(\left(\sqrt{2}\right)^4\right)^n\right)$$

$$4^n = \boxed{} \left(3^n \cdot 2^n\right) \qquad 4^n = \boxed{o} \left(3^n \cdot 2^n\right)$$

$$4^n = \boxed{} \left(3^n + 2^n\right) \qquad 4^n = \boxed{\omega} \left(3^n + 2^n\right)$$

Problem 276 Arrange the following fifteen functions into a table, such that any two functions, say $f(n)$ and $g(n)$, appear *in the same horizontal row* if and only if $f(n) = \Theta(g(n))$, while $f(n)$ appears *above* $g(n)$ in the table if and only if $f(n) = o(g(n))$.

$$n^2 + 3n^3 + 2^n \quad 2^{n+1} \quad n^3 + \lg^2 n \qquad 12n \qquad \lg \lg n$$

$$n^2 \lg \lg n \qquad n^2 \lg^2 n \quad 15n + n^2 + 1 \qquad \lg n \qquad n^3 + 3^n$$

$$n/(\lg n) \qquad (1.5)^n \quad (n-1)(n^2+1) \quad \log(\log^3 \sqrt{n}) \quad 3 \cdot 2^n + \log n$$

Answer:

$\lg \lg n$	$\log \left(\log^3 \sqrt{n}\right)$	
$\lg n$		
$n/\left(\lg n\right)$		
$12n$		
$15n + n^2 + 1$		
$n^2 \lg \lg n$		
$n^2 \lg^2 n$		
$n^3 + \lg^2 n$	$(n-1)\left(n^2+1\right)$	
$(1.5)^n$		
$n^2 + 3n^3 + 2^n$	2^{n+1}	$3 \cdot 2^n + \log n$
$n^3 + 3^n$		

Problem 277 Arrange the following fifteen functions into a table, such that any two functions, say $f(n)$ and $g(n)$, appear *in the same horizontal row* if and only if $f(n) = \Theta(g(n))$, while $f(n)$ appears *above* $g(n)$ in the table if and only if $f(n) = o(g(n))$.

$$\log \log(n^3) \quad \log(n^3) \quad n^3 + \log n \qquad 5^n + (3/2)^n \quad n^2 \log n$$

$$n^3 + 5^n \quad n! + n^3 \quad \log^3 n \qquad \log \log n \qquad \log^2 n + \log n + n^2$$

$$12n^5 \qquad \sqrt{n^3} \quad 11n^3 + 3n^5 + n^3 \log n \quad 5n^n + n^3 \qquad n \log n$$

Answer:

$\log \log(n^3)$	$\log \log n$
$\log(n^3)$	
$\log^3 n$	
$n \log n$	
$\sqrt{n^3}$	
$\log^2 n + \log n + n^2$	
$n^2 \log n$	
$n^3 + \log n$	
$12n^5$	$11n^3 + 3n^5 + n^3 \log n$
$n^3 + 5^n$	$5^n + (3/2)^n$
$n! + n^3$	
$5n^n + n^3$	

Problem 278   Arrange the following fifteen functions into a table, such that any two functions, say $f(n)$ and $g(n)$, appear *in the same horizontal row* if and only if $f(n) = \Theta(g(n))$, while $f(n)$ appears *above* $g(n)$ in the table if and only if $f(n) = o(g(n))$.

$$\log\log(n^5) \quad n^5 + 5^n \quad (\sqrt{5})^n \quad \log(n^5) \quad 5n^n$$
$$n\log n \quad n! + n^5 \quad n^5 + \log n \quad n^5 + (3/2)^n \quad n^2 + \log n$$
$$\log^5 n \quad \sqrt{n^5} \quad (\log\log n)^5 \quad \log^2 n + \log n + n^2 \quad 11n^2 + 3n^5 + \log\log n$$

Answer:

$\log\log(n^5)$	
$(\log\log n)^5$	
$\log(n^5)$	
$\log^5 n$	
$n\log n$	
$n^2 + \log n$	$\log^2 n + \log n + n^2$
$\sqrt{n^5}$	
$n^5 + \log n$	$11n^2 + 3n^5 + \log\log n$
$n^5 + (3/2)^n$	
$(\sqrt{5})^n$	
$n^5 + 5^n$	
$n! + n^5$	
$5n^n$	

Problem 279   Arrange the following fifteen functions into a table, such that any two functions, say $f(n)$ and $g(n)$, appear *in the same horizontal row* if and only if $f(n) = \Theta(g(n))$, while $f(n)$ appears *above* $g(n)$ in the table if and only if $f(n) = o(g(n))$.

$$0.5 \quad \log^2 n \quad n^{1.5}\log n \quad (n+3)! \quad n^3 + \log^3 n$$
$$\log\log n \quad \log n \quad n^{0.5} \quad 3n + (n+1)^3 \quad \log^2 n + 0.5n$$
$$\log\log n + n^3 \quad 3^n + n^n + n \quad \sqrt{n^3} \quad 3^{n+1} \quad 3^n + n$$

Answer:

0.5		
$\log\log n$		
$\log n$		
$\log^2 n$		
$n^{0.5}$		
$\log^2 n + 0.5n$		
$\sqrt{n^3}$		
$n^{1.5}\log n$		
$n^3 + \log^3 n$	$3n + (n+1)^3$	$\log\log n + n^3$
3^{n+1}	$3^n + n$	
$(n+3)!$		
$3^n + n^n + n$		

Problem 280   Arrange the following eighteen functions into a table, such that any two functions, say $f(n)$ and $g(n)$, appear *in the same horizontal row* if and only if $f(n) = \Theta(g(n))$, while $f(n)$ appears *above* $g(n)$ in the table if and only if $f(n) = o(g(n))$.

$$
\begin{array}{llllll}
4^{n+4} & (n-1)(n+2) & 5 & n^{0.01} & (n+1)^2 + 2n & 0.001n \\
\sqrt{n^5} & 5\log n + 5^n + 5 & (n-\sqrt{n})(n+\log n) & n^2\log n & (n+1)! & n + \log^2 n \\
4^n & n^2 + \log\log n & \log\log\log n & \log n & (\log n)^3 & n + \sqrt{\log n}
\end{array}
$$

Answer:

5			
$\log\log\log n$			
$\log n$			
$(\log n)^3$			
$n^{0.01}$			
$0.001n$	$n + \log^2 n$	$n + \sqrt{\log n}$	
$(n+1)^2 + 2n$	$n^2 + \log\log n$	$(n-1)(n+2)$	$(n-\sqrt{n})(n+\log n)$
$n^2\log n$			
$\sqrt{n^5}$			
4^{n+4}	4^n		
$5\log n + 5^n + 5$			
$(n+1)!$			

Problem 281   Arrange the following eighteen functions into a table, such that any two functions, say $f(n)$ and $g(n)$, appear *in the same horizontal row* if and only if $f(n) = \Theta(g(n))$, while $f(n)$ appears *above* $g(n)$ in the table if and only if $f(n) = o(g(n))$.

$$
\begin{array}{llllll}
5\cdot 3^n + 2^{n+1} & 3^n + 3 & (n-5)(n-4) & \lg(3n^5) & (n+1)! & 1 \\
n/(\lg n) + n^2 & 2^{n+2} + n^2 & \lg n & n^2 + \lg n & \lg\lg n + \lg n & \sqrt{n^3} + n^2 \\
n/(\lg^2 n) & 3^{n-3} + n! & n^3 + 3^n & 3^{n-3} + n^3 & \lg\lg n & n^n + 2^n
\end{array}
$$

Answer:

1			
$\lg\lg n$			
$\lg\lg n + \lg n$	$\lg n$	$\lg(3n^5)$	
$n/(\lg^2 n)$			
$n/(\lg n) + n^2$	$n^2 + \lg n$	$\sqrt{n^3} + n^2$	$(n-5)(n-4)$
$2^{n+2} + n^2$			
$n^3 + 3^n$	$3^{n-3} + n^3$	$5\cdot 3^n + 2^{n+1}$	$3^n + 3$
$3^{n-3} + n!$			
$(n+1)!$			
$n^n + 2^n$			

Problem 282   Arrange the following eighteen functions into a table, such that any two functions, say $f(n)$ and $g(n)$, appear *in the same horizontal row* if and only if $f(n) = \Theta(g(n))$, while $f(n)$ appears *above* $g(n)$ in the table if and only if $f(n) = o(g(n))$.

$$3^{n-3} + n^3 \quad \log_2(n^3) \quad \log^2 n \quad n/(\log_2 \log_2 n) + \log_2 n \quad 2^{n+2} + (3n)^2 \quad \log_{10}(n^2)$$

$$n! + 2^n \quad n! + \sqrt{n^3} \quad n^{3/2} + 5 \quad n^2/(\log_{10} n) + 3n \quad n^3 + 4^n \quad (2n)^2 + \log_2 n$$

$$n! + n^n \quad \sqrt{n^3} + 3n \quad 2\log^3 n \quad n/(\log_2^2 n) + \log_2 n \quad (3/2)^n \quad n^2/(\sqrt{n}) + 2n$$

Answer:

$\log_2(n^3)$	$\log_{10}(n^2)$	
$\log^2 n$		
$2\log^3 n$		
$n/(\log_2^2 n) + \log_2 n$		
$n/(\log_2 \log_2 n) + \log_2 n$		
$n^{3/2} + 5$	$\sqrt{n^3} + 3n$	$n^2/(\sqrt{n}) + 2n$
$n^2/(\log_{10} n) + 3n$		
$(2n)^2 + \log_2 n$		
$(3/2)^n$		
$2^{n+2} + (3n)^2$		
$3^{n-3} + n^3$		
$n^3 + 4^n$		
$n! + \sqrt{n^3}$	$n! + 2^n$	
$n! + n^n$		

Problem 283   In each of the following cases, construct a function $f(n)$ that satisfies the specified constraints, or state that it is impossible.

Answer:

$f(n) = O\left(\sqrt{n} + \lg\lg n\right)$ and $f(n) = \omega\left((\lg\lg n)^2\right)$	$f(n) = \sqrt{n}$
$f(n) = \omega\left(n^3 + 3^n\right)$ and $f(n) = o\left(n^4 + 3^{n+1}\right)$	Impossible.
$f(n) = \Omega\left(n^3 + n^2 + n\right)$ and $f(n) = o\left(n^4\right)$	$f(n) = n^3$
$f(n) = \Omega\left(3n \log n\right)$ and $f(n) = O\left(n \log(n^2)\right)$	$f(n) = n \log n$
$f(n) = \omega\left(4n^3\right)$ and $f(n) = o\left(2n^4\right)$	$f(n) = n^{3.5}$
$f(n) = \omega\left(2^n/n^2\right)$ and $f(n) = o\left(3^n/n^3\right)$	$f(n) = 2^n$
$f(n) = O\left(\log_{10} n\right)$ and $f(n) = \Omega\left(\log_5 n\right)$	$f(n) = \log_2 n$
$f(n) = \Omega\left(n^n\right)$ and $f(n) = \Omega\left(n!\right)$	$f(n) = n^n$
$f(n) = \Omega\left(4n^2\right)$ and $f(n) = O\left(3n^2\right)$	$f(n) = n^2$
$f(n) = \omega\left(n^2 \lg n\right)$ and $f(n) = o\left(n^2 \lg^2 n\right)$	$f(n) = n^2 \lg^{1.5} n$
$f(n) = O\left(n^{0.1}\right)$ and $f(n) = \omega\left(\lg n\right)$	$f(n) = \lg^2 n$
$f(n) = \Omega\left(n^3\right)$ and $f(n) = o\left(n^3 + (3/2)^n\right)$	$f(n) = (5/4)^n$
$f(n) = \Theta\left(n^2\right)$ and $f(n) = O\left(n^2 + n\right)$	$f(n) = n^2$

Problem 284 Arrange the following eighteen functions into a table, such that any two functions, say $f(n)$ and $g(n)$, appear *in the same horizontal row* if and only if $f(n) = \Theta(g(n))$, while $f(n)$ appears *above* $g(n)$ in the table if and only if $f(n) = o(g(n))$.

$$
\begin{array}{cccccc}
n^2 & n^3 & \log n & n^{1/3} + \log n & (\log n)^2 & n! \\
(1/3)^n & (3/2)^n & 6 & \log n + 5 & n/(\log n) & \log\log n \\
\sqrt{n} & n & 2^n & n\log n & n + n^3 + 7n^5 & n^2 + \log n
\end{array}
$$

Answer:

$(1/3)^n$	
6	
$\log\log n$	
$\log n + 5$	$\log n$
$(\log n)^2$	
$n^{1/3} + \log n$	
$\sqrt{n}$	
$n/(\log n)$	
n	
$n\log n$	
n^2	$n^2 + \log n$
n^3	
$n + n^3 + 7n^5$	
$(3/2)^n$	
2^n	
$n!$	

Problem 285 In each of the following cases, construct a function $f(n)$ that satisfies the specified constraints, or state that it is impossible.

Answer:

$f(n) = \Theta\left(n^2/\lg n\right)$ and $f(n) = o\left(n^2\right)$	$f(n) = \left(n^2/\lg n\right)$
$f(n) = O\left(4 \cdot n^n\right)$ and $f(n) = \omega\left(3 \cdot n!\right)$	$f(n) = n^n$
$f(n) = o\left(4 \cdot n^n\right)$ and $f(n) = \Omega\left(3 \cdot n!\right)$	$f(n) = n!$
$f(n) = \omega\left(n \cdot \sqrt{n}\right)$ and $f(n) = o\left(n^2\right)$	$f(n) = n^{(7/4)}$
$f(n) = \omega\left(\left(\sqrt{n}\right)^{3/2}\right)$ and $f(n) = \Omega(n)$	$f(n) = n$
$f(n) = \omega\left(n \cdot \lg n\right)$ and $f(n) = o\left(n^2\right)$	$f(n) = n \cdot \sqrt{n}$
$f(n) = O\left(\lg n\right)$ and $f(n) = \omega\left(\lg\lg n\right)$	$f(n) = \lg n$
$f(n) = \omega\left(3^n\right)$ and $f(n) = o\left(4^n\right)$	$f(n) = (3.5)^n$
$f(n) = O\left(\lg n\right)$ and $f(n) = \Omega\left(\lg^2 n\right)$	Impossible.
$f(n) = \omega\left(n!\right)$	$f(n) = n^n$
$f(n) = o\left(n^3/\left(\lg n\right)\right)$ and $f(n) = \omega\left(n^2 \cdot \lg n\right)$	$f(n) = n^2 \cdot \lg^2 n$
$f(n) = \Theta\left(\log_2 n^2\right)$ and $f(n) = \Theta\left(\log_{10} n^{10}\right)$	$f(n) = \lg n$

Problem 286   Arrange the following eighteen functions into a table, such that any two functions, say $f(n)$ and $g(n)$, appear *in the same horizontal row* if and only if $f(n) = \Theta(g(n))$, while $f(n)$ appears *above* $g(n)$ in the table if and only if $f(n) = o(g(n))$.

$$\sqrt{n^3} \qquad n \qquad 3^n \qquad n^n \qquad\qquad n + (\log n)^{3/2} \quad n^2 + \log n$$

$$n^2 + (1/4) \quad n^3 \qquad \log n \quad n^{3/2} + \log n \quad (\log n)^2 \qquad\qquad n!$$

$$(1/4)^n \qquad (3/2)^n \quad 1/4 \qquad \log n + n^3 \quad n/((\log n)^2) \qquad \log \log n$$

Answer:

$(1/4)^n$	
$1/4$	
$\log \log n$	
$\log n$	
$(\log n)^2$	
$n/((\log n)^2)$	
n	$n + (\log n)^{3/2}$
$n^{3/2} + \log n$	$\sqrt{n^3}$
$n^2 + (1/4)$	$n^2 + \log n$
n^3	$\log n + n^3$
$(3/2)^n$	
3^n	
$n!$	
n^n	

Problem 287   In each of the following cases, construct a function $f(n)$ that satisfies the specified constraints, or state that it is impossible.

	Answer:
$f(n) = \Omega(n)$ and $f(n) = o(n^2)$	$f(n) = n$
$f(n) = \Omega(n^2)$ and $f(n) = \Omega(n)$	$f(n) = n^2$
$f(n) = \Theta(n)$ and $f(n) = o(n)$	Impossible.
$f(n) = O(\lg n)$ and $f(n) = \omega(\lg \lg n)$	$f(n) = \sqrt{\lg n}$
$f(n) = \omega(2^n)$ and $f(n) = o(5^n)$	$f(n) = 3^n$
$f(n) = O(\lg n)$ and $f(n) = \Omega(\lg^2 n)$	Impossible.
$f(n) = \omega(3^n)$	$f(n) = 4^n$
$f(n) = o(n^3)$ and $f(n) = \omega(n^2)$	$f(n) = n^2 \sqrt{n}$
$f(n) = \Theta(\log_2 n)$ and $f(n) = \Theta(\log_{10} n)$	$f(n) = \log_2 n$
$f(n) = \omega(n/\lg^2 n)$ and $f(n) = o(n)$	$f(n) = n/(\lg n)$
$f(n) = \omega(n/\lg n)$ and $f(n) = o(n)$	$f(n) = n/(\lg \lg n)$
$f(n) = \Theta(n^2)$ and $f(n) = O(n^2 + n)$	$f(n) = n^2$
$f(n) = \Theta(n^3)$ and $f(n) = o(n^3 + n)$	Impossible.

Problem 288   Arrange the following eighteen functions into a table, such that any two functions, say $f(n)$ and $g(n)$, appear *in the same horizontal row* if and only if $f(n) = \Theta(g(n))$, while $f(n)$ appears *above* $g(n)$ in the table if and only if $f(n) = o(g(n))$.

$\log(n!)$	$n \log n$	$\log(n^n)$	$n^2 \log n$	$(1/2)n^n$	$n/(\log n)$
$\sqrt{n}$	5^n	$\log\log n$	$n^5 + \log n + 2\log^5 n$	$n^5 + (3/2)^n$	$\log^2 n + 2\log\log n$
$5n + \sqrt{n}$	$n! + n^5$	$n^5 + 5n$	$(\log\log n)^2$	$n^2 + \log n$	$3n^2 + 11n + 5\log^3 n$

Answer:

$\log\log n$		
$(\log\log n)^2$		
$\log^2 n + 2\log\log n$		
$\sqrt{n}$		
$n/(\log n)$		
$5n + \sqrt{n}$		
$\log(n!)$	$n \log n$	$\log(n^n)$
$n^2 + \log n$	$3n^2 + 11n + 5\log^3 n$	
$n^2 \log n$		
$n^5 + 5n$	$n^5 + \log n + 2\log^5 n$	
$n^5 + (3/2)^n$		
5^n		
$n! + n^5$		
$(1/2)n^n$		

Problem 289   In each of the following cases, construct a function $f(n)$ that satisfies the specified constraints, or state that it is impossible.

Answer:

$f(n) = o(n \cdot \sqrt{n})$ and $f(n) = \omega(n)$ — $f(n) = n^{5/4}$

$f(n) = O(n^2)$ and $f(n) = o(n^3)$ — $f(n) = n^2$

$f(n) = \Theta(n)$ and $f(n) = O(\sqrt{n})$ — Impossible.

$f(n) = \omega(\log_2 n)$ and $f(n) = o(n)$ — $f(n) = \sqrt{n}$

$f(n) = \Omega(n^n)$ and $f(n) = O(n!)$ — Impossible.

$f(n) = \omega(n/(\lg^2 n))$ and $f(n) = o(n/(\log n))$ — $f(n) = n/\left(\lg^{3/2} n\right)$

$f(n) = \Omega(n \cdot \lg n)$ and $f(n) = o(n^2 \cdot \lg n)$ — $f(n) = n^2$

$f(n) = \Theta\left(\sqrt{n^5}\right)$ and $f(n) = \omega\left(\sqrt{n^3}\right)$ — $f(n) = n^{5/2}$

$f(n) = O(3^n)$ and $f(n) = \Omega(3^{n+1})$ — $f(n) = 3^n$

$f(n) = \omega(n^2)$ and $f(n) = o(2^n)$ — $f(n) = n^3$

$f(n) = \Omega(n^2/(\lg n))$ and $f(n) = O(n^2)$ — $f(n) = n^2$

$f(n) = \omega(n^2 \lg n)$ and $f(n) = o(n^2)$ — Impossible.

$f(n) = \Omega(n\sqrt{n})$ and $f(n) = o(n^2)$ — $f(n) = n\sqrt{n}$

Problem 290   Arrange the following eighteen functions into a table, such that any two functions, say $f(n)$ and $g(n)$, appear *in the same horizontal row* if and only if $f(n) = \Theta(g(n))$, while $f(n)$ appears *above* $g(n)$ in the table if and only if $f(n) = o(g(n))$.

$$n/(\lg n) + \lg^2 n \quad 2^{n-2} + n^2 \quad \lg(n^2) \qquad n^2 + 12n \quad n + \sqrt{n} \quad n^{1/2} + 15$$

$$n^2/(\lg^2 n) + 12n \quad (3/2)^n \quad n/(\lg^2 n) + \lg n \quad 3^{n-3} + n^3 \quad \lg n \qquad \lg^2 n$$

$$n^2/(\lg n) + 11n \quad n^3 + 11n \quad n! + 2^n \qquad n^n + 4^n \quad \sqrt{n^3} + n \quad 2\lg^3 n$$

Answer:

$\lg n$	$\lg(n^2)$
$\lg^2 n$	
$2\lg^3 n$	
$n^{1/2} + 15$	
$n/(\lg^2 n) + \lg n$	
$n/(\lg n) + \lg^2 n$	
$n + \sqrt{n}$	
$\sqrt{n^3} + n$	
$n^2/(\lg^2 n) + 12n$	
$n^2/(\lg n) + 11n$	
$n^2 + 12n$	
$n^3 + 11n$	
$(3/2)^n$	
$2^{n-2} + n^2$	
$3^{n-3} + n^3$	
$n! + 2^n$	
$n^n + 4^n$	

Problem 291   In each of the following cases, construct a function $f(n)$ that satisfies the specified constraints, or state that it is impossible.

Answer:

$f(n) = \omega(3^n)$ and $f(n) = o(4^n)$ $\qquad$ $f(n) = (3.5)^n$

$f(n) = \omega\left(n \cdot \sqrt{n^3}\right)$ and $f(n) = o(n^3)$ $\qquad$ $f(n) = n^{(11/4)}$

$f(n) = \omega\left((\log\log n)^2\right)$ and $f(n) = o(\log n)$ $\qquad$ $f(n) = \sqrt{\log n}$

$f(n) = \Omega(\log^2 n)$ and $f(n) = O(\log\log n)$ $\qquad$ Impossible.

$f(n) = \Omega((7/6)^n)$ and $O((9/8)^n)$ $\qquad$ Impossible.

$f(n) = \Omega(n!)$ and $f(n) = O(n^n)$ $\qquad$ $f(n) = n!$

$f(n) = O(n/(\lg n))$ and $f(n) = \Omega(n/(\log\log n))$ $\qquad$ Impossible.

$f(n) = o(n/(\lg\lg n))$ and $f(n) = \omega(n/(\log n))$ $\qquad$ $f(n) = n/(\sqrt{\log n})$

$f(n) = \Theta(3^n)$ and $f(n) = \Theta(3^{n+1})$ $\qquad$ $f(n) = 3^n$

$f(n) = \Theta(n^n)$ and $f(n) = \Theta(n^{n+1})$ $\qquad$ Impossible.

$f(n) = \omega(n^3)$ and $f(n) = o\left((n+1)^3\right)$ $\qquad$ Impossible.

Problem 292   In each of the following cases, construct a function $f(n)$ that satisfies the specified constraints, or state that it is impossible.

Answer:

$f(n) = \Theta(\lg n)$ and $f(n) = o\left(\sqrt{n}\right)$	$f(n) = \lg n$
$f(n) = O(\log_{10} n)$ and $f(n) = \Omega(10^n)$	Impossible.
$f(n) = \Theta(\log\log n)$ and $f(n) = \omega(1)$	$f(n) = \log\log n$
$f(n) = \omega(n^n)$ and $f(n) = O(n!)$	Impossible.
$f(n) = \omega(n \cdot \sqrt{n})$ and $f(n) = o(n)$	Impossible.
$f(n) = O(2^n + 3^n)$ and $f(n) = \Omega(4^n)$	Impossible.
$f(n) = \omega\left(15n^5 + 3n^3 + n\right)$ and $f(n) = o\left(n^6\right)$	$f(n) = n^{5.5}$
$f(n) = \omega\left(n^2/\lg n\right)$ and $f(n) = o\left(n^2\right)$	$f(n) = \left(n^2/\lg\lg n\right)$
$f(n) = O(3 \cdot n^n)$ and $f(n) = \omega(15 \cdot n!)$	$f(n) = n^n$
$f(n) = O(5 \cdot n!)$ and $f(n) = \omega(3 \cdot 4^n)$	$f(n) = 5^n$

Problem 293   In each of the following cases, construct a function $f(n)$ that satisfies the specified constraints, or state that it is impossible.

Answer:

$f(n) = \omega(n \cdot \sqrt{n})$ and $f(n) = o(n \cdot \lg n)$	Impossible.
$f(n) = \Omega(n^2)$ and $f(n) = \omega(n^3)$	$f(n) = n^{3.1}$
$f(n) = \Omega(n^2)$ and $f(n) = o\left(\sqrt{n^5}\right)$	$f(n) = n^2$
$f(n) = \Omega(\log_2 n)$ and $f(n) = O(\log_3(n+1))$	$f(n) = \log_2 n$
$f(n) = \Omega(n^n)$ and $f(n) = \Theta(n!)$	Impossible.
$f(n) = o(n/(\lg\lg n))$ and $f(n) = \omega\left(n/(\log^2 n)\right)$	$f(n) = n/(\lg n)$
$f(n) = \omega(n \cdot 3^n)$ and $f(n) = o\left(n^2 \cdot 4^n\right)$	$f(n) = n \cdot 4^n$
$f(n) = \Omega\left(\sqrt{n^5} \cdot (1/2)^n\right)$ and $f(n) = O\left(\sqrt{n}\right)$	$f(n) = 1$
$f(n) = \Omega(3^n \cdot (1/2)^n)$ and $f(n) = O(3^n)$	$f(n) = 2^n$
$f(n) = O(n^2)$ and $f(n) = \Omega(2^n)$	Impossible.

Problem 294   In each of the following cases, construct a function $f(n)$ that satisfies the specified constraints, or state that it is impossible.

Answer:

$f(n) = \omega(n!)$	$f(n) = n^n$
$f(n) = o(\lg n)$	$f(n) = \lg\lg n$
$f(n) = \omega(\lg n)$ and $f(n) = O(n/(\lg n))$	$f(n) = n/(\lg n)$
$f(n) = \omega(n\lg n)$ and $f(n) = o\left(n^2\right)$	$f(n) = n^{1.01}$
$f(n) = \Omega(n \cdot \sqrt{n})$ and $f(n) = o\left(n^2\right)$	$f(n) = n \cdot \sqrt{n}$
$f(n) = O(2^n + 3^n)$ and $f(n) = \omega(2^n)$	$f(n) = 3^n$
$f(n) = \omega\left(15n^5 + 3n^3 + n\right)$ and $f(n) = o\left(n^5\right)$	Impossible.
$f(n) = \omega\left(n^2/\lg n\right)$ and $f(n) = o\left(n^3\right)$	$f(n) = n^2$
$f(n) = \Omega(3 \cdot n^n)$ and $f(n) = O(15 \cdot n!)$	Impossible.
$f(n) = O\left(4^n \cdot \sqrt{n}\right)$ and $f(n) = \omega(3 \cdot 4^n)$	$f(n) = 4^n \cdot \sqrt{n}$

Problem 295   In each of the following cases, construct a function $f(n)$ that satisfies the specified constraints, or state that it is impossible.

Answer:

$f(n) = \omega(n^3)$ and $f(n) = o(n^2)$	Impossible.
$f(n) = \Omega(n^4)$ and $f(n) = O(n^3)$	Impossible.
$f(n) = \Theta(\lg n)$ and $f(n) = o(\lg\lg n)$	Impossible.
$f(n) = o(\lg n)$ and $f(n) = \Theta(\lg\lg n)$	$f(n) = \lg\lg n$
$f(n) = o(n \cdot \sqrt{n})$ and $f(n) = \omega(n)$	$f(n) = n \cdot \lg n$
$f(n) = \Omega(2^n + 3^n)$ and $f(n) = O(4^n)$	$f(n) = 3^n$
$f(n) = \Omega(15n^5 + 3n^3 + n)$ and $f(n) = o(n^6)$	$f(n) = n^5$
$f(n) = \omega(n^2/\lg^2 n)$ and $f(n) = o(n^2)$	$f(n) = (n^2/\lg n)$
$f(n) = \Omega(3 \cdot n^n)$ and $f(n) = \omega(15 \cdot n!)$	$f(n) = n^n$
$f(n) = O(5 \cdot n!)$ and $f(n) = o(3 \cdot 4^n)$	$f(n) = 3^n$

Problem 296   In each of the following cases, construct a function $f(n)$ that satisfies the specified constraints, or state that it is impossible.

Answer:

$f(n) = \Omega(n/\sqrt{n})$ and $f(n) = o(n)$	$f(n) = \sqrt{n}$
$f(n) = \Theta(3^n)$ and $f(n) = \Omega(4^n)$	Impossible.
$f(n) = \Theta(\lg n)$ and $f(n) = \Omega(\sqrt{n})$	Impossible.
$f(n) = O(\log_{10} n)$ and $f(n) = \Omega(\log_2 n)$	$f(n) = \log_2 n$
$f(n) = O(n^n)$ and $f(n) = \Omega(n!)$	$f(n) = n!$
$f(n) = \Omega(n^3/(\lg n))$ and $f(n) = O(n^2 \log n)$	Impossible.
$f(n) = \omega(n^2 \lg n)$ and $f(n) = o(n^2/\lg n)$	Impossible.
$f(n) = o(\sqrt{n^7})$ and $f(n) = \omega(\sqrt{n^5})$	$f(n) = n^3$
$f(n) = O(n^3)$ and $f(n) = o(n^3 + n)$	$f(n) = n^2$
$f(n) = \Omega(n^2)$ and $f(n) = O(n^2 + n)$	$f(n) = n^2$

Problem 297   In each of the following cases, construct a function $f(n)$ that satisfies the specified constraints, or state that it is impossible.

Answer:

$f(n) = \omega(n)$ and $f(n) = O(n^2)$	$f(n) = n^2$
$f(n) = \Theta(n^2)$ and $f(n) = \Omega(n)$	$f(n) = n^2$
$f(n) = \omega(n)$ and $f(n) = o(n)$	Impossible.
$f(n) = O(3^n)$ and $f(n) = \omega(2^n)$	$f(n) = 3^n$
$f(n) = \omega(2^n)$ and $f(n) = \Theta(4^n)$	$f(n) = 4^n$
$f(n) = O(3^n)$ and $f(n) = \Omega(4^n)$	Impossible.
$f(n) = \omega(\log n)$	$f(n) = \log^2 n$
$f(n) = o(n)$	$f(n) = \sqrt{n}$
$f(n) = \Omega(\lg n)$ and $f(n) = O(n)$	$f(n) = n$
$f(n) = \omega(n \lg n)$ and $f(n) = o(n)$	Impossible.

Problem 298   In each of the following cases, construct a function $f(n)$ that satisfies the specified constraints, or state that it is impossible.

Answer:

$f(n) = \Theta(n \lg n)$ and $f(n) = O(n)$	Impossible.
$f(n) = \Theta(\log_{10} n)$ and $f(n) = \Theta(\log_2 n)$	$f(n) = \log_{39} n^{27}$
$f(n) = \omega(2^n)$ and $f(n) = \Omega((3/2)^n)$	$f(n) = 3^n$
$f(n) = O(n^2 \lg n)$ and $f(n) = \Omega(n^2)$	$f(n) = n^2$
$f(n) = o(n^2 \lg n)$ and $f(n) = \omega(n^2)$	$f(n) = n^2 \cdot \sqrt{\lg n}$
$f(n) = o(n)$ and $f(n) = \omega(1)$	$f(n) = \lg n$
$f(n) = o((n + \lg n)^2)$ and $f(n) = \omega(n^2 + \lg^2 n)$	Impossible.
$f(n) = o((2^n + n)(n^2 + 1))$ and $f(n) = \omega((2^n + n^2)(n+1))$	$f(n) = 2^n \cdot n\sqrt{n}$
$f(n) = o(3^n)$ and $f(n) = \omega(2^n)$	$f(n) = \left(\dfrac{5}{2}\right)^n$
$f(n) = o(\lg(3^n))$ and $f(n) = \omega(\lg(2^n))$	Impossible.

Problem 299   In each of the following cases, construct a function $f(n)$ that satisfies the specified constraints, or state that it is impossible.

Answer:

$f(n) = \Omega(2n)$ and $f(n) = O(n)$	$f(n) = n$
$f(n) = \omega(n^2)$ and $f(n) = o(n)$	Impossible.
$f(n) = \Theta(n \cdot \sqrt{n} \cdot \log_2 n)$ and $f(n) = O\left(n^{(3/2)}\right)$	Impossible.
$f(n) = \omega(\log_2 n)$ and $f(n) = o(\log_2^2 n)$	$f(n) = \log_2^{(3/2)} n$
$f(n) = \Theta(\log_2 n)$ and $f(n) = \Theta(\log_{10}(n^2))$	$f(n) = \lg n$
$f(n) = \Omega(\log_2 n)$ and $f(n) = O(\log_{10}(n^3))$	$f(n) = \lg n$
$f(n) = o(\log_2 n / \log_2 \log_2 n)$	$f(n) = \lg \lg n$
$f(n) = o(n)$ and $f(n) = \omega(n / \log_2 n)$	$f(n) = n / \lg \lg n$
$f(n) = \omega(\log_2 n)$ and $f(n) = O(\log_{10}^2(3n))$	$f(n) = \lg^2 n$
$f(n) = \omega(3^n)$ and $f(n) = o(4^n)$	$f(n) = (3.5)^n$

4.2   Divide & Conquer Recurrences

Problem 300   Find asymptotic estimates for the positive, increasing functions f, g, h, ℓ, p, t, defined by the following recurrence relations, and explan your answers briefly.

(a) $f(n) = 8 \cdot f\left(\dfrac{n}{2}\right) + n^3$

   Answer: This is an instance of the Master Recurrence, whose form is:

$$f(n) = af\left(\frac{n}{b}\right) + n^d$$

and whose solution is determined according to the result of the comparison between the values of a and b^d. In this instance:

$$a = 8, \ b = 2, \ d = 3$$

yielding:
$$b^d = 2^3 = 8 = a$$

Hence:
$$f(n) = \Theta\left(n^d \cdot \lg n\right) = \Theta\left(n^3 \cdot \lg n\right)$$

(b) $g(n) = 8 \cdot g\left(\dfrac{n}{3}\right) + n^2$

Answer: This is an instance of the Master Recurrence, with:
$$a = 8,\ b = 3,\ d = 2$$

yielding:
$$b^d = 3^2 = 9 > 8 = a$$

Hence:
$$g(n) = \Theta\left(n^d\right) = \Theta\left(n^2\right)$$

(c) $h(n) = 9 \cdot h\left(\dfrac{n}{3}\right) + n^2$

Answer: This is an instance of the Master Recurrence, with:
$$a = 9,\ b = 3,\ d = 2$$

yielding:
$$b^d = 3^2 = 9 = a$$

Hence:
$$h(n) = \Theta\left(n^d \cdot \lg n\right) = \Theta\left(n^2 \cdot \lg n\right)$$

(d) $\ell(n) = 9 \cdot \ell\left(\dfrac{n}{3}\right) + n$

Answer: This is an instance of the Master Recurrence, with:
$$a = 9,\ b = 3,\ d = 1$$

yielding:
$$b^d = 3^1 = 3 < 9 = a$$

Hence:
$$\ell(n) = \Theta\left(n^{\log_b a}\right) = \Theta\left(n^{\log_3 9}\right) = \Theta\left(n^2\right)$$

(e) $p(n) = 3 \cdot p\left(\dfrac{n}{4}\right) + n$

Answer: This is an instance of the Master Recurrence, with:
$$a = 3,\ b = 4,\ d = 1$$

yielding:
$$b^d = 4^1 = 4 > 3 = a$$

Hence:
$$p(n) = \Theta\left(n^d\right) = \Theta\left(n^1\right) = \Theta(n)$$

(f) $t(n) = 4 \cdot t\left(\dfrac{n}{2}\right) + n$

    Answer: This is an instance of the Master Recurrence, with:

$$a = 4, \; b = 2, \; d = 1$$

    yielding:

$$b^d = 2^1 = 2 < 4 = a$$

    Hence:

$$t(n) = \Theta\left(n^{\log_b a}\right) = \Theta\left(n^{\log_2 4}\right) = \Theta\left(n^2\right)$$

Problem 301   Find asymptotic estimates for the positive, increasing functions f, g, h, ℓ, p, t, defined by the following recurrence relations, and explan your answers briefly.

(a) $f(n) = 3 \cdot f\left(\dfrac{3n}{5}\right) + n^2$

    Answer: This is an instance of the Master Recurrence, with:

$$a = 3, \; b = \dfrac{5}{3}, \; d = 2$$

    yielding:

$$b^d = \left(\dfrac{5}{3}\right)^2 = \dfrac{25}{9} < \dfrac{27}{9} = 3 = a$$

    Hence:

$$f(n) = \Theta\left(n^{\log_b a}\right) = \Theta\left(n^{\log_{(5/3)} 3}\right)$$

(b) $g(n) = 6 \cdot g\left(\dfrac{n}{4}\right) + n$

    Answer: This is an instance of the Master Recurrence, with:

$$a = 6, \; b = 4, \; d = 1$$

    yielding:

$$b^d = 4^1 = 4 < 6 = a$$

    Hence:

$$g(n) = \Theta\left(n^{\log_b a}\right) = \Theta\left(n^{\log_4 6}\right)$$

(c) $h(n) = h\left(\dfrac{n}{5}\right) + 1$

    Answer: This is an instance of the Master Recurrence, with:

$$a = 1, \; b = 5, \; d = 0$$

    yielding:

$$b^d = 5^0 = 1 = a$$

    Hence:

$$h(n) = \Theta\left(n^d \cdot \lg n\right) = \Theta\left(n^0 \cdot \lg n\right) = \Theta\left(\lg n\right)$$

(d) $\ell(n) = 16 \cdot \ell\left(\dfrac{n}{2}\right) + n^4$

    Answer: This is an instance of the Master Recurrence, with:

$$a = 16,\ b = 2,\ d = 4$$

    yielding:

$$b^d = 2^4 = 16 = a$$

    Hence:

$$\ell(n) = \Theta\left(n^d \cdot \lg n\right) = \Theta\left(n^4 \cdot \lg n\right)$$

(e) $p(n) = 25 \cdot p\left(\dfrac{5n}{6}\right) + n^2$

    Answer: This is an instance of the Master Recurrence, with:

$$a = 25,\ b = \dfrac{6}{5},\ d = 2$$

    yielding:

$$b^d = \left(\dfrac{6}{5}\right)^2 = \dfrac{36}{25} < \dfrac{50}{25} = 2 < 25 = a$$

    Hence:

$$p(n) = \Theta\left(n^{\log_b a}\right) = \Theta\left(n^{\log_{(6/5)} 25}\right)$$

(f) $t(n) = 16 \cdot t\left(\dfrac{n}{4}\right) + n$

    Answer: This is an instance of the Master Recurrence, with:

$$a = 16,\ b = 4,\ d = 1$$

    yielding:

$$b^d = 4^1 = 4 < 16 = a$$

    Hence:

$$t(n) = \Theta\left(n^{\log_b a}\right) = \Theta\left(n^{\log_4 16}\right) = \Theta\left(n^2\right)$$

Problem 302   Find asymptotic estimates for the positive, increasing functions f, g, h, ℓ, p, t, defined by the following recurrence relations, and explan your answers briefly.

(a) $f(n) = 3 \cdot f\left(\dfrac{2n}{3}\right) + n^2$

    Answer: This is an instance of the Master Recurrence, with:

$$a = 3,\ b = \dfrac{3}{2},\ d = 2$$

    yielding:

$$b^d = \left(\dfrac{3}{2}\right)^2 = \dfrac{9}{4} < \dfrac{12}{4} = 3 = a$$

    Hence:

$$f(n) = \Theta\left(n^{\log_b a}\right) = \Theta\left(n^{\log_{(3/2)} 3}\right)$$

(b) $g(n) = 6 \cdot g\left(\dfrac{2n}{5}\right) + n^2$

Answer: This is an instance of the Master Recurrence, with:

$$a = 6, \ b = \frac{5}{2}, \ d = 2$$

yielding:

$$b^d = \left(\frac{5}{2}\right)^2 = \frac{25}{4} > \frac{24}{4} = 6 = a$$

Hence:

$$g(n) = \Theta\left(n^d\right) = \Theta\left(n^2\right)$$

(c) $h(n) = 9 \cdot h\left(\dfrac{n}{3}\right) + n^3$

Answer: This is an instance of the Master Recurrence, with:

$$a = 9, \ b = 3, \ d = 3$$

yielding:

$$b^d = 3^3 = 27 > 9 = a$$

Hence:

$$h(n) = \Theta\left(n^d\right) = \Theta\left(n^3\right)$$

(d) $\ell(n) = 5 \cdot \ell\left(\dfrac{n}{5}\right) + n^2$

Answer: This is an instance of the Master Recurrence, with:

$$a = 5, \ b = 5, \ d = 2$$

yielding:

$$b^d = 5^2 = 25 > a$$

Hence:

$$\ell(n) = \Theta\left(n^d\right) = \Theta\left(n^2\right)$$

(e) $p(n) = 5 \cdot p\left(\dfrac{n}{5}\right) + n$

Answer: This is an instance of the Master Recurrence, with:

$$a = 5, \ b = 5, \ d = 1$$

yielding:

$$b^d = 5^1 = 5 = a$$

Hence:

$$p(n) = \Theta\left(n^d \cdot \lg n\right) = \Theta\left(n^1 \cdot \lg n\right) = \Theta\left(n \cdot \lg n\right)$$

(f) $t(n) = 5 \cdot t\left(\dfrac{n}{5}\right) + 1$

Answer: This is an instance of the Master Recurrence, with:

$$a = 5, \; b = 5, \; d = 0$$

yielding:

$$b^d = 5^0 = 1 < 5 = a$$

Hence:

$$t(n) = \Theta\left(n^{\log_b a}\right) = \Theta\left(n^{\log_5 5}\right) = \Theta\left(n^1\right) = \Theta(n)$$

Problem 303   Find asymptotic estimates for the positive, increasing functions f, g, h, ℓ, p, t, defined by the following recurrence relations, and explan your answers briefly.

(a) $f(n) = 8 \cdot f\left(\dfrac{n}{4}\right) + \sqrt{n}$

Answer: This is an instance of the Master Recurrence, with:

$$a = 8, \; b = 4, \; d = \frac{1}{2}$$

yielding:

$$b^d = 4^{(1/2)} = 2 < 8 = a$$

Hence:

$$f(n) = \Theta\left(n^{\log_b a}\right) = \Theta\left(n^{\log_4 8}\right) = \Theta\left(n^{(\lg 8/\lg 4)}\right) = \Theta\left(n^{(3/2)}\right) = \Theta\left(\sqrt{n^3}\right)$$

(b) $g(n) = 6 \cdot g\left(\dfrac{n}{2}\right) + \sqrt{n^5}$

Answer: This is an instance of the Master Recurrence, with:

$$a = 6, \; b = 2, \; d = \frac{5}{2}$$

yielding:

$$b^d = 2^{(5/2)} = \sqrt{2^5} = \sqrt{32} < \sqrt{36} = 6 = a$$

Hence:

$$g(n) = \Theta\left(n^{\log_b a}\right) = \Theta\left(n^{\log_2 6}\right)$$

(c) $h(n) = 27 \cdot h\left(\dfrac{n}{3}\right) + \sqrt{n^5}$

Answer: This is an instance of the Master Recurrence, with:

$$a = 27, \; b = 3, \; d = \frac{5}{2}$$

yielding:

$$b^d = 3^{(5/2)} = \sqrt{3^5} = \sqrt{243} < \sqrt{256} = 16 < 27 = a$$

Hence:

$$h(n) = \Theta\left(n^{\log_b a}\right) = \Theta\left(n^{\log_3 27}\right) = \Theta\left(n^3\right)$$

(d) $\ell(n) = 3 \cdot \ell\left(\dfrac{2n}{5}\right) + n \cdot \sqrt{n}$

Answer: This is an instance of the Master Recurrence, with:

$$a = 3, \; b = \frac{5}{2}, \; d = \frac{3}{2}$$

yielding:

$$b^d = \left(\frac{5}{2}\right)^{(3/2)} = \sqrt{\left(\frac{5}{2}\right)^3} = \sqrt{\frac{125}{8}} = \sqrt{15 + \frac{5}{8}} > \sqrt{9} = 3 = a$$

Hence:

$$\ell(n) = \Theta\left(n^d\right) = \Theta\left(n^{3/2}\right) = \Theta\left(n \cdot \sqrt{n}\right)$$

(e) $p(n) = 3.375 \cdot p\left(\dfrac{2n}{3}\right) + n^3$

Answer: This is an instance of the Master Recurrence, with:

$$a = 3.375, \; b = \frac{3}{2}, \; d = 3$$

yielding:

$$b^d = \left(\frac{3}{2}\right)^3 = \frac{27}{8} = 3 + \frac{3}{8} = 3.375 = a$$

Hence:

$$p(n) = \Theta\left(n^d \cdot \lg n\right) = \Theta\left(n^3 \cdot \lg n\right)$$

(f) $t(n) = 5 \cdot t\left(\dfrac{2n}{3}\right) + n^4$

Answer: This is an instance of the Master Recurrence, with:

$$a = 5, \; b = \frac{3}{2}, \; d = 4$$

yielding:

$$b^d = \left(\frac{3}{2}\right)^4 = \frac{81}{16} > \frac{80}{16} = 5 = a$$

Hence:

$$t(n) = \Theta\left(n^d\right) = \Theta\left(n^4\right)$$

Problem 304   Find asymptotic estimates for the positive, increasing functions f, g, h, defined by the following recurrence relations, and explain your answers briefly.

(a) $f(n) = 4 \cdot f\left(\dfrac{n}{2}\right) + \lg n$

Answer: Since the recurrence which defines f does not honor the form of the Master Recurrence, we proceed to obtain an estimate of the asymptotic order of f. Observe that:

$$1 \le \lg n \le n \; \text{ whenever } n \ge 2$$

Hence, for a sufficiently large n, function f satisfies the following pair of inequalities:

$$f_1(n) \le f(n) \le f_2(n)$$

where functions f_1 and f_2 are governed by instances of the Master Recurrence:

$$f_1(n) = 4 \cdot f_1\left(\frac{n}{2}\right) + 1$$
$$f_2(n) = 4 \cdot f_2\left(\frac{n}{2}\right) + n$$

To determine the asymptotic orders of f_1 and f_2, observe that:

$$2^0 = 1 < 4$$
$$2^1 = 2 < 4$$

yielding:

$$f_1(n) = \Theta\left(n^{\log_2 4}\right) = \Theta\left(n^2\right)$$
$$f_2(n) = \Theta\left(n^{\log_2 4}\right) = \Theta\left(n^2\right)$$

Since:

$$f(n) = \Omega\left(f_1(n)\right) = \Omega\left(n^2\right)$$
$$f(n) = O\left(f_2(n)\right) = O\left(n^2\right)$$

we conclude that:

$$f(n) = \Theta(n^2)$$

(b) $g(n) = 4 \cdot g\left(\dfrac{3n}{5}\right) + n^2 \cdot \lg n$

Answer: Since the recurrence which defines g does not honor the form of the Master Recurrence, we proceed to obtain an estimate of the asymptotic order of g. Observe that:

$$1 \le \lg n \le \sqrt{n} \text{ whenever } n \ge 2$$

Hence, for a sufficiently large n, function g satisfies the following pair of inequalities:

$$g_1(n) \le g(n) \le g_2(n)$$

where functions g_1 and g_2 are governed by instances of the Master Recurrence:

$$g_1(n) = 4 \cdot g_1\left(\frac{3n}{5}\right) + n^2$$
$$g_2(n) = 4 \cdot g_2\left(\frac{3n}{5}\right) + n^2 \cdot \sqrt{n}$$

To determine the asymptotic orders of g_1 and g_2, observe that:

$$\left(\frac{5}{3}\right)^2 = \frac{25}{9} < \frac{27}{9} = 3 < 4$$
$$\left(\frac{5}{3}\right)^{(5/2)} = \left(\frac{5}{3}\right)^2 \cdot \sqrt{\frac{5}{3}} < \left(\frac{5}{3}\right)^2 \cdot \frac{13}{10} = \frac{5 \cdot 13}{9 \cdot 2} = \frac{65}{18} < \frac{72}{18} = 4$$

where a part of the calculation is based on the following fact:

$$\left(\frac{13}{10}\right)^2 = \frac{169}{100} = \frac{507}{300} > \frac{500}{300} = \frac{5}{3}$$

The asymptotic orders of g_1 and g_2 are:

$$g_1(n) = \Theta\left(n^{\log_{(5/3)} 4}\right)$$
$$g_2(n) = \Theta\left(n^{\log_{(5/3)} 4}\right)$$

Since:

$$g(n) = \Omega\left(g_1(n)\right) = \Omega\left(n^{\log_{(5/3)} 4}\right)$$
$$g(n) = O\left(g_2(n)\right) = O\left(n^{\log_{(5/3)} 4}\right)$$

we conclude that:

$$g(n) = \Theta(n^{\log_{(5/3)} 4})$$

(c) $h(n) = 2 \cdot h\left(\dfrac{2n}{3}\right) + n \cdot \lg n$

Answer: Since the recurrence which defines h does not honor the form of the Master Recurrence, we proceed to obtain an estimate of the asymptotic order of h. Observe that:

$$1 \leq \lg n \leq n^{(1/2)} \text{ for a sufficiently large } n$$

Hence, for a sufficiently large n, function h satisfies the following pair of inequalities:

$$h_1(n) \leq h(n) \leq h_2(n)$$

where functions h_1 and h_2 are governed by instances of the Master Recurrence:

$$h_1(n) = 2 \cdot h_1\left(\frac{2n}{3}\right) + n$$
$$h_2(n) = 2 \cdot h_2\left(\frac{2n}{3}\right) + n^{(3/2)}$$

To determine the asymptotic orders of h_1 and h_2, observe that:

$$\left(\frac{3}{2}\right)^1 < 2$$
$$\left(\frac{3}{2}\right)^{(3/2)} = \sqrt{\frac{27}{8}} < \sqrt{\frac{32}{8}} = \sqrt{4} = 2$$

The asymptotic orders of h_1 and h_2 are:

$$h_1(n) = \Theta\left(n^{\log_{(3/2)} 2}\right)$$
$$h_2(n) = \Theta\left(n^{\log_{(3/2)} 2}\right)$$

Since:

$$h(n) = \Omega\left(h_1(n)\right) = \Omega\left(n^{\log_{(3/2)} 2}\right)$$
$$h(n) = O\left(h_2(n)\right) = O\left(n^{\log_{(3/2)} 2}\right)$$

we conclude that:

$$h(n) = \Theta(n^{\log_{(3/2)} 2})$$

Problem 305   Find asymptotic estimates for the positive, increasing functions f, g, h, ℓ, defined by the following recurrence relations, and explain your answers briefly.

(a) $f(n) = f\left(\frac{n}{2}\right) + n \cdot \lg n$

> **Answer:** Since the recurrence which defines f does not honor the form of the Master Recurrence, we proceed to obtain an estimate of the asymptotic order of f. Straightforwardly, we conclude from the recurrence that:
>
> $$f(n) = \Omega\left(n \cdot \lg n\right)$$
>
> since:
>
> $$f\left(\frac{n}{2}\right) > 0$$
>
> Furthermore, for a sufficiently large n:
>
> $$\lg n \le n^\varepsilon \text{ for any } \varepsilon > 0$$
>
> Hence, for a sufficiently large n, function f satisfies the following inequality:
>
> $$f(n) \le f_2(n)$$
>
> where the function f_2 is governed by an instance of the Master Recurrence:
>
> $$f_2(n) = f_2\left(\frac{n}{2}\right) + n^{1+\varepsilon}$$
>
> where $\varepsilon > 0$ is an arbitrarily small positive real constant. To determine the asymptotic order of f_2, observe that:
>
> $$2^{1+\varepsilon} = 2 \cdot 2^\varepsilon > 2 \cdot 1 > 1 \text{ whenever } \varepsilon > 0$$
>
> yielding:
>
> $$f_2(n) = \Theta\left(n^{1+\varepsilon}\right) = \Theta\left(n \cdot n^\varepsilon\right)$$
>
> Since: $f(n) = \Omega\left(n \cdot \lg n\right)$ and $f(n) = O\left(f_2(n)\right)$, we conclude that the asymptotic order of the function f is bounded as follows:
>
> $$f(n) = \Omega(n \cdot \lg n) \text{ but } f(n) = O\left(n \cdot n^\varepsilon\right)$$
>
> for an arbitrarily small positive real constant $\varepsilon > 0$. (The actual asymptotic order of $f(n)$ is $\Theta\left(n \cdot \lg n\right)$.)

(b) $g(n) = g\left(\frac{n}{3}\right) + \lg n$

> **Answer:** Since the recurrence which defines g does not honor the form of the Master Recurrence, we proceed to obtain an estimate of the asymptotic order of g. Straightforwardly, we conclude from the recurrence that:
>
> $$g(n) = \Omega\left(\lg n\right)$$
>
> since:
>
> $$g\left(\frac{n}{3}\right) > 0$$
>
> Furthermore, for a sufficiently large n:
>
> $$\lg n \le n^\varepsilon \text{ for any } \varepsilon > 0$$

Hence, for a sufficiently large n, function g satisfies the following inequality:

$$g(n) \le g_2(n)$$

where the function g_2 is governed by an instance of the Master Recurrence:

$$g_2(n) = g_2\left(\frac{n}{3}\right) + n^\varepsilon$$

To determine the asymptotic order of g_2, observe that:

$$3^\varepsilon > 1 \text{ whenever } \varepsilon > 0$$

yielding:

$$g_2(n) = \Theta\left(n^\varepsilon\right)$$

Since: $g(n) = \Omega\left(\lg n\right)$ and $g(n) = O\left(g_2(n)\right)$, we conclude that the asymptotic order of the function g is bounded as follows:

$$g(n) = \Omega(\lg n) \text{ but } g(n) = O\left(n^\varepsilon\right)$$

for an arbitrarily small positive real constant $\varepsilon > 0$. (The actual asymptotic order of $g(n)$ is $\Theta\left(\lg^2 n\right)$.)

(c) $h(n) = 3 \cdot h\left(\dfrac{n}{2}\right) + \dfrac{n^2}{\lg^3 n}$

Answer: Since the recurrence which defines h does not honor the form of the Master Recurrence, we proceed to obtain an estimate of the asymptotic order of h. Straightforwardly, we conclude from the recurrence that:

$$h(n) = \Omega\left(\frac{n^2}{\lg^3 n}\right)$$

since:

$$h\left(\frac{n}{2}\right) > 0$$

Furthermore, for a sufficiently large n:

$$1 \le \lg^3 n$$

Hence, for a sufficiently large n, function h satisfies the following inequality:

$$h(n) \le h_2(n)$$

where the function h_2 is governed by an instance of the Master Recurrence:

$$h_2(n) = 3 \cdot h_2\left(\frac{n}{2}\right) + n^2$$

To determine the asymptotic order of h_2, observe that:

$$2^2 = 4 > 3$$

yielding:

$$h_2(n) = \Theta\left(n^2\right)$$

Since: $h(n) = \Omega\left(n^2/\lg^3 n\right)$ and $h(n) = O(h_2(n))$, we conclude that the asymptotic order of the function h is bounded as follows:

$$h(n) = \Omega\left(\frac{n^2}{\lg^3 n}\right) \text{ but } h(n) = O(n^2)$$

(The actual asymptotic order of $h(n)$ is $\Theta\left(\frac{n^2}{\lg^3 n}\right)$.)

(d) $\ell(n) = 8 \cdot \ell\left(\frac{n}{2}\right) + \frac{n^3}{\lg n}$

Answer: Since the recurrence which defines ℓ does not honor the form of the Master Recurrence, we proceed to obtain an estimate of the asymptotic order of ℓ. Observe that, for a sufficiently large n:

$$1 \leq \lg n \leq n^\varepsilon \text{ for any } \varepsilon > 0$$

Hence, for a sufficiently large n, function ℓ satisfies the following pair of inequalities:

$$\ell_1(n) \leq \ell(n) \leq \ell_2(n)$$

where functions ℓ_1 and ℓ_2 are governed by instances of the Master Recurrence:

$$\ell_1(n) = 8 \cdot \ell_1\left(\frac{n}{2}\right) + n^{3-\varepsilon}$$
$$\ell_2(n) = 8 \cdot \ell_2\left(\frac{n}{2}\right) + n^3$$

To determine the asymptotic orders of ℓ_1 and ℓ_2, observe that:

$$2^{3-\varepsilon} = \frac{2^3}{2^\varepsilon} = \frac{8}{2^\varepsilon} < \frac{8}{1} = 8 \text{ whenever } \varepsilon > 0$$
$$2^3 = 8$$

The asymptotic orders of ℓ_1 and ℓ_2 are:

$$\ell_1(n) = \Theta\left(n^{\log_2 8}\right) = \Theta\left(n^3\right)$$
$$\ell_2(n) = \Theta\left(n^3 \cdot \lg n\right)$$

Since: $\ell(n) = \Omega(\ell_1(n))$ and $\ell(n) = O(\ell_2(n))$, we conclude that the asymptotic order of the function ℓ is bounded as follows:

$$\ell(n) = \Omega\left(n^3\right) \text{ but } \ell(n) = O\left(n^3 \cdot \lg n\right)$$

for an arbitrarily small positive real constant $\varepsilon > 0$. (The actual asymptotic order of $\ell(n)$ is $\Theta\left(n^3 \cdot \lg\lg n\right)$.)

Problem 306   Find asymptotic estimates for the positive, increasing functions f, g, h, ℓ, defined by the following recurrence relations, and explain your answers briefly.

(a) $f(n) = 6 \cdot f\left(\frac{2n}{5}\right) + n^2 \cdot \lg^2 n$

Answer: Since the recurrence which defines f does not honor the form of the Master Recurrence, we proceed to obtain an estimate of the asymptotic order of f. Straightforwardly, we conclude from the recurrence that:

$$f(n) = \Omega\left(n^2 \cdot \lg^2 n\right)$$

since:

$$f\left(\frac{2n}{5}\right) > 0$$

Furthermore, for a sufficiently large n:

$$\lg^2 n \leq n^\varepsilon \text{ for any } \varepsilon > 0$$

Hence, for a sufficiently large n, function f satisfies the following inequality:

$$f(n) \leq f_2(n)$$

where the function f_2 is governed by an instance of the Master Recurrence:

$$f_2(n) = 6 \cdot f_2\left(\frac{2n}{5}\right) + n^{2+\varepsilon}$$

where $\varepsilon > 0$ is an arbitrarily small positive real constant. To determine the asymptotic order of f_2, observe that:

$$\left(\frac{5}{2}\right)^{2+\varepsilon} = \left(\frac{5}{2}\right)^2 \cdot \left(\frac{5}{2}\right)^\varepsilon > \left(\frac{5}{2}\right)^2 \cdot 1 = \frac{25}{4} > \frac{24}{4} = 6 \text{ whenever } \varepsilon > 0$$

yielding:

$$f_2(n) = \Theta\left(n^{2+\varepsilon}\right) = \Theta\left(n^2 \cdot n^\varepsilon\right)$$

Since: $f(n) = \Omega\left(n^2 \cdot \lg^2 n\right)$ and $f(n) = O\left(f_2(n)\right)$, we conclude that the asymptotic order of the function f is bounded as follows:

$$f(n) = \Omega\left(n^2 \cdot \lg^2 n\right) \text{ but } f(n) = O\left(n^2 \cdot n^\varepsilon\right)$$

for an arbitrarily small positive real constant $\varepsilon > 0$. (The actual asymptotic order of $f(n)$ is $\Theta\left(n^2 \cdot \lg^2 n\right)$.)

(b) $g(n) = 7 \cdot g\left(\frac{2n}{5}\right) + n^2 \cdot \lg^2 n$

Answer: Since the recurrence which defines g does not honor the form of the Master Recurrence, we proceed to obtain an estimate of the asymptotic order of g. Observe that, for a sufficiently large n:

$$1 \leq \lg^2 n \leq n^\varepsilon \text{ for any } \varepsilon > 0$$

Hence, for a sufficiently large n, function g satisfies the following pair of inequalities:

$$g_1(n) \leq g(n) \leq g_2(n)$$

where functions g_1 and g_2 are governed by instances of the Master Recurrence:

$$g_1(n) = 7 \cdot g_1\left(\frac{2n}{5}\right) + n^2$$

$$g_2(n) = 7 \cdot g_2\left(\frac{2n}{5}\right) + n^{2+\varepsilon}$$

To determine the asymptotic orders of g_1 and g_2, observe that:

$$\left(\frac{5}{2}\right)^2 = \frac{25}{4} < \frac{28}{4} = 7$$

$$\left(\frac{5}{2}\right)^{2+\varepsilon} = \left(\frac{5}{2}\right)^2 \cdot \left(\frac{5}{2}\right)^\varepsilon = \frac{25}{4} \cdot \left(\frac{5}{2}\right)^\varepsilon < \frac{25}{4} \cdot \frac{28}{25} = \frac{28}{4} = 7 \text{ whenever } 0 < \varepsilon < \log_{(5/2)} \frac{28}{25}$$

The asymptotic orders of g_1 and g_2 are:

$$g_1(n) = \Theta\left(n^{\log_{(5/2)} 7}\right)$$
$$g_2(n) = \Theta\left(n^{\log_{(5/2)} 7}\right)$$

Since: $g(n) = \Omega\left(n^{\log_{(5/2)} 7}\right)$ and $g(n) = O\left(n^{\log_{(5/2)} 7}\right)$, we conclude that:

$$g(n) = \Theta\left(n^{\log_{(5/2)} 7}\right)$$

(c) $h(n) = 2 \cdot h\left(\frac{n}{4}\right) + \sqrt{n \cdot \lg n}$

Answer: Since the recurrence which defines h does not honor the form of the Master Recurrence, we proceed to obtain an estimate of the asymptotic order of h. Observe that, for a sufficiently large n:

$$1 \leq \sqrt{\lg n} \leq n^\varepsilon \text{ for any } \varepsilon > 0$$

Hence, for a sufficiently large n, function h satisfies the following pair of inequalities:

$$h_1(n) \leq h(n) \leq h_2(n)$$

where functions h_1 and h_2 are governed by instances of the Master Recurrence:

$$h_1(n) = 2 \cdot h_1\left(\frac{n}{4}\right) + n^{(1/2)}$$
$$h_2(n) = 2 \cdot h_2\left(\frac{n}{4}\right) + n^{(1/2)+\varepsilon}$$

To determine the asymptotic orders of h_1 and h_2, observe that:

$$4^{(1/2)} = \sqrt{4} = 2$$
$$4^{(1/2)+\varepsilon} = \sqrt{4} \cdot 4^\varepsilon = 2 \cdot 4^\varepsilon > 2 \cdot 1 = 2 \text{ whenever } \varepsilon > 0$$

The asymptotic orders of h_1 and h_2 are:

$$h_1(n) = \Theta\left(n^{(1/2)} \cdot \lg n\right) = \Theta\left(\sqrt{n} \cdot \lg n\right)$$
$$h_2(n) = \Theta\left(n^{(1/2)+\varepsilon}\right) = \Theta\left(\sqrt{n} \cdot n^\varepsilon\right)$$

Since: $h(n) = \Omega\left(h_1(n)\right)$ and $h(n) = O\left(h_2(n)\right)$, we conclude that the asymptotic order of the function h is bounded as follows:

$$h(n) = \Omega\left(\sqrt{n} \cdot \lg n\right) \text{ but } h(n) = O\left(\sqrt{n} \cdot n^\varepsilon\right)$$

for an arbitrarily small positive real constant $\varepsilon > 0$. (The actual asymptotic order of $h(n)$ is $\Theta\left(\sqrt{n} \cdot \lg^{(3/2)} n\right)$.)

(d) $\ell(n) = 2 \cdot \ell\left(\dfrac{n}{4}\right) + \sqrt{\dfrac{n}{\lg n}}$

Answer: Since the recurrence which defines ℓ does not honor the form of the Master Recurrence, we proceed to obtain an estimate of the asymptotic order of ℓ. Observe that, for a sufficiently large n:

$$1 \le \sqrt{\lg n} \le n^{\varepsilon} \text{ for any } \varepsilon > 0$$

Hence, for a sufficiently large n, function ℓ satisfies the following pair of inequalities:

$$\ell_1(n) \le \ell(n) \le \ell_2(n)$$

where functions ℓ_1 and ℓ_2 are governed by instances of the Master Recurrence:

$$\ell_1(n) = 2 \cdot \ell_1\left(\frac{n}{4}\right) + n^{(1/2)-\varepsilon}$$
$$\ell_2(n) = 2 \cdot \ell_2\left(\frac{n}{4}\right) + n^{(1/2)}$$

To determine the asymptotic orders of ℓ_1 and ℓ_2, observe that:

$$4^{(1/2)-\varepsilon} = \frac{\sqrt{4}}{4^{\varepsilon}} = \frac{2}{4^{\varepsilon}} < \frac{2}{1} = 2 \text{ whenever } \varepsilon > 0$$
$$4^{(1/2)} = \sqrt{4} = 2$$

The asymptotic orders of ℓ_1 and ℓ_2 are:

$$\ell_1(n) = \Theta\left(n^{\log_4 2}\right) = \Theta\left(n^{(1/\log_2 4)}\right) = \Theta\left(n^{(1/2)}\right) = \Theta\left(\sqrt{n}\right)$$
$$\ell_2(n) = \Theta\left(n^{(1/2)} \cdot \lg n\right) = \Theta\left(\sqrt{n} \cdot \lg n\right)$$

Since: $\ell(n) = \Omega\left(\ell_1(n)\right)$ and $\ell(n) = O\left(\ell_2(n)\right)$, we conclude that the asymptotic order of the function ℓ is bounded as follows:

$$\ell(n) = \Omega\left(\sqrt{n}\right) \text{ but } \ell(n) = O\left(\sqrt{n} \cdot \lg n\right)$$

(The actual asymptotic order of $\ell(n)$ is $\Theta\left(\sqrt{n \cdot \lg n}\right)$.)

Problem 307 Consider a positive, increasing function $f(n)$, defined recursively as follows:

$$f(n) = 8 \cdot f\left(\frac{n}{2}\right) + n^{\tau}$$

where τ is a rational constant.

(a) Select a value for τ such that:

$$f(n) = \omega\left(n^4\right)$$

and explain your answer. If such a value does not exist, explain why.

Answer: This is an instance of the Master Recurrence, with:

$$a = 8, \, b = 2, \, d = \tau$$

By setting:

$$\tau = 5$$

we obtain:

$$b^d = 2^{\tau} = 2^5 = 32 > 8 = a$$

and the asymptotic order of f is:

$$f(n) = \Theta\left(n^d\right) = \Theta\left(n^5\right) = \omega\left(n^4\right)$$

(b) Select a value for τ such that:

$$f(n) = \omega\left(n^3\right)$$

and at the same time:

$$f(n) = o\left(n^{3+\varepsilon}\right) \text{ for every } \varepsilon > 0$$

and explain your answer. If such a value does not exist, explain why.

Answer: By setting:

$$\tau = 3$$

we obtain:

$$b^d = 2^\tau = 2^3 = 8 = a$$

and the asymptotic order of f is:

$$f(n) = \Theta\left(n^d \cdot \lg n\right) = \Theta\left(n^3 \cdot \lg n\right)$$

which is indeed simultaneously $\omega(n^3)$ and $o(n^{3+\varepsilon})$ for any real constant $\varepsilon > 0$.

(c) Select a value for τ such that:

$$f(n) = o\left(n^2\right)$$

and explain your answer. If such a value does not exist, explain why.

Answer: By the Master Theorem, the asymptotic order of f is always:

$$f(n) = \Omega\left(n^{(\log_b a)}\right) = \Omega\left(n^{(\log_2 8)}\right) = \Omega\left(n^3\right)$$

Hence, no value of τ can bring $f(n)$ down to $o(n^2)$.

Problem 308   Let β, γ, δ be positive rational constants.

(a) Select a value for β such that the recurrence relation:

$$f(n) = 9 \cdot f\left(\frac{n}{\beta}\right) + n^3$$

defines a function $f(n)$ such that:

$$f(n) = \omega(n^3) \text{ and } f(n) = o(n^5)$$

Explain your answer. If such a value does not exist, explain why.

Answer: This is a Master Recurrence with:

$$a = 9, \, b = \beta, \, d = 3$$

By setting:

$$\beta = 2$$

we obtain:

$$b^d = 2^3 = 8 < 9 = a$$

and the asymptotic order of f is:

$$f(n) = \Theta\left(n^{\log_b a}\right) = \Theta\left(n^{\log_2 9}\right)$$

Since:

$$2^3 = 8 < 9 < 16 = 2^4$$

or, equivalently:

$$3 < \log_2 9 < 4$$

we conclude that:

$$f(n) = \omega\left(n^3\right) \quad \text{and} \quad f(n) = o\left(n^4\right)$$

which satisfies the requirements.

(b) Select a value for γ such that the recurrence relation:

$$g(n) = 5 \cdot g\left(\frac{n}{\gamma}\right) + n^3$$

defines a function $g(n)$ such that:

$$g(n) = \Theta\left(n^3\right)$$

Explain your answer. If such a value does not exist, explain why.

Answer: This is a Master Recurrence with:

$$a = 5, \; b = \gamma, \; d = 3$$

By setting:

$$\gamma = 2$$

we obtain:

$$b^d = 2^3 = 8 > 5 = a$$

and the asymptotic order of f is:

$$f(n) = \Theta\left(n^d\right) = \Theta\left(n^3\right)$$

(c) Select a value for δ such that the recurrence relation:

$$h(n) = 2 \cdot h\left(\frac{n}{\delta}\right) + \sqrt{n}$$

defines a function $h(n)$ such that:

$$h(n) = \omega\left(\sqrt{n}\right) \quad \text{and} \quad h(n) = o(n)$$

Explain your answer. If such a value does not exist, explain why.

Answer: This is a Master Recurrence with:

$$a = 2, \; b = \delta, \; d = \frac{1}{2}$$

By setting:

$$\delta = 4$$

we obtain:

$$b^d = 4^{(1/2)} = \sqrt{4} = 2 = a$$

and the asymptotic order of f is:

$$f(n) = \Theta\left(n^d \cdot \lg n\right) = \Theta\left(\sqrt{n} \cdot \lg n\right)$$

which satisfies the requirements.

Problem 309   Let $\alpha, \beta, \gamma, \delta$ be positive rational constants.

(a) Consider a positive, increasing function $f_1(n)$, defined by the following recurrence:

$$f_1(n) = \alpha \cdot f_1\left(\frac{3n}{4}\right) + n^2$$

Select a value for α, such that $\alpha \leq 3$ and:

$$f_1(n) = \Theta\left(n^2\right)$$

and explain your answer. If this is impossible, explain why.

Answer: This is an instance of the Master Recurrence, with:

$$a = \alpha, \; b = \frac{4}{3}, \; d = 2$$

To obtain:

$$f_1(n) = \Theta\left(n^d\right) = \Theta\left(n^2\right)$$

we need:

$$a < b^d$$

which yields:

$$\alpha < \left(\frac{4}{3}\right)^2 = \frac{16}{9} = 1 + \frac{7}{9}$$

Hence, one correct choice is:

$$\alpha = 1$$

(b) Consider a positive, increasing function $f_2(n)$, defined by the following recurrence:

$$f_2(n) = \beta \cdot f_2\left(\frac{3n}{4}\right) + n^2$$

Select a value for β, such that $\beta \leq 3$ and:

$$f_2(n) = \omega\left(n^2\right)$$

and explain your answer. If this is impossible, explain why.

Answer: The analysis given in the answer to part (a) shows that the requirement is:

$$f_2(n) = \omega\left(n^d\right) = \omega\left(n^2\right)$$

for which we need:

$$a = \beta \geq b^d = \left(\frac{4}{3}\right)^2 = 1 + \frac{7}{9}$$

One correct choice is:

$$\beta = 2$$

yielding:

$$f_2(n) = \Theta\left(n^{\log_{(4/3)} 2}\right)$$

(c) Consider a positive, increasing function $f_3(n)$, defined by the following recurrence:

$$f_3(n) = 8 \cdot f_3\left(\frac{n}{2}\right) + n^\gamma$$

Select a value for γ, such that:

$$f_3(n) = o\left(n^3\right)$$

and explain your answer. If this is impossible, explain why.

Answer: To see that this is impossible, observe that this is an instance of the Master Recurrence, with:

$$a = 8, \ b = 2, \ d = \gamma$$

We recognize three cases:

Case (1): $a > b^d$

whence: $f_3(n) = \Theta\left(n^{\log_2 8}\right) = \Theta\left(n^3\right)$.

Case (2): $a = b^d$, or equivalently: $d = \gamma = \log_2 8 = 3$ whence: $f_3(n) = \Theta\left(n^d \log n\right) = \Theta\left(n^3 \log n\right) = \omega\left(n^3\right)$.

Case (3): $a < b^d$, or equivalently: $d = \gamma > \log_2 8 = 3$ whence: $f_3(n) = \Theta\left(n^d\right) = \omega\left(n^3\right)$.

(d) Consider a positive, increasing function $f_4(n)$, defined by the following recurrence:

$$f_4(n) = 8f_4\left(\frac{n}{2}\right) + n^\delta$$

Select a value for δ, such that:

$$f_4(n) = \omega\left(n^3\right)$$

and explain your answer. If this is impossible, explain why.

Answer: The analysis given in the answer to part (c) shows that the requirement is:

$$f_4(n) = \omega\left(n^3\right)$$

for which we need:

$$d = \delta \geq \log_b a = 3$$

One correct choice is:

$$\delta = 4$$

yielding:

$$a = 8 < 2^4 = b^d$$

whence:

$$f_4(n) = \Theta\left(n^d\right) = \Theta\left(n^4\right)$$

Problem 310   Dangerous Professor gives a test to her students at the beginning of every lecture. However, students are escaping rapidly from the class, so that there is nobody left in the class after a certain number of tests. In fact, after every test, $(1/6)$ of the current student population drop the course and never appear again.

Initially, there are n students in the class. At the beginning of a test, every student receives a problem set with a number of questions. At the end of the test, each student hands in the problem set.

The Professor prepares a variety of different questions, so that individual problem sets differ among themselves (and those who copy their neighbor's answer are sure to fail.) Her rule is: if n students show up for an exam, then a total of about $\sqrt{n}+5$ questions are administered on the n problem sets in that exam.

To contribute to the clarity and neatness of students' answers, the Professor gives them a lot of scratch paper, in the form of blue notebooks. Usually, she gives 5 blue notebooks to each student in each exam.

(a) Let $t(n)$ be the number of tests administered before the last student drops the course. Write a recurrence relation that governs the asymptotic order of $t(n)$.

   Answer: After 1 test is administered to n students, $(5n/6)$ students remain:

$$t(n) = t\left(\frac{5n}{6}\right) + 1$$

(b) Let $p(n)$ be the total number of problem sets administered in all the exams, before the last student drops the course. Write a recurrence relation that governs the asymptotic order of $p(n)$.

   Answer: After n problem sets are administered to n students, $(5n/6)$ students remain:

$$p(n) = p\left(\frac{5n}{6}\right) + n$$

(c) Let $q(n)$ be the total number of questions administered in all the exams, before the last student drops the course. Write a recurrence relation that governs the asymptotic order of $q(n)$.

   Answer: After $\sqrt{n}+5 = \Theta(\sqrt{n})$ questions are administered to n students, $(5n/6)$ students remain:

$$q(n) = q\left(\frac{5n}{6}\right) + \sqrt{n}$$

(d) Let $\ell(n)$ be the total number of blue books given in all the exams, before the last student drops the course. Write a recurrence relation that governs the asymptotic order of $\ell(n)$.

   Answer: After $5n = \Theta(n)$ blue books are administered to n students, $(5n/6)$ students remain:

$$\ell(n) = \ell\left(\frac{5n}{6}\right) + n$$

(e) Find an asymptotic estimate for $t(n)$, and explain your answer.

   Answer: This is a Master Recurrence with:

$$a = 1, b = \frac{6}{5}, d = 0$$

   Since:

$$b^d = \left(\frac{6}{5}\right)^0 = 1 = a$$

   the asymptotic order of t is:

$$t(n) = \Theta\left(n^d \cdot \lg n\right) = \Theta(\lg n)$$

(f) Find an asymptotic estimate for $p(n)$, and explain your answer.

    Answer: This is a Master Recurrence with:

$$a = 1, \; b = \frac{6}{5}, \; d = 1$$

Since:

$$b^d = \left(\frac{6}{5}\right)^1 = \left(\frac{6}{5}\right) > 1 = a$$

the asymptotic order of p is:

$$p(n) = \Theta\left(n^d\right) = \Theta(n)$$

(g) Find an asymptotic estimate for $q(n)$, and explain your answer.

    Answer: This is a Master Recurrence with:

$$a = 1, \; b = \frac{6}{5}, \; d = \frac{1}{2}$$

Since:

$$b^d = \sqrt{\left(\frac{6}{5}\right)} > 1 = a$$

the asymptotic order of q is:

$$q(n) = \Theta\left(n^d\right) = \Theta\left(\sqrt{n}\right)$$

(h) Find an asymptotic estimate for $b(n)$, and explain your answer.

    Answer: This is a Master Recurrence with:

$$a = 1, \; b = \frac{6}{5}, \; d = 1$$

Since:

$$b^d = \left(\frac{6}{5}\right)^1 = \frac{6}{5} > 1 = a$$

the asymptotic order of ℓ is:

$$\ell(n) = \Theta\left(n^d\right) = \Theta(n)$$

Problem 311   The CLUB OF HAPPY NATURE LOVERS is on an outing, planting a tree. The actual honor of planting the tree belongs to the winner of the following courteous competition.

Initially, each member of the Club writes his own list, which contains the names of all competitors (including himself.) Next, he puts on a clean pair of socks, and ties safely his sneakers. After this, all competitors go for a fast cross-country walking race; $(1/3)$ of the walkers that finish latest are eliminated, while the $(2/3)$ of the competitors that arrive first remain in the tree-planting competition. Each of these competitors again gets a new pair of socks and writes a new list, which contains the names of all members still remaining in the game. Next, those still in the game go for another walking race, where $(1/3)$ of them are again eliminated. This procedure continues until only one participant is left, who becomes the winner and plants the tree.

(a) Let $\alpha(n)$ be the total number of names written while selecting the winner from n Club members. Write a recurrence relation that governs the asymptotic order of $\alpha(n)$.

    Answer: After n^2 names are written, $(2n/3)$ members remain:

$$\alpha(n) = \alpha\left(\frac{2n}{3}\right) + n^2$$

(b) Let $s(n)$ be the total number of pairs of socks used while selecting the winner from n Club members. Write a recurrence relation that governs the asymptotic order of $s(n)$.

Answer: After n pairs are used, $(2n/3)$ members remain:

$$s(n) = s\left(\frac{2n}{3}\right) + n$$

(c) Let $w(n)$ be the total number of walking races arranged while selecting the winner from n Club members. Write a recurrence relation that governs the asymptotic order of $w(n)$.

Answer: After 1 race is completed, $(2n/3)$ members remain:

$$w(n) = w\left(\frac{2n}{3}\right) + 1$$

(d) Find an asymptotic estimate for $\alpha(n)$, and explain your answer.

Answer: This is a Master Recurrence with:

$$a = 1, \, b = (3/2), \, d = 2$$

Since:

$$a = 1 < \left(\frac{9}{4}\right) = \left(\frac{3}{2}\right)^2 = b^d$$

the asymptotic order of α is:

$$\alpha(n) = \Theta\left(n^d\right) = \Theta\left(n^2\right)$$

(e) Find an asymptotic estimate for $s(n)$, and explain your answer.

Answer: This is a Master Recurrence with:

$$a = 1, \, b = \left(\frac{3}{2}\right), \, d = 1$$

Since:

$$a = 1 < \left(\frac{3}{2}\right)^1 = b^d$$

the asymptotic order of t is:

$$s(n) = \Theta\left(n^d\right) = \Theta(n)$$

(f) Find an asymptotic estimate for $w(n)$, and explain your answer.

Answer: This is a Master Recurrence with:

$$a = 1, \, b = \left(\frac{3}{2}\right), \, d = 0$$

Since:

$$a = 1 = \left(\frac{3}{2}\right)^0 = b^d$$

the asymptotic order of w is:

$$w(n) = \Theta\left(n^d \lg n\right) = \Theta(\lg n)$$

Problem 312   FINE COMPARISON is a problem solving competition, where the winner is selected as follows.

First, each of the n participants receives one clean notebook and a copy of the problem set. All the n copies of the problem set are identical. The participants write their solutions in the notebooks for an hour, after which each notebook is handed to one of the graders. The grading is organized so that each grader grades exactly two notebooks, and selects the participant whose work received the higher score. After finishing the grading, the grader goes to the proctor to receive a dollar in wages for this grading session, while each of the selected participants receives a new clean notebook and a copy of a new problem set; after an hour of work the notebooks are again organized in pairs, each pair is assigned to one grader, and each commissioned grader selects the author of the better of the two assigned notebooks for the next round, earning a dollar for the work. This procedure continues until only one participant is left, who becomes the winner.

(a) Let $\delta(n)$ be the total amount of dollars paid to the graders while selecting the winner from n participants. Write a recurrence relation that governs the asymptotic order of $\delta(n)$.

Answer: After $(n/2) = \Theta(n)$ dollars are paid, $(n/2)$ participants remain:

$$\delta(n) = \delta\left(\frac{n}{2}\right) + n$$

(b) Let $h(n)$ be the total number of hours that the winner has spent working on the problems while competing among n participants. Write a recurrence relation that governs the asymptotic order of $h(n)$.

Answer: After 1 hour, $(n/2)$ participants remain:

$$h(n) = h\left(\frac{n}{2}\right) + 1$$

(c) Find an asymptotic estimate for $\delta(n)$, and explain your answer.

Answer: This is a Master Recurrence with:

$$a = 1, b = 2, d = 1$$

Since:

$$a = 1 < 2 = 2^1 = b^d$$

the asymptotic order of δ is:
$$\delta(n) = \Theta\left(n^d\right) = \Theta\left(n\right)$$

(d) Find an asymptotic estimate for $h(n)$, and explain your answer.

Answer: This is a Master Recurrence with:

$$a = 1, b = 2, d = 0$$

Since:

$$a = 1 = 2^0 = b^d$$

the asymptotic order of h is:

$$h(n) = \Theta\left(n^d \cdot \lg n\right) = \Theta\left(\lg n\right)$$

Problem 313   The THOUGHTFUL STUDENT CHESS CLUB is awarding a prize to the best player. Their chess competition proceeds as follows.

Initially, in the first round, each player plays one match against every other player. When all the matches in the first round are over, the referee compiles a list that ranks all the players in sequence. One quarter of the players (that are ranked the highest) proceed into the second round, where again each of them plays a match against every other player that is still in the competition, after which the referee again selects the best quarter of them for the third round. This procedure continues until only one player is left, who wins the prize.

(a) Let $t(n)$ be the total number of matches played while selecting the winner from n Club members. Write a recurrence relation that governs the asymptotic order of $t(n)$.

Answer: After $\begin{pmatrix} n \\ 2 \end{pmatrix} = \Theta(n^2)$ matches, $(n/4)$ players remain:

$$t(n) = t\left(\frac{n}{4}\right) + n^2$$

(b) Let $r(n)$ be the total number of rounds held while selecting the winner from n Club members. Write a recurrence relation that governs the asymptotic order of $h(n)$.

Answer: After 1 round, $(n/4)$ players remain:

$$r(n) = r\left(\frac{n}{4}\right) + 1$$

(c) Let $\ell(n)$ be the total length of all ranking lists written while selecting the winner from n Club members. (The length of a list is the number of names on it.) Write a recurrence relation that governs the asymptotic order of $\ell(n)$.

Answer: After n players are listed, $(n/4)$ of them remain:

$$\ell(n) = \ell\left(\frac{n}{4}\right) + n$$

(d) Find an asymptotic estimate for $t(n)$, and explain your answer.

Answer: This is a Master Recurrence with:

$$a = 1,\, b = 4,\, d = 2$$

Since:

$$a = 1 < 16 = 4^2 = b^d$$

the asymptotic order of t is:

$$t(n) = \Theta\left(n^d\right) = \Theta\left(n^2\right)$$

(e) Find an asymptotic estimate for $r(n)$, and explain your answer.

Answer: This is a Master Recurrence with:

$$a = 1,\, b = 4,\, d = 0$$

Since:

$$a = 1 = 4^0 = b^d$$

the asymptotic order of h is:

$$h(n) = \Theta\left(n^d \cdot \lg n\right) = \Theta\left(\lg n\right)$$

(f) Find an asymptotic estimate for $\ell(n)$, and explain your answer.

> **Answer:** This is a Master Recurrence with:

$$a = 1,\ b = 4,\ d = 1$$

Since:

$$a = 1 < 4 = 4^1 = b^d$$

the asymptotic order of ℓ is:

$$\ell(n) = \Theta\left(n^d\right) = \Theta\left(n\right)$$

Problem 314 The management of HIGH BALCONY HOTEL is running a promotion campaign. As the final event of the campaign, the management is awarding a prize to one of the n guests. The winner is chosen as follows. First, all n candidates gather in one of the dining rooms, and they all sit down so that exactly four persons occupy each table. The four occupants of each table then play one short lottery game, in which exactly one of these four persons wins. All winners then proceed to the next dining room, sit down again in groups of four, and each group selects one winner by playing a lottery game; all the winners then move into a new dining room. This procedure continues until only one winner is left, who receives the prize.

(a) Let $r(n)$ be the total number of dining rooms visited while selecting the winner from n guests. Write a recurrence relation that governs the asymptotic order of $r(n)$.

> **Answer:** After 1 room is visited, $(n/4)$ guests remain:

$$r(n) = r\left(\frac{n}{4}\right) + 1$$

(b) Let $g(n)$ be the total number of lottery games played while selecting the winner from n guests. Write a recurrence relation that governs the asymptotic order of $g(n)$.

> **Answer:** After $(n/4) = \Theta(n)$ games are over, $(n/4)$ guests remain:

$$g(n) = g\left(\frac{n}{4}\right) + n$$

(c) Find an asymptotic estimate for $r(n)$, and explain your answer.

> **Answer:** This is a Master Recurrence with:

$$a = 1,\ b = 4,\ d = 0$$

Since:

$$a = 1 = 4^0 = b^d$$

the asymptotic order of r is:

$$r(n) = \Theta\left(n^d \cdot \lg n\right) = \Theta\left(\lg n\right)$$

(d) Find an asymptotic estimate for $g(n)$, and explain your answer.

> **Answer:** This is a Master Recurrence with:

$$a = 1,\ b = 4,\ d = 1$$

Since:

$$a = 1 < 4 = 4^1 = b^d$$

the asymptotic order of g is:

$$g(n) = \Theta\left(n^d\right) = \Theta\left(n\right)$$

Problem 315   MASSIVE KNOWLEDGE is a problem solving competition, where the winner is selected as follows. First, each of the n participants receives a clean notebook and a copy of the problem set. All the n copies of the problem set are identical. They work for an hour, after which the graders grade all the n participants, and select $(n/3)$ of them that scored the highest. Each of these selected participants then receives a new clean notebook and a copy of a new problem set; after an hour of work the graders again select one third of them for the next round. This procedure continues until only one participant is left, who becomes the winner.

(a) Let $p(n)$ be the total number of different problem sets prepared while selecting the winner from n participants. Write a recurrence relation that governs the asymptotic order of $p(n)$.

   Answer: After 1 problem set is solved, $(n/3)$ participants remain:

$$p(n) = p\left(\frac{n}{3}\right) + 1$$

(b) Let $\beta(n)$ be the total number of notebooks distributed while selecting the winner from n participants. Write a recurrence relation that governs the asymptotic order of $\beta(n)$.

   Answer: After n notebooks are distributed, $(n/3)$ participants remain:

$$\beta(n) = \beta\left(\frac{n}{3}\right) + n$$

(c) Find an asymptotic estimate for $p(n)$, and explain your answer.

   Answer: This is a Master Recurrence with:

$$a = 1, b = 3, d = 0$$

   Since:

$$a = 1 = 3^0 = b^d$$

   the asymptotic order of p is:

$$p(n) = \Theta\left(n^d \cdot \lg n\right) = \Theta\left(\lg n\right)$$

(d) Find an asymptotic estimate for $\beta(n)$, and explain your answer.

   Answer: This is a Master Recurrence with:

$$a = 1, b = 3, d = 1$$

   Since:

$$a = 1 < 3 = 3^1 = b^d$$

   the asymptotic order of β is:

$$\beta(n) = \Theta\left(n^d\right) = \Theta\left(n\right)$$

Problem 316   The SMART STUDENT PROGRAMMING CLUB is awarding a prize to the best programmer. Their programming competition proceeds as follows. Initially, in the first round, each programmer is given one new computer. Then, each programmer enters a series of games, where a game is a short coding duel against one of the other programmers. In the first round, every programmer is scheduled for exactly one game against every other programmer. In each game, the Judge administers one coding problem to every pair of competing programmers; no problem is ever administered more than once in the entire competition. When all the games in the first round are over, the Judge ranks the programmers. One third of the programmers that are ranked the highest proceed into the second round, where again each of them receives a new computer (since the one from the previous round is already outdated) and plays one game against every other programmer that is still in the competition. At the end of the round, the Judge selects the upper third for the next round. This procedure continues until only one programmer is left, who wins the prize.

(a) Let $p(n)$ be the total number of problems administered while selecting the winner from n Club members. Write a recurrence relation that governs the asymptotic order of $p(n)$.

Answer:

$$p(n) = p\left(\frac{n}{3}\right) + n^2$$

(b) Let $r(n)$ be the total number of rounds held while selecting the winner from n Club members. Write a recurrence relation that governs the asymptotic order of $r(n)$.

Answer:

$$r(n) = r\left(\frac{n}{3}\right) + 1$$

(c) Let $c(n)$ be the total number of computers given away while selecting the winner from n Club members. Write a recurrence relation that governs the asymptotic order of $c(n)$.

Answer:

$$c(n) = c\left(\frac{n}{3}\right) + n$$

(d) Find an asymptotic estimate for $p(n)$, and explain your answer.

Answer: This is a Master Recurrence with:

$$a = 1, \, b = 3, \, d = 2$$

Since:

$$a = 1 < 9 = 3^2 = b^d$$

the asymptotic order of p is:

$$p(n) = \Theta\left(n^d\right) = \Theta\left(n^2\right)$$

(e) Find an asymptotic estimate for $r(n)$, and explain your answer.

Answer: This is a Master Recurrence with:

$$a = 1, \, b = 3 \, d = 0$$

Since:

$$a = 1 = 3^0 = b^d$$

the asymptotic order of r is:

$$r(n) = \Theta\left(n^d \cdot \lg n\right) = \Theta\left(\lg n\right)$$

(f) Find an asymptotic estimate for $c(n)$, and explain your answer.

Answer: This is a Master Recurrence with:

$$a = 1, \, b = 3, \, d = 1$$

Since:

$$a = 1 < 3 = 3^1 = b^d$$

the asymptotic order of c is:

$$c(n) = \Theta\left(n^d\right) = \Theta\left(n\right)$$

Problem 317   The members of the SOCIETY OF LUCKY FELLOWS are electing their Chairman as follows.

First, the n members group themselves into pairs. Next, the Judge gives one coin to each pair. At the sound of the Judge's whistle, all pairs simultaneously toss their coins. If a coin lands heads, the older of the two members of that pair wins. If a coin lands tails, the younger of the two members of that pair wins. All losers then quit, taking the coins with them as a consolation prize. All winners again form pairs, each pair again receives a coin, waits for the whistle, tosses the coin, and determines the winner. This procedure continues until only one winner is left, who becomes the Chairman.

(a) Let $w(n)$ be the total number of times the whistle is blown while electing the Chairman from n members. Write a recurrence relation that governs the asymptotic order of $w(n)$.

   Answer:
$$w(n) = w\left(\frac{n}{2}\right) + 1$$

(b) Let $c(n)$ be the total number of coins distributed while electing the Chairman from n members. Write a recurrence relation that governs the asymptotic order of $c(n)$.

   Answer:
$$c(n) = c\left(\frac{n}{2}\right) + n$$

(c) Find an asymptotic estimate for $w(n)$, and explain your answer.

   Answer: This is a Master Recurrence with:
$$a = 1,\, b = 2,\, d = 0$$

   Since:
$$a = 1 = 2^0 = b^d$$

   the asymptotic order of w is:
$$w(n) = \Theta\left(n^d \cdot \lg n\right) = \Theta\left(\lg n\right)$$

(d) Find an asymptotic estimate for $c(n)$, and explain your answer.

   Answer: This is a Master Recurrence with:
$$a = 1,\, b = 2,\, d = 1$$

   Since:
$$a = 1 < 2 = 2^1 = b^d$$

   the asymptotic order of c is:
$$c(n) = \Theta\left(n^d\right) = \Theta\left(n\right)$$

4.3   Recursive Algorithms

Problem 318   You are negotiating with a sibyl who is guarding access to a treasury.

The sibyl has placed before you a basket containing n coins. To gain access to the treasury, you have to recognize the one magic coin among the n coins in the basket. Coins are labeled by distinct natural numbers. Apart from the labeling, the coins are indistinguishable, and you have no way of telling which one is magic.

The sibyl knows which coin is magic, however. She is willing to help you, by answering your questions. Only the following two types of questions are acceptable.

Question Type 1: If two coins are shown to the sibyl, she will say which one of them, if any, is magic. However, once she answers a question of type 1, she will not talk to you any more. If you have not discovered the magic coin after you hear her answer to this question, you lose forever.

Question Type 2: If any sequence of coins, sorted by label, is shown to the sibyl, she will remove one-third of these coins, guaranteeing that the magic coin is not removed. Then, she will mix thoroughly the remaining coins, so that they are not sorted any more. You may ask her this question as often as you like.

Find an algorithm for entering the treasury, and determine a function that governs its asymptotic worst-case running time, counting only label lookup operations.

Answer: The algorithm for entering the treasury is as follows.

while there are more than 2 coins in the basket

{ sort the coins,
ask the second question; }

with 2 coins

ask the first question.

To calculate the running time, recall that n elements are sorted with $t_s(n) = \Theta(n \lg n)$ comparisons. Furthermore, each sorting is followed by a question of type 2, which reduces the number of coins in the game to two-thirds of its original value. Hence, the running time of our algorithm is governed by the following recurrence:

$$t(n) = t_s(n) + t\left(\frac{2n}{3}\right)$$

which is asymptotically equivalent to:

$$t(n) = t\left(\frac{2n}{3}\right) + n \lg n$$

This is an instance of a generalized Master Recurrence, whose form is:

$$t(n) = a\, t\left(\frac{n}{b}\right) + f(n)$$

where $f(n)$ is a positive, increasing function. Recall that the solution of this recurrence is a function $t(n) = \Theta(f(n))$ if the function $f(n)$ satisfies the following two properties:

There exists an $\epsilon > 0$ such that $f(n) = \Omega\left(n^{(\log_b a)+\epsilon}\right)$.

There exists a $c < 1$ such that $a f\left(\frac{n}{b}\right) \le c f(n)$ for a sufficiently large n.

To verify that this case is indeed in effect, observe that:

$$a = 1,\ b = \left(\frac{3}{2}\right),\ f(n) = n \lg n,\ \log_b a = 0$$

yielding the two required conditions:

$$f(n) = n \lg n \ge n = n^{0+1} = n^{\log_b a + 1}$$

$$a f\left(\frac{n}{b}\right) = f\left(\frac{2n}{3}\right) = \left(\frac{2n}{3}\right) \cdot \lg\left(\frac{2n}{3}\right) < \left(\frac{2n}{3}\right) \cdot \lg n = \left(\frac{2}{3}\right) \cdot (n \lg n) = \left(\frac{2}{3}\right) \cdot f(n)$$

Hence, we conclude that $t(n) = \Theta(n \cdot \lg n)$.

Problem 319   The sibyl from Problem 318 offers you another riddle to solve, to gain access to the same treasury.

This time you have to find 2 magic coins in a basket containing n coins. The coins are unmarked and indistinguishable from one another. You may ask the sibyl only one type of a question:

> *Question:*   If two coins are shown to the sibyl and both of these two coins are magic, she will say "yes"; otherwise she will say "no". You may ask her this question as often as you like.

Find an algorithm for entering the treasury, and determine a function that governs its asymptotic worst-case running time, counting questions. If you were in a hurry and there were many coins in the basket, would you prefer the riddle described in this problem to that solved in Problem 318?

Answer: To win with n coins, we may have to show the sibyl every possible pair of coins, and therefore ask as many as $n(n-1)/2$ questions in the worst case. Assuming that a label lookup operation, whose executions were counted in Problem 318, and the question operation, counted in the present algorithm, are performed within comparable time intervals, it is appropriate to compare the functions that express the running time of the two procedures. Since:

$$\frac{n(n-1)}{2} = \omega(n \cdot \lg n)$$

we conclude that for a sufficiently large number of coins n, it takes longer to enter the treasury according to the rules of this game than that described in Problem 318.

Problem 320   Consider the algorithm *guess*, defined as follows.

algorithm guess (n: natural number) **returns** integer
if $(n = 0)$ **then**
    $m \longleftarrow n + 1$; **return** (m)
else
    $m \longleftarrow$ guess $(n - 1) +$ guess $(n - 1) + 1$; **return** (m)
endif

(In the algorithm pseudo-code $\longleftarrow$ denotes assignment.)

(a) Let $g(n)$ be the output of the algorithm *guess*, invoked on input n. Write a recursive definition of $g(n)$.

Answer:
$$g(0) = 1$$
$$g(n) = g(n-1) + g(n-1) + 1, \ \text{ for } n \geq 1$$

(b) Write a closed form for $g(n)$, and prove that it indeed defines the function which you have proposed in part **(a)**.

Answer: We use induction to prove that the closed form for $g(n)$ is:

$$g(n) = 2^{n+1} - 1 \ \ \text{ for } n \geq 0$$

The base case occurs for $n = 0$, and follows from the base case of the recursive definition of the function g. Indeed:

$$g(0) = 1 = 2^{0+1} - 1$$

For the inductive step, assume that: $g(n) = 2^{n+1} - 1$, for some $n \geq 0$. We have to prove that: $g(n+1) = 2^{(n+1)+1} - 1$. Indeed:

$$g(n+1) =$$
$$\bullet \quad = g((n+1) - 1) + g((n+1) - 1) + 1 = g(n) + g(n) + 1 = 2g(n) + 1$$
$$* \quad = 2\left(2^{n+1} - 1\right) + 1 = 2^{n+2} - 2 + 1 = 2^{(n+1)+1} - 1$$

The equality $\bullet$ follows from the recursive definition of $g(n+1)$ for $n + 1 \geq 1$; the equality $*$ is obtained by an application of the inductive hypothesis to g_n; and the other equalities are obtained by arithmetic transformations.

Problem 321   Recall the algorithm *guess*, defined in Problem 320.

(a) Let $a(n)$ be the number of times the algorithm *guess* performs an addition (the operation "+") when invoked on input n. Write a recursive definition of $a(n)$.

Answer: By inspection of the code, we conclude that *guess* (0) performs one addition. Whenever $n > 0$, *guess* (n) performs an addition $a(n-1)$ times within each of the two invocations of *guess* $(n-1)$, and performs two more additions to calculate m as a sum. Hence, the recurrence for $a(n)$ is as follows.

$$a(0) = 1$$
$$a(n) = 2a(n-1) + 2, \text{ for } n \geq 1$$

(b) Write a closed form for $a(n)$, and prove that it indeed defines the function which you have proposed in part **(a)**.

Answer: We use induction to prove that the closed form for $a(n)$ is:

$$a(n) = 3 \cdot 2^n - 2 \quad \text{for } n \geq 0$$

The base case occurs for $n = 0$, and follows from the base case of the recursive definition of the function a. Indeed:

$$a(0) = 1 = 3 \cdot 2^0 - 2$$

For the inductive step, assume that: $a(n) = 3 \cdot 2^n - 2$, for some $n \geq 0$. We have to prove that $a(n+1) = 3 \cdot 2^{n+1} - 2$. Indeed:

$$a(n+1) = 2a((n+1) - 1) + 2 = 2a(n) + 2 = 2(3 \cdot 2^n - 2) + 2 = 6 \cdot 2^n - 2 = 3 \cdot 2^{n+1} - 2$$

The first equality follows from the recursive definition of $a(n+1)$ for $n + 1 \geq 1$; the third equality is obtained by an application of the inductive hypothesis to a_n; and the other equalities are obtained by arithmetic transformations.

(c) Let $t(n)$ be the number of times the test in the **if**-statement returns the value of true, if the algorithm *guess* is invoked on input n. Write a recursive definition of $t(n)$.

Answer: By inspection of the code, we conclude that the **if**-statement evaluates a true result if executed by *guess* (0). In an invocation of *guess* (n) for $n > 0$, this happens $t(n-1)$ times within each of the two invocations of *guess* $(n-1)$, whence the recurrence:

$$t(0) = 1$$
$$t(n) = 2t(n-1), \text{ for } n \geq 1$$

(d) Write a closed form for $t(n)$, and prove that it indeed defines the function which you have proposed in part **(c)**.

Answer: We use induction to prove that the closed form for $t(n)$ is:

$$t(n) = 2^n \quad \text{for } n \geq 0$$

The base case occurs for $n = 0$, and follows from the base case of the recursive definition of the function t. Indeed:

$$t(0) = 1 = 2^0$$

For the inductive step, assume that $t(n) = 2^n$, for some $n \geq 0$.
We have to prove that $t(n + 1) = 2^{n+1}$. Indeed:

$$t(n + 1) = 2t((n + 1) - 1) = 2t(n) = 2 \cdot 2^n = 2^{n+1}$$

The first equality follows from the recursive definition of $t(n + 1)$ for $n + 1 \geq 1$; the third equality is obtained by an application of the inductive hypothesis to t_n; and the other equalities are obtained by arithmetic transformations.

(e) Let $f(n)$ be the number of times the test in the **if**-statement returns the value false, if the algorithm *guess* is invoked on input n. Write a recursive definition of $f(n)$.

Answer: By inspection of the code, we conclude that the **if**-statement does not evaluates a false result if executed by *guess* (0). In an invocation of *guess* (n) for $n > 0$, the **if**-statement executed prior to the recursive calls returns the value of false. Additionally, this happens $f(n - 1)$ times within each of the two invocations of *guess* $(n - 1)$, whence the recurrence:

$$f(0) = 0$$
$$f(n) = 2f(n - 1) + 1, \quad \text{for } n \geq 1$$

(f) Write a closed form for $f(n)$, and prove that it indeed defines the function which you have proposed in part **(e)**.

Answer: We use induction to prove that the closed form for $f(n)$ is:

$$f(n) = 2^n - 1 \quad \text{for } n \geq 0$$

The base case occurs for $n = 0$, and follows from the base case of the recursive definition of the function t. Indeed:

$$f(0) = 0 = 2^0 - 1$$

For the inductive step, assume that: $f(n) = 2^n - 1$, for some $n \geq 0$. We have to prove that: $f(n + 1) = 2^{n+1} - 1$. Indeed:

$$f(n + 1) = 2f(n) + 1 = 2(2^n - 1) + 1 = 2^{n+1} - 1$$

The first equality follows from the recursive definition of $f(n + 1)$ for $n + 1 \geq 1$; the third equality is obtained by an application of the inductive hypothesis to f_n; and the other equalities are obtained by arithmetic transformations.

(g) Let $c(n)$ be the total number of times the algorithm *guess* performs an addition (the operation "+") or a test (the **if**-statement) when invoked on input n. Write a closed form for $a(n)$ and prove that your answer is correct.

Answer: By summing the counts obtained in the answers to parts **(b)**, **(d)** and **(f)**, we obtain:

$$c(n) = a(n) + t(n) + f(n) = (3 \cdot 2^n - 2) + (2^n) + (2^n - 1) = 5 \cdot 2^n - 3$$

Problem 322 Recall the algorithm *guess*, defined in Problem 320.

(a) Construct an algorithm *straightGuess* to compute efficiently the function $g(n) = 2^{n+1} - 1$, computed by the algorithm *guess*. The argument list of *straightGuess* must be identical to that of *guess*, and its arithmetic capabilities must be limited to addition (and subtraction), as is the case in the algorithm *guess*. If the algorithm *guess* is reasonably efficient, so that the construction of *straightGuess* is not justifiable, explain why.

Answer: The running time of the algorithm *guess* is exponential in the value of its input—hence, *guess* cannot be deemed efficient.

algorithm straightGuess $(n$: natural number$)$ **returns** integer
$m \longleftarrow 2;$
$\ell \longleftarrow 0;$
while $(\ell < n)$ **do**
$\quad m \longleftarrow m + m;$
$\quad \ell \longleftarrow \ell + 1$
endwhile
$r \longleftarrow m - 1;$
return (r)

(b) Prove that the algorithm which you constructed in the answer to part **(a)** is correct in that it indeed computes the function $g(n)$.

Answer: Let $\mu(n)$ be the final value of the variable m in an invocation of *straightGuess* (n). The value computed by *straightGuess* (n) is equal to $\mu(n) - 1$, as is verified by inspection of the code. We have to prove that $\mu(n) - 1 = g(n)$ for all $n \geq 0$, which is equivalent to the claim that $\mu(n) = g(n) + 1 = 2^{n+1}$.

By inspection of the code, we obtain the following recurrence for $\mu(n)$.

$$\mu(0) = 2$$
$$\mu(n) = \mu(n-1) + \mu(n-1) = 2\mu(n-1)$$

To prove that the closed form of $\mu(n)$ is indeed as required, observe that, in the base case:

$$\mu(0) = 2 = 2^{0+1}$$

Inductively, assuming that $\mu(n-1) = 2^{(n-1)+1} = 2^n$, we conclude:

$$\mu(n) = 2\mu(n-1) = 2 \cdot 2^n = 2^{n+1}$$

whence the claim.

(c) Prove that your algorithm is indeed more efficient than *guess*, by calculating the total count of additions and tests performed by it, as a function of n.

Answer: Let $\gamma(n)$ be total number of additions and tests performed by *straightGuess* (n). By inspection of the code, we find that each pass through the **while**-loop entails one test prior to it and two additions within it; the loop is executed exactly n times, but attempted $n + 1$ times. Upon the exit from the loop, one more addition (subtraction) is performed. These observations lead to the closed form for $\gamma(n)$:

$$\gamma(n) = (n + 1) + 2n + 1 = 3n + 2$$

(d) Compare the asymptotic orders of the functions that express the running time of the algorithms *guess* and *straightGuess*.

Answer: We conclude from the answers given in the part (g) of Problem 321 and part (c) of this problem that the running time $c(n)$ of the algorithm *guess* and the running time $\gamma(n)$ of the algorithm *straightGuess* have the following properties.

$$c(n) = \Theta(2^n) \text{ and } \gamma(n) = \Theta(n)$$

Hence:

$$\gamma(n) = \Theta(\lg c(n))$$

indicating that *straightGuess* is dramatically (exponentially) faster than *guess*.

(e) Compare the total counts of additions and tests, performed by *guess* (n) and *straightGuess* (n), for $n = 0, 1, 2, 5, 10, 20, 30$.

Answer: We compute the following values, based on the answers given in the part (g) of Problem 321 and part (c) of this problem.

n	guess (n) $c(n) = 5 \cdot 2^n - 3$	straightGuess (n) $\gamma(n) = 3n + 2$
0	2	2
1	7	5
2	17	8
5	157	17
10	5,117	32
20	5,242,877	62
30	5,368,709,117	92

Problem 323   Consider the algorithm *press*, defined as follows.

algorithm press $(s$: binary string$)$ **returns** integer
if $(s = \lambda)$ **then**
    $m \longleftarrow 0$; **return** (m)
else
    $m \longleftarrow (2 \cdot \text{press}(s^+) + s[0])$; **return** (m)
endif

In the algorithm pseudo-code $\longleftarrow$ denotes assignment, λ denotes the empty string, $s[0]$ denotes the rightmost bit of the string s, and s^+ denotes the string obtained from s by a removal of $s[0]$.

(a) Let $g(s)$ be the output of *press* (s). Calculate $g(s)$ for $s = 1, 10, 111, 101$.
Answer:

s	$g(s)$
1	1
10	2
111	7
101	5

(b) Let $a(n)$ be the number of times the algorithm *press* performs any of the following six operations: addition, multiplication, assignment, access to $s[0]$, access to s^+, equality test, when invoked on an input string of length n. Determine a closed form of the function $a(n)$, and explain your answer.

Answer: By inspection of the code, we find that *press* performs one test and one assignment if invoked on an empty string. When invoked on an input of length equal to n, where $n > 0$, *press* performs a total of $a(n-1)$ operations inside the recursive call to itself on an input of length $n-1$. Moreover, it performs one test and one string truncation prior to this recursive call, as well as one multiplication, one access to the last bit, one addition and one assignment, after the return from the recursive call. Hence, $a(n)$ is governed by the following recurrence.

$$a(0) = 2$$
$$a(n) = 6 + a(n-1) \text{ for } n \geq 1$$

It is straightforward to prove by induction that the closed form for $a(n)$ is:

$$a(n) = 6n + 2$$

(c) Which function is performed by the algorithm *press*? In other words, what is the relationship between an input to the algorithm and the corresponding output?

Answer: An invocation of *press* (s) returns the value of the natural number whose binary representation is the string s. In other words, if s is a bit string, say $s = s_{n-1}s_{n-2} \ldots s_1 s_0$, where $s_k \in \{0, 1\}$ for all k such that $0 \leq k < n$, then *press* (s) returns the number:

$$g(s) = \sum_{k=0}^{n-1} s_k \cdot 2^k$$

(d) Compare the asymptotic orders of the input length n, the output value $g(n)$, and the runtime cost $a(n)$ for the algorithm *press*.

Answer: We conclude the following from the answers given in the parts **(b)** and **(c)**.

$$a(n) = \Theta(n), \quad g(n) = \Theta(2^n), \quad a(n) = \Theta(\lg g(n))$$

Problem 324 Recall the algorithm *press*, defined in Problem 323.

(a) Consider the algorithm *hardPress*, which computes the natural number $g(s)$ from its binary representation $s = s_{n-1}s_{n-2} \ldots s_1 s_0$, performing the sequence of additions and multiplications specified by the following expression, in the order determined by the parentheses.

$$g(s) = s_0 + \sum_{k=1}^{n-1} \left(s_k \cdot \left(\prod_{i=1}^{k} (2) \right) \right)$$

Let $b(n)$ be the total number of additions and multiplications performed by *hardPress* (s), where n is the length of the input string s. Determine the asymptotic order of the function $b(n)$.

Answer: By inspection of the procedure, we conclude that the computation of $s_k \cdot 2^k$ requires k multiplications, for $0 \leq k < n$. Calculating the sum of these n products entails a total of $n-1$ additions. Hence, the total operation count is:

$$b(n) = (n-1) + \sum_{k=0}^{n-1} k = (n-1) + \frac{n(n-1)}{2} = \frac{(n-1)(n+2)}{2}$$

whence the conclusion that $b(n) = \Theta(n^2)$.

(b) Compare the asymptotic orders of the functions that express the running time of the algorithms *press* and *hardPress*.

Answer: We conclude from the answers given in the part **(b)** of Problem 323 and the part **(a)** of this problem:

$$b(n) = \Theta\left((a(n))^2\right) \quad \text{or} \quad a(n) = \Theta\left(\sqrt{b(n)}\right)$$

which indicates that *hardPress* is significantly (polynomially) slower than *press*.

Problem 325   Consider the algorithm *stress*, defined as follows.

algorithm stress (*s*: binary string) **returns** integer
if $(s = \lambda)$ **then**
     $m \longleftarrow 0;$ **return** (m)
else
     $m \longleftarrow \left(\text{stress}(s^-) + s^{[L]}\right);$ **return** (m)
endif

In the algorithm pseudo-code $\longleftarrow$ denotes assignment, λ denotes the empty string, $s^{[L]}$ denotes the leftmost bit of the string s, and s^- denotes the string obtained from s by a removal of $s^{[L]}$.

(a) Let $g(s)$ be the output of *stress* (s). Calculate $g(s)$ for $s = 1, 10, 111, 110$.
Answer:

s	$g(s)$
1	1
10	1
111	3
110	2

(b) Let $a(n)$ be the number of times the algorithm *stress* performs any of the following five operations: assignment, addition, access to $s^{[L]}$, access to s^-, equality test, when invoked on an input string of length n. Determine a closed form of the function $a(n)$, and explain your answer.

Answer: By inspection of the code, we find that *stress* performs one test and one assignment if invoked on an empty string. When invoked on an input of length equal to n, where $n > 0$, *stress* performs $a(n-1)$ operations inside the recursive call to itself on input s^-, whose length is equal to $n-1$. Apart from this, *stress* performs one test, one assignment, one addition, one accesses to the last bit, and one string truncation. Hence, $a(n)$ is governed by the following recurrence.

$$a(0) = 2$$
$$a(n) = 5 + a(n-1) \text{ for } n \geq 1$$

It is straightforward to prove by induction that the closed form for $a(n)$ is:

$$a(n) = 5n + 2$$

(c) Which function is performed by the algorithm *stress*? In other words, what is the relationship between an input s to the algorithm and the corresponding output $g(s)$?

Answer: An invocation of *stress* (s) returns the number of occurrences of the symbol 1 in the input bit string s.

Problem 326   Recall the algorithm *stress*, defined in Problem 325.

(a) Assume that the computation space is managed as follows.

(A) Each letter of the input string occupies one word of the computation space. Every integer occupies four words of this space.

(B) The space occupied by all program variables prior to a recursive call remains occupied for the duration of the call.

(C) The space occupied by all program variables is released upon returning from the program, with the exception of the result returned by the program, whose space becomes a part of the space allocated to the calling program.

(D) The arguments of a recursive call are copied prior to the call. The space required for these copies is allocated by the calling program, and re-claimed after the return from the recursive call.

Let $p(n)$ be the size (in words) of the computation space required for the completion of an invocation of the algorithm *stress* on an input string s of length n, not counting (i.e., in addition to) the space occupied by the input s itself. Determine a closed form for $p(n)$.

Answer: When invoked on an empty input string, *stress* assigns a value to only one program variable, namely m. The assigned value is an integer, which (by assumption) occupies four words. When *stress* is invoked with an input string of length $n > 0$, a recursive call is entered, whose argument is string s^-, of length equal to $n - 1$. This copy requires additional space, of size equal to $n - 1$ words. Moreover, once the recursive call to *stress* (s^-) has been completed, the returned value is assigned to the integer variable m, which requires four words of additional space. Hence, the computation space of the algorithm *stress* is governed by the recurrence:

$$p(0) = 4$$
$$p(n) = (n - 1) + p(n - 1) + 4 = p(n - 1) + (n + 3) \text{ for } n \geq 1$$

whence the closed form:

$$p(n) = p(0) + \sum_{k=1}^{n}(k + 3) = 4 + \sum_{k=1}^{n}3 + \sum_{k=1}^{n}k = 4 + 3n + \frac{n(n + 1)}{2} = \frac{n^2}{2} + \frac{7n}{2} + 4$$

(b) Assume that the runtime environment has been modified so that it is addresses (pointers) rather than entire argument values that are copied prior to a call. Precisely, the last (only) of the rules given in the part **(a)** is modified so as to become:

(D) The address of each argument of a recursive call is generated prior to the call. The space required for these addresses is allocated by the calling program, and re-claimed after the return from the recursive call. Each address occupies two words.

Let $q(n)$ be the size (in words) of the computation space required for the completion of an invocation of the algorithm *stress* on input string s of length n under these assumptions, not counting (i.e., in addition to) the space occupied by the input s itself. Determine a closed form for $q(n)$.

Answer: When invoked on an empty input string, *stress* does not perform any recursive calls— hence, $q(0) = p(0) = 4$. When *stress* is invoked with an input string of length $n > 0$, a recursive call with one argument is entered. The address of this arguments is generated prior to the call, and two words of the computation space are expended thereby. Once the recursive call to *stress* (s^-) has been completed, the returned value is assigned to the integer variable m, which requires four words of additional space. Consequently, the recurrence that governs the computation space of the algorithm *stress* in this runtime environment is as follows:

$$q(0) = 4$$
$$q(n) = 2 + 4 + q(n-1) = 6 + q(n-1) \text{ for } n \geq 1$$

It is straightforward to prove by induction that the closed form for $q(n)$ is:

$$q(n) = 6n + 4$$

(c) Compare the asymptotic orders of $p(n)$ and $q(n)$.

Answer: We conclude from the answers given in the parts **(a)** and **(b)** that:

$$p(n) = \Theta\left((q(n))^2\right) \text{ or } q(n) = \Theta\left(\sqrt{p(n)}\right)$$

which indicates that the space utilization in the runtime environment described in the part **(b)** is significantly (polynomially) more efficient than that assumed in the part **(a)**.

Problem 327   Consider the algorithm *mess*, defined as follows.

algorithm mess (s: binary string) **returns** binary string
if $(s = \lambda)$ **then**
    $w \longleftarrow s$; **return** (w)
else
    $w \longleftarrow \left(s^{[L]} \mid \text{mess} (s^-) \mid s^{[L]}\right)$; **return** (w)
endif

In the algorithm pseudo-code $\longleftarrow$ denotes assignment, λ denotes the empty string, $|\ldots|$ denotes concatenation of (any finite number of) strings, $s^{[L]}$ denotes the leftmost bit of the string s, and s^- denotes the string obtained from s by a removal of $s^{[L]}$.

(a) Let $g(s)$ be the output of *mess* (s). Calculate $g(s)$ for $s = 1, 01, 001, 1000$.
Answer:

s	$g(s)$
1	11
01	0110
001	001100
1000	10000001

(b) Let $a(n)$ be the number of times the algorithm *mess* performs any of the following five operations: assignment, string concatenation, access to $s^{[L]}$, access to s^-, equality test, when invoked on an input string of length n. Determine a closed form of the function $a(n)$, and explain your answer.

Answer: By inspection of the code, we find that *mess* performs one test and one assignment if invoked on an empty string. When invoked on an input of length equal to n, where $n > 0$, *mess* performs a total of $a(n-1)$ operations inside the recursive call to itself on an input of length $n-1$, in addition to one test, one assignment, one string concatenation, two accesses to the last bit, and one string truncation. Hence, $a(n)$ is governed by the following recurrence.

$$a(0) = 2$$
$$a(n) = 6 + a(n-1) \text{ for } n \geq 1$$

It is straightforward to prove by induction that the closed form for $a(n)$ is:

$$a(n) = 6n + 2$$

(c) Let $c(n)$ be the number of times the algorithm *mess* performs a string concatenation, when invoked on an input string of length n. Determine a closed form of the function $c(n)$, and compare the asymptotic orders of $a(n)$ and $c(n)$.

Answer: The argument given in the answer to part **(b)** implies that *mess* performs about five other operations together with each string concatenation. Hence, $a(n)$ and $c(n)$ differ by a constant factor only, meaning:

$$c(n) = \Theta\left((a(n))\right)$$

To determine the exact closed form for $c(n)$, observe that *mess* does not perform any string concatenations when invoked on an empty string as input. When invoked on an input string of length equal to n, where $n > 0$, *mess* performs a total of $c(n-1)$ string concatenations inside the recursive call to itself on an input of length $n-1$, in addition to the string concatenation performed after the return from the recursive call. Hence, $c(n)$ is governed by the following recurrence.

$$c(0) = 0$$
$$c(n) = 1 + c(n-1) \text{ for } n \geq 1$$

It is straightforward to prove by induction that the closed form for $c(n)$ is:

$$c(n) = n$$

(d) Which function is performed by the algorithm *mess*? In other words, what is the relationship between an input s to the algorithm and the corresponding output $g(s)$?

Answer: An invocation of *mess* (s) returns the concatenation of the input string s and its reversal s^R:

$$g(s) = ss^R$$

Problem 328   Recall the algorithm *mess*, defined in Problem 327.

(a) Assume that *mess* is executed in a runtime environment where the computation space is managed as is described in the part **(a)** of the Problem 326, with the following additional rule.

(E) The result of a string concatenation is written into a new segment of the computation space, unoccupied before. This segment is reserved after the arguments of the concatenation are evaluated.

Let $p(n)$ be the size (in words) of the computation space required for the completion of an invocation of the algorithm *mess* on an input string s of length n under these assumptions, not counting (i.e., in addition to) the space occupied by the input s itself. Determine a closed form for $p(n)$.

Answer: When invoked on an input string of length equal to 0, *mess* assigns a value to only one program variable, namely w. The assigned value is the empty string, which (by assumption) does not occupy any computation space. When *mess* is invoked with an input string of length $n > 0$, it enters a recursive call *mess* (s^-). The argument of this call requires an segment of space of size equal to $n-1$ words, since the length of s^- is equal to $n-1$. Next, the recursive invocation of *mess* (s^-) employs additional space of size equal to $p(n-1)$. However, the space allocated for the arguments and the execution of the recursive call is released after the return from the call, contributing to the space of the calling program only the result returned by this call, which amounts to $2(n-1)$ words, according to the answer given in the part **(d)** of Problem 327. It is only after the space reserved for the recursive call is reclaimed that a new allocation is performed, which reserves the $2n$ words required for the result of the concatenation.

Hence, the computation space ought to be large enough to enable one allocation of $(n-1)+p(n-1)$ words first, and then a second allocation of $2(n-1)+2n$ words, but not both of these allocations simultaneously. This means that the higher of these two requirements ought to be met (so that the smaller reservation (re)uses the space reserved by the larger one.) Consequently, the recurrence that governs the computation space of the algorithm *mess* in this runtime environment is as follows:

$$p(0) = 0$$
$$p(n) = \max\left((n-1+p(n-1)),\ 4n-2\right)\ \text{ for } n \geq 1$$

We use induction to prove that:

$$p(n) = \frac{n(n-1)}{2} + 8\ \text{ for } n \geq 4$$

The base case occurs when $n = 4$, and is verified as follows.

$$p(1) = \max\left(0 + p(0),\ 4\cdot 1 - 2\right) = \max\left(0,\ 2\right) = 2$$
$$p(2) = \max\left(1 + p(1),\ 4\cdot 2 - 2\right) = \max\left(3,\ 6\right) = 6$$
$$p(3) = \max\left(2 + p(2),\ 4\cdot 3 - 2\right) = \max\left(8,\ 10\right) = 10$$
$$p(4) = \max\left(3 + p(3),\ 4\cdot 4 - 2\right) = \max\left(13,\ 14\right) = 14 = \frac{4(4-1)}{2} + 8$$

For the inductive step, assuming that $p(n-1) = \dfrac{(n-1)(n-2)}{2} + 8$ for some $n \geq 4$, the recursive definition mandates that:

$$p(n) = \max\left((n-1+p(n-1)),\ 4n-2\right) = \max\left(\left(n-1+\frac{(n-1)(n-2)}{2}+8\right),\ 4n-2\right)$$
$$= \max\left(\frac{(n-1)(n-2+2)}{2}+8,\ 4n-2\right) = \max\left(\frac{n(n-1)}{2}+8,\ 4n-2\right) = \frac{n(n-1)}{2}+8$$

since $n(n-1)/2 + 8 \geq 4n - 2$ whenever $n \geq 4$.

(b) Assume that the runtime environment has been modified so as to comply with the rules given in the part **(b)** of the Problem 326, along with the additional rule given in the part **(a)** of this problem. Let $q(n)$ be the size (in words) of computation space required for the completion of an invocation of the algorithm *mess* on input string s of length n under these assumptions, not counting (i.e., in addition to) the space occupied by the input s itself. Determine a closed form for $q(n)$.

Answer: When invoked on an empty input string, *mess* does not perform any recursive calls—hence, $q(0) = p(0) = 0$. When *mess* is invoked with an input string of length $n > 0$, a recursive call with one argument is entered. The address of this arguments is generated prior to the call, and two words of the computation space are allocated for it. Next, the recursive invocation of *mess* (s^-) employs additional space of size equal to $q(n-1)$. After this space is reclaimed, the returned result still occupies space of size equal to $2(n-1)$ words. Next, a new allocation is performed, which reserves the $2n$ words required for the result of the concatenation. Hence, the computation space ought to be large enough to enable one allocation of $2 + q(n-1)$ words first, and then a second allocation of $2(n-1) + 2n$ words, whence the recurrence:

$$q(0) = 0$$
$$q(n) = \max\left(2 + q(n-1),\ 4n-2\right)\ \text{ for } n \geq 1$$

We use induction to prove that:

$$q(n) = 4n - 2, \text{ for } n \geq 1$$

The base case occurs when $n = 1$, and evidently holds, since:

$$q(1) = \max(2 + q(0), 4 - 2) = \max(2, 2) = 2 = 4 \cdot 1 - 2$$

For the inductive step, assuming that $q(n-1) = 4(n-1) - 2$ for some $n > 0$, the recursive definition mandates that:

$$\begin{aligned} q(n) &= \max(2 + q(n-1), 4n - 2) \\ &= \max(2 + 4(n-1) - 2, 4n - 2) = \max(4n - 4, 4n - 2) = 4n - 2 \end{aligned}$$

whence the claim.

(c) Compare the asymptotic orders of $p(n)$ and $q(n)$.

Answer: We conclude from the answers given in the parts (a) and (b) that:

$$p(n) = \Theta\left((q(n)^2)\right) \text{ or } q(n) = \Theta\left(\sqrt{p(n)}\right)$$

which indicates that the space utilization in the runtime environment described in the part (f) is significantly (polynomially) more efficient than that assumed in the part (e).

Problem 329 Consider the algorithm *dress*, defined as follows.

algorithm dress (*s*: binary string) **returns** binary string
if $(s = \lambda)$ **then**
    $w \longleftarrow s;$ **return** (w)
else
    $w \longleftarrow (\text{dress}(s^+) \,|\, s[0] \,|\, s[0] \,|\, s[0]);$ **return** (w)
endif

In the algorithm pseudo-code $\longleftarrow$ denotes assignment, λ denotes the empty string, $|\ldots|$ denotes concatenation of (any finite number of) strings, $s[0]$ denotes the rightmost bit of the string s, and s^+ denotes the string obtained from s by a removal of $s[0]$.

(a) Let $g(s)$ be the output of *dress* (s). Calculate $g(s)$ for $s = 1, 01, 101, 0100$.

Answer:

s	$g(s)$
1	111
01	000111
101	111000111
0100	000111000000

(b) Let $c(n)$ be the number of operations of string concatenation, performed by *dress*, when invoked on an input string of length n. Determine a closed form of the function $c(n)$, and explain your answer.

Answer: By inspection of the code, we find that *dress* does not perform any concatenations if invoked on an empty string. When invoked on an input of length equal to n, where $n > 0$, *dress* performs a total of $c(n-1)$ string concatenations inside the recursive call to itself on an input of length $n - 1$, in addition to one concatenation performed after the return from the call. Hence, $c(n)$ is governed by the following recurrence.

$$c(0) = 0$$
$$c(n) = 1 + c(n-1) \text{ for } n \geq 1$$

It is straightforward to prove by induction that the closed form for $c(n)$ is:

$$c(n) = n$$

(c) Which function is performed by the algorithm *dress*? In other words, what is the relationship between an input s to the algorithm and the corresponding output $g(s)$?

Answer: An invocation of *dress* (s) returns an "expanded" copy of the input string s, in which every bit, say α, is substituted by the string $\alpha\alpha\alpha$, which consists of three consecutive copies of the original bit.

Problem 330   Recall the algorithm *dress*, defined in Problem 329.

(a) Assume that *dress* is executed in a runtime environment managed as is described in the part **(a)** of the Problem 328, with the following additional rule.

(F) The execution of an assignment (copy) of a variable that occupies ℓ words of space requires exactly ℓ units or time.

Let $t(n)$ be the computation time required for the completion of an invocation of the algorithm *dress* on an input string s of length n under these assumptions. Determine a closed form for $t(n)$.

Answer: When invoked on an input string of length equal to 0, *dress* assigns a value to only one program variable, namely w. Since the assigned value is the empty string, which does not occupy any space, the computation time required for this assignment is (by assumption) negligible. When *dress* is invoked with an input string of length $n > 0$, it enters a recursive call *dress* (s^+). Generating a copy of the argument of this call requires $n - 1$ units of time, since the length of s^+ is equal to $n - 1$. Next, the recursive invocation of *dress* (s^+) requires $t(n-1)$ units of time. The size of the result of the final concatenation is equal to $3n$ words—hence, its generation requires additional $3n$ units of time. Finally, the recurrence that governs the running time of the algorithm *dress* in this runtime environment is as follows:

$$t(0) = 0$$
$$t(n) = n - 1 + t(n-1) + 3n = 4n - 1 + t(n-1) \text{ for } n \geq 0$$

We use induction to prove that:

$$p(n) = 2n^2 + n$$

The base case occurs when $n = 0$, and evidently holds, since $t(0) = 0 = 2 \cdot 0^2 + 0$. For the inductive step, assuming that $t(n-1) = 2(n-1)^2 + (n-1)$ for some $n > 0$, the recursive definition mandates that:

$$t(n) \quad = 4n - 1 + t(n-1) = 4n - 1 + 2(n-1)^2 + (n-1)$$
$$= 2n^2 + n$$

whence the claim.

(b) Assume that the runtime environment has been modified so as to comply with the rules given in the part **(b)** of the Problem 328, along with the additional rule given in the part **(a)** of this problem.

Let $r(n)$ be the computation time required for the completion of an invocation of the algorithm *dress* on input string of length n under these assumptions. Determine a closed form for $r(n)$.

Answer: When invoked on an empty input string, *dress* assigns a value to only one program variable, namely w. Since the assigned value is the empty string, which does not occupy any space, the computation time required for this assignment is (by assumption) negligible. When *dress* is invoked with an input string of length $n > 0$, a recursive call with one argument is entered. The two-word address of this arguments is generated prior to the call, and two units of time are required for writing it. The running time of the recursive invocation of *dress* (s^+) is equal to $r(n-1)$. Next, additional $3n$ units of time are required to generate the result of the final concatenation, which is a string of length equal to $3n$ words. Hence, the recurrence which governs the running time of *dress* on an input string of length equal to n is as follows.

$$r(0) = 0$$
$$r(n) = 2 + r(n-1) + 3n = 3n + 2 + r(n-1) \text{ for } n \geq 1$$

We use induction to prove that:

$$r(n) = \frac{3n^2 + 7n}{2}$$

The base case occurs when $n = 0$, and evidently holds, since $r(0) = 0 = (3 \cdot 0^2 + 7 \cdot 0)/2$. For the inductive step, assuming that $r(n-1) = (3(n-1)^2 + 7(n-1))/2$ for some $n > 0$, the recursive definition mandates that:

$$r(n) = 3n + 2 + r(n-1) = 3n + 2 + \frac{3(n-1)^2 + 7(n-1)}{2}$$
$$= \frac{(6n+4) + (3n^2 - 6n + 3) + (7n - 7)}{2} = \frac{3n^2 + 7n}{2}$$

whence the claim.

(c) Compare the asymptotic orders of the functions that govern the number of concatenations and the running time of the algorithm in the two proposed environment models.

Answer: We conclude from the answers given in the part **(b)** of Problem 329 and the parts **(a)** and **(b)** of this problem that the running time of the algorithm *dress* in both of the two environment models is of the same asymptotic order, which is the order of the square of the number of concatenations. The time spent on argument copying is asymptotically insignificant in both cases, since it is dominated by the time required for generation of concatenation results.

$$t(n) = \Theta\left(r(n)\right) = \Theta\left(\left(c(n)\right)^2\right) = \Theta\left(n^2\right)$$

Problem 331 Consider the algorithms *digress* and *regress*, defined as follows. (In the algorithm pseudo-code $\longleftarrow$ denotes assignment, and all variables assume integer values.)

algorithm digress (n: positive integer) **returns** integer
if $(n = 1)$ **then**
      $y \longleftarrow 1$; **return** (y)
else
      $w \longleftarrow$ digress $(n-1)$;
      $y \longleftarrow n \cdot w$;
      return (y)
endif

```
algorithm regress (n: positive integer) returns integer
if (n = 1) then
        y ⟵ 1;  return (y)
else
        w ⟵ regress (n − 1);
        ℓ ⟵ 1;
        y ⟵ w;
        while (ℓ < n) do
                y ⟵ y + w;
                ℓ ⟵ ℓ + 1;
        endwhile
        return  (y)
endif
```

Let $f(n)$ be the output of *digress* (n), and let $g(n)$ be the output of *regress* (n). Prove that $f(n) = g(n)$ for every $n \geq 0$.

Answer: We prove that $f(n) = g(n) = n!$. Observe that the output $f(n)$ of *digress* (n) is governed by the recurrence:

$$f(1) = 1$$
$$f(n) = n \cdot f(n − 1)$$

which is a standard recursive definition of the factorial function. Furthermore, the output $g(n)$ of *regress* (n) is governed by the recurrence:

$$g(1) = 1$$
$$g(n) = \sum_{k=1}^{n} g(n − 1), \text{ for } n \geq 1$$

whose closed form is: $g(n) = n!$, as is proved in Problem 170.

Problem 332   Recall the algorithms *digress* and *regress*, defined in Problem 331. Observe that *digress* performs only multiplications, while *regress* performs only additions.

Let $\xi(n)$ be the number of multiplications performed by an invocation of *digress* (n), and let $\delta(n)$ be the number of additions performed by an invocation of *regress* (n). Determine the closed forms for $\xi(n)$ and $\delta(n)$, and compare their asymptotic orders.

Answer: We conclude by inspection of the code that *digress* (1) does not perform any multiplications. For $n > 1$, *digress* (n) performs $\xi(n − 1)$ multiplications within the recursive call and one more multiplication after the return from the call. This yields the recurrence:

$$\xi(1) = 0$$
$$\xi(n) = n + \xi(n − 1) \text{ for } n \geq 1$$

It is straightforward to prove by induction that the closed form for $\xi(n)$ is given by:

$$\xi(n) = n − 1$$

By inspection of the code, we conclude that *regress* (1) does not perform any additions. For $n > 1$, *regress* (n) performs $\delta(n − 1)$ additions within the recursive call. After the return from the call, the body of the ℓ-loop, which contains two additions, is executed $n - 1$ times. Hence, the number of additions performed by *regress* (n) is governed by the following recurrence.

$$\delta(1) = 0$$
$$\delta(n) = \delta(n - 1) + 2(n - 1), \text{ for } n \geq 1$$

whence the closed form:

$$\delta(n) = \delta(1) + \sum_{k=2}^{n} 2(k - 1) = 0 + \left(\sum_{k=1}^{n-1}(2k)\right) = 2\left(\sum_{k=1}^{n-1} k\right) = 2\left(\frac{n(n - 1)}{2}\right)$$
$$= n(n - 1)$$

The number of multiplications $\xi(n)$, performed by *digress* (n), and the number of additions $\delta(n)$, performed by *regress* (n), are related as follows.

$$\xi(n) = \Theta(n) = \Theta\left(\sqrt{\delta(n)}\right) \text{ and } \delta(n) = \Theta(n^2) = \Theta\left(((\xi(n))^2)\right)$$

Problem 333 Consider the algorithms *express* and *impress*, defined as follows. (In the algorithm pseudo-code $\longleftarrow$ denotes assignment, and all variables assume integer values.)

algorithm express (b: natural number, n: integer) **returns** integer
if $(n = 0)$ **then**
        $y \longleftarrow b$; **return** (y)
else
        $x \longleftarrow 1$; $\ell \longleftarrow 0$
        while $(\ell < n)$ **do**
               $\ell \longleftarrow \ell + 1$;
               $x \longleftarrow x \cdot 2$
        endwhile
        $y \longleftarrow 1$; $j \longleftarrow 0$
        while $(j < x)$ **do**
               $j \longleftarrow j + 1$;
               $y \longleftarrow y \cdot b$
        endwhile
        return (y)
endif

algorithm impress (b: natural number, n: integer) **returns** integer
if $(n = 0)$ **then**
        $y \longleftarrow b$; **return** (y)
else
        $x \longleftarrow$ impress $(b, n - 1)$;
        $y \longleftarrow x \cdot x$; **return** (y)
endif

Let $f(n)$ be the output of *express* (b, n), and let $g(n)$ be the output of *impress* (b, n). Prove that $f(n) = g(n)$ for every $n \geq 0$.

Answer: We prove that $f(n) = g(n) = b^{2^n}$. To this end, we formulate and prove a sequence of claims as follows.

Claim (1): After k traversals of the ℓ-loop in the algorithm *express*, the value of the variable x is equal to 2^k. In particular, the final value of x is equal to 2^n.

The claim is proved by induction on k. The base case occurs for $k = 0$ and corresponds to the value of x prior to the first traversal of the loop, which is indeed equal to $1 = 2^0$. Assuming that the value of x after k traversals of the loop is equal to 2^k, we find that the assignment within the $(k+1)$st traversal of the loop modifies this value to $2^k \cdot 2 = 2^{k+1}$. Since the loop is traversed n times, the final value of x is indeed equal to 2^n.

Claim (2): After k traversals of the j-loop in the algorithm *express*, the value of the variable y is equal to b^k. In particular, the final value of y, which is equal to $f(n)$, the value returned by the algorithm, is equal to b^{2^n}.

The proof of this claim is analogous to the proof of the *Claim (1)*, with understanding that the multiplication within the loop introduces a factor of b, and that the loop is executed 2^n times.

Claim (3): The value returned by *impress* (b, n) is governed by the following recurrence:

$$g(0) = b$$
$$g(n) = (g(n-1))^2, \text{ for } n \geq 1$$

The claim is verified straightforwardly, by inspection of the code.

Claim (4): The closed form of the value returned by *impress* (b, n) is $g(n) = b^{2^n}$.

The claim is proved by induction on n. When $n = 0$, the base case of the recursive definition of $g(n)$ yields: $g(0) = b = b^{2^0}$. Assuming that $g(n) = b^{2^n}$ for some $n \geq 0$, the recursive rule of the definition of $g(n)$ implies:

$$g(n+1) = (g(n))^2 = \left(b^{2^n}\right)^2 = b^{2^n \cdot 2} = b^{2^{n+1}}$$

whence the claim.

Problem 334   Recall the algorithms *express* and *impress*, defined in Problem 333.

(a) Let $\xi(n)$ be the number of multiplications performed by an invocation of *express* (b, n), and let $\mu(n)$ be the number of multiplications performed by an invocation of *impress* (b, n). Determine the closed forms for $\xi(n)$ and $\mu(n)$, and compare their asymptotic orders.

Answer: We conclude by inspection of the code that *express* (b, n) performs n multiplications within the ℓ-loop and 2^n multiplications within the j-loop, whence the count:

$$\xi(n) = 2^n + n, \text{ for } n \geq 1$$

The number of multiplications performed by *impress* (b, n) is governed by the following recurrence.

$$\mu(0) = 0$$
$$\mu(n) = \mu(n-1) + 1, \text{ for } n \geq 1$$

whence the closed form:

$$\mu(n) = n$$

Hence, the number of multiplications performed by the first algorithm is exponentially greater than the number of multiplications performed by the second algorithm—precisely:

$$\xi(n) = \Theta\left(2^n\right) \quad \text{and} \quad \mu(n) = \Theta\left(n\right) = \Theta\left(\lg \xi(n)\right)$$

(b) Assume that the two algorithms are executed in a runtime environment where the execution time of a multiplication $x \cdot y$ is equal to $(\lg x)(\lg y)$ time units. Let $t(n)$ be the running time expended on multiplication by an invocation of *express* (b, n), and let $r(n)$ be the running time expended on multiplication by an invocation of *impress* (b, n). Determine the closed form for $t(n)$ and $r(n)$, and compare their asymptotic orders.

Answer: We conclude from the analysis presented in the answer to Problem 333 that the multiplication time of the ℓ-loop in the algorithm *express* is as follows.

$$t_\ell(n) = \sum_{\ell=1}^{n} \left(\lg\left(2^\ell\right) \cdot \lg 2\right) = \sum_{\ell=1}^{n} (\ell \cdot 1) = \frac{n(n+1)}{2}$$

The multiplication time of the j-loop in the algorithm *express* is as follows.

$$t_j(n) = \sum_{j=1}^{2^n} \left(\lg\left(2^j\right) \cdot \lg b\right) = \sum_{j=1}^{2^n} (j \cdot \lg b) = (\lg b)\left(\frac{2^n(2^n+1)}{2}\right) = 4^n \cdot \left(\frac{\lg b}{2}\right) + 2^{n-1} \cdot \lg b$$

Hence, an invocation of *express* (b, n) spends the following number of time units on multiplication.

$$t(n) = t_\ell(n) + t_j(n) = 4^n \cdot \left(\frac{\lg b}{2}\right) + 2^{n-1} \cdot \lg b + \frac{n(n+1)}{2}$$

The running time spent by an invocation of *impress* (b, n) on multiplication is governed by the following recurrence.

$$r(0) = 0$$
$$r(n) = r(n-1) + \left(\lg\left(b^{2^{n-1}}\right) \cdot \lg\left(b^{2^{n-1}}\right)\right), \quad \text{for } n \geq 1$$

whence the closed form:

$$
\begin{aligned}
r(n) \quad &= r(n-1) + \left(\lg^2\left(b^{2^{n-1}}\right)\right) = r(n-1) + \left(\frac{2^{n-1}}{\lg b}\right)^2 = r(n-1) + \frac{2^{2(n-1)}}{\lg^2 b} \\
&= r(n-1) + \frac{4^{n-1}}{\lg^2 b} = r(0) + \sum_{k=1}^{n} \frac{4^{k-1}}{\lg^2 b} = 0 + \frac{1}{\lg^2 b}\left(\sum_{k=0}^{n-1} 4^k\right) = \frac{4^n - 1}{3 \lg^2 b} \\
&= 4^n \cdot \frac{1}{3 \lg^2 b} - \frac{1}{3 \lg^2 b}
\end{aligned}
$$

Observe that the cost model adopted in this part of the problem indicates that the multiplication operation is implemented by a sequence of bit additions—the functions $t(n)$ and $r(n)$ correspond to the number of bit additions performed by *express* (b, n) and *impress* (b, n), respectively. While the number of multiplications performed by *impress* (b, n) is dramatically (exponentially) smaller than the number of multiplications performed by *express* (b, n), the running time spent on performing these multiplications is of the same order in both algorithms, if measured in this cost model. Informally, the multiplication cost in this model grows with the length of the binary representation of the factors. The many multiplications in *express* (b, n) are fast because they are performed on small numbers, while the far fewer multiplications in *impress* (b, n) are slow because they are performed on large numbers.

In summary:

$$t(n) = \Theta\left(4^n\right) \quad \text{and} \quad r(n) = \Theta\left(4^n\right)$$

$$\frac{t(n)}{r(n)} \approx \frac{3\lg^3 b}{2}$$

Problem 335   Consider the algorithm *bell*, whose original code is defined as follows.

algorithm bell (n: natural number)
if $(n = 0)$ **then**
    return
else
    $x \leftarrow 1; \ y \leftarrow 2; \ z \leftarrow 1;$
    $\ell \leftarrow 0$
    bell $\left(\dfrac{n}{2y}\right)$; bell $\left(\dfrac{n}{2y}\right)$; bell $\left(\dfrac{n}{2y}\right)$;

    while $(\ell < n^z)$ **do**
        $\ell \leftarrow \ell + 1;$
        RING
    endwhile
    $j \leftarrow 0$
    while $(j < x)$ **do**
        bell $\left(\dfrac{n}{2y}\right)$

    endwhile
    return
endif

In the algorithm pseudo-code $\leftarrow$ denotes assignment. Variables x, y, z assume integer values. Let $t(n)$ be the number of times the operation RING is performed by the algorithm *bell*, when invoked on input n.

(a) Determine the asymptotic order of $t(n)$.

Answer: We find by inspection of the code that an invocation *bell* (n) executes $(3 + x)$ recursive calls to *bell* $(n/(2y))$, in addition to n^z repetitions of the operation RING. Hence, the asymptotic order of $t(n)$ is governed by the recurrence:

$$t(n) = (x + 3) \cdot t\left(\frac{n}{2y}\right) + n^z$$

This is an instance of the Master Recurrence, with:

$$a = x + 3 = 1 + 3 = 4$$
$$b = 2y = 2 \cdot 2 = 4$$
$$d = z = 1$$

yielding:

$$b^d = 4^1 = 4 = a$$

Hence:
$$t(n) = \Theta\left(n \cdot \lg n\right)$$

(b) Edit (write, remove, or alter) one symbol in the code of *bell* so that the asymptotic order of $t(n)$ becomes $\Theta(n^2)$.

Answer: The anlysis presented in the answer to the part **(a)** implies that the condition $t(n) = \Theta(n^2)$ is attained if:
$$b^d < a \quad \text{and} \quad \log_b a = 2$$
which is fulfilled when $a = 4$, $b = 2$, and $d = 1$. These parameters differ from those given in the part **(a)** only in the value of b, which mandates a modification of the value of y, yielding:
$$y = \frac{b}{2} = \frac{2}{2} = 1$$

Therefore, we modify the code which assigns a value to the variable y to read: $y \longleftarrow \boxed{1}$.

(c) Edit (write, remove, or alter) one symbol in the code of *bell* so that the asymptotic order of $t(n)$ becomes $\omega(n^3)$ but $o(n^4)$.

Answer: The anlysis presented in the answer to the part **(a)** implies that the condition $t(n) = \omega(n^3)$ and $t(n) = o(n^4)$ is attained if:
$$b^d < a \quad \text{and} \quad 3 < \log_b a < 4$$
which is fulfilled when $64 < a < 256$, $b = 4$, and $d = 1$. These parameters differ from those given in the part **(a)** only in the value of a, which mandates a modification of the value of x, yielding:
$$64 < x + 3 < 256 \quad \text{or, equivalently} \quad 61 < x < 253$$

which is satisfied by, say, $x = 71$. Therefore, we modify the code which assigns a value to the variable x to read: $x \longleftarrow \boxed{7}\, 1$.

(d) Edit (write, remove, or alter) one symbol in the code of *bell* so that the asymptotic order of $t(n)$ becomes $\Theta(n)$.

Answer: The anlysis presented in the answer to the part **(a)** implies that the condition $t(n) = \Theta(n)$ is attained if:
$$b^d > a \quad \text{and} \quad d = 1$$
which is fulfilled when $a = 3$, $b = 4$, and $d = 1$. These parameters differ from those given in the part **(a)** only in the value of a, which mandates a modification of the value of x, yielding:
$$x + 3 = 3 \quad \text{or, equivalently} \quad x = 0$$

Therefore, we modify the code which assigns a value to the variable x to read: $x \longleftarrow \boxed{0}$.

(e) Edit (write, remove, or alter) one symbol in the code of *bell* so that the asymptotic order of $t(n)$ becomes $\Theta(n^5)$.

Answer: The anlysis presented in the answer to the part **(a)** implies that the condition $t(n) = \Theta(n^5)$ is attained if:
$$b^d > a \quad \text{and} \quad d = 5$$
which is fulfilled when $a = 4$, $b = 4$, and $d = 5$. These parameters differ from those given in the part **(a)** only in the value of d, which mandates a modification of the value of z, yielding:
$$z = d = 5$$

Therefore, we modify the code which assigns a value to the variable z to read: $z \longleftarrow \boxed{5}$.

Problem 336   Consider the algorithm *well*, whose original code is defined as follows.

algorithm well (n: natural number)
if ($n \leq 1$) **then**
        return ($4 \cdot n + 1$)
else
        return ($10 \cdot$ well ($n - 1$) $- 24 \cdot$ well ($n - 2$))
endif

In the algorithm pseudo-code $\longleftarrow$ denotes assignment.

Let $f(n)$ be the value returned by an invocation of *well* (n). You are allowed to edit (write, remove, or alter) no more than one symbol in the original code of the program. For each of the following properties, describe precisely the editing action (if any) required for the output function $f(n)$ to attain the given property, and prove that your answer is correct.

(a) $f(n) = \Theta(4^n)$.

Answer: Modify the first **return** statement so as to read:

$$\textbf{return } (\boxed{3} \cdot n + 1)$$

We find by inspection of the code that $f(n)$ is defined by the recurrence:

$$f(0) = 3 \cdot 0 + 1 = 1$$
$$f(1) = 3 \cdot 1 + 1 = 4$$
$$f(n) = 10f(n - 1) - 24f(n - 2)$$

This is a linear homogeneous recurrence relation of second order with constant coefficients, whose characteristic equation is:

$$x^2 - 10x + 24 = 0 \text{ or, equivalently: } (x - 4)(x - 6) = 0$$

which yields two distinct roots: $x_1 = 4$, $x_2 = 6$. The solutions are of the form: $f(n) = \alpha \cdot 4^n + \beta \cdot 6^n$, where α and β are calculated from the initial conditions:

$$f(0) = \alpha + \beta = 1$$
$$f(1) = 4\alpha + 6\beta = 4$$

Hence, $\beta = 0$ and $\alpha = 1$, yielding: $f(n) = 4^n$.

(b) $f(n) = \Theta(5^n)$.

Answer: Modify the second **return** statement so as to read:

$$\textbf{return } (10 \cdot \text{ well } (n - 1) - 2\boxed{5} \cdot \text{ well } (n - 2))$$

This modification leads to a new recurrence, whose characteristic equation is:

$$x^2 - 10x + 25 = 0 \text{ or, equivalently: } (x - 5)(x - 5) = 0$$

which yields a single root: $x_0 = 5$. The solutions are of the form: $f(n) = \alpha \cdot 5^n + \beta \cdot n \cdot 5^n$, where α and β are calculated from the initial conditions:

$$f(0) = \alpha = 1$$
$$f(1) = 5\alpha + 5\beta = 5$$

Hence, $\beta = 0$ and $\alpha = 1$, yielding: $f(n) = 5^n$.

(c) $f(n) = \Theta(6^n)$.

Answer: No editing should be performed. The analysis presented in the answer to the part **(a)** implies that the closed form for $f(n)$ is given as $f(n) = \alpha \cdot 4^n + \beta \cdot 6^n$, where α and β are calculated from the initial conditions:

$$f(0) = \alpha + \beta = 1$$
$$f(1) = 4\alpha + 6\beta = 5$$

Hence, $\beta = \dfrac{1}{2}$ and $\alpha = \dfrac{1}{2}$, yielding: $f(n) = 2 \cdot 4^{n-1} + 3 \cdot 6^{n-1}$.

(d) $f(n) = \Theta(7^n)$.

Answer: Modify the second **return** statement so as to read:

$$\textbf{return } (10 \cdot \text{ well } (n-1) - 2\boxed{1} \cdot \text{ well } (n-2))$$

This modification leads to a new recurrence, whose characteristic equation is:

$$x^2 - 10x + 21 = 0 \text{ or, equivalently: } (x-3)(x-7) = 0$$

which yields two distinct roots: $x_1 = 3$, $x_2 = 7$. The solutions are of the form: $f(n) = \alpha \cdot 3^n + \beta \cdot 7^n$, where α and β are calculated from the initial conditions:

$$f(0) = \alpha + \beta = 1$$
$$f(1) = 3\alpha + 7\beta = 5$$

Hence, $\beta = \dfrac{1}{2}$ and $\alpha = \dfrac{1}{2}$, yielding: $f(n) = \dfrac{1}{2}(3^n + 7^n)$.

(e) $f(n) = \Theta(8^n)$.

Answer: Modify the second **return** statement so as to read:

$$\textbf{return } (1\boxed{1} \cdot \text{ well } (n-1) - 24 \cdot \text{ well } (n-2))$$

This modification leads to a new recurrence, whose characteristic equation is:

$$x^2 - 11x + 24 = 0 \text{ or, equivalently: } (x-3)(x-8) = 0$$

which yields two distinct roots: $x_1 = 3$, $x_2 = 8$. The solutions are of the form: $f(n) = \alpha \cdot 3^n + \beta \cdot 8^n$, where α and β are calculated from the initial conditions:

$$f(0) = \alpha + \beta = 1$$
$$f(1) = 3\alpha + 8\beta = 5$$

Hence, $\beta = \dfrac{2}{5}$ and $\alpha = \dfrac{3}{5}$, yielding: $f(n) = \dfrac{1}{5}(2 \cdot 3^n + 3 \cdot 8^n)$.

(f) $f(n) = \Theta(12^n)$.

Answer: Modify the second **return** statement so as to read:

$$\textbf{return } (1\boxed{4} \cdot \text{ well } (n-1) - 24 \cdot \text{ well } (n-2))$$

This modification leads to a new recurrence, whose characteristic equation is:

$$x^2 - 14x + 24 = 0 \text{ or, equivalently: } (x - 2)(x - 12) = 0$$

which yields two distinct roots: $x_1 = 2$, $x_2 = 12$. The solutions are of the form: $f(n) = \alpha \cdot 2^n + \beta \cdot 12^n$, where α and β are calculated from the initial conditions:

$$f(0) = \alpha + \beta = 1$$
$$f(1) = 2\alpha + 12\beta = 5$$

Hence, $\beta = \dfrac{3}{10}$ and $\alpha = \dfrac{7}{10}$, yielding: $f(n) = \dfrac{1}{5}(7 \cdot 2^n + 3 \cdot 12^n)$.

(g) $f(n)$ oscillates between positive and negative values, changing sign for every n such that $n > 1$.
Answer: Modify the second **return** statement so as to read:

$$\textbf{return } (\boxed{-}\ 10 \cdot \text{ well } (n - 1) - 24 \cdot \text{ well } (n - 2))$$

This modification leads to a new recurrence, whose characteristic equation is:

$$x^2 + 10x + 24 = 0 \text{ or, equivalently: } (x + 4)(x + 6) = 0$$

which yields two distinct roots: $x_1 = -4$, $x_2 = -6$. The solutions are of the form:
$f(n) = \alpha \cdot (-4)^n + \beta \cdot (-6)^n$, where α and β are calculated from the initial conditions:

$$f(0) = \alpha + \beta = 1$$
$$f(1) = -4\alpha - 6\beta = 5$$

Hence, $\beta = -\dfrac{9}{2}$ and $\alpha = \dfrac{11}{2}$, yielding:

$f(n) = \dfrac{1}{2}(11 \cdot (-4)^n - 9 \cdot (-6)^n)$, or equivalently: $f(n) = \dfrac{1}{2}(-1)^n (11 \cdot 4^n - 9 \cdot 6^n)$.

To prove that $f(n)$ changes sign whenever $n > 1$, observe that:

$$11 \cdot 4^n - 9 \cdot 6^n = 6^n \left(11 \cdot \left(\frac{2}{3}\right)^n - 9\right) < 0 \text{ whenever } n > 0$$

Hence, the sign of $f(n)$ is equal to the sign of $(-1)^{n+1}$ whenever $n \geq 1$.

Chapter 5

Relations and Graphs

5.1 Relations

Problem 337   Binary relation ρ on the set $S = \{a, b, c, d, e\}$ is defined as follows:

$$\rho = \{(a, c), (a, e), (b, a), (e, d)\}$$

(a) Write the matrix representation of ρ.

> **Answer:**

	a	b	c	d	e
a	0	0	1	0	1
b	1	0	0	0	0
c	0	0	0	0	0
d	0	0	0	0	0
e	0	0	0	1	0

(b) Is ρ a reflexive relation? Explain your answer.

> **Answer:** Relation ρ is not reflexive. For ρ to be reflexive, $(x, x) \in \rho$ would have to hold for any element $x \in S$. However, for instance:
>
> $$(a, a) \notin \rho$$
>
> which violates the condition of reflexivity.

(c) Write the matrix representing the reflexive closure of ρ. If this is impossible, explain why.

> **Answer:**

	a	b	c	d	e
a	1	0	1	0	1
b	1	1	0	0	0
c	0	0	1	0	0
d	0	0	0	1	0
e	0	0	0	1	1

(d) Is ρ a symmetric relation? Explain your answer.

Answer: Relation ρ is not symmetric. For ρ to be symmetric, $(y, x) \in \rho$ would have to hold whenever $(x, y) \in \rho$ holds, for any pair of elements $x, y \in S$. However, for instance:

$$(b, a) \in \rho \text{ but } (a, b) \notin \rho$$

which violates the condition of symmetry.

(e) Write the matrix representing the symmetric closure of ρ. If this is impossible, explain why.

Answer:

	a	b	c	d	e
a	0	1	1	0	1
b	1	0	0	0	0
c	1	0	0	0	0
d	0	0	0	0	1
e	1	0	0	1	0

(f) Is ρ a transitive relation? Explain your answer.

Answer: Relation ρ is not transitive. For ρ to be transitive, $(x, z) \in \rho$ would have to hold whenever $(x, y) \in \rho$ and $(y, z) \in \rho$ are simultaneously true, for any triplet of elements $x, y, z \in S$. However, for instance:

$$(b, a) \in \rho \text{ and } (a, c) \in \rho \text{ but } (b, c) \notin \rho$$

(g) Is ρ an antisymmetric relation? Explain your answer.

Answer: Relation ρ is antisymmetric. Antisymmetry means that $(x, y) \in \rho$ and $(y, x) \in \rho$ cannot hold simultaneously, unless $x = y$, for any pair of elements $x, y \in S$. By inspection, we conclude that ρ does not violate the condition of antisymmetry.

(h) Is ρ a function? Explain your answer.

Answer: Relation ρ is not a function. For ρ to be a function, every element of S would have to have exactly one image. However, ρ assigns two images to the element a, since:

$$(a, c) \in \rho \text{ and } (a, e) \in \rho$$

Furthermore, elements c and d do not have any image under ρ, since:

$$(\forall x \in S)((c, x) \notin \rho \land (d, x) \notin \rho)$$

(i) Is any subset of ρ a function? Explain your answer.

Answer: Relation ρ and all of its subsets leave elements c and d unassigned. Hence, no subset of ρ is a function.

Problem 338   Binary relation ρ on the set $S = \{1, 2, 3, 4\}$ is defined as follows:

$$\rho = \{(1, 2), (1, 4), (2, 3), (3, 1), (4, 2)\}$$

(a) List all elements of the reflexive closure of ρ, or explain why ρ does not have a reflexive closure.

Answer: The reflexive closure of ρ is the set:

$$\rho \cup \{(1, 1), (2, 2), (3, 3), (4, 4)\}$$

(b) List all elements of a reflexive relation $\xi \in S \times S$ that contains ρ but is not equal to the reflexive closure of ρ. If such a relation does not exist, explain why.

Answer:
$$\xi = \rho \cup \{(1,1),(2,2),(3,3),(4,4),(3,2)\}$$

Every reflexive relation on S that contains ρ also contains the reflexive closure of ρ, and possibly other pairs.

(c) List all elements of the symmetric closure of ρ, or explain why ρ does not have a symmetric closure.

Answer: The symmetric closure of ρ is the set:

$$\rho \cup \{(2,1),(4,1),(3,2),(1,3),(2,4)\}$$

(d) List all elements of a symmetric relation $\sigma \in S \times S$ that contains ρ but is not equal to the symmetric closure of ρ. If such a relation does not exist, explain why.

Answer:
$$\sigma = \rho \cup \{(2,1),(4,1),(3,2),(1,3),(2,4),(3,4),(4,3)\}$$

Every symmetric relation on S that contains ρ also contains the symmetric closure of ρ, and possibly other pairs.

(e) Is ρ antisymmetric? Explain your answer.

Answer: Relation ρ is antisymmetric. For every pair $(x,y) \in \rho$ such that $x \neq y$, we conclude by inspection that $(y,x) \notin \rho$, which by definition implies that ρ is antisymmetric.

(f) Is ρ a function? Explain your answer.

Answer: Relation ρ is not a function. For ρ to be a function, every element of S would have to have exactly one image. However, ρ assigns two images to the element 1, since:

$$(1,2) \in \rho \text{ and } (1,4) \in \rho$$

(g) Is any subset of ρ a function? Explain your answer.

Answer: Relation ρ indeed contains as proper subsets the following two functions:

$$\varphi_1 = \{(1,2),(2,3),(3,1),(4,2)\}$$
$$\varphi_2 = \{(1,4),(2,3),(3,1),(4,2)\}$$

Problem 339 In each of the following cases, complete the matrix by writing a symbol into each empty box, so as to obtain a representation of a binary relation R on a 5-element set that satisfies the stated constraints. If this is impossible, explain why.

(a) R is symmetric.

Answer :

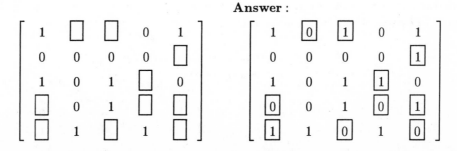

(b) R is antisymmetric.

Answer :

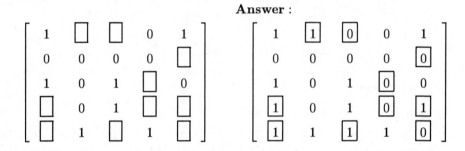

$$\begin{bmatrix} 1 & \Box & \Box & 0 & 1 \\ 0 & 0 & 0 & 0 & \Box \\ 1 & 0 & 1 & \Box & 0 \\ \Box & 0 & 1 & \Box & \Box \\ \Box & 1 & \Box & 1 & \Box \end{bmatrix} \qquad \begin{bmatrix} 1 & \boxed{1} & \boxed{0} & 0 & 1 \\ 0 & 0 & 0 & 0 & \boxed{0} \\ 1 & 0 & 1 & \boxed{0} & 0 \\ \boxed{0} & 0 & 1 & \boxed{1} & \boxed{0} \\ \boxed{0} & 1 & \boxed{0} & 1 & \boxed{0} \end{bmatrix}$$

(c) R is neither symmetric nor antisymmetric.

Answer :

$$\begin{bmatrix} 1 & \Box & \Box & 0 & 1 \\ 0 & 0 & 0 & 0 & \Box \\ 1 & 0 & 1 & \Box & 0 \\ \Box & 0 & 1 & \Box & \Box \\ \Box & 1 & \Box & 1 & \Box \end{bmatrix} \qquad \begin{bmatrix} 1 & \boxed{1} & \boxed{0} & 0 & 1 \\ 0 & 0 & 0 & 0 & \boxed{0} \\ 1 & 0 & 1 & \boxed{0} & 0 \\ \boxed{1} & 0 & 1 & \boxed{0} & \boxed{1} \\ \boxed{1} & 1 & \boxed{1} & 1 & \boxed{0} \end{bmatrix}$$

(d) R is reflexive.

Answer : Impossible.

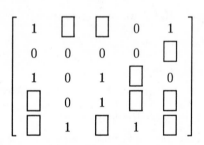

$$\begin{bmatrix} 1 & \Box & \Box & 0 & 1 \\ 0 & 0 & 0 & 0 & \Box \\ 1 & 0 & 1 & \Box & 0 \\ \Box & 0 & 1 & \Box & \Box \\ \Box & 1 & \Box & 1 & \Box \end{bmatrix}$$

For a matrix to represent a reflexive relation, all of its elements on the main diagonal must be equal to 1. This matrix, however, has one 0 on the main diagonal.

Problem 340   Let $A = \{1, 2, 3, 4\}$.

(a) Write matrices representing three different symmetric but not reflexive binary relations on the set A, or explain why this is impossible.

Answer:

R_1	1	2	3	4
1	0	0	0	0
2	0	1	0	0
3	0	0	1	0
4	0	0	0	1

R_2	1	2	3	4
1	1	0	0	1
2	0	0	0	0
3	0	0	1	0
4	1	0	0	1

R_3	1	2	3	4
1	1	1	0	0
2	1	1	1	0
3	0	1	0	0
4	0	0	0	1

(b) Write matrices representing three different antisymmetric and reflexive binary relations on the set A, or explain why this is impossible.

Answer:

R_1	1	2	3	4
1	1	0	1	0
2	1	1	0	1
3	0	1	1	0
4	1	0	1	1

R_2	1	2	3	4
1	1	0	1	1
2	0	1	0	1
3	0	0	1	0
4	0	0	0	1

R_3	1	2	3	4
1	1	0	0	0
2	1	1	0	0
3	1	1	1	0
4	1	1	1	1

(c) Write matrices representing three different binary relations on the set A, which are not reflexive, not symmetric and not antisymmetric, or explain why this is impossible.

Answer:

R_1	1	2	3	4
1	0	1	0	1
2	1	1	0	0
3	1	0	1	0
4	1	0	0	1

R_2	1	2	3	4
1	1	1	0	1
2	0	0	0	1
3	0	0	1	1
4	1	1	1	1

R_3	1	2	3	4
1	1	0	0	0
2	1	1	0	0
3	1	1	0	1
4	0	1	1	1

(d) Write matrices representing three different antisymmetric and symmetric but not reflexive binary relations on the set A, or explain why this is impossible.

Answer:

R_1	1	2	3	4
1	0	0	0	0
2	0	1	0	0
3	0	0	1	0
4	0	0	0	1

R_2	1	2	3	4
1	1	0	0	0
2	0	0	0	0
3	0	0	1	0
4	0	0	0	1

R_3	1	2	3	4
1	1	0	0	0
2	0	1	0	0
3	0	0	0	0
4	0	0	0	1

(e) Write matrices representing all different antisymmetric, symmetric, and reflexive binary relations on the set A, or explain why this is impossible.

Answer:

R_1	1	2	3	4
1	1	0	0	0
2	0	1	0	0
3	0	0	1	0
4	0	0	0	1

Problem 341 Sets A, B, C are defined as follows:

$$A = \{2, 3\}$$
$$B = \{14, 15, 16, 17, 18\}$$
$$C = \left\{ z \mid (\exists x \in A)(\exists y \in B)(z = \frac{y}{x}) \right\}$$
$$D = \{ z \mid (z \in C) \wedge (\exists y \in N)(z = 3y + 1) \}$$

All variables assume values from the set of natural numbers $N = \{0, 1, \ldots\}$.

(a) Write matrices representing three different reflexive and symmetric binary relations on the set C, or explain why this is impossible.

Answer:

R_1	5	6	7	8	9
5	1	0	0	0	0
6	0	1	0	0	0
7	0	0	1	0	0
8	0	0	0	1	0
9	0	0	0	0	1

R_2	5	6	7	8	9
5	1	0	0	1	0
6	0	1	0	0	0
7	0	0	1	0	1
8	1	0	0	1	0
9	0	0	1	0	1

R_3	5	6	7	8	9
5	1	0	0	0	0
6	0	1	0	0	0
7	0	0	1	0	0
8	0	0	0	1	1
9	0	0	0	1	1

(b) Write matrices representing three different reflexive and symmetric binary relations on the set D, or explain why this is impossible.

Answer: To see that this is impossible, recall that every binary relation on the set D is a subset of $D \times D$. However, by definition of D:

$$D = \{7\} \implies |D| = 1 \implies |D \times D| = 1 \cdot 1 = 1 \implies |\mathcal{P}(D \times D)| = 2^1 = 2$$

Hence, there are only 2 binary relations on the set D, whose matrix representation is as follows:

R_1	7
7	0

R_2	7
7	1

Both R_1 and R_2 are symmetric; R_2 is also reflexive.

(c) Write matrices representing three different antisymmetric but not reflexive binary relations on the set C, or explain why this is impossible.

Answer:

R_1	5	6	7	8	9
5	0	0	0	0	0
6	0	1	0	0	0
7	0	0	1	0	0
8	0	0	0	1	0
9	0	0	0	0	1

R_2	5	6	7	8	9
5	0	0	0	1	0
6	0	0	0	0	0
7	0	0	1	0	1
8	0	0	0	0	0
9	0	0	0	0	0

R_3	5	6	7	8	9
5	0	0	0	0	0
6	1	0	0	0	0
7	1	1	0	0	0
8	1	1	1	0	0
9	1	1	1	1	0

(d) Write matrices representing three different reflexive but not symmetric and not antisymmetric binary relations on the set C, or explain why this is impossible.

Answer:

R_1	5	6	7	8	9
5	1	1	1	1	0
6	1	1	0	0	0
7	1	0	1	0	0
8	1	0	0	1	0
9	1	0	0	0	1

R_2	5	6	7	8	9
5	1	0	1	1	0
6	0	1	0	0	0
7	1	0	1	0	1
8	0	0	0	1	0
9	0	0	0	0	1

R_3	5	6	7	8	9
5	1	0	0	0	1
6	1	1	0	0	0
7	1	1	1	0	0
8	1	1	1	1	0
9	1	1	1	1	1

(e) Write matrices representing three different antisymmetric and symmetric but not reflexive binary relations on the set C, or explain why this is impossible.

Answer:

R_1	5	6	7	8	9
5	0	0	0	0	0
6	0	1	0	0	0
7	0	0	1	0	0
8	0	0	0	1	0
9	0	0	0	0	1

R_2	5	6	7	8	9
5	0	0	0	0	0
6	0	0	0	0	0
7	0	0	1	0	0
8	0	0	0	1	0
9	0	0	0	0	1

R_3	5	6	7	8	9
5	0	0	0	0	0
6	0	0	0	0	0
7	0	0	0	0	0
8	0	0	0	1	0
9	0	0	0	0	1

(f) Write matrices representing three different antisymmetric, symmetric, and reflexive binary relations on the set C, or explain why this is impossible.

Answer: This is impossible—there exists only one such relation on the set C, represented by the folowing matrix:

R_1	5	6	7	8	9
5	1	0	0	0	0
6	0	1	0	0	0
7	0	0	1	0	0
8	0	0	0	1	0
9	0	0	0	0	1

Problem 342   Let $A = \{1, 2, 3, 4\}$.

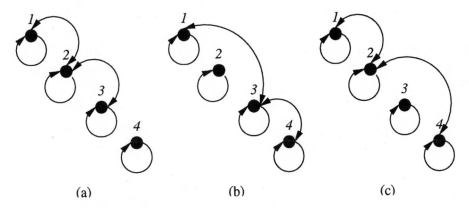

(a)   (b)   (c)

Figure 5.1: Reflexive, symmetric, and not transitive relations: (a) R_1; (b) R_2; (c) R_3

(a) Write matrices and draw graphs representing three different reflexive, symmetric but not transitive binary relations on the set A, or explain why this is impossible.

Answer: See Figure 5.1 for graphs of the relations represented by the following matrices.

R_1	1	2	3	4
1	1	1	0	0
2	1	1	1	0
3	0	1	1	0
4	0	0	0	1

R_2	1	2	3	4
1	1	0	1	0
2	0	1	0	0
3	1	0	1	1
4	0	0	1	1

R_3	1	2	3	4
1	1	1	0	0
2	1	1	0	1
3	0	0	1	0
4	0	1	0	1

(b) Write matrices and draw graphs representing three different reflexive, antisymmetric and transitive binary relations on the set A, or explain why this is impossible.

Answer: See Figure 5.2 for graphs of the relations represented by the following matrices.

R_1	1	2	3	4
1	1	1	1	1
2	0	1	1	1
3	0	0	1	1
4	0	0	0	1

R_2	1	2	3	4
1	1	0	0	0
2	0	1	0	0
3	0	0	1	0
4	0	0	0	1

R_3	1	2	3	4
1	1	0	0	0
2	1	1	0	0
3	1	1	1	0
4	1	1	1	1

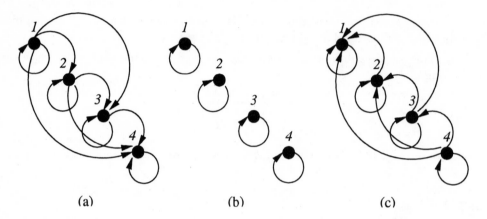

(a)   (b)   (c)

Figure 5.2: Reflexive, antisymmetric, and transitive relations: (a) R_1; (b) R_2; (c) R_3

(c) Write matrices and draw graphs representing three different binary relations on the set A, which are not reflexive, not symmetric, not antisymmetric, and not transitive, or explain why this is impossible.

Answer: See Figure 5.3 for graphs of the relations represented by the following matrices.

R_1	1	2	3	4
1	0	1	0	0
2	0	0	0	1
3	0	0	1	0
4	0	1	0	0

R_2	1	2	3	4
1	1	1	0	0
2	1	0	0	0
3	0	0	1	0
4	0	0	1	1

R_3	1	2	3	4
1	0	1	0	0
2	1	0	1	0
3	1	0	0	0
4	0	0	1	1

(d) Write matrices and draw graphs representing three different symmetric and transitive binary relations on the set A, or explain why this is impossible.

Answer: See Figure 5.4 for graphs of the relations represented by the following matrices.

R_1	1	2	3	4
1	1	1	1	1
2	1	1	1	1
3	1	1	1	1
4	1	1	1	1

R_2	1	2	3	4
1	1	0	0	0
2	0	1	0	0
3	0	0	1	0
4	0	0	0	1

R_3	1	2	3	4
1	1	0	0	0
2	0	1	0	1
3	0	0	1	0
4	0	1	0	1

Problem 343   Binary relation ρ on the set $S = \{1, 2, 3, 4, 5, 6\}$ is defined as follows:

$$\rho = \{(1, 2), (1, 3), (3, 4), (4, 5)\}$$

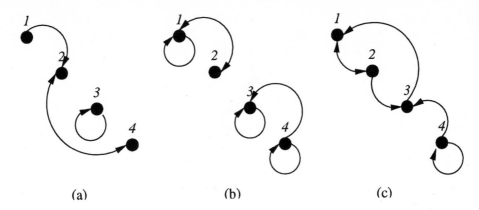

Figure 5.3: Not reflexive, symmetric, antisymmetric, or transitive: (a) R_1; (b) R_2; (c) R_3

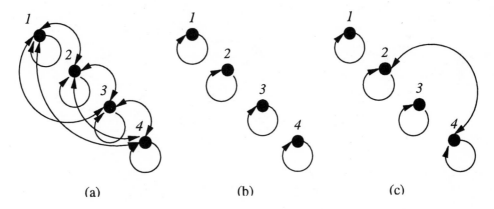

Figure 5.4: Symmetric and transitive relations: (a) R_1; (b) R_2; (c) R_3

Construct three different relations, R_1, R_2, and R_3 on the set S, which are transitive and contain ρ.
Answer: See Figure 5.5 for a graph that represents ρ, and see Figure 5.6 for graphs of the following relations.

$$R_1 = \rho \cup \{(1,4),(1,5),(3,5)\}$$
$$R_2 = R_1 \cup \{(6,6)\}$$
$$R_3 = R_1 \cup \{(2,2)\}$$

Observe that R_1 is the transitive closure of ρ, since R_1 is the minimal transitive relation that contains ρ. Every transitive relation on S that contains ρ also contains R_1.

Problem 344   Let $D \subseteq N^+$ be a subset of the set of positive integers $N^+ = \{1, 2, \ldots\}$. Let $R \subseteq D \times D$ be an equivalence relation on the set D. Sets $S \subseteq R$ and $T \subseteq R$ are two subsets of the relation R, defined as follows:

$$S = \{(1,2),(1,4),(3,5),(4,5)\}$$
$$T = \{(x,y) \mid y = x + 15\}$$

All variables assume values from the set of positive integers $N^+ = \{1, 2, \ldots\}$.

Determine the truth value of each of the following propositions, and justify your answer briefly.

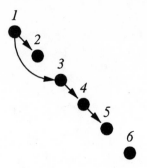

Figure 5.5: A relation that is not transitive: ρ

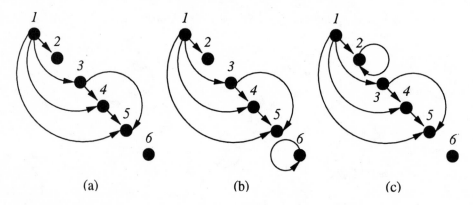

Figure 5.6: Transitive relations that contains ρ: (a) R_1; (b) R_2; (c) R_3

(a) $1 \in R$

   Answer: This is false. R is a set of pairs—it cannot contain a number as an element.

(b) $1 \in R \wedge 2 \in R$

   Answer: This is false. R is a set of pairs—it cannot contain a number as an element.

(c) $1 \in D$

   Answer: This is true. Since $(1,2) \in S$ and $S \subseteq R \subseteq D \times D$, it follows that $(1,2) \in D \times D$, whence the claim.

(d) $(1,1) \in R$

   Answer: This is true, since $1 \in D$, and R is reflexive.

(e) $(4,2) \in R$

   Answer: This is true. Since $(1,4),(1,2) \in S$ and $S \subseteq R$, it follows that $(1,4),(1,2) \in R$. Since R is symmetric, we conclude that $(4,1) \in R$. Finally, since R is transitive, $(4,2) \in R$ holds because $(4,1),(1,2) \in R$ holds.

(f) $(2,5) \in R$

> **Answer:** This is true. Since $(4,2) \in R$ holds (by the answer to the previous part) and R is symmetric, it follows that $(2,4) \in R$ also holds. Next, $(4,5) \in S$ and $S \subseteq R$, yielding: $(4,5) \in R$. Finally, since R is transitive, $(2,5) \in R$ holds because $(2,4), (4,5) \in R$ holds.

(g) $(\forall x)(\exists y)(x < y \wedge (x,y) \in R)$

> **Answer:** This is true. By definition of T, for every $x \in N^+$ it is true that $(x, x+15) \in T$. Since $T \subseteq R$, it is also true that $(x, x+15) \in R$ for every $x \in N^+$. Evidently: $x < x + 15$.

(h) $(\forall x > 15)(\exists y)(x > y \wedge (y,x) \in R)$

> **Answer:** This is true. By definition of T, for every $y \in N^+$ it is true that $(y, y+15) \in T$. Since $T \subseteq R$, it is also true that $(y, y+15) \in R$ for every $y \in N^+$. Let $x = y + 15$; evidently: $x > 15$, $x - 15 < x$, and $(x - 15, x) \in R$.

(i) $(\exists x)(\exists y)(x > 15 \wedge (x,y) \in R \wedge (y,x) \notin R)$

> **Answer:** This is false. R is symmetric—$(x,y) \in R$ and $(y,x) \in R$ are simultaneously true or simultaneously false.

(j) $(\forall x)((x,3) \notin S)$

> **Answer:** This is true. S is a finite set containing four elements, each of which is an ordered pair whose second component is different from 3.

(k) $(\forall x)((x,3) \notin R)$

> **Answer:** This is true. Since $(3,5) \in S$ and $S \subseteq R$, it follows that $(3,5) \in R$. Since R is symmetric, $(5,3) \in R$ is also true, providing a counterexample to the false claim.

Problem 345 Let $R \subseteq N \times N$ be a binary relation on the set of natural numbers, defined as follows:

$$(x,y) \in R \iff x + y \geq 18$$

(a) Is R reflexive? Prove your answer.

> **Answer:** Relation R is not reflexive. For R to be reflexive, $(x,x) \in R$ has to hold for every $x \in N$. However:
> $$x + x \geq 18$$
> is false whenever $x < 9$.

(b) Is R symmetric? Prove your answer.

> **Answer:** Relation R is symmetric. Symmetry means that $(x,y) \in R$ holds just when $(y,x) \in R$ holds, for any pair of numbers x, y. Indeed:
> $$x + y = y + x$$
> since addition is commutative. Hence:
> $$x + y \geq 18 \iff y + x \geq 18$$
> whence the claim.

(c) Is R antisymmetric? Prove your answer.

> **Answer:** Relation R is not antisymmetric. For R to be antisymmetric, $(x,y) \in R$ and $(y,x) \in R$ cannot be true simultaneously for any pair (x,y), unless $x = y$. However:

$$(20,10) \in R \text{ since } 20 + 10 \geq 18$$
$$(10,20) \in R \text{ since } 10 + 20 \geq 18$$

but still:

$$10 \neq 20$$

(d) Is R transitive? Prove your answer.

> **Answer:** Relation R is not transitive. For R to be transitive, $(x,z) \in R$ cannot be false when both $(x,y) \in R$ and $(y,z) \in R$ are true, for any choice of numbers $x, y, z \in N$. However:

$$(1,20) \in R \text{ since } 1 + 20 \geq 18$$
$$(20,2) \in R \text{ since } 20 + 2 \geq 18$$

but still:

$$(1,2) \notin R \text{ since } 1 + 2 < 18$$

Problem 346   Let $R \subseteq (\mathcal{P}(N) \setminus \emptyset) \times (\mathcal{P}(N) \setminus \emptyset)$ be a binary relation on the set of non-empty subsets of the set N of natural numbers, defined as follows:

$$(A,B) \in R \Longleftrightarrow A \cap B \neq \emptyset$$

(a) Is R reflexive? Prove your answer.

> **Answer:** Relation R is reflexive. Reflexivity means that $(A,A) \in R$ holds for every non-empty set $A \subseteq N$. Indeed:
>
> $$A \cap A \neq \emptyset \text{ whenever } A \neq \emptyset$$

(b) Is R symmetric? Prove your answer.

> **Answer:** Relation R is symmetric. Recall that symmetry means that $(A,B) \in R$ holds just when $(B,A) \in R$ holds, for any pair of sets $A, B \subseteq N \setminus \emptyset$. Indeed:
>
> $$A \cap B = B \cap A$$

since the operation of set intersection is commutative. Hence:

$$A \cap B \neq \emptyset \Longleftrightarrow B \cap A \neq \emptyset$$

whence the claim.

(c) Is R an equivalence relation? Prove your answer.

> **Answer:** Relation R is not an equivalence relation because it is not transitive. For R to be transitive, $(A,C) \in R$ cannot be false when both $(A,B) \in R$ and $(B,C) \in R$ are true, for any choice of sets $A, B, C \subseteq N \setminus \emptyset$. However:

$$(\{1,2\},\{2,3\}) \in R \text{ since } \{1,2\} \cap \{2,3\} = \{2\} \neq \emptyset$$
$$(\{2,3\},\{3,4\}) \in R \text{ since } \{2,3\} \cap \{3,4\} = \{3\} \neq \emptyset$$

but still:

$$(\{1,2\},\{3,4\}) \notin R \text{ since } \{1,2\} \cap \{3,4\} = \emptyset$$

Problem 347   Let $R \subseteq (\mathcal{P}(N) \setminus \emptyset) \times (\mathcal{P}(N) \setminus \emptyset)$ be a binary relation on the set of non-empty subsets of the set N of natural numbers, defined as follows:

$$(A, B) \in R \Longleftrightarrow 1 \in A \cup B$$

(a) Is R reflexive? Prove your answer.

Answer: Relation R is not reflexive. For R to be reflexive, $(A, A) \in R$ has to hold for every non-empty set $A \subseteq N$. However, $\{2\} \neq \emptyset$ but still:

$$(\{2\}, \{2\}) \notin R \text{ since } 1 \notin \{2\} \cup \{2\}$$

(b) Is R symmetric? Prove your answer.

Answer: Relation R is symmetric. Recall that symmetry means that $(A, B) \in R$ holds just when $(B, A) \in R$ holds, for any pair of sets $A, B \subseteq N \setminus \emptyset$. Indeed:

$$A \cup B = B \cup A$$

since the operation of set union is commutative. Hence:

$$1 \in A \cup B \Longleftrightarrow 1 \in B \cup A$$

whence the claim.

(c) Is R transitive? Prove your answer.

Answer: Relation R is not transitive. For R to be transitive, $(A, C) \in R$ cannot be false when both $(A, B) \in R$ and $(B, C) \in R$ are true, for any choice of sets $A, B, C \subseteq N \setminus \emptyset$. However:

$$(\{0, 2\}, \{1, 3\}) \in R \text{ since } 1 \in \{0, 2\} \cup \{1, 3\}$$
$$(\{1, 3\}, \{2, 4\}) \in R \text{ since } 1 \in \{1, 3\} \cup \{2, 4\}$$

but still:

$$(\{0, 2\}, \{2, 4\}) \notin R \text{ since } 1 \notin \{0, 2\} \cup \{2, 4\}$$

(d) Is R antisymmetric? Prove your answer.

Answer: Relation R is not antisymmetric. For R to be antisymmetric, $(A, B) \in R$ and $(B, A) \in R$ cannot be true simultaneously for any pair of sets $A, B \subseteq N \setminus \emptyset$, unless $A = B$. However:

$$(\{1\}, \{2\}) \in R \text{ since } 1 \in \{1\} \cup \{2\}$$
$$(\{2\}, \{1\}) \in R \text{ since } 1 \in \{2\} \cup \{1\}$$

but still:

$$\{1\} \neq \{2\}$$

(e) Is R a relation of partial order? Prove your answer.

Answer: Relation R is not a relation of partial order because it is not antisymmetric, as is proved in the answer to part **(d)**.

(f) Is R an equivalence relation? Prove your answer.

Answer: Relation R is not an equivalence relation because it is not reflexive, as is proved in the answer to part **(a)**.

Problem 348   Let $R \subseteq N \times N$ be a binary relation on the set of natural numbers, defined as follows:

$$(x, y) \in R \iff x \leq 3y$$

(a) Is R reflexive? Prove your answer.

   Answer: Relation R is reflexive. Reflexivity means that $(x, x) \in R$ holds for every number $x \in N$. Indeed:

$$x \leq 3x$$

   for any $x \in N$.

(b) Is R symmetric? Prove your answer.

   Answer: Relation R is not symmetric. For R to be symmetric, $(x, y) \in R$ cannot be false when $(y, x) \in R$ is true, for any pair (x, y). However:

$$(1, 5) \in R \text{ since } 1 \leq 3 \cdot 5$$

   but

$$(5, 1) \notin R \text{ since } 5 > 3 \cdot 1$$

(c) Is R antisymmetric? Prove your answer.

   Answer: Relation R is not antisymmetric. For R to be antisymmetric, $(x, y) \in R$ and $(y, x) \in R$ cannot be true simultaneously for any pair (x, y), unless $x = y$. However:

$$(1, 2) \in R \text{ since } 1 \leq 3 \cdot 2$$
$$(2, 1) \in R \text{ since } 2 \leq 3 \cdot 1$$

   but still:

$$1 \neq 2$$

(d) Is R transitive? Prove your answer.

   Answer: Relation R is not transitive. For R to be transitive, $(x, z) \in R$ cannot be false when both $(x, y) \in R$ and $(y, z) \in R$ are true, for any choice of $x, y, z \in N$. However:

$$(5, 2) \in R \text{ since } 5 \leq 3 \cdot 2$$
$$(2, 1) \in R \text{ since } 2 \leq 3 \cdot 1$$

   but still:

$$(5, 1) \notin R \text{ since } 5 > 3 \cdot 1$$

(e) Is R a relation of partial order? Prove your answer.

   Answer: Relation R is not a relation of partial order because it is not antisymmetric, as is proved in the answer to part **(c)**. Moreover, R is not transitive, as is proved in the answer to part **(d)**.

(f) Is R an equivalence relation? Prove your answer.

   Answer: Relation R is not an equivalence relation. For R to be an equivalence relation, it would have to be symmetric and transitive. However, by the argument given in the answer to part **(b)**, R is not symmetric; by the argument given in the answer to part **(d)**, R is not transitive.

Problem 349   For a natural number $n > 1$, let $\Psi(n)$ be the set of all primes that divide n. For instance:

$$\Psi(15) = \{3, 5\} \qquad \text{since } 15 = 3 \cdot 5$$
$$\Psi(14000) = \{2, 5, 7\} \quad \text{since } 14000 = 2^4 \cdot 5^3 \cdot 7$$
$$\Psi(5) = \{5\}$$

Let $R \subseteq N \times N$ be a binary relation on the set of natural numbers, defined as follows:

$$(x, y) \in R \iff \Psi(x) \subseteq \Psi(y)$$

(a) Is R reflexive? Prove your answer.

Answer: Relation R is reflexive. Reflexivity means that $(x, x) \in R$ holds for every number $x \in N$. Indeed:

$$\Psi(x) \subseteq \Psi(x)$$

for any $x \in N$, since the relation of set inclusion is reflexive.

(b) Is R symmetric? Prove your answer.

Answer: Relation R is not symmetric. For R to be symmetric, $(x, y) \in R$ cannot be false when $(y, x) \in R$ is true, for any pair (x, y). However:

$$(2, 6) \in R \text{ since } \Psi(2) = \{2\} \subseteq \{2, 3\} = \Psi(6)$$

but

$$(6, 2) \notin R \text{ since } \Psi(6) = \{2, 3\} \not\subseteq \{2\} = \Psi(2)$$

(c) Is R antisymmetric? Prove your answer.

Answer: Relation R is not antisymmetric. For R to be antisymmetric, $(x, y) \in R$ and $(y, x) \in R$ cannot be true simultaneously for any pair (x, y), unless $x = y$. However:

$$(3, 9) \in R \text{ since } \Psi(3) = \{3\} \subseteq \{3\} = \Psi(9)$$
$$(9, 3) \in R \text{ since } \Psi(9) = \{3\} \subseteq \{3\} = \Psi(3)$$

but still:

$$3 \neq 9$$

(d) Is R transitive? Prove your answer.

Answer: Relation R is transitive. Transitivity means that $(x, z) \in R$ is true whenever both both $(x, y) \in R$ and $(y, z) \in R$ are true, for any choice of numbers $x, y, z \in N$. Indeed:

$$\Psi(x) \subseteq \Psi(y) \wedge \Psi(y) \subseteq \Psi(z) \implies \Psi(x) \subseteq \Psi(z)$$

since the relation of set inclusion is transitive.

(e) Is R a relation of partial order? Prove your answer.

Answer: Relation R is not a relation of partial order because it is not antisymmetric, as is proved in the answer to part **(c)**.

(f) Is R an equivalence relation? Prove your answer.

Answer: Relation R is not an equivalence relation because it is not symmetric, as is proved in the answer to part **(b)**.

Problem 350   For a natural number $n > 1$, let $\Psi(n)$ be the set of all primes that divide n (as is defined in Problem 349.) Let $R \subseteq N \times N$ be a binary relation on the set of natural numbers, defined as follows:

$$(x, y) \in R \Longleftrightarrow \Psi(x) = \Psi(y)$$

(a) Is R reflexive? Prove your answer.

   Answer: Relation R is reflexive. Reflexivity means that $(x, x) \in R$ holds for every number $x \in N$. Indeed:

$$\Psi(x) = \Psi(x)$$

   for any $x \in N$, since the relation of set equality is reflexive.

(b) Is R symmetric? Prove your answer.

   Answer: Relation R is symmetric. Symmetry means that $(x, y) \in R$ holds just when $(y, x) \in R$ holds, for any pair of numbers x, y. Indeed:

$$(x, y) \in R \Longleftrightarrow \Psi(x) = \Psi(y) \Longleftrightarrow \Psi(y) = \Psi(x) \Longleftrightarrow (y, x) \in R$$

   where the second equality holds because the relation of set equality is symmetric.

(c) Is R antisymmetric? Prove your answer.

   Answer: Relation R is not antisymmetric. For R to be antisymmetric, $(x, y) \in R$ and $(y, x) \in R$ cannot be true simultaneously for any pair (x, y), unless $x = y$. However:

$$(3, 9) \in R \text{ since } \Psi(3) = \{3\} = \Psi(9)$$
$$(9, 3) \in R \text{ since } \Psi(9) = \{3\} = \Psi(3)$$

   but still:

$$3 \neq 9$$

(d) Is R transitive? Prove your answer.

   Answer: Relation R is transitive. Transitivity means that $(x, z) \in R$ is true whenever both both $(x, y) \in R$ and $(y, z) \in R$ are true, for any choice of numbers $x, y, z \in N$. Indeed:

$$\Psi(x) = \Psi(y) \wedge \Psi(y) = \Psi(z) \Longrightarrow \Psi(x) = \Psi(z)$$

   since the relation of set equality is transitive.

(e) Is R a relation of partial order? Prove your answer. If your answer is "yes", find one element that precedes 30 and one element that follows 30 in R, and explain your answer.

   Answer: Relation R is not a relation of partial order because it is not antisymmetric, as is proved in the answer to part **(c)**.

(f) Is R an equivalence relation? Prove your answer. If your answer is "yes", define the equivalence class that contains 30.

   Answer: Relation R is an equivalence relation because it is reflexive, symmetric, and transitive, as is proved in the answers to parts **(a)**, **(b)** and **(d)**. The equivalence class of 30 contains all numbers divisible by $2, 3, 5$ and by no other primes:

$$[30]_R = \{2^j 3^k 5^\ell \mid j, k, l \in N^+\}$$

   The relation R has infinitely many equivalence classes, and each equivalence class is an infinite set.

Problem 351   Let $R \subseteq (N^+ \times N^+) \times (N^+ \times N^+)$ be a binary relation on the set of ordered pairs (2-tuples) of positive natural numbers, defined as follows:

$$(\langle x, y \rangle, \langle z, w \rangle) \in R \iff (x + y < z + w) \vee (x + y = z + w \wedge x \leq z)$$

(a) Is R reflexive? Prove your answer.

Answer: Relation R is reflexive. Reflexivity means that $(\langle x, y \rangle, \langle x, y \rangle) \in R$ holds for every 2-tuple of positive natural numbers $\langle x, y \rangle \in N^+ \times N^+$. Indeed:

$$x + y = x + y \wedge x \leq x$$

for any $x, y \in N^+$, whence the claim.

(b) Is R symmetric? Prove your answer.

Answer: Relation R is not symmetric. For R to be symmetric, $(\langle x, y \rangle, \langle z, w \rangle) \in R$ must hold if $(\langle z, w \rangle, \langle x, y \rangle) \in R$ is true, for any pair of tuples $\langle x, y \rangle$ and $\langle z, w \rangle$. However:

$$(\langle 1, 2 \rangle, \langle 3, 4 \rangle) \in R \text{ since } 1 + 2 < 3 + 4$$

but

$$(\langle 3, 4 \rangle, \langle 1, 2 \rangle) \notin R \text{ since } 3 + 4 > 1 + 2$$

(c) Is R antisymmetric? Prove your answer.

Answer: Relation R is antisymmetric. Antisymmetry means that $(\langle x, y \rangle, \langle z, w \rangle) \in R$ and $(\langle z, w \rangle, \langle x, y \rangle) \in R$ cannot be true simultaneously for any pair of tuples of positive natural numbers $\langle x, y \rangle \in N^+ \times N^+$ and $\langle z, w \rangle \in N^+ \times N^+$ unless $\langle x, y \rangle = \langle z, w \rangle$. Indeed, assuming that $(\langle x, y \rangle, \langle z, w \rangle) \in R$, we distinguish between two cases:

Case (1): $x + y < z + w$

In this case, it is impossible that $(\langle z, w \rangle, \langle x, y \rangle) \in R$, since this would require that:

$$z + w \leq x + y$$

which is a contradiction.

Case (2): $x + y = z + w \wedge x \leq z$

In this case, $(\langle z, w \rangle, \langle x, y \rangle) \in R$ requires:

$$z + w = x + y \wedge z \leq x$$

However, $x \leq z$ and $z \leq x$ hold simultaneously only if $x = z$, since the relation "$\leq$" is antisymmetric. Finally, $x = z$ and $x + y = z + w$ imply: $y = w$, yielding the claim: $\langle x, y \rangle = \langle z, w \rangle$.

(d) Is R transitive? Prove your answer.

Answer: Relation R is transitive. Recall that transitivity means that $(\langle x, y \rangle, \langle z, w \rangle) \in R$ is true whenever both $(\langle x, y \rangle, \langle u, v \rangle)$ and $(\langle u, v \rangle, \langle z, w \rangle)$ belong to R, for any choice of numbers $x, y, u, v, z, w \in N^+$. Indeed, assuming that $(\langle x, y \rangle, \langle u, v \rangle) \in R$ and $(\langle u, v \rangle, \langle z, w \rangle) \in R$, we distinguish between four cases, each of which leads to the conclusion: $(\langle x, y \rangle, \langle z, w \rangle) \in R$.

Case (1): $x + y < u + v \wedge u + v < z + w$

By the transitivity of the relation "$<$": $x + y < z + w$.

Case (2): $(x + y = u + v \wedge x \leq u) \wedge u + v < z + w$

By a substitution into the second inequality: $x + y < z + w$.

Case (3): $x + y < u + v \wedge (u + v = z + w \wedge u \leq z)$

By a substitution into the first inequality: $x + y < z + w$.

Case (4): $(x + y = u + v \wedge x \leq u) \wedge (u + v = z + w \wedge u \leq z)$

By the transitivity of the relations "$=$" and "$\leq$": $x + y = z + w$ and $x \leq z$.

(e) Is R a relation of partial order? Prove your answer. If your answer is "yes", find one element that precedes $\langle 3, 5 \rangle$ in R and one element that follows $\langle 3, 5 \rangle$ in R, and explain your answer.

Answer: Relation R is a relation of partial order because it is reflexive, antisymmetric, and transitive, as is proved in the answers to parts **(a)**, **(c)** and **(d)**. In R, $\langle 6, 1 \rangle$ precedes $\langle 3, 5 \rangle$, and $\langle 4, 4 \rangle$ follows $\langle 3, 5 \rangle$. To verify this observe:

$$6 + 1 < 3 + 5 \implies (\langle 6, 1 \rangle, \langle 3, 5 \rangle) \in R$$
$$(3 + 5 = 4 + 4 \wedge 3 \leq 4) \implies (\langle 3, 5 \rangle, \langle 4, 4 \rangle) \in R$$

(f) Is R an equivalence relation? Prove your answer. If your answer is "yes", define the equivalence class that contains $\langle 3, 5 \rangle$.

Answer: Relation R is not an equivalence relation because it is not symmetric, as is proved in the answer to part **(b)**.

Problem 352   Let $R \subseteq (N^+ \times N^+) \times (N^+ \times N^+)$ be a binary relation on the set of ordered pairs (2-tuples) of positive natural numbers, defined as follows:

$$(\langle x, y \rangle, \langle z, w \rangle) \in R \iff x + y = z + w$$

(a) Is R reflexive? Prove your answer.

Answer: Relation R is reflexive. Reflexivity means that $(\langle x, y \rangle, \langle x, y \rangle) \in R$ holds for every 2-tuple of positive natural numbers $\langle x, y \rangle \in N^+ \times N^+$. Indeed:

$$x + y = x + y$$

for any $x, y \in N^+$, whence the claim.

(b) Is R symmetric? Prove your answer.

Answer: Relation R is symmetric. Recall that symmetry means that $(\langle x, y \rangle, \langle x', y' \rangle) \in R$ and $(\langle x', y' \rangle, \langle x, y \rangle) \in R$ are simultaneously true or simultaneously false, for any pair of 2-tuples of positive natural numbers $\langle x, y \rangle \in N^+ \times N^+$ and $\langle x', y' \rangle \in N^+ \times N^+$ Indeed:

$$(\langle x, y \rangle, \langle x', y' \rangle) \in R \iff x + y = x' + y' \iff x' + y' = x + y \iff (\langle x', y' \rangle, \langle x, y \rangle) \in R$$

The second equivalence holds because the relation of equality in numbers is symmetric.

(c) Is R antisymmetric? Prove your answer.

> **Answer:** Relation R is not antisymmetric. For R to be antisymmetric, $(\langle x, y\rangle, \langle z, w\rangle) \in R$ and $(\langle z, w\rangle, \langle x, y\rangle) \in R$ cannot be true simultaneously for any pair of 2-tuples of positive natural numbers $\langle x, y\rangle \in N^+ \times N^+$, and $\langle z, w\rangle \in N^+ \times N^+$ unless $\langle x, y\rangle = \langle z, w\rangle$. However:
>
> $$(\langle 1, 5\rangle, \langle 4, 2\rangle) \in R \text{ since } 1 + 5 = 4 + 2$$
> $$(\langle 4, 2\rangle, \langle 1, 5\rangle) \in R \text{ since } 4 + 2 = 1 + 5$$

but still:

$$\langle 1, 5\rangle \neq \langle 4, 2\rangle$$

(d) Is R transitive? Prove your answer.

> **Answer:** Relation R is transitive. Recall that transitivity means that $(\langle x, y\rangle, \langle z, w\rangle) \in R$ is true whenever both $(\langle x, y\rangle, \langle u, v\rangle)$ and $(\langle u, v\rangle, \langle z, w\rangle)$ belong to R, for any choice of numbers $x, y, u, v, z, w \in N^+$. Assuming that $(\langle x, y\rangle, \langle u, v\rangle) \in R$ and $(\langle u, v\rangle, \langle z, w\rangle) \in R$, we conclude:
>
> $$x + y = u + v = z + w \implies (\langle x, y\rangle, \langle z, w\rangle) \in R$$
>
> since the relation of equality on the set N^+ is transitive.

(e) Is R a relation of partial order? Prove your answer. If your answer is "yes", find one element that precedes $\langle 3, 5\rangle$ and one element that follows $\langle 3, 5\rangle$ in R, and explain your answer.

> **Answer:** Relation R is not a relation of partial order because it is not antisymmetric, as is proved in the answer to part **(c)**.

(f) Is R an equivalence relation? Prove your answer. If your answer is "yes", define the equivalence class that contains $\langle 3, 5\rangle$.

> **Answer:** Relation R is an equivalence relation because it is reflexive, symmetric, and transitive, as is proved in the answers to parts **(a)**, **(b)** and **(d)**. The equivalence class of $\langle 3, 5\rangle$ contains seven tuples:
>
> $$[\langle 3, 5\rangle]_R = \{\langle 1, 7\rangle,\ \langle 2, 6\rangle,\ \langle 3, 5\rangle,\ \langle 4, 4\rangle,\ \langle 5, 3\rangle,\ \langle 6, 2\rangle,\ \langle 7, 1\rangle\}$$
>
> Relation R has infinitely many equivalence classes. Each equivalence class is a finite set of tuples. For every number $n \geq 2$, there exists an equivalence class of R, say R_n, which contains exactly $n - 1$ tuples:
>
> $$R_n = \{\langle x, y\rangle \mid x + y = n\} = \{(1, n - 1),\ (2, n - 2),\ \ldots,\ (n - 1, 1)\}$$
>
> See Figure 5.7 for an illustration of the set $N^+ \times N^+$ with the equivalence relation R. Solid lines correspond to equivalence classes of R.

Problem 353   Let $R \subseteq (N^+ \times N^+) \times (N^+ \times N^+)$ be a binary relation on the set of ordered pairs (2-tuples) of positive natural numbers, defined as follows:

$$(\langle x, y\rangle, \langle z, w\rangle) \in R \iff x - y = z - w$$

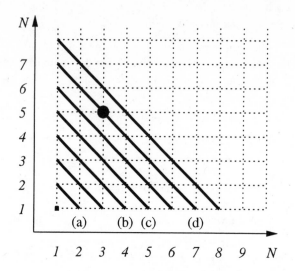

Figure 5.7: $R_n = \{\langle x, y\rangle \mid x + y = n\}$; (a) R_3; (b) R_5; (c) R_6; (d) R_8 with tuple $\langle 3, 5\rangle$ marked.

(a) Is R reflexive? Prove your answer.

Answer: Relation R is reflexive. Reflexivity means that $(\langle x, y\rangle, \langle x, y\rangle) \in R$ holds for every 2-tuple of positive natural numbers $\langle x, y\rangle \in N^+ \times N^+$. Indeed:

$$x - y = x - y$$

for any $x, y \in N^+$, whence the claim.

(b) Is R symmetric? Prove your answer.

Answer: Relation R is symmetric. Recall that symmetry means that $(\langle x, y\rangle, \langle x', y'\rangle) \in R$ and $(\langle x', y'\rangle, \langle x, y\rangle) \in R$ are simultaneously true or simultaneously false, for any pair of 2-tuples of positive natural numbers $\langle x, y\rangle \in N^+ \times N^+$ and $\langle x', y'\rangle \in N^+ \times N^+$ Indeed:

$$(\langle x, y\rangle, \langle x', y'\rangle) \in R \Longleftrightarrow x - y = x' - y' \Longleftrightarrow x' - y' = x - y \Longleftrightarrow (\langle x', y'\rangle, \langle x, y\rangle) \in R$$

The second equivalence holds because the relation of equality in numbers is symmetric.

(c) Is R antisymmetric? Prove your answer.

Answer: Relation R is not antisymmetric. For R to be antisymmetric, $(\langle x, y\rangle, \langle z, w\rangle) \in R$ and $(\langle z, w\rangle, \langle x, y\rangle) \in R$ cannot be true simultaneously for any pair of 2-tuples of positive natural numbers $\langle x, y\rangle \in N^+ \times N^+$, and $\langle z, w\rangle \in N^+ \times N^+$ unless $\langle x, y\rangle = \langle z, w\rangle$. However:

$$(\langle 2, 1\rangle, \langle 4, 3\rangle) \in R \text{ since } 2 - 1 = 4 - 3$$
$$(\langle 4, 3\rangle, \langle 2, 1\rangle) \in R \text{ since } 4 - 3 = 2 + 1$$

but still:

$$\langle 2, 1\rangle \neq \langle 4, 3\rangle$$

(d) Is R transitive? Prove your answer.

Answer: Relation R is transitive. Recall that transitivity means that $(\langle x, y\rangle, \langle z, w\rangle) \in R$ is true whenever both $(\langle x, y\rangle, \langle u, v\rangle)$ and $(\langle u, v\rangle, \langle z, w\rangle)$ belong to R, for any choice of numbers

$x, y, u, v, z, w \in N^+$. Indeed, assuming that $(\langle x, y \rangle, \langle u, v \rangle) \in R$ and $(\langle u, v \rangle, \langle z, w \rangle) \in R$, we conclude:

$$x - y = u - v = z - w \implies (\langle x, y \rangle, \langle z, w \rangle) \in R$$

since the relation of equality on the set N^+ is transitive.

(e) Is R a relation of partial order? Prove your answer. If your answer is "yes", find one element that precedes $\langle 3, 5 \rangle$ and one element that follows $\langle 3, 5 \rangle$ in R, and explain your answer.

Answer: Relation R is not a relation of partial order because it is not antisymmetric, as is proved in the answer to part **(c)**.

(f) Is R an equivalence relation? Prove your answer. If your answer is "yes", define the equivalence class that contains $\langle 3, 5 \rangle$.

Answer: Relation R is an equivalence relation because it is reflexive, symmetric, and transitive, as is proved in the answers to parts **(a)**, **(b)** and **(d)**. The equivalence class of $\langle 3, 5 \rangle$ contains infinitely many tuples:

$$[\langle 3, 5 \rangle]_R = \{\langle x, x + 2 \rangle \mid x \in N^+\}$$

Relation R has infinitely many equivalence classes, and each equivalence class is an infinite set of tuples. For every integer ℓ, there exists an equivalence class of R, say R_ℓ, which contains infinitely many tuples, precisely all tuples of the form:

$$R_\ell = \{\langle x, x + \ell \rangle \mid x \in N^+, \ell \in Z\}$$

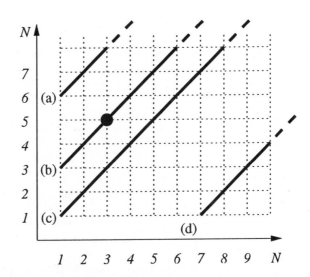

Figure 5.8: $R_\ell = \{\langle x, y \rangle \mid x - y = \ell\}$; (a) R_5; (b) R_2 with tuple $\langle 3, 5 \rangle$ marked; (c) R_0; (d) $R_{(-6)}$.

See Figure 5.8 for an illustration of the set $N^+ \times N^+$ with the equivalence relation R. Bold lines indicate the structure of equivalence classes of R.

Problem 354 Construct a binary equivalence relation $R \subseteq (N^+ \times N^+) \times (N^+ \times N^+)$ on the set of ordered pairs (2-tuples) of positive natural numbers, whose equivalence classes correspond to the bold lines comprising (the straightforward infinite extension of) the graph given on Figure 5.9.

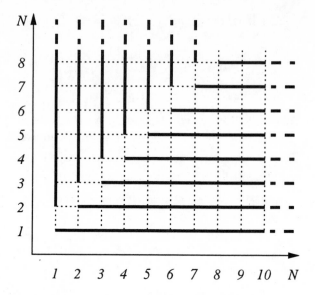

Figure 5.9: Equivalence classes of a relation $R \subseteq (N^+ \times N^+) \times (N^+ \times N^+)$

Answer: The representation of R given on Figure 5.9 indicates that R has infinitely many equivalence classes, each consisting of infinitely many points. For every positive integer n, there exist two equivalence classes of R, say R_n (a horizontal line) and and R'_n (a vertical line), defined as follows:

$$R_n = \{\langle x + n, n \rangle \mid x \in N\}$$
$$R'_n = \{\langle n, y + n + 1 \rangle \mid y \in N\}$$

The equivalence relation that generates these equivalence classes is defined as follows.

$$(\langle x, y \rangle, \langle z, w \rangle) \in R \iff ((x \geq y \wedge z \geq w \wedge y = w) \vee (y > x \wedge w > z \wedge x = z))$$

Problem 355   Construct a binary equivalence relation $R \subseteq (N^+ \times N^+) \times (N^+ \times N^+)$ on the set of ordered pairs (2-tuples) of positive natural numbers, whose equivalence classes correspond to the closed curves comprising (the straightforward infinite extension of) the graph given on Figure 5.10.

Answer: The representation of R given on Figure 5.10 indicates that R has infinitely many equivalence classes, each consisting of exactly four points, which form a unit square, whose lower left node has both coordinates odd. Precisely, for every pair of natural numbers $k, n \geq 0$, there exists an equivalence class of R, say $R_{k,n}$, defined as follows:

$$R_{k,n} = \{\langle 2k + 1, 2n + 1 \rangle, \ \langle 2k + 2, 2n + 1 \rangle, \ \langle 2k + 1, 2n + 2 \rangle, \ \langle 2k + 2, 2n + 2 \rangle\}$$

The equivalence relation that generates these equivalence classes is defined as follows.

$$(\langle x, y \rangle, \langle z, w \rangle) \in R \iff \left(\left\lfloor \frac{x + 1}{2} \right\rfloor = \left\lfloor \frac{z + 1}{2} \right\rfloor \wedge \left\lfloor \frac{y + 1}{2} \right\rfloor = \left\lfloor \frac{w + 1}{2} \right\rfloor \right)$$

Problem 356   Let $R \subseteq (N^+ \times N^+) \times (N^+ \times N^+)$ be a binary relation on the set of ordered pairs (2-tuples) of positive natural numbers, defined as follows:

$$(\langle x, y \rangle, \langle z, w \rangle) \in R \iff (\exists k, \ell \in N^+)(z = kx \wedge w = \ell y)$$

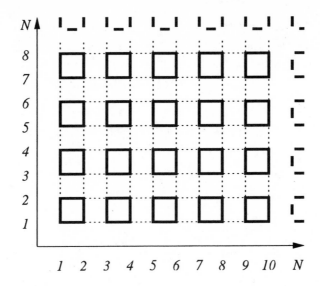

Figure 5.10: Equivalence classes of a relation $R \subseteq (N^+ \times N^+) \times (N^+ \times N^+)$

(a) Is R reflexive? Prove your answer.

Answer: Relation R is reflexive. Reflexivity means that $(\langle x, y \rangle, \langle x, y \rangle) \in R$ holds for every 2-tuple of positive natural numbers $\langle x, y \rangle \in N^+ \times N^+$. Indeed:

$$x = 1 \cdot x \wedge y = 1 \cdot y$$

for any $x, y \in N^+$, whence the claim.

(b) Is R symmetric? Prove your answer.

Answer: Relation R is not symmetric. For R to be symmetric, $(\langle x, y \rangle, \langle z, w \rangle) \in R$ must hold if $(\langle z, w \rangle, \langle x, y \rangle) \in R$ is true, for any pair (x, y). However:

$$(\langle 2, 3 \rangle, \langle 10, 12 \rangle) \in R \text{ since } 10 = 5 \cdot 2 \wedge 12 = 4 \cdot 3$$

but

$$(\langle 10, 12 \rangle, \langle 2, 3 \rangle) \notin R \text{ since } (\forall k, \ell \in N^+)(2 \neq 10k \wedge 3 \neq 12\ell)$$

(c) Is R antisymmetric? Prove your answer.

Answer: Relation R is antisymmetric. Antisymmetry means that $(\langle x, y \rangle, \langle z, w \rangle) \in R$ and $(\langle z, w \rangle, \langle x, y \rangle) \in R$ cannot be true simultaneously for any pair of tuples of positive natural numbers $\langle x, y \rangle \in N^+ \times N^+$ and $\langle z, w \rangle \in N^+ \times N^+$ unless $\langle x, y \rangle = \langle z, w \rangle$. Indeed, assuming that $(\langle x, y \rangle, \langle z, w \rangle) \in R$ and $(\langle z, w \rangle, \langle x, y \rangle) \in R$, we find:

$$z = k_1 x \wedge x = k_2 z \text{ for some } k_1, k_2 \geq 1$$
$$w = \ell_1 y \wedge y = \ell_2 w \text{ for some } \ell_1, \ell_2 \geq 1$$

yielding:

$$z = (k_1 k_2) z \text{ for some } k_1 k_2 \geq 1$$
$$w = (\ell_1 \ell_2) w \text{ for some } \ell_1 \ell_2 \geq 1$$

which is possible only if $k_1 = k_2 = \ell_1 = \ell_2 = 1$, yielding: $x = z$ and $y = w$, whence the claim that $\langle x, y \rangle = \langle z, w \rangle$.

(d) Is R transitive? Prove your answer.

   Answer: Relation R is transitive. Recall that transitivity means that $(\langle x,y\rangle, \langle z,w\rangle) \in R$ is true whenever both $(\langle x,y\rangle, \langle u,v\rangle)$ and $(\langle u,v\rangle, \langle z,w\rangle)$ belong to R, for any choice of numbers $x,y,u,v,z,w \in N^+$. Indeed, assuming that $(\langle x,y\rangle, \langle u,v\rangle) \in R$ and $(\langle u,v\rangle, \langle z,w\rangle) \in R$, we find:

$$u = k_1 x \wedge z = k_2 u \text{ for some } k_1, k_2 \geq 1$$
$$v = \ell_1 y \wedge w = \ell_2 v \text{ for some } \ell_1, \ell_2 \geq 1$$

   yielding:

$$z = (k_1 k_2)x \wedge w = (\ell_1 \ell_2)y \text{ for some } k_1 k_2 \geq 1 \text{ and } \ell_1 \ell_2 \geq 1$$

   whence the claim that $(\langle x,y\rangle, \langle z,w\rangle) \in R$.

(e) Prove that R a relation of partial order.

   Answer: Relation R is a relation of partial order because it is reflexive, antisymmetric, and transitive, as is proved in the answers to parts **(a)**, **(c)** and **(d)**.

(f) Find one element that precedes $\langle 3,5\rangle$ in R and one element that follows $\langle 3,5\rangle$ in R, and explain your answer. If this is impossible, prove it.

   Answer: Tuple $\langle 1,1\rangle$ precedes $\langle 3,5\rangle$ in R, and tuple $\langle 6,20\rangle$ follows $\langle 3,5\rangle$ in R. To verify this, observe:

$$(\langle 1,1\rangle, \langle 3,5\rangle) \in R \text{ since } 3 = 3 \cdot 1 \wedge 5 = 5 \cdot 1$$
$$(\langle 3,5\rangle, \langle 6,20\rangle) \in R \text{ since } 6 = 2 \cdot 3 \wedge 20 = 4 \cdot 5$$

(g) Find four distinct elements different from $\langle 3,5\rangle$ that precede $\langle 3,5\rangle$ in R, and explain your answer. If this is impossible, prove it.

   Answer: This is impossible, since there exist exactly three other tuples that precede $\langle 3,5\rangle$ in R. To verify this, observe that:

$$(\langle x,y\rangle, \langle 3,5\rangle) \in R \Longleftrightarrow x \,|\, 3 \wedge y \,|\, 5$$

   Since both 3 and 5 are prime, the only tuples that precede $\langle 3,5\rangle$ are:

$$\langle 1,1\rangle, \ \langle 3,1\rangle, \ \langle 1,5\rangle$$

(h) Find two incomparable elements that follow $\langle 3,5\rangle$ in R, and explain your answer. If this is impossible, prove it.

   Answer: Tuples $\langle 21,10\rangle$ and $\langle 12,15\rangle$ follow $\langle 3,5\rangle$ in R.

$$(\langle 3,5\rangle, \langle 21,10\rangle) \in R \text{ since } 21 = 7 \cdot 3 \wedge 10 = 2 \cdot 5$$
$$(\langle 3,5\rangle, \langle 12,15\rangle) \in R \text{ since } 12 = 4 \cdot 3 \wedge 15 = 3 \cdot 5$$

   However, $\langle 21,10\rangle$ and $\langle 12,15\rangle$ are incomparable, since:

$$(\forall k,\ell,i,j \in N^+)\,((12 \neq k \cdot 21 \wedge 15 \neq \ell \cdot 10) \wedge (21 \neq i \cdot 12 \wedge 10 \neq j \cdot 15))$$

(i) Find the least element of the set $N^+ \times N^+$ with the partial order R. If this is impossible, prove it.

   Answer: The least element of $N^+ \times N^+$ under R is $\langle 1,1\rangle$, since for an arbitrary tuple $\langle x,y\rangle \in N^+ \times N^+$:

$$(1 \,|\, x \wedge 1 \,|\, y) \Longrightarrow (\langle 1,1\rangle, \langle x,y\rangle) \in R$$

(j) Find the greatest element of the set $N^+ \times N^+$ with the partial order R. If this is impossible, prove it.

Answer: This is impossible. If the set $N^+ \times N^+$ with the partial order R had the greatest element, say $\langle x, y \rangle$, then $\langle 2x, 2y \rangle$ would follow $\langle x, y \rangle$ in R, causing a contradiction.

Problem 357   Let $R \subseteq (N^+ \times N^+) \times (N^+ \times N^+)$ be a binary relation on the set of ordered pairs (2-tuples) of positive natural numbers, defined as follows:

$$(\langle x, y \rangle, \langle z, w \rangle) \in R \iff (\exists k, \ell \in N^+)(z + 1 = k(x + 1) \wedge w + 1 = \ell(y + 1))$$

(a) Is R reflexive? Prove your answer.

Answer: Relation R is reflexive. Reflexivity means that $(\langle x, y \rangle, \langle x, y \rangle) \in R$ holds for every 2-tuple of positive natural numbers $\langle x, y \rangle \in N^+ \times N^+$. Indeed:

$$x + 1 = 1 \cdot (x + 1) \wedge y + 1 = 1 \cdot (y + 1)$$

for any $x, y \in N^+$, whence the claim.

(b) Is R symmetric? Prove your answer.

Answer: Relation R is not symmetric. For R to be symmetric, $(\langle x, y \rangle, \langle z, w \rangle) \in R$ must hold if $(\langle z, w \rangle, \langle x, y \rangle) \in R$ is true, for any pair (x, y). However:

$$(\langle 2, 3 \rangle, \langle 8, 19 \rangle) \in R \text{ since } 8 + 1 = 3 \cdot (2 + 1) \wedge 19 + 1 = 5 \cdot (3 + 1)$$

but

$$(\langle 8, 19 \rangle, \langle 2, 3 \rangle) \notin R \text{ since } (\forall k, \ell \in N^+)(2 + 1 \neq k(8 + 1) \wedge 3 + 1 \neq \ell(19 + 1))$$

(c) Is R antisymmetric? Prove your answer.

Answer: Relation R is antisymmetric. Antisymmetry means that $(\langle x, y \rangle, \langle z, w \rangle) \in R$ and $(\langle z, w \rangle, \langle x, y \rangle) \in R$ cannot be true simultaneously for any pair of tuples of positive natural numbers $\langle x, y \rangle \in N^+ \times N^+$ and $\langle z, w \rangle \in N^+ \times N^+$ unless $\langle x, y \rangle = \langle z, w \rangle$. Indeed, assuming that $(\langle x, y \rangle, \langle z, w \rangle) \in R$ and $(\langle z, w \rangle, \langle x, y \rangle) \in R$, we find:

$$z + 1 = k_1(x + 1) \wedge x + 1 = k_2(z + 1) \text{ for some } k_1, k_2 \geq 1$$
$$w + 1 = \ell_1(y + 1) \wedge y + 1 = \ell_2(w + 1) \text{ for some } \ell_1, \ell_2 \geq 1$$

yielding:

$$z + 1 = (k_1 k_2)(z + 1) \text{ for some } k_1 k_2 \geq 1$$
$$w + 1 = (\ell_1 \ell_2)(w + 1) \text{ for some } \ell_1 \ell_2 \geq 1$$

which is possible only if $k_1 = k_2 = \ell_1 = \ell_2 = 1$, yielding: $x = z$ and $y = w$, whence the claim that $\langle x, y \rangle = \langle z, w \rangle$.

(d) Is R transitive? Prove your answer.

Answer: Relation R is transitive. Recall that transitivity means that $(\langle x, y \rangle, \langle z, w \rangle) \in R$ is true whenever both $(\langle x, y \rangle, \langle u, v \rangle)$ and $(\langle u, v \rangle, \langle z, w \rangle)$ belong to R, for any choice of numbers $x, y, u, v, z, w \in N^+$. Indeed, assuming that $(\langle x, y \rangle, \langle u, v \rangle) \in R$ and $(\langle u, v \rangle, \langle z, w \rangle) \in R$, we find:

$$u + 1 = k_1(x + 1) \wedge z + 1 = k_2(u + 1) \text{ for some } k_1, k_2 \geq 1$$
$$v + 1 = \ell_1(y + 1) \wedge w + 1 = \ell_2(v + 1) \text{ for some } \ell_1, \ell_2 \geq 1$$

yielding:

$$z + 1 = (k_1 k_2)(x + 1) \wedge w + 1 = (\ell_1 \ell_2)(y + 1) \text{ for some } k_1 k_2 \geq 1 \text{ and } \ell_1 \ell_2 \geq 1$$

whence the claim that $(\langle x, y \rangle, \langle z, w \rangle) \in R$.

(e) Prove that R a relation of partial order.

Answer: Relation R is a relation of partial order because it is reflexive, antisymmetric, and transitive, as is proved in the answers to parts **(a)**, **(c)** and **(d)**.

(f) Find one element different from $\langle 2, 4 \rangle$ that precedes $\langle 2, 4 \rangle$ in R, and one element different from $\langle 2, 4 \rangle$ that follows $\langle 2, 4 \rangle$ in R. Explain your answer. If this is impossible, prove it.

Answer: Tuple $\langle 11, 34 \rangle$ follows $\langle 2, 4 \rangle$ in R. To verify this, observe:

$$(\langle 2, 4 \rangle, \langle 11, 34 \rangle) \in R \text{ since } 11 + 1 = 4(2 + 1) \wedge 34 + 1 = 7(4 + 1)$$

However, $\langle 2, 4 \rangle$ does not follow any other tuple in R. To verify this, assume that there exists a tuple $\langle x, y \rangle \in N^+ \times N^+$ such that $\langle x, y \rangle \neq \langle 2, 4 \rangle$ and $(\langle x, y \rangle, \langle 2, 4 \rangle) \in R$. This means that:

$$(x + 1) \,|\, (2 + 1) \wedge (y + 1) \,|\, (4 + 1) \text{ for some } x, y \geq 1$$

Observe that $x + 1 > 1$ and $y + 1 > 1$. Since 3 and 5 are prime, each of them has only one divisor greater than one, namely itself. Hence:

$$(x + 1 = 3 \wedge y + 1 = 5) \Longleftrightarrow (x = 2 \wedge y = 4) \Longleftrightarrow (\langle x, y \rangle = \langle 2, 4 \rangle)$$

(g) Find two incomparable elements that follow $\langle 2, 4 \rangle$ in R, and explain your answer. If this is impossible, prove it.

Answer: Tuples $\langle 5, 14 \rangle$ and $\langle 11, 9 \rangle$ follow $\langle 2, 4 \rangle$:

$$(\langle 2, 4 \rangle, \langle 5, 14 \rangle) \in R \text{ since } 5 + 1 = 2(2 + 1) \wedge 14 + 1 = 3(4 + 1)$$
$$(\langle 2, 4 \rangle, \langle 11, 9 \rangle) \in R \text{ since } 11 + 1 = 4(2 + 1) \wedge 9 + 1 = 2(4 + 1)$$

However, $\langle 5, 14 \rangle$ and $\langle 11, 9 \rangle$ are incomparable, since:

$$(\forall \ell \in N^+)\, (9 + 1 \neq \ell(14 + 1))$$
$$(\forall \ell \in N^+)\, (14 + 1 \neq \ell(9 + 1))$$

(h) Find the least element of the set $N^+ \times N^+$ with the partial order R. If this is impossible, prove it.

Answer: This is impossible. The set $N^+ \times N^+$ with the partial order R has infinitely many minimal elements. Precisely, the set μ of minimal elements is defined as follows:

$$\mu = \{\langle p - 1, q - 1 \rangle \mid p - 1, q - 1 \in N^+, \text{ and } p, q \text{ are prime } \}$$

To verify that all elements of μ are indeed minimal in $N^+ \times N^+$ under R, consider an arbitrary tuple $\langle p - 1, q - 1 \rangle \in \mu$, and assume that there exists a tuple $\langle x, y \rangle \in N^+ \times N^+$ such that $\langle x, y \rangle \neq \langle p - 1, q - 1 \rangle$ and $(\langle x, y \rangle, \langle p - 1, q - 1 \rangle) \in R$. By definition of R, we find that:

$$(x + 1) \,|\, p \wedge (y + 1) \,|\, q \text{ where } x + 1 > 1, y + 1 > 1, \text{ and } p, q \text{ are prime}$$

This is possible only if $x + 1 = p$ and $y + 1 = q$, meaning that $(\langle x, y \rangle, \langle p - 1, q - 1 \rangle) \in R$, whence the contradiction.

(i) Find the greatest element of the set $N^+ \times N^+$ with the partial order R. If this is impossible, prove it.

Answer: This is impossible. If the set $N^+ \times N^+$ with the partial order R had the greatest element, say $\langle x, y \rangle$, then $\langle 2x + 1, 2y + 1 \rangle$ would follow $\langle x, y \rangle$ in R, causing a contradiction.

Problem 358   Set S has n elements. Answer each of the following questions and explain your answer.

(a) How many elements are there in $S \times S$?

Answer: n^2.

(b) How many binary relations are there on S?

Answer: $2^{(n^2)}$, since every binary relation on S corresponds to a matrix of type $n \times n$, whose entries are equal to 0 or 1.

(c) How many binary relations on S are not reflexive?

Answer: $(2^n - 1) \cdot 2^{(n^2 - n)}$. To verify this, first observe that, in the matrix of a relation that is not reflexive, only one of the 2^n possible assignments of values to the n entries on the main diagonal is forbidden, precisely that in which all these entries are equal to 1. The remaining $n^2 - n$ entries may assume any values.

(d) How many binary relations on S are functions?

Answer: n^n. In the matrix of a function, exactly one of the n entries in each of the n rows is equal to 1, while all the other entries are equal to 0.

(e) How many reflexive binary relations on S are functions?

Answer: 1. This is the identity function on the set S, which corresponds to a matrix in which exactly the entries on the main diagonal are equal to 1, while all the other entries are equal to 0.

(f) How many binary relations on S are reflexive and antisymmetric?

Answer: $\sqrt{3^{(n^2 - n)}}$. To verify this, first observe that the matrix of a reflexive relation is constrained to have all entries on its main diagonal equal to 1. The remaining $n^2 - n$ entries should be considered as $(n^2 - n)/2$ pairs, where each pair consists of two entries whose positions are symmetric with respect to the main diagonal. Antisymmetry constrains each such pair to one of the following three values: $(0, 0)$, $(0, 1)$, $(1, 0)$.

(g) How many binary relations on S are antisymmetric but not reflexive?

Answer: $(2^n - 1) \cdot 3^{((n^2 - n)/2)}$. To verify this, first observe that, in the matrix of a relation that is not reflexive, only one of the 2^n possible assignments of values to the n entries on the main diagonal is forbidden, precisely that in which all these entries are equal to 1. The remaining $n^2 - n$ entries should be considered as $(n^2 - n)/2$ pairs, where each pair consists of two entries whose positions are symmetric with respect to the main diagonal. Antisymmetry constrains each such pair to one of the following three values: $(0, 0)$, $(0, 1)$, $(1, 0)$.

(h) How many binary relations on S are symmetric but not reflexive?

Answer: $(2^n - 1) \cdot 2^{((n^2 - n)/2)}$. To verify this, first observe that, in the matrix of a relation that is not reflexive, only one of the 2^n possible assignments of values to the n entries on the main diagonal is forbidden, precisely that in which all these entries are equal to 1. The remaining $n^2 - n$ entries should be considered as $(n^2 - n)/2$ pairs, where each pair consists of two entries whose positions are symmetric with respect to the main diagonal. Symmetry constrains each such pair to one of the following two values: $(0,0)$, $(1,1)$.

(i) How many binary relations on S are symmetric and antisymmetric but not reflexive?

Answer: $(2^n - 1)$. To verify this, first observe that, in the matrix of a relation that is not reflexive, only one of the 2^n possible assignments of values to the n entries on the main diagonal is forbidden, precisely that in which all these entries are equal to 1. The remaining $n^2 - n$ entries should be considered as $(n^2 - n)/2$ pairs, where each pair consists of two entries whose positions are symmetric with respect to the main diagonal. Simultaneous properties of symmetry and antisymmetry constrain each of these pairs to only one possible value: $(0,0)$.

(j) How many binary relations on S are symmetric or antisymmetric but not reflexive?

Answer: $(2^n - 1) \cdot \left(3^{((n^2-n)/2)} + 2^{((n^2-n)/2)} - 1\right)$, as follows from the answers given in the parts **(g)**, **(h)**, **(i)** and the Inclusion-Exclusion Theorem.

5.2   Graphs and Trees

Problem 359   Let $\mathcal{G}$ be the set of all simple undirected graphs that have 4 nodes.

(a) Draw all non-isomorphic members of $\mathcal{G}$ that consist of 4 connected components.
Answer: The single graph that satisfies the requirements is given on Figure 5.11 (a).

(a)   (b)

Figure 5.11: 4-node graphs with many connected components: (a) 4; (b) 3

(b) Draw all non-isomorphic members of $\mathcal{G}$ that consist of 3 connected components.
Answer: The single graph that satisfies the requirements is given on Figure 5.11 (b).

(c) Draw all non-isomorphic members of $\mathcal{G}$ that consist of 2 connected components.
Answer: There exist three such graphs and they are represented on Figure 5.12.

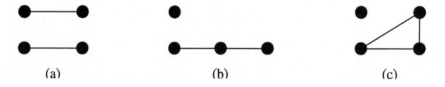

(a)   (b)   (c)

Figure 5.12: 4-node graphs with two connected components

(d) Draw all non-isomorphic members of $\mathcal{G}$ that are connected.
Answer: There exist six such graphs and they are represented on Figure 5.13.

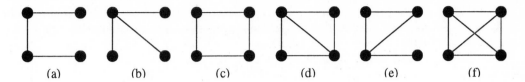

Figure 5.13: 4-node connected graphs

(e) Identify all trees and forests that are members of $\mathcal{G}$.

Answer: There exist two 4-node trees and they are represented on Figure 5.13 (a) and (b). Apart from these trees, there are four more forests. Precisely, with the single exception of the graph given on Figure 5.12 (c), every 4-node graph with at least two connected components is a forest.

(f) Identify all articulation points (if any) of all graphs that belong to $\mathcal{G}$.

Answer: Recall that a node is an articulation point if its removal (together with the edges incident to it) produces a graph with more connected components that the original graph. No members of $\mathcal{G}$ with 3 or 4 connected components have an articulation point. The graph given on Figure 5.12 (b) is the only member of $\mathcal{G}$ with two connected components that has an articulation point: the node whose degree is equal to 2. The only connected graphs in $\mathcal{G}$ that have articulation points are the two trees; their articulation points are those nodes whose degree is greater than 1.

(g) Identify all cut edges (if any) of all graphs that belong to $\mathcal{G}$.

Answer: Recall that an edge is a cut edge if its removal produces a graph with more connected components that the original graph. The graphs given on Figure 5.11 (a), Figure 5.12 (c), and Figure 5.13 (c,d,f) do not have any cut edges. The graph given on Figure 5.13 (e) has one cut edge, which is incident to the node whose degree is equal to 1. In all other graphs that belong to $\mathcal{G}$ every edge is a cut edge.

Problem 360 Let $\mathcal{T}$ be the set of all undirected (non-rooted) trees that have 6 nodes.

(a) Draw four pairwise non-isomorphic rooted trees that would be isomorphic to the same member of $\mathcal{T}$ if they were not rooted.

Answer: See Figure 5.14 (b,c,d,e) for four rooted trees that would be isomorphic to the tree given on Figure 5.14 (a) if they were not rooted. To verify that no two of the four rooted trees are isomorphic, observe, for instance, that any two of them differ in the number of levels or in the number of nodes at level 1.

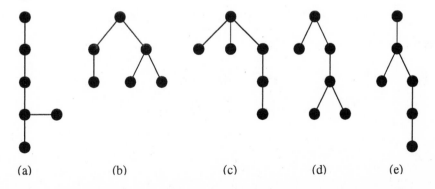

Figure 5.14: (a) A non-rooted 6-node tree T_1; (b–e) Four distinct rooted versions of T_1

(b)   Draw four pairwise non-isomorphic rooted trees which, if they were not rooted, would be isomorphic to the same member of $\mathcal{T}$, but different from the one constructed in the answer to part **(a)**.

Answer: See Figure 5.15 (b,c,d,e) for four rooted trees that would be isomorphic to the tree given on Figure 5.15 (a) if they were not rooted. To verify that no two of the four rooted trees are isomorphic, observe, for instance, that any two of them differ in the number of levels or in the number of nodes at level 1.

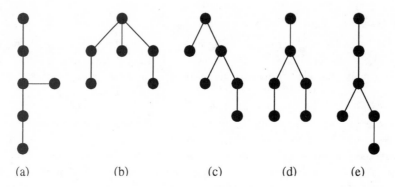

(a)   (b)   (c)   (d)   (e)

Figure 5.15: (a) A non-rooted 6-node tree T_2; (b–e) Four distinct rooted versions of T_2

To verify that the two corresponding undirected trees are not isomorphic, observe that T_1 contains two different paths of length 4 (with 5 nodes) as subgraphs, while T_2 has only one such path.

Problem 361   Directed graph J is represented by its adjacency matrix:

$$
M : \quad
\begin{array}{c|ccccccc}
 & A & B & C & D & E & F & G \\
\hline
A & 0 & 1 & 0 & 0 & 1 & 0 & 1 \\
B & 0 & 0 & 1 & 0 & 0 & 1 & 0 \\
C & 0 & 0 & 0 & 0 & 0 & 0 & 1 \\
D & 0 & 0 & 1 & 0 & 0 & 0 & 0 \\
E & 1 & 1 & 0 & 1 & 0 & 0 & 0 \\
F & 0 & 0 & 1 & 1 & 0 & 0 & 1 \\
G & 1 & 1 & 0 & 1 & 0 & 1 & 0 \\
\end{array}
$$

Directed graph H is represented by its adjacency lists:

source	destination
a	b, g
b	e
c	d, f
d	b, e
e	a, c, g
f	a, c
g	

$L :$

(a)   What is the out-degree of node F in J?

Answer: The out-degree of node F in J is equal to the sum of entries in the row of the adjacency matrix M which corresponds to F. Hence, the out-degree of F is equal to **3**.

(b) What is the in-degree of node F in J?

Answer: The in-degree of node F in J is equal to the sum of entries in the column of the adjacency matrix M which corresponds to F. Hence, the in-degree of F is equal to 2.

(c) What is the out-degree of node g in H?

Answer: The out-degree of node g in H is equal to the number of entries in the adjacency list which corresponds to g. Hence, the out-degree of g is equal to zero.

(d) What is the in-degree of node g in H?

Answer: The in-degree of node g in H is equal to the number of occurrences of g in the adjacency lists of H. Hence, the in-degree of g is equal to 2.

(e) How many arcs are there in J?

Answer: The number of arcs of J is equal to the sum of all entries of the adjacency matrix M. Hence, J has 17 arcs.

(f) How many arcs are there in H?

Answer: The number of arcs of H is equal to the number of all entries in the adjacency lists of H. Hence, H has 12 arcs.

(g) Draw graphs J and H.

Answer: See Figure 5.16.

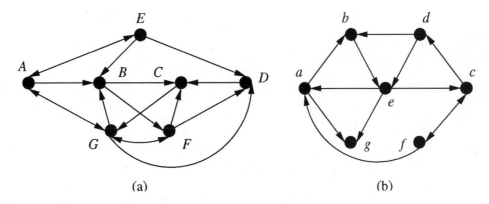

Figure 5.16: (a) Graph J; (b) Graph H

(h) Find a graph S with exactly 6 nodes and 8 arcs such that each of the two graphs J and H contains a subgraph isomorphic to S. Find these subgraphs and prove that they are isomorphic.

Answer: See Figure 5.17. Graph S is given in the part (a), while parts (b) and (c) identify the isomorphism bijections between the nodes of S and the nodes of the corresponding subgraphs of J and H.

(i) Let K and I be the simple undirected graphs obtained from J and H, respectively, by a conversion of every arc into an edge. Does the graphs K contain a subgraph isomorphic to I? Prove your answer.

Answer: The graph I, represented on Figure 5.18 (b) is isomorphic to the subgraph of K represented on Figure 5.18 (a). The node names on the drawing identify the isomorphism bijection.

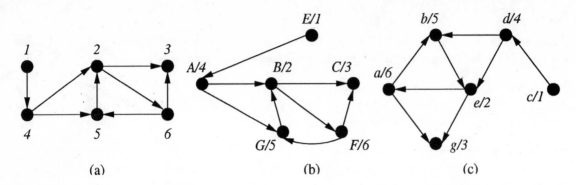

Figure 5.17: Isomorphic graphs: (a) S; (b) a subgraph of J; (c) a subgraph of H

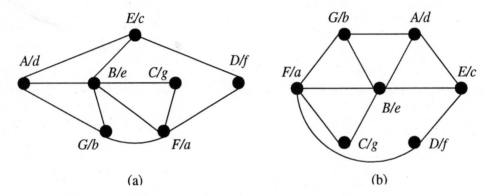

Figure 5.18: Isomorphic graphs: (a) S; (b) a subgraph of J; (c) a subgraph of H

Problem 362   Directed graph H is represented by the following adjacency matrix.

$$\mathbf{A}:\quad
\begin{array}{c|ccccccc}
  & a & b & c & d & e & f & g \\
\hline
a & 0 & 1 & 1 & 0 & 0 & 0 & 0 \\
b & 0 & 0 & 1 & 1 & 1 & 0 & 0 \\
c & 0 & 0 & 0 & 0 & 0 & 1 & 1 \\
d & 0 & 0 & 0 & 0 & 1 & 0 & 0 \\
e & 0 & 0 & 0 & 0 & 0 & 1 & 0 \\
f & 0 & 0 & 0 & 0 & 0 & 0 & 1 \\
g & 0 & 0 & 0 & 0 & 0 & 0 & 0 \\
\end{array}$$

Graph G is a simple undirected graph obtained from H by a conversion of every arc into an edge.

(a) Draw graphs H and G.

Answer: See Figure 5.19.

(b) Let $\overline{G}$ be the (simple, undirected) complement of G. How many edges does $\overline{G}$ have? Prove your answer.

Answer: Recall that a complete graph is the edge-disjoint union of any one of its subgraphs and the complement of that subgraph. The number of edges of an n-node complete graph is:

$$\binom{n}{2}$$

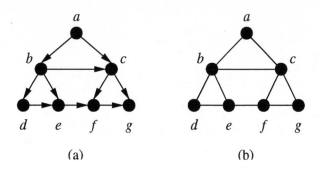

Figure 5.19: (a) Directed graph H; (b) Undirected graph G

In this case, $n = 7$. Hence, the total number of edges of G and $\overline{G}$ together is:

$$\binom{7}{2} = 21$$

Since G has 10 edges, we conclude that $\overline{G}$ has 11 edges.

(c) What is the distance between nodes a and g in the graph G? Prove your answer.

Answer: The distance between nodes a and g in the graph G is equal to 2. To verify this, observe that this distance has to be at least 2, since there is no edge (a, g). To see that this distance is not greater than 2, note the path $a \to c \to g$ of length 2.

(d) List the nodes (in sequence) of a simple path of length 3 from a to g in the graph G, or prove that such a path does not exist.

Answer: $a \to b \to c \to g$

(e) List the nodes (in sequence) of a simple path of length 4 from a to g in the graph G, or prove that such a path does not exist.

Answer: $a \to b \to e \to f \to g$

(f) List the nodes (in sequence) of a simple path of length 5 from a to g in the graph G, or prove that such a path does not exist.

Answer: $a \to b \to d \to e \to f \to g$

(g) List the nodes (in sequence) of a simple path of length 6 from a to g in the graph G, or prove that such a path does not exist.

Answer: $a \to c \to b \to d \to e \to f \to g$

Problem 363   The complete (undirected) bipartite graph $K_{\ell,m}$ is defined as follows:

$$K_{\ell,m} = (V, E)$$

The node set V of $K_{\ell,m}$ is:

$$V = A \cup B$$

where:

$$A = \{a_1, a_2, \ldots, a_\ell\}$$
$$B = \{b_1, b_2, \ldots, b_m\}$$

and

$$A \cap B = \emptyset$$

The edge set of $K_{\ell,m}$ is:

$$E = A \times B$$

(a) Draw $K_{3,4}$ and the complement of $K_{3,4}$.

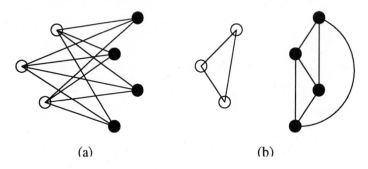

(a)   (b)

Figure 5.20: (a) $K_{3,4}$; (b) $\overline{K_{3,4}}$

Answer: See Figure 5.20.

(b) How many connected components does the complement of $K_{3,4}$ have? Explain your answer.

Answer: The complement $\overline{K_{3,4}}$ of the graph $K_{3,4}$ has two connected components: one containing the nodes a_1, a_2, a_3, and the other containing the nodes b_1, b_2, b_3, b_4.

(c) How many nodes and how many edges are there in each connected component of the complement of $K_{\ell,m}$? Explain your answer.

Answer: The complement $\overline{K_{\ell,m}}$ of the graph $K_{\ell,m}$ has two connected components: an ℓ-node complete graph K_ℓ on the nodes of the set A, and an m-node complete graph K_m on the nodes of the set B. The complete graph K_ℓ has $\ell(\ell-1)/2$ edges, and the complete graph K_m has $m(m-1)/2$ edges.

To see this, observe that the complete bipartite graph $K_{\ell,m}$ contains exactly those edges whose endpoints belong to two different node subsets, A and B. Hence, its complement $\overline{K_{\ell,m}}$ contains exactly those edges whose endpoints belong to the same node subset, A or B.

Problem 364   In each of the following cases, construct a simple, undirected, connected graph G, which has exactly 8 nodes and 12 edges, and satisfies the stated constraints. Justify your construction briefly. If such a graph G does not exist, explain why.

(a) G has a Hamilton cycle and an Euler tour.

Answer: The graph represented on Figure 5.21 (a) has a Hamilton cycle, which visits the graph nodes in the increasing order of their names. Since the degree of every node is even, this graph also has an Euler tour.

(b) G has a Hamilton cycle, but does not have an Euler tour and does not have an Euler path.

Answer: The graph represented on Figure 5.21 (b) has a Hamilton cycle, which visits the graph nodes in the increasing order of their names. Since it has as many as six nodes with an odd degree (precisely, nodes $1, 2, 3, 4, 5, 8$) this graph cannot have an Euler tour or an Euler path.

(c) G has an Euler tour, but does not have a Hamilton cycle.

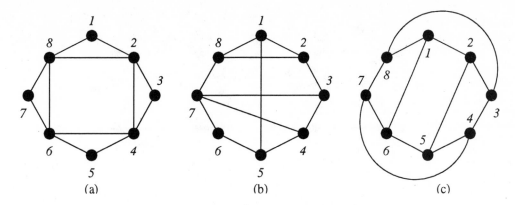

Figure 5.21: Some instances of graphs with 8 nodes, 12 edges, and a Hamilton cycle

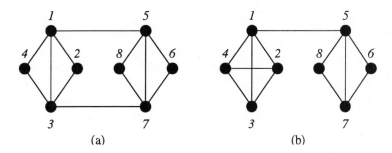

Figure 5.22: Some instances of graphs with 8 nodes, 12 edges, and no Hamilton cycle

Answer: The graph represented on Figure 5.22 (a) has an Euler tour, since every node has an even degree. To verify that it cannot have a Hamilton cycle, observe that the two edges $(1,5)$ and $(3,7)$ form a cut set of the graph. Every cut set has to contribute an even number of edges to a Hamilton cycle, since the cycle, starting in one of the cut components, has to re-enter that component as many times as it exits from the component. Hence, a purported Hamilton cycle in this graph has to contain edges $(1,5)$ and $(3,7)$, and cannot be closed unless nodes 5 and 7 are joined. However, nodes 5 and 7 can be joined by exactly one of three possible paths: $(5,6,7)$, $(5,8,7)$, and $(5,7)$. In each of the three cases, the supposed Hamilton cycle fails to visit some node(s) of the graph. Precisely, node 8 is left out in the first and third case, while node 6 is left out in the second and third case.

(d) G does not have a Hamilton cycle, does not have an Euler tour, and does not have an Euler path.

Answer: The graph represented on Figure 5.22 (b) cannot have an Euler tour or an Euler path, since it has as many as four nodes with an odd degree (precisely, nodes $2,3,4,7$). It cannot have a Hamilton cycle, since it has a cut set consisting of a single edge $(1,5)$.

Problem 365   In each of the following cases, construct a simple, undirected, connected graph G, which has exactly 8 nodes and 12 edges, and satisfies the stated constraints. Justify your construction briefly. If such a graph G does not exist, explain why.

(a) G has a Hamilton cycle and an Euler path, and exactly three of its nodes have a degree equal to 2.

Answer: The graph represented on Figure 5.23 (a) has a Hamilton cycle, which visits the graph nodes in the increasing order of their names. Since it has exactly two nodes with an odd degree, it

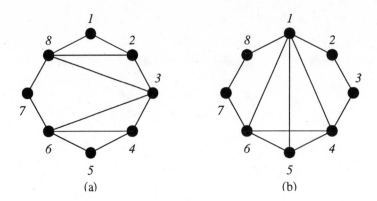

Figure 5.23: Some instances of graphs with 8 nodes, 12 edges, and a Hamilton path

has an Euler path whose endpoints are these two nodes: 2 and 4. Evidently, nodes $1, 5, 7$ are the only three nodes with a degree equal to 2.

(b) G has a Hamilton cycle and an Euler path, and exactly four of its nodes have a degree equal to 2.

Answer: The graph represented on Figure 5.23 (b) has a Hamilton cycle, which visits the graph nodes in the increasing order of their names. Since it has exactly two nodes with an odd degree, it has an Euler path whose endpoints are these two nodes: 1 and 5. Evidently, nodes $2, 3, 7, 8$ are the only four nodes with a degree equal to 2.

(c) G has a Hamilton cycle, but does not have an Euler cycle nor an Euler path; exactly five of its nodes have a degree equal to 2.

Answer: To see that this is impossible, recall that the number of nodes with an odd degree is even in every graph. In the graph G, the number of nodes with an odd degree cannot be equal to zero, lest G would have an Euler cycle. Likewise, the number of nodes with an odd degree cannot be equal to 2, lest G would have an Euler path. Hence, G has at least four nodes with an odd degree. However, since G has 8 nodes and exactly 5 of these nodes have a degree equal to 2, there are exactly 3 nodes whose degree is different from 2. This contradicts the previous conclusion that G has at least four nodes with an odd degree.

(d) G is regular, planar, and has a Hamilton cycle.

Answer: The graph represented by a planar drawing given on Figure 5.21 (c) is a degree-3 regular graph, with a Hamilton cycle which visits the graph nodes in the increasing order of their names.

Problem 366   Recall that the direct product $G_1 \times G_2$ of two simple undirected graphs $G_1 = (V_1, E_1)$ and $G_2 = (V_2, E_2)$ is a graph (V, E) whose node set is the product of the node sets of the two original graphs:

$$V = V_1 \times V_2$$

while the edge set E is defined as follows. Given two nodes of the product graph $G_1 \times G_2$, say $\langle a, x \rangle$ and $\langle b, y \rangle$, where $a, b \in V_1$ and $x, y \in V_2$, these nodes are connected by an edge $(\langle a, x \rangle, \langle b, y \rangle)$ in G if and only if one of the following conditions hold:

$$a = b \wedge (x, y) \in E_2$$
$$x = y \wedge (a, b) \in E_1$$

(a) Draw the direct product of a 3-node cycle and a 4-node cycle.

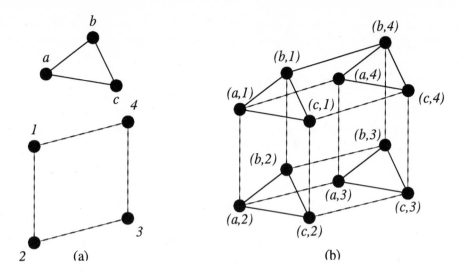

Figure 5.24: Direct product of cycles: (a) C_3 and C_4; (b) $C_3 \times C_4$

Answer: See Figure 5.24.

(b) If G_1 has n_1 nodes and e_1 edges, and G_2 has n_2 nodes and e_2 edges, how many nodes and edges does $G_1 \times G_2$ have?

Answer: The product graph $G_1 \times G_2$ has $n_1 n_2$ nodes, and $(n_1 e_2 + n_2 e_1)$ edges.

(c) Is the direct product of two connected graphs connected? Prove your answer.

Answer: To verify that it is connected, consider two nodes of the product graph $G_1 \times G_2$, say $\langle a, x \rangle$ and $\langle b, y \rangle$, where $a, b \in V_1$ and $x, y \in V_2$. Since G_1 and G_2 are connected, there exist the following two paths:

$$a \to d_0 \to \ldots \to d_m \to b \text{ in } G_1, \quad \text{for some } m \geq 0$$
$$x \to z_0 \to \ldots \to z_n \to y \text{ in } G_2, \quad \text{for some } n \geq 0$$

To construct a path from $\langle a, x \rangle$ to $\langle b, y \rangle$, employ the known path in G_1 to connect $\langle a, x \rangle$ with $\langle b, x \rangle$, and then follow the known path in G_2 from $\langle b, x \rangle$ to $\langle b, y \rangle$. The required path in the product graph $G_1 \times G_2$ has the following node sequence.

$$\langle a, x \rangle \to \langle d_0, x \rangle \to \ldots \to \langle d_m, x \rangle \to \langle b, x \rangle \to \langle b, z_0 \rangle \to \ldots \to \langle b, z_n \rangle \to \langle b, y \rangle$$

Problem 367 Consider two simple undirected graphs $G_1 = (V_1, E_1)$ and $G_2 = (V_2, E_2)$, and their direct product $G_1 \times G_2$.

(a) If each of the two component graphs G_1 and G_2 has an Euler tour, does $G_1 \times G_2$ have an Euler tour? Prove your answer.

Answer: To verify that such a product graph has an Euler tour, first recall that a graph has an Euler tour if and only the degree of every node in the graph is even. Consider an arbitrary node $\langle a, x \rangle$ of the graph $G_1 \times G_2$. This node is adjacent to all nodes of the form $\langle \beta, x \rangle$, where $(a, \beta) \in E_1$, and to all nodes of the form $\langle a, \tau \rangle$, where $(x, \tau) \in E_2$. Hence, the degree of a node in the product graph is equal to the sum of the degrees of the component nodes in their component graphs:

$$d_{G_1 \times G_2}(\langle a, x \rangle) = d_{G_1}(a) + d_{G_2}(x)$$

Since both $d_{G_1}(a)$ and $d_{G_2}(x)$ are even, their sum is also even, whence the claim.

(b) If neither of the two component graphs G_1 and G_2 has an Euler tour, is it possible for $G_1 \times G_2$ to have an Euler tour? Prove your answer.

Answer: See Figure 5.25 to confirm that this is indeed possible.

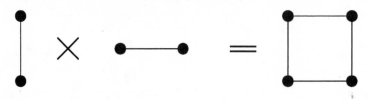

Figure 5.25: Direct product: $P_2 \times P_2 = C_4$.

(c) If each of the two component graphs G_1 and G_2 has an Euler path, does $G_1 \times G_2$ have an Euler path? Prove your answer.

Answer: Such a product graph need not have an Euler path. A graph has Euler path just when exactly two of its nodes have an odd degree. See Figure 5.26 for an illustration of two graphs with an Euler path whose direct product has as many as four nodes of odd degree.

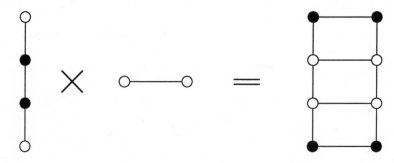

Figure 5.26: Direct product of two paths: many nodes with an odd degree.

Problem 368   A directed simple graph $G = (V, A)$ (possibly with self-loops) has n nodes, numbered from 0 to $n - 1$, and a arcs. The undirected graph G' obtained from G by a conversion of every arc into an edge is connected and simple.

Consider two representations for G: adjacency matrix and adjacency lists. Assume that node names are stored as bit strings of equal length. All operations are performed on bits. In particular, an access to an entry, say (j, k) of the adjacency matrix constitutes one bit operation, once the bit strings representing j and k are available. Furthermore, an access to the (first bit) of the first entry of the adjacency list of node j constitutes one bit operation, once the bit string representing j is available; an access to each subsequent bit in the adjacency list (in sequence) constitutes one bit operation.

(a) Estimate the memory size in bits $\mu(n)$, required for representing the arc set A of the graph G using the adjacency matrix representation.

Answer: The adjacency matrix is a bit matrix of size $n \times n$, whence the required memory size: $\mu(n) = n^2$ bits.

(b) Estimate the memory size in bits $\tau(n)$, required for representing the arc set A of the graph G using the adjacency list representation.

Answer: The adjacency list of a node, say node x, is a sequence of names of destination nodes of those arcs that are incident out of x. To see that each of these node names requires $\lceil \lg n \rceil$ bits of storage, recall that there are exactly 2^ℓ distinct binary strings of any length $\ell \geq 1$. If ℓ is selected so as to to be the minimum length of a binary string that can assume at least n distinct values, then it must be that $2^{\ell-1} < n \leq 2^\ell$. Therefore, $\ell = \lceil \lg n \rceil$. The total number of nodes that appear in all adjacency lists is equal to the number of arcs a. Since G is simple, the upper bound on the number of its arcs is: $a \leq n^2$. Since G' is connected, the lower bound on the number of arcs of G is: $a \geq n - 1$. Hence, the required memory size in bits is:

$$(n-1) \cdot \lg n \leq \tau(n) \leq n^2 \cdot \lg n$$

(c) Compare the (asymptotic orders of) sizes of the storage space $\mu(n)$ and $\tau(n)$ in the case when graph G is a directed tree.

Answer: If G is a directed tree, the number of its arcs is: $a = n - 1$. Recall that each arc contributes $\lceil \lg n \rceil$ bits to $\tau(n)$. Hence, $\tau(n) = \Theta(n \cdot \lg n)$, while $\mu(n) = \Theta\left(n^2\right)$, yielding: $\tau(n) = o(\mu(n))$.

(d) Compare the (asymptotic orders of) sizes of the storage space $\mu(n)$ and $\tau(n)$ in the case when the undirected graph G' contains a subgraph isomorphic to $K_{n/2, n/2}$, the complete bipartite graph on $(n/2) + (n/2)$ nodes.

Answer: The complete bipartite graph $K_{n/2, n/2}$ has $n^2/8$ edges, which means that a lower bound on the number of arcs of G is: $a \geq n^2/8$. Each arc contributes $\lceil \lg n \rceil$ bits to $\tau(n)$. Hence, $\tau(n) = \Theta\left(n^2 \cdot \lg n\right)$, while $\mu(n) = \Theta\left(n^2\right)$, yielding: $\mu(n) = o(\tau(n))$.

Problem 369   Consider the representations of the directed simple graph $G = (V, A)$, described in Problem 368.

(a) Estimate the number of bit operations required to determine whether (x, y) is an arc of G, using the adjacency matrix representation.

Answer: Once the binary representations of the node names x and y are given, a lookup of the entry (x, y) in the adjacency matrix of G constitutes one bit operation, by the assumptions of the programming model. Since $(x, y) \in A$ exactly when the corresponding entry in the adjacency matrix is equal to 1, we conclude that a single bit operation (or a constant number of them) suffices to determine whether (x, y) is an arc of G.

(b) Estimate the number of bit operations required to determine whether (x, y) is an arc of G, using the adjacency list representation.

Answer: Once the binary representation of the node name x is given, the first bit in the first node name in the adjacency list of node x is accessible at the cost of a single bit lookup. Recall that $(x, y) \in A$ holds exactly when y is found in the adjacency list of x. Hence, it may be necessary to read the entire adjacency list of x in order to determine whether it contains the node y. Since there are no restrictions on the structure of the graph G, the worst case occurs when there is an arc incident out of node x into every node of G. In this case, the length of the adjacency list of x is equal to $n \cdot \lceil \lg n \rceil$ bits, and we conclude that the asymptotic order of the number of bit operations required to determine whether (x, y) is an arc of G is estimated to be $\Theta(n \cdot \lg n)$ in the worst case.

(c) Assuming that the out-degree of G is bounded by a constant, estimate the number of bit operations required to determine whether (x, y) is an arc of G, using the adjacency list representation.

Answer: This case is different from the one analyzed in the previous part in that the number of entries in the adjacency list of x is limited by a constant, say d. Consequently, the length of this adjacency list has an upper bound of $d \cdot \lceil \lg n \rceil$ bits. Hence, the asymptotic order of the required number of bit operations is estimated to be $\Theta(\lg n)$ in the worst case.

Problem 370   Consider the representations of the directed simple graph $G = (V, A)$, described in Problem 368.

(a) Estimate the number of bit operations required to determine the out-degree of an arbitrary node x of the graph G, using the adjacency matrix representation.

Answer: The out-degree of x is equal to the sum of values in the row x of the adjacency matrix. Since there are n bits to be added together, the required number of bit operations is estimated to be $\Theta(n)$.

(b) Estimate the number of bit operations required to determine the out-degree of an arbitrary node x of the graph G, using the adjacency list representation.

Answer: The out-degree of x, say Δ, is equal to the number of nodes whose names appear in the adjacency list of x. If there is no way to determine the length of the adjacency list other than by reading it in its entirety, then an access is required to each of the $\Delta \cdot \lceil \lg n \rceil$ bits in the list. In the absence of any assumptions about the maximum out-degree of x, the worst case occurs when $\Delta = n$. Hence, the required number of bit operations is estimated to be $\Theta(n \cdot \lg n)$.

(c) Assuming that the out-degree of G is bounded by a constant, estimate the number of bit operations required to determine the out-degree of an arbitrary node x of the graph G, using the adjacency matrix representation.

Answer: The calculation of the out-degree of x still requires a computation of the sum of all n entries in the corresponding row of the adjacency matrix. Hence, this case is not different from that analyzed in the part **(a)**, and the required number of bit operations is estimated to be $\Theta(n)$.

(d) Assuming that the out-degree of G is bounded by a constant, estimate the number of bit operations required to determine the out-degree of an arbitrary node x of the graph G, using the adjacency list representation.

Answer: This case is different from the one analyzed in the part **(b)** in that the number of entries in the adjacency list of x is limited by a constant, say d, yielding an upper bound of $d \cdot \lceil \lg n \rceil$ bits on the length of an individual adjacency list. Hence, the asymptotic order of the required number of bit operations is estimated to be $\Theta(\lg n)$ in the worst case.

Problem 371   Consider the representations of the directed simple graph $G = (V, A)$, described in Problem 368.

(a) Estimate the number of bit operations required to determine the in-degree of an arbitrary node x of the graph G, using the adjacency matrix representation.

Answer: The in-degree of x is equal to the sum of values in the column x of the adjacency matrix. Since there are n bits to be added together, the required number of bit operations is estimated to be $\Theta(n)$.

(b) Assuming that the in-degree of G is bounded by a constant, estimate the number of bit operations required to determine the in-degree of an arbitrary node x of the graph G, using the adjacency matrix representation.

Answer: The calculation of the in-degree of x still requires a computation of the sum of all n entries in the corresponding column of the adjacency matrix. Hence, this case is not different from that analyzed in the previous part, and the required number of bit operations is estimated to be $\Theta(n)$.

(c) Estimate the number of bit operations required to determine the in-degree of an arbitrary node x of the graph G, using the adjacency list representation.

Answer: The in-degree of x is equal to the number of occurrences of x in the adjacency lists of all nodes of G. To obtain this count, all adjacency lists have to be read. The total number of nodes in all adjacency lists of G is equal to the number of arcs a, and the total length of all lists is equal to $a \cdot \lceil \lg n \rceil$ bits. Recall that a is bounded by: $n - 1 \le a \le n^2$, since the undirected graph G' is connected and simple. Hence, the total number of required bit accesses is never smaller than $(n - 1) \cdot \lceil \lg n \rceil$, and may be as large as $n^2 \cdot \lceil \lg n \rceil$. We conclude that the required number of bit operations is estimated to be $\Theta(n^2 \cdot \lg n)$.

(d) Assuming that the in-degree of G is bounded by a constant, estimate the number of bit operations required to determine the in-degree of an arbitrary node x of the graph G, using the adjacency list representation.

Answer: The calculation of the in-degree of x still requires access to each bit in the adjacency lists of all nodes of G. However, this case is different from the one analyzed in the previous part in that the number of arcs is limited by: $a \le dn$, for some constant d. Hence, the total number of required bit accesses has an upper bound of $dn \cdot \lceil \lg n \rceil$, and the required number of bit operations is estimated to be $\Theta(n \cdot \lg n)$ in the worst case.

Problem 372 G is a directed graph represented by its adjacency matrix A as follows.

$$A = \begin{bmatrix} 0 & 1 & 1 & 1 & 0 & 0 & 0 \\ 0 & 0 & 1 & 0 & 1 & 1 & 0 \\ 0 & 1 & 0 & 1 & 0 & 1 & 0 \\ 0 & 1 & 1 & 0 & 1 & 0 & 0 \\ 0 & 0 & 1 & 1 & 0 & 1 & 0 \\ 0 & 0 & 1 & 1 & 0 & 0 & 1 \\ 1 & 1 & 1 & 0 & 0 & 0 & 0 \end{bmatrix}$$

(a) Draw the graph G.

Answer: See Figure 5.27 (a). The nodes of the graph are labeled so that the lexicographic order of labels corresponds to the left-to-right and top-to-bottom order of the entries in the matrix A.

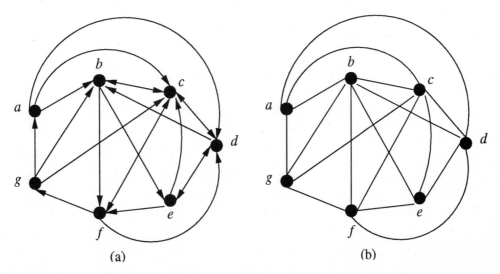

Figure 5.27: (a) Directed graph G; (b) Undirected graph H

(b) For all possible pairs of nodes (x, y) of G, calculate the number of different paths of length 2 from x to y. Explain your answer.

Answer: The number of different paths of length 2 between node x and node y is equal to the entry corresponding to the pair (x, y) in the matrix A^2, which is computed as follows:

$$A^2 = A \times A = \begin{bmatrix} 0 & 1 & 1 & 1 & 0 & 0 & 0 \\ 0 & 0 & 1 & 0 & 1 & 1 & 0 \\ 0 & 1 & 0 & 1 & 0 & 1 & 0 \\ 0 & 1 & 1 & 0 & 1 & 0 & 0 \\ 0 & 0 & 1 & 1 & 0 & 1 & 0 \\ 0 & 0 & 1 & 1 & 0 & 0 & 1 \\ 1 & 1 & 1 & 0 & 0 & 0 & 0 \end{bmatrix} \times \begin{bmatrix} 0 & 1 & 1 & 1 & 0 & 0 & 0 \\ 0 & 0 & 1 & 0 & 1 & 1 & 0 \\ 0 & 1 & 0 & 1 & 0 & 1 & 0 \\ 0 & 1 & 1 & 0 & 1 & 0 & 0 \\ 0 & 0 & 1 & 1 & 0 & 1 & 0 \\ 0 & 0 & 1 & 1 & 0 & 0 & 1 \\ 1 & 1 & 1 & 0 & 0 & 0 & 0 \end{bmatrix} = \begin{bmatrix} 0 & 2 & 2 & 1 & 2 & 2 & 0 \\ 0 & 1 & 2 & 3 & 0 & 2 & 1 \\ 0 & 1 & 3 & 1 & 2 & 1 & 1 \\ 0 & 1 & 2 & 2 & 1 & 3 & 0 \\ 0 & 2 & 2 & 2 & 1 & 1 & 1 \\ 1 & 3 & 2 & 1 & 1 & 1 & 0 \\ 0 & 2 & 2 & 2 & 1 & 2 & 0 \end{bmatrix}$$

(c) For all possible pairs of nodes (x, y) of G, calculate the number of different paths of length 3 from x to y. Explain your answer.

Answer: The number of different paths of length 3 between node x and node y is equal to the entry corresponding to the pair (x, y) in the matrix A^3, which is computed as follows:

$$A^3 = A^2 \times A = \begin{bmatrix} 0 & 2 & 2 & 1 & 2 & 2 & 0 \\ 0 & 1 & 2 & 3 & 0 & 2 & 1 \\ 0 & 1 & 3 & 1 & 2 & 1 & 1 \\ 0 & 1 & 2 & 2 & 1 & 3 & 0 \\ 0 & 2 & 2 & 2 & 1 & 1 & 1 \\ 1 & 3 & 2 & 1 & 1 & 1 & 0 \\ 0 & 2 & 2 & 2 & 1 & 2 & 0 \end{bmatrix} \times \begin{bmatrix} 0 & 1 & 1 & 1 & 0 & 0 & 0 \\ 0 & 0 & 1 & 0 & 1 & 1 & 0 \\ 0 & 1 & 0 & 1 & 0 & 1 & 0 \\ 0 & 1 & 1 & 0 & 1 & 0 & 0 \\ 0 & 0 & 1 & 1 & 0 & 1 & 0 \\ 0 & 0 & 1 & 1 & 0 & 0 & 1 \\ 1 & 1 & 1 & 0 & 0 & 0 & 0 \end{bmatrix} = \begin{bmatrix} 0 & 3 & 7 & 6 & 3 & 6 & 2 \\ 1 & 6 & 7 & 4 & 4 & 3 & 2 \\ 1 & 5 & 6 & 6 & 2 & 6 & 1 \\ 0 & 4 & 7 & 6 & 3 & 4 & 3 \\ 1 & 5 & 7 & 4 & 4 & 5 & 1 \\ 0 & 4 & 7 & 5 & 4 & 6 & 1 \\ 0 & 4 & 7 & 5 & 4 & 5 & 2 \end{bmatrix}$$

(d) Find a pair (x, y) of nodes (if any) in the graph G, such that there exists an arc (x, y), but there are no paths of length equal to 2 or 3 from x to y.

Answer: There exists an arc (g, a), but there are no paths of length equal to 2 or 3 from g to a, since the corresponding entries in the matrices A^2 and A^3 are equal to 0.

(e) Find all paths of length equal to 2 in the graph G from node c to node a.

Answer: The entry of the matrix A^2 which corresponds to the pair (c, a) is equal to 0, indicating that there are no paths of length equal to 2 from node c to node a.

(f) Find all paths of length equal to 2 in the graph G from node a to node c.

Answer: The entry of the matrix A^2 which corresponds to the pair (a, c) is equal to 2, indicating that there are two paths of length equal to 2 from node a to node c. These paths are:

$$a \to b \to c \quad \text{and} \quad a \to d \to c$$

(g) Find all paths of length equal to 3 in the graph G from node c to node a.

Answer: The entry of the matrix A^3 which corresponds to the pair (c, a) is equal to 1, indicating that there exists one path of length equal to 3 from node c to node a. This path is:

$$c \to f \to g \to a$$

(h) Find all paths of length equal to 3 in the graph G from node a to node c.

Answer: The entry of the matrix A^3 which corresponds to the pair (a, c) is equal to 7, indicating that there are seven paths of length equal to 3 from node a to node c. These paths are:

$$
\begin{array}{lll}
a \to b \to e \to c & a \to d \to b \to c & a \to c \to b \to c \\
a \to b \to f \to c & a \to d \to e \to c & a \to c \to d \to c \\
& & a \to c \to f \to c
\end{array}
$$

Problem 373 Consider the undirected graph G, defined in Problem 372. Let graph H be obtained from the graph G by a conversion of every arc into an edge.

(a) Draw the graph H.

Answer: See Figure 5.27 (b).

(b) Construct the adjacency matrix B of the graph H.

Answer: The adjacency matrix B of the graph H is the matrix of the symmetric closure of the relation represented by the matrix A, defined in Problem 372, whence the answer:

$$
B = \begin{bmatrix}
0 & 1 & 1 & 1 & 0 & 0 & 1 \\
1 & 0 & 1 & 1 & 1 & 1 & 1 \\
1 & 1 & 0 & 1 & 1 & 1 & 1 \\
1 & 1 & 1 & 0 & 1 & 1 & 0 \\
0 & 1 & 1 & 1 & 0 & 1 & 0 \\
0 & 1 & 1 & 1 & 1 & 0 & 1 \\
1 & 1 & 1 & 0 & 0 & 1 & 0
\end{bmatrix}
$$

(c) For all possible pairs of nodes (x, y) of H, calculate the number of different paths of length 2 from x to y. Explain your answer.

Answer: The number of different paths of length 2 between node x and node y is equal to the entry corresponding to the pair (x, y) in the matrix B^2, which is computed as follows:

$$
B^2 = B \times B = \begin{bmatrix}
0 & 1 & 1 & 1 & 0 & 0 & 1 \\
1 & 0 & 1 & 1 & 1 & 1 & 1 \\
1 & 1 & 0 & 1 & 1 & 1 & 1 \\
1 & 1 & 1 & 0 & 1 & 1 & 0 \\
0 & 1 & 1 & 1 & 0 & 1 & 0 \\
0 & 1 & 1 & 1 & 1 & 0 & 1 \\
1 & 1 & 1 & 0 & 0 & 1 & 0
\end{bmatrix} \times \begin{bmatrix}
0 & 1 & 1 & 1 & 0 & 0 & 1 \\
1 & 0 & 1 & 1 & 1 & 1 & 1 \\
1 & 1 & 0 & 1 & 1 & 1 & 1 \\
1 & 1 & 1 & 0 & 1 & 1 & 0 \\
0 & 1 & 1 & 1 & 0 & 1 & 0 \\
0 & 1 & 1 & 1 & 1 & 0 & 1 \\
1 & 1 & 1 & 0 & 0 & 1 & 0
\end{bmatrix} = \begin{bmatrix}
4 & 3 & 3 & 2 & 3 & 4 & 2 \\
3 & 6 & 5 & 4 & 3 & 4 & 3 \\
3 & 5 & 6 & 4 & 3 & 4 & 3 \\
2 & 4 & 4 & 5 & 3 & 3 & 4 \\
3 & 3 & 3 & 3 & 4 & 3 & 3 \\
4 & 4 & 4 & 3 & 3 & 5 & 2 \\
2 & 3 & 3 & 4 & 3 & 2 & 4
\end{bmatrix}
$$

(d) For all possible pairs of nodes (x, y) of H, calculate the number of different paths of length 3 from x to y. Explain your answer.

Answer: The number of different paths of length 3 between node x and node y is equal to the entry corresponding to the pair (x, y) in the matrix B^3, which is computed as follows:

$$
B^3 = B^2 \times B = \begin{bmatrix}
10 & 18 & 18 & 17 & 12 & 13 & 14 \\
18 & 22 & 23 & 21 & 19 & 21 & 18 \\
18 & 23 & 22 & 21 & 19 & 21 & 18 \\
17 & 21 & 21 & 16 & 16 & 20 & 13 \\
12 & 19 & 19 & 16 & 12 & 16 & 12 \\
13 & 21 & 21 & 20 & 16 & 16 & 17 \\
14 & 18 & 18 & 13 & 12 & 17 & 10
\end{bmatrix}
$$

(e) Find all paths of length equal to 2 in the graph H from node c to node a.

Answer: The entry of the matrix $\boldsymbol{B}^2$ which corresponds to the pair (c, a) is equal to 3, indicating that there are three paths of length equal to 2 from node c to node a. These paths are:

$$c \to b \to a, \quad c \to d \to a, \quad c \to g \to a$$

(f) Find all paths of length equal to 3 in the graph H from node c to node a.

Answer: The entry of the matrix $\boldsymbol{B}^3$ which corresponds to the pair (c, a) is equal to 18, indicating that there are eighteen paths of length equal to 3 from node c to node a. These paths are:

$$
\begin{array}{lll}
c \to a \to b \to a & c \to b \to c \to a & c \to d \to b \to a \\
c \to a \to c \to a & c \to b \to d \to a & c \to d \to c \to a \\
c \to a \to d \to a & c \to b \to g \to a & \\
c \to a \to g \to a & & \\
c \to e \to b \to a & c \to f \to b \to a & c \to g \to b \to a \\
c \to e \to c \to a & c \to f \to c \to a & c \to g \to c \to a \\
c \to e \to d \to a & c \to f \to d \to a & \\
& c \to f \to g \to a & \\
\end{array}
$$

Problem 374   Graph G is a simple undirected graph, represented on Figure 5.28.

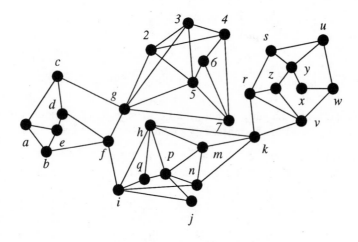

Figure 5.28: Graph G

(a) Graph G' is a graph obtained from G by a removal of five edges, so as to comprise exactly four connected components. Identify the removed edges and the connected components of G'.

Answer: See Figure 5.29 for an illustration of the graph G', obtained from G by a removal of the edges: $(c, g), (f, g), (f, i), (k, r), (k, v)$.

(b) Determine the number of edges of the graph representing the transitive closure of the relation represented by the graph G, and explain your answer.

Answer: The transitive closure of the relation that corresponds to G is a complete graph on the node set of G. To see this, recall that in a graph of a transitive relation there exists an edge between any two nodes that are connected by a path. Since G is connected, there exists a path in G between every pair of its nodes, implying the existence of an edge between these nodes in the graph of the

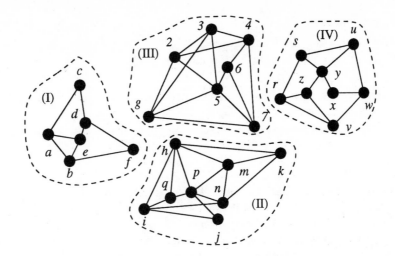

Figure 5.29: After a removal of five edges from G: graph G' with 4 connected components

transitive closure. Since G has 29 nodes, the complete graph on its node set has: $\begin{pmatrix} 29 \\ 2 \end{pmatrix} = 406$ edges.

(c) Determine the number of edges of the graph representing the transitive closure of the relation represented by the graph G', and explain your answer.

Answer: Construction of the graph of the transitive closure corresponds to a conversion of every path in the original graph into an edge, which transforms every connected graph into a complete graph. Since G' has exactly four conncted components, the graph of the transitive closure is the union of four pairwise unconnected complete graphs, each containing the node set of one of the four connected components. The number of edges in these four complete graphs is:

$$\begin{pmatrix} 6 \\ 2 \end{pmatrix} + \begin{pmatrix} 8 \\ 2 \end{pmatrix} + \begin{pmatrix} 7 \\ 2 \end{pmatrix} + \begin{pmatrix} 8 \\ 2 \end{pmatrix} = 92$$

(d) Consider an arbitrary set V, with an equivalence relation ρ, represented by a graph H. Let x and y be arbitrary nodes of H. What are the possible values of the distance between x and y in H? Explain your answer.

Answer: The graph of any equivalence relation is a union of pairwise unconnected complete graphs, which correspond to the equivalence classes of the represented equivalence relation. Hence, if $(x, y) \in \rho$ then nodes x and y belong to the same connected component of H (equivalence class of ρ), and the distance between x and y is equal to 1, since this is the distance between any two nodes of a complete graph. Otherwise, if $(x, y) \notin \rho$, there is no path between nodes x and y.

Problem 375   Directed simple graph $G = (V, A)$ (possibly with self-loops) has n nodes, and is represented by its adjacency matrix A.

For each of the following claims, provide a criterion equivalent to the claim but expressed in terms of the matrix A, and explain your answer.

(a) G represents a transitive relation.

Answer: Recall that a relation is transitive if and only if it is equal to its transitive closure. To calculate the matrix of the transitive closure of the relation represented by the matrix A, first compute the following matrix.

$$B = \sum_{k=0}^{n-1} A^k$$

where $A^0 = I$ is the identity matrix. Finally, construct a binary matrix C from the matrix B by a substitution of 1 for every nonzero entry of B. Precisely:

$$C[i,j] = \begin{cases} 1 & \text{if} \quad B[i,j] > 0 \\ 0 & \text{if} \quad B[i,j] = 0 \end{cases}$$

To verify the construction, recall that the entry $A^k[i,j]$ is equal to the number of distinct paths of length k from node i to node j. Moreover, if there exists a path from i to j in G, then there exists a simple path of no more than n nodes (and no more than $n-1$ edges). To see this, observe that every path of more than n nodes must visit some node twice—the loop between the two visits to this node can be eliminated, thus shortening the path. Hence, there exists at least one value of k such that $1 \le k \le n$ and $A^k[i,j] > 0$ if and only if there exists a path (of length equal to k) from i to j in G. However, in that case:

$$(\exists k)(A^k[i,j] > 0) \iff B[i,j] > 0 \iff C[i,j] = 1$$

Hence, there exists a path from i to j in G if and only if $C[i,j] = 1$, which confirms that C is the matrix of the transitive closure of the relation represented by A.

Finally, the relation represented by A is transitive if and only if its matrix is equal to the matrix of its transitive closure, which means if and only if $A = C$.

(Note that an efficient algorithm for construction of the matrix C entails fewer operations than is suggested by the description of C presented in this answer.)

(b) G represents an equivalence relation.

Answer: Recall that a relation is an equivalence relation if and only if it is reflexive, symmetric and transitive. In other words, the relation represented by A is an equivalence relation if and only if the following three conditions hold.

reflexivity	$(\forall k, 1 \le k \le n) (A[k,k] = 1)$
symmetry	$(\forall i,j, 1 \le i,j \le n) (A[i,j] = A[j,i])$
transitivity	$A = C$

where C is the matrix of the transitive closure of A, computed in the answer to the part **(a)**.

(c) G represents a relation of partial order.

Answer: Recall that a relation is a partial order if and only if it is reflexive, antisymmetric and transitive. In other words, the relation represented by A is a partial order if and only if the following three conditions hold.

reflexivity	$(\forall k, 1 \le k \le n) (A[k,k] = 1)$
antisymmetry	$(\forall i,j, 1 \le i,j \le n) (A[i,j] \cdot A[j,i] = 0)$
transitivity	$A = C$

where C is the matrix of the transitive closure of A, computed in the answer to the part **(a)**.

(d) G is strongly connected.

Answer: Recall that G is strongly connected if and only if there exists a directed path between every pair of its nodes. For an arbitrary pair of nodes i, j, there exists a path from i to j in G if and only if the corresponding entry of the matrix C of the transitive closure of A is equal to 1, according to the analysis given in the answer to the part **(a)**. Hence, G is strongly connected if and only if:

$$(\forall i, j,\ 1 \leq i, j \leq n)\,(C[i,j] = 1)$$

(e) G is undirected.

Answer: Recall that G is undirected if and only if its matrix is symmetric, which means if and only if:

$$(\forall i, j,\ 1 \leq i, j \leq n)\,(A[i,j] = A[j,i])$$

Problem 376   Consider the directed simple graph $G = (V, A)$, defined in Problem 375. Let G^+ be the undirected graph obtained from G by a conversion of every arc into an edge.

For each of the following claims, provide a criterion equivalent to the claim but expressed in terms of the adjacency matrix A of the graph G, and explain your answer.

(a) G^+ is connected.

Answer: The adjacency matrix A^+ of the undirected graph G^+ is computed as follows.

$$(\forall i, j,\ 1 \leq i, j \leq n)\,(A^+[i,j] = \max(A[i,j],\ A[j,i]))$$

We proceed to compute the matrix B^+, which counts the simple paths in G^+, and the matrix C^+, which is the matrix of the connectivity relation on the nodes of G^+. The construction is analogous to that explained in the answer to part **(a)** of Problem 375.

$$B^+ = \sum_{k=0}^{n-1} \left(A^+\right)^k$$

$$C^+[i,j] = \begin{cases} 1 & \text{if} \quad B^+[i,j] > 0 \\ 0 & \text{if} \quad B^+[i,j] = 0 \end{cases}$$

Finally, G^+ is connected if and only if there are no zero entries in the matrix C^+.

$$(\forall i, j,\ 1 \leq i, j \leq n)\,(C^+[i,j] = 1)$$

(Note that an efficient algorithm for construction of the matrix C^+ entails fewer operations than is suggested by the description of C presented in this answer.)

(b) The relation represented by the matrix C^+, computed in the answer to the previous part, is an equivalence relation.

Answer: This claim is always true, regardless of the original matrix A. In fact, the connectedness relation in the undirected version of the graph G^+ (represented by the matrix C^+) is the reflexive, symmetric, and transitive closure of the adjacency relation (represented by the matrix A.)

To verify that C^+ represents a reflexive relation, observe that, by construction:

$$B^+ = I + \sum_{k=1}^{n-1} \left(A^+\right)^k$$

Since $I[k,k]=1$ for any k such that $1 \le k \le n$, we conclude that $B^+[k,k] > 0$ and, consequently: $C^+[k,k]=1$.

To verify that C^+ represents a symmetric relation, observe that A^+ is symmetric, by construction. Since the sum and product of two symmetric matrices is symmetric, and B^+ is equal to a sum of products of symmetric matrices (recall that I is also symmetric), we conclude that B^+ is symmetric. Consequently, C^+ is symmetric.

To verify that C^+ represents a transitive relation, recall the argument presented in the answer to the part (a) of Problem 375.

(c) A subset S of nodes of G comprises an equivalence class of the equivalence relation, represented by the matrix C^+, described in the previous part.

Answer: Select an arbitrary node $x \in S$, and let X be the subset of nodes of G that are related to x under the relation represented by C^+:

$$X = \{y \mid C^+[x,y] = 1\}$$

The set X is exactly the equivalence class of x in the equivalence relation represented by C^+, whence the criterion that S is an equivalence class if and only if $S = X$.

The equivalence classes of C^+ are precisely the connected components of the undirected graph G^+.

Problem 377   Graph H is the undirected graph represented on Figure 5.27 (b).

(a) Is the graph H planar? Prove your answer.

Answer: Graph H is not planar, since it contains as a subgraph $K_{3,3}$, the complete bipartite graph on $3+3$ nodes whose node set is $\{a, e, f\} \cup \{b, c, d\}$, as is demonstrated on Figure 5.30.

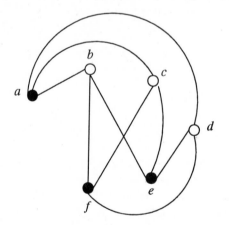

Figure 5.30: $K_{3,3}$ as a subgraph of the graph H

(b) Let η be an integer such that there exists a set of exactly η edges of the graph H whose removal renders the graph planar, while any subgraph of H obtained by a removal of fewer than η edges is not planar. Determine the value of η and identify the edges whose removal leaves a planar subgraph of the graph H. Prove your answer.

Answer: $\eta = 2$. We need to demonstrate that there exists a pair of edges whose removal from the graph H produces a graph H', which is planar. Furthermore, we have to show that H does not become planar after a removal of any single edge.

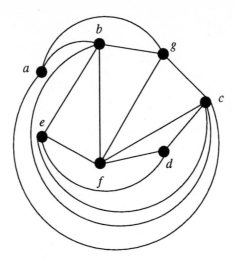

Figure 5.31: After a removal of $\{(a,d),\ (b,d)\}$ from H: a planar drawing of H'

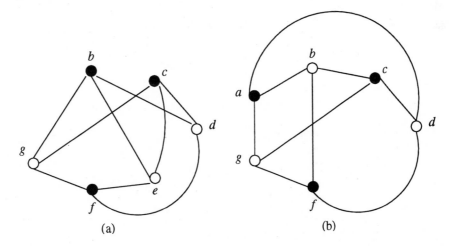

(a)   (b)

Figure 5.32: Copies of $K_{3,3}$ in H after a removal of edges incident to: (a) node a; (b) node e

To construct H', remove from H the pair of edges: $\{(a,d),(b,d)\}$. The planar drawing of the resulting subgraph H' is given on Figure 5.31.

To prove that a removal of any single edge, say (x,y), is insufficient to render the graph H planar, we distinguish among a number of cases, according to the choice of the edge (x,y). Evidently, if the removed edge (x,y) is not any of the nine edges that form the copy of $K_{3,3}$ in H given on Figure 5.30, then the removal of this edge leaves the subgraph $K_{3,3}$ intact, guaranteeing that the graph which contains it is not planar. Hence, we assume that the removed edge (x,y) is one of the nine edges represented on Figure 5.30, and we recognize the following three cases.

   Case (1): $(x,y) \in \{(a,b),\,(a,c),\,(a,d)\}$

In this case, the remaining subgraph of H contains a copy of $K_{3,3}$, given on Figure 5.32 (a), whose node set is $\{b,c,f\} \cup \{d,e,g\}$.

   Case (2): $(x,y) \in \{(e,b),\,(e,c),\,(e,d)\}$

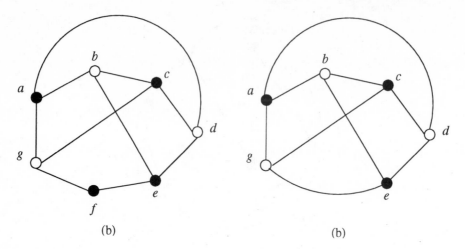

(b)   (b)

Figure 5.33: After a removal of (f, b), (f, c), (f, d): (a) a subgraph homeomorphic to $K_{3,3}$; (b) $K_{3,3}$

In this case, the remaining subgraph of H contains a copy of $K_{3,3}$, given on Figure 5.32 (b), whose node set is $\{a, c, f\} \cup \{b, d, g\}$.

> *Case (3):* $(x, y) \in \{(f, b), (f, c), (f, d)\}$

In this case, the remaining subgraph of H contains a subgraph given on Figure 5.33 (a), which is homeomorphic to a copy of $K_{3,3}$, given on Figure 5.33 (b).

Problem 378   Graphs considered in this problem are simple (without self-loops), undirected, and connected.

Given a graph $G = (V, E)$, recall that a node subset $D \subseteq V$ is a dominating set for the graph G if every node outside D is adjacent to some node in D. Furthermore, a node subset $C \subseteq V$ is a node cover for the graph G if at least one endpoint of every edge of G belongs to C.

Let $\Delta(G)$ be the minimum of the sizes of all dominating sets of G, and let $\Upsilon(G)$ be the minimum of the sizes of all node covers of G.

(a) Prove that every node cover is a dominating set.

Answer: Let C be a node cover for a graph $G = (V, E)$. To prove that C is a dominating set for G, assume the opposite, and let $x \in V \setminus C$ be a node that is not adjacent to any node in C. Since G is connected, there exists at least one edge, say (x, y) adjacent to the node x (unless x is the only node of G, in which case the claim holds vacuously.) Since C is a node cover, at least one of the two endpoints x, y of the edge (x, y) must belong to C. By assumption, $x \notin C$. Hence, $y \in C$, in contradiction with the assumption that there are no nodes in C that are adjacent to x.

(b) Construct a graph that has a dominating set which is not a node cover.

Answer: Every node of a 3-node cycle constitutes a dominating set, since the other two nodes are adjacent to it. However, each of the three nodes is adjacent to only two edges—hence, every node cover of the 3-node cycle contains at least two nodes.

(c) Construct an infinite family of graphs whose every dominating set is a node cover.

Answer: Define the family of n-beam stars, for all $n \geq 1$, as follows. The node set of the n-beam star consists of the center, labeled with 0, and n peripheral nodes, labeled with integers 1 through n. Every peripheral node is connected by an edge to the center.

The n-beam star has exactly two dominating sets that are minimal in that they do not contain a proper subset that is also a dominating set. One of these sets is the singleton containing the center, while the other is the n-element set consisting of all the peripheral nodes. Each of these sets is also a node cover, since every edge has exactly one endpoint in each of the two sets.

Problem 379 Graphs considered in this problem are simple (without self-loops), undirected, and connected.

Given a graph $G = (V, E)$, let $\Delta(G)$ be the minimum of the sizes of all dominating sets of G, and let $\Upsilon(G)$ be the minimum of the sizes of all node covers of G.

Construct an infinite family of graphs whose every m-node instance G has the property that:

$$\Upsilon(G) - \Delta(G) \geq 0.9\, m$$

Answer: Consider the n-node complete graph K_n. We claim that: $\Delta(K_n) = 1$ and $\Upsilon(K_n) = n - 1$.

Evidently, any node of the complete graph constitutes a dominating set, since all the other nodes are adjacent to it. To see that the n-node complete graph cannot have a node cover with fewer than $n - 1$ nodes, assume the opposite, and let K_m be a complete graph with a node cover C of size equal to $m - 2$. Let x and y be the two nodes that do not belong to C. Since any pair of nodes are adjacent in a complete graph, (x, y) is an edge. However, neither of the two endpoints x, y of the edge (x, y) belongs to C, contradicting the assumption that C is a node cover.

Finally, calculate the difference: $\Upsilon(K_n) - \Delta(K_n) = (n - 1) - 1 = n - 2$. To conclude the proof, note that $n - 2 \geq (9n/10)$ whenever $n \geq 20$. Hence, in every complete graph with no fewer than 20 nodes, the minimum size of a node cover exceeds the minimum size of a dominating set by at least 90% of the size of the graph.

Problem 380 Graphs considered in this problem are simple (without self-loops), undirected, and connected.

Given a graph $G = (V, E)$, let $\Delta(G)$ be the minimum of the sizes of all dominating sets of G, and let $\Upsilon(G)$ be the minimum of the sizes of all node covers of G.

Construct an infinite family of graphs whose degree is bounded by a certain constant, and whose every m-node instance G has the property that:

$$\Upsilon(G) - \Delta(G) \geq (m/2)$$

Answer: We construct a family R_n, such that R_n has $6n$ nodes and a maximum degree of 7, while $\Upsilon(R_n) - \Delta(R_n) = 3n$. Informally, R_n is a collection of 5-beam wheels, whose centers are connected into an n-node path. See Figure 5.34 for an illustration of R_4.

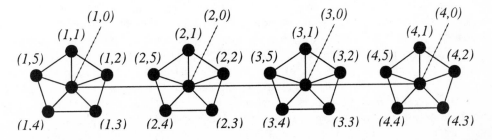

Figure 5.34: Degree-7 graph whose any node cover is much larger than a certain dominating set

Precisely, $R_n = (V_n, E_n)$ is defined as follows. The node set V_n is a set of $6n$ ordered pairs:

$$V_n = \{\langle x, y \rangle \mid 1 \leq x \leq n, \, 0 \leq y \leq 5\}$$

The edge set E_n is a union of three types of edges: path edges P_n, star edges S_n, and cycle edges C_n:

$$E_n = P_n \cup S_n \cup C_n$$

where each edge subset is defined as follows. The path edges connect all nodes whose last component is equal to 0 into an n-node path, in the sequence of their first components:

$$P_n = \{(\langle k, 0 \rangle, \langle k+1, 0 \rangle) \mid 1 \leq k < n\}$$

The star edges connect every node whose last component is equal to 0 with the five nodes with which it agrees in the value of the first component:

$$S_n = \{(\langle k, 0 \rangle, \langle k, \ell \rangle) \mid 1 \leq k \leq n, \, 1 \leq \ell \leq 5\}$$

The cycle edges connect each group of five nodes that agree in the value of the first component but their second component is not equal to zero into a cycle, in the sequence of their second components:

$$C_n = \{(\langle k, 1 \rangle, \langle k, 2 \rangle), (\langle k, 2 \rangle, \langle k, 3 \rangle), (\langle k, 3 \rangle, \langle k, 4 \rangle), (\langle k, 4 \rangle, \langle k, 5 \rangle), (\langle k, 5 \rangle, \langle k, 1 \rangle) \mid 1 \leq k \leq n\}$$

For brevity, refer to the group of six nodes whose first component is equal to k as the kth wheel, to the node $\langle k, 0 \rangle$ as the center of the kth wheel, and to the remaining five nodes as the periphery of the kth wheel.

We claim that $\Delta(R_n) = n$. To see this, observe that the set of centers of the wheels:

$$D_n = \{(k, 0) \mid 1 \leq k \leq n\}$$

is a dominating set for R_n, since every other node is adjacent to some node in D_n via a star edge. Furthermore, no dominating set can have fewer than n nodes, since every wheel has to contribute at least one node to any dominating set. To see this, note that if any wheel fails to contribute its center to the dominating set, the periphery of that wheel cannot be adjacent to any nodes in the dominating set unless such node(s) are designated within the periphery itself, whence the claim.

To verify that $\Upsilon(R_n) = 4n$, first observe that the set:

$$C_n = \{(k, 0), (k, 1), (k, 3), (k, 5) \mid 1 \leq k \leq n\}$$

is a node cover for R_n, since the wheel centers cover the path edges and the star edges, while the three peripheral nodes selected from every wheel cover the cycle edges of that wheel. To verify that no node cover can have fewer than $4n$ nodes, first observe that the periphery of any wheel has to contribute at least 3 nodes to C_n, since as many as 5 peripheral edges have to be covered only by the peripheral nodes of the wheel, while each such node covers no more than 2 distinct edges. Next, we claim that each wheel has to contribute at least one more node to C_n, in addition to these 3 peripheral nodes. To verify the claim, recall that the five star edges in each wheel have five distinct peripheral endpoints. Unless the wheel center is in C_n (as the fourth node from that wheel), all the 5 peripheral nodes have to be in C_n in order to cover the star edges.

Finally, we calculate the difference: $\Upsilon(R_n) - \Delta(R_n) = 4n - n = 3n = \dfrac{6n}{2} = \dfrac{|V_n|}{2}$.

Problem 381 Consider two simple undirected graphs $G_1 = (V_1, E_1)$ and $G_2 = (V_2, E_2)$, and their direct product $G_1 \times G_2$. Let $n_j = |V_j|$, for $j = 1, 2$, be the sizes of node sets of the two graphs. Let $D_j \subseteq V_j$ be a dominating set for G_j of size $\Delta_j = |D_j|$.

(a) Determine whether $G_1 \times G_2$ has a dominating set of size equal to:

$$\min(\Delta_1 \cdot n_2, \Delta_2 \cdot n_1,)$$

and prove your answer.

Answer: We prove that each of the two sets $D_1 \times V_2$ and $V_1 \times D_2$ is a dominating set for $G_1 \times G_2$.

To prove that $D_1 \times V_2$ is a dominating set for $G_1 \times G_2$, consider an arbitrary node $\langle a, x \rangle$ of $G_1 \times G_2$. Assuming that $\langle a, x \rangle \notin D_1 \times V_2$, we have to prove that $\langle a, x \rangle$ is adjacent to some node in $D_1 \times V_2$. Observe that $\langle a, x \rangle \notin D_1 \times V_2$ implies that $a \notin D_1$. Since D_1 is a dominating set for G_1, there exists a node $b \in D_1$ adjacent to a. This means that node $\langle a, x \rangle$ is adjacent to node $\langle b, x \rangle$, which in turn belongs to $D_1 \times V_2$, meaning that $D_1 \times V_2$ is indeed a dominating set.

The proof that $D_2 \times V_1$ is analogous, provided that the adjacencies in the second component are considered, while the first component is fixed.

(b) Assume that the minimum size of a dominating set for G_j is equal to Δ_j, for $j = 1, 2$. Determine whether the minimum size of a dominating set for $G_1 \times G_2$ is equal to that obtained in part **(a)**. Prove your answer.

Answer: In general, the dominating set produced by the construction presented in the part **(a)** does not have the minimum size. For instance, consider the direct product of two 4-node paths, illustrated on Figure 5.35. Evidently, the two internal nodes of a 4-node path constitute a dominating set. Furthermore, it is one of minimum size, since no single node in a degree-2 graph can be adjacent to as many as 3 other nodes. Hence, $\Delta_1 = \Delta_2 = 2$ and $n_1 = n_2 = 4$, whence:

$$\min(\Delta_1 \cdot n_2, \Delta_2 \cdot n_1,) = \min(8, 8) = 8$$

However, the direct product of two 4-node paths has a dominating set of size equal to 6, as is illustrated on Figure 5.35.

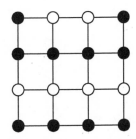

Figure 5.35: A dominating set of size 6 for the product of two 4-node paths

Problem 382 Consider two simple undirected graphs $G_1 = (V_1, E_1)$ and $G_2 = (V_2, E_2)$, and their direct product $G_1 \times G_2$. Let $n_j = |V_j|$, for $j = 1, 2$, be the sizes of node sets of the two graphs. Let $C_j \subseteq V_j$ be a node cover for G_j of size $\Upsilon_j = |C_j|$.

Determine whether $G_1 \times G_2$ has a node cover of size equal to:

$$\min(\Upsilon_1 \cdot n_2, \Upsilon_2 \cdot n_1,)$$

and prove your answer.

Answer: In general, $G_1 \times G_2$ does not have a node cover of the specified size. In fact, $G_1 \times G_2$ need not have a node cover of size equal to $\max(\Upsilon_1 \cdot n_2, \Upsilon_2 \cdot n_1,)$, and neither of the two node sets $C_1 \times V_2$ and $V_1 \times C_2$ need be a node cover. For instance, consider the direct product of two 3-node paths. The middle node of each of the two paths constitutes a node cover of size equal to 1, which is evidently optimal. Hence, $\Upsilon_1 = \Upsilon_2 = 1$ and $n_1 = n_2 = 3$, whence:

$$\min(\Upsilon_1 \cdot n_2, \Upsilon_2 \cdot n_1,) = \max(\Upsilon_1 \cdot n_2, \Upsilon_2 \cdot n_1,) = 3$$

However, the direct product of two 3-node paths has 12 edges, one node of degree 4, four nodes of degree 3, and four nodes of degree 2. This graph cannot have a node cover consisting of only three nodes, since the sum of degrees of any three of its nodes is less than 12. It has a node cover comprising the four nodes with the degree equal to 3, as is illustrated on Figure 5.36.

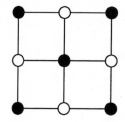

Figure 5.36: An optimal node cover for the product of two 3-node paths

Problem 383   In each of the following cases, construct three pairwise non-isomorphic rooted binary trees that satisfy the stated constraints.

(a) The pre-order traversal of each of the three trees produces the same list of nodes, say: $1, 2, 3, 4, 5, 6, 7, 8, 9$, while the root and all of its children have identical labels in all of the three trees.

Answer: See Figure 5.37 for a representation of the trees T_1, T_2, and T_3. To verify that they are indeed pairwise non-isomorphic, observe, for instance, that they have distinct heights: 3, 4, and 5, respectively.

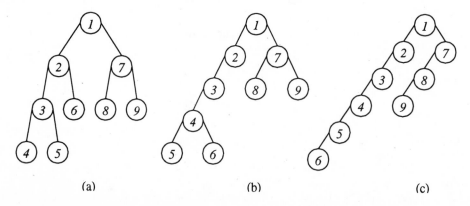

Figure 5.37: Identical pre-order traversal lists: (a) T_1; (b) T_2; (c) T_3

(b) The in-order traversal of each of the three trees produces the same list of nodes, say: $1, 2, 3, 4, 5, 6, 7, 8, 9$, while the root and all of its children have identical labels in all of the three trees.

Answer: See Figure 5.38 for a representation of the trees T_1, T_2, and T_3. To verify that they are indeed pairwise non-isomorphic, observe, for instance, that the nodes whose degree is equal to 2 are located at levels $0, 3$ in T_1, $0, 2, 3$ in T_2, and $0, 2$ in T_3.

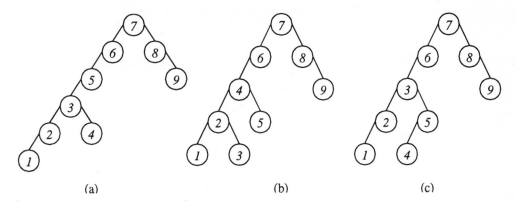

(a)   (b)   (c)

Figure 5.38: Identical in-order traversal lists: (a) T_1; (b) T_2; (c) T_3

Problem 384   Your baby brother Jerry likes a mathematical computer game, which allows him to play with various graphs. Right now, he is constructing complete, undirected, rooted trees. The parameters of his construction are: the height of the tree h, the number of leaves n, and the number of children of internal nodes b (the branch factor, as Jerry calls it). Jerry is often puzzled by the effects of various parameter changes on the appearance of his trees.

(a) Jerry says: "My tree has too few leaves—I would like it to have twice as many as now, for the same height. Therefore, I am doubling the branch factor." What do you say?

Answer: The number of leaves of a height-h complete b-ary tree is: $n = b^h$. If the new branch factor is $b_1 = 2b$, the new number of leaves is going to be:

$$n_1 = b_1^h = (2b)^h = 2^h \cdot b^h = 2^h \cdot n$$

Unless the tree has only one level, Jerry's is estimate is going to be inaccurate—precisely, he is underestimating the resulting number of leaves by a factor equal to:

$$\frac{n_1}{2n} = \frac{2^h \cdot n}{2n} = 2^{h-1}$$

(b) Jerry says: "The height of my tree is equal to 5. I would like to have a few hundred leaves. I wonder what my branch factor should be." What do you say?

Answer: The number of leaves n of a complete, height-5 tree is calculated for a few initial values of the branch factor b as follows.

b	$n = b^5$
2	32
3	243
4	1024

Hence, the only branch factor that fits Jerry's specification of "a few hundred leaves" is: $b = 3$.

(c) Jerry says: "My branch factor is equal to 4 only. However, I would like to have a million leaves. I guess I will need a very tall tree." What do you say?

Answer: The height of a complete b-ary tree with at least n leaves is calculated as follows.

$$h = \lceil \log_b n \rceil = \left\lceil \frac{\lg n}{\lg b} \right\rceil = \left\lceil \frac{\lg 10^6}{\lg 4} \right\rceil \leq \left\lceil \frac{\lg 2^{20}}{\lg 4} \right\rceil = \frac{20}{2} = 10$$

Hence, ten levels, in addition to the root, suffice.

(d) Jerry says: "Some of my trees have too many leaves to be drawn. Once I know my branch factor and I get to see the first few levels, say one half of the tree height, I wish I could guess how wide the leaf level is going to be." What do you say?

Answer: The number of nodes at a level ℓ of a complete, height-h, b-ary tree is equal to b^ℓ. Hence, the number of nodes, say m, that are located at the level $h/2$ is calculated as follows:

$$m = b^{(h/2)} = \left(b^h\right)^{(1/2)} = \sqrt{n}$$

or, equivalently, $n = m^2$. This means that the size of the population of the tree level halfway between the root and the leaves is about the square root of the number of leaves. Moreover, this relationship is independent from the branch factor of the tree.

Problem 385   Your baby brother Jerry is playing the game described in Problem 384.

(a) Jerry says: "My branch factor is equal to 4, and the height of my tree is equal to 10. I just colored all the leaves, and the program took a while to do so. I would like very much to color also the internal nodes of my tree, but am not even thinking of it—if it took so long to color only the tenth level, I cannot even imagine how long it would take to color all the levels." What do you say?

Answer: The number of nodes of a height-h complete b-ary tree is:

$$N = \frac{b^{h+1} - 1}{b - 1}$$

The number of nodes is greater than the number of leaves by the following factor.

$$\frac{N}{n} = \frac{b^{h+1} - 1}{(b-1) \cdot b^h} = \frac{b}{b-1} - \frac{1}{b^h(b-1)} < \frac{b}{b-1} = \frac{4}{4-1} = 1 + \frac{1}{3}$$

Hence, $N \leq (1 + 1/3)n$, meaning that the number of internal nodes is:

$$N - n \leq \left(1 + \frac{1}{3}\right) n - n = \frac{n}{3}$$

indicating that the number of internal nodes is less than one-third of the number of leaves—for a modest overhead of one-third in the running time of his program, Jerry could color the entire tree.

(b) Jerry says: "Does there exist a tree where 90% of nodes are leaves?" What do you say?

Answer: If at least a fraction of $\alpha < 1$ nodes are leaves, then the analysis given in the answer to the previous part implies:

$$\frac{N}{n} = \frac{b}{b-1} - \frac{1}{b^h(b-1)} \leq \frac{1}{\alpha}$$

Since $b^h(b-1) \gg 1$ even for relatively small values of h, a good approximation of the last inequality is:

$$\frac{b}{b-1} \leq \frac{1}{\alpha} \iff \alpha b \leq b - 1 \iff b \geq \frac{1}{1 - \alpha}$$

Observe that the approximation is on the safe side, since every value of b that satisfies the approximation also satisfies the original inequality. In Jerry's case: $\alpha = 9/10$, whence the branch factor:

$$b \geq \frac{1}{1 - (9/10)} = 10$$

We conclude that, in a complete tree where every internal node has exactly 10 children, at least 90% of nodes are leaves.

Problem 386   Your baby brother Jerry is playing the game described in Problem 384.

Jerry says: "I have $h + 1$ levels in my tree, but these levels are very unevenly populated—the root is alone at level 0, but at level h there are very many leaves. I would like to pack my tree into a rectangle with the same number of levels, $h + 1$, but so that all levels of this rectangle have the same width. I wonder what the width is going to be—perhaps something half-way between the width of my tree at the root (1, that is) and the width at the leaf level (b^h). I am going to try it for a binary tree whose height is equal to 20." What do you say?

Answer: If all the nodes of a complete, height h, b-ary tree are put in a rectangular box with $h + 1$ levels, the number of nodes w at such an individual level is an integer that satisfies the following pair of constraints.

$$\left\lfloor \frac{N}{h+1} \right\rfloor \leq w \leq \left\lceil \frac{N}{h+1} \right\rceil$$

To estimate the width w, recall the calculation given in the answer to part **(b)** of Problem 385, which implies that the number of nodes is:

$$N = \left(\frac{n}{b-1} \right) \left(b - \frac{1}{b^h} \right) = \left(\frac{bn}{b-1} \right) \left(1 - \frac{1}{b^{h+1}} \right)$$

Since $b^{h+1} \geq 2^1 = 2$, the number of nodes N honors the following bounds:

$$\frac{bn}{2(b-1)} < N < \frac{bn}{b-1}$$

while the width of the rectangle is estimated as follows.

$$w \approx \frac{N}{h+1} = \left(\frac{bn}{(h+1)(b-1)} \right) \left(1 - \frac{1}{b^{h+1}} \right)$$

In Jerry's tree: $b = 2$, $h = 20$, $n = b^h = 2^{20}$, and $b^{h+1} = 2^{21} \gg 1$, meaning that the term $(1/b^{h+1})$ vanishes next to 1. This leads to a good approximation of the required width:

$$w \approx \frac{bn}{(h+1)(b-1)} = \frac{2n}{21} = \frac{2^{21}}{21}$$

Jerry's estimate of the width is: $w_J = (n+1)/2$, which is over five times the actual value. In any case:

$$w < \frac{bn}{(h+1)(b-1)}$$

implying that Jerry over-estimates the width by a factor not less than the following.

$$\frac{w_J}{w} > \left(\frac{n+1}{n} \right) \left(\frac{b-1}{b} \right) \left(\frac{h+1}{2} \right) = \left(1 + \frac{1}{n} \right) \left(1 - \frac{1}{b} \right) \left(\frac{h+1}{2} \right) > \left(1 - \frac{1}{b} \right) \left(\frac{h+1}{2} \right)$$

Problem 387   Your baby brother Jerry is playing the game described in Problem 384.

(a) Jerry says: "My beautiful tree has over one billion leaves, but its height is too big: 19. I wish the height was less than 10, but I don't want to lose any leaves. I am going to double the branch factor—I hope this will cut the height to one half of its present value." What do you say?

Answer: The height of a complete, b-ary tree with n leaves is calculated as follows.

$$h = \log_b n = \frac{\lg n}{\lg b}$$

If Jerry's new branch factor is: $B = 2b$, the height assumes the smallest integer value that satisfies the condition:

$$h_1 \geq \log_B n = \frac{\lg n}{\lg B} = \frac{\lg n}{\lg (2b)} = \frac{\lg n}{1 + \lg b}$$

The ratio of the new height and the old height is equal to:

$$\frac{h_1}{h} = h_1 \cdot \frac{1}{h} = \left(\frac{\lg n}{1 + \lg b} \right) \left(\frac{\lg b}{\lg n} \right) = \frac{\lg b}{1 + \lg b} \geq \frac{\lg b}{\lg b + \lg b} = \frac{1}{2}$$

When $\lg b = 1$ or, equivalently, $b = 2$, the new height h_1 may attain a value as low as $h/2$. However, for any branch factor b such that $b > 2$, the new height h_1 is greater than $h/2$. Note that this ratio is independent from the number of leaves n.

The branch factor of Jerry's tree is equal to 3. To verify this, observe that $3^{19} \approx 1.16 \cdot 10^9$, which matches Jerry's quote of over one billion. (The nearest two candidates for the branch factor, 2 and 4, would give vastly incorrect number of leaves for a height-19 tree: about 0.5 million and 0.27 trillion, respectively.) To estimate the new height, recall that:

$$2^{(3/2)} = \left(2^3 \right)^{(1/2)} = \sqrt{8} < \sqrt{9} = 3 \implies \lg 3 > \frac{3}{2}$$

yielding:

$$h_1 \geq h \cdot \left(\frac{\lg b}{1 + \lg b} \right) = h \cdot \left(\frac{1}{(1/\lg b) + 1} \right) > h \cdot \left(\frac{1}{(2/3) + 1} \right) = \frac{3h}{5} = \frac{3 \cdot 19}{5} > 11$$

Hence, the new height exceeds one-half of the old height and does not meet Jerry's expectations to fit into a single digit. (Indeed, $h_1 = 12$ is the smallest height that allows for the required number of leaves with the branch factor $b = 6$. To verify this, note that $6^{12} \approx 2.18 \cdot 10^9$.)

(b) Jerry says: "Does there exists a tree where a twofold increase in the branch factor reduces the height by no more than 10%?" What do you say?

Answer: Select the branch factor b such that: $\lg b \geq \alpha/(1 - \alpha)$, for some $0 < \alpha < 1$. The analysis given in the answer to the previous part implies:

$$h_1 \geq h \cdot \left(\frac{1}{(1/\lg b) + 1} \right) = h \cdot \left(\frac{1}{((1 - \alpha)/\alpha) + 1} \right) = h \cdot \left(\frac{\alpha}{(1 - \alpha) + \alpha} \right) = \alpha \cdot h$$

Jerry's instance mandates that: $\alpha \geq 9/10$ (or, equivalently: $1 - \alpha \leq 1/10$), which implies that:

$$\lg b \geq \frac{\alpha}{1 - \alpha} \geq 9 \implies b \geq 2^9 = 512$$

In other words, if a complete tree, say T_1, has a branch factor equal to 512 and a height equal to h, then any complete tree T_2 whose branch factor is equal to 1024 must have a height of at least $0.9\,h$ in order to attain the number of leaves of T_1—regardless of the number of leaves or the height of the tree.

Problem 388 Your baby brother Jerry is playing the game described in Problem 384.

Jerry says: "I have tried many different values of height and branch factor, but I have not been able to construct a complete tree with exactly $1,852,200$ leaves, unless its height is equal to 1." What do you say?

Answer: The number of leaves n is an integer of the form: $n = b^h$, for integers b and h. Let p be any prime factor of b, such that $p^k \mid b$. Then, $p^{kh} \mid b^h$. On the other hand, if q is any prime factor of b^h, then $q \mid b$. Hence, the number n can be represented in the form $n = b^h$ if and only if there exists a positive integer h that divides all the exponents of primes in the factorization of n. However:

$$1,852,200 = 2^3 \cdot 3^3 \cdot 5^2 \cdot 7^3$$

Hence, the only positive integer that divides all the exponents in this prime factorization is 1, leaving the trivial possibility of a unit height as the only option.

Problem 389 Prove that there are infinitely many natural numbers, say k, such that there does not exist a simple, undirected, k-node graph (without self-loops) that is isomorphic to its complement.

Answer: Recall that the complement $\overline{G} = (V, \overline{E})$ of a simple, undirected graph $G = (V, E)$ has a node set identical to that of the node set of G, and exactly those edges that are absent in G. Precisely, define R to be the set of all possible pairs of distinct nodes of G (the set of all possible edges):

$$R = (V \times V) \setminus \{(s, s) \mid s \in V\}$$

The edge sets E and $\overline{E}$ are related as follows.

$$(\forall s, t \in R)((s, t) \in E \iff (s, t) \notin \overline{E})$$

Hence, $E \cap \overline{E} = \emptyset$ and $E \cup \overline{E} = R$. Assuming that G has k nodes, the sum $|E| + |\overline{E}|$ is equal to the number of elements of R: $k(k-1)/2$. A necessary condition for two graphs, G and $\overline{G}$, to be isomorphic is that they agree in the number of edges. Hence:

$$|E| + |\overline{E}| = 2 \cdot |E| = \frac{k(k-1)}{2} \implies |E| = \frac{k(k-1)}{4}$$

Since $|E|$ is certainly an integer, and exactly one of the numbers k and $k-1$ is odd (and thereby not divisible by 2), we conclude that exactly one of the numbers k and $k-1$ is divisible by 4. This means that a necessary condition for the existence of a k-node graph isomorphic to its complement is that $k = 4n$ or $k = 4n + 1$, for some natural number n. Consequently, whenever k is of the form: $k = 4n + 2$ or $k = 4n + 3$, there does not exist a k-node graph isomorphic to its complement.

Problem 390 Construct an infinite family B_n (for $n \geq 1$) of simple, undirected graphs without self-loops, such that B_n has $4n$ nodes and is isomorphic to its complement. Demonstrate the isomorphism.

Answer: The family of graphs $B_n = (V_n, E_n)$ and the family of complements $\overline{B_n} = (V_n, \overline{E_n})$ are defined as follows. For a given $n \geq 1$, define $\eta(n)$ to be the natural number that satisfies the following condition.

$$2^{\eta(n)-1} < n \leq 2^{\eta(n)}$$

In other words, there exist at least n distinct binary strings of length equal to $\eta(n)$, but there are fewer than n distinct binary strings of any smaller length. Let $C(n)$ be any subset of the set of all $2^{\eta(n)}$ length-$\eta(n)$ binary strings, such that $C(n)$ contains exactly n distinct strings.

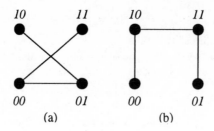

Figure 5.39: Isomorphism with the complement: (a) B_1; (b) $\overline{B_1}$

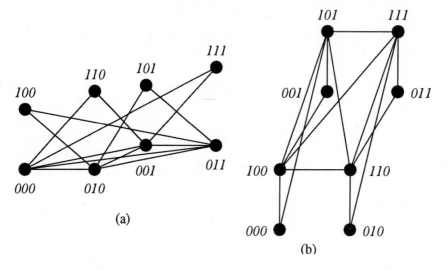

Figure 5.40: Isomorphism with the complement: (a) B_2; (b) $\overline{B_2}$

The node set of B_n contains the $4n$ binary strings of length equal to $\eta(n) + 2$, whose length-$\eta(n)$ suffix belongs to $\mathcal{C}(n)$. Precisely:

$$V_n = \{\alpha\beta x \mid \alpha, \beta \in \{0,1\} \wedge x \in \mathcal{C}(n)\}$$

Any individual pair of distinct nodes $(\alpha_1\beta_1 x, \alpha_2\beta_2 y) \in V_n \times V_n$ belongs to exactly one of the edge sets E_n and $\overline{E_n}$. The membership is decided according to the pair $(\alpha_1\beta_1, \alpha_2\beta_2)$ of 2-bit prefixes of the node names, disregarding the pair (x, y) of their $\eta(n)$-bit suffixes (selected from the suffix pool $\mathcal{C}(n)$). Precisely:

$$E_n = \{(\alpha_1\beta_1 x, \alpha_2\beta_2 y) \in V_n \times V_n \mid ((\alpha_1\beta_1, \alpha_2\beta_2) \in \mathcal{E} \wedge x, y \in \mathcal{C}(n))\}$$
$$\overline{E_n} = \{(\alpha_1\beta_1 x, \alpha_2\beta_2 y) \in V_n \times V_n \mid ((\alpha_1\beta_1, \alpha_2\beta_2) \in \overline{\mathcal{E}} \wedge x, y \in \mathcal{C}(n))\}$$

where the sets of pairs $\mathcal{E}$ and $\overline{\mathcal{E}}$ are defined as follows.

$$\mathcal{E} = \{(00,00), (00,01), (00,11), (01,00), (01,01), (01,10), (10,01), (11,00)\}$$
$$\overline{\mathcal{E}} = \{(00,10), (01,11), (10,00), (10,10), (10,11), (11,01), (11,10), (11,11)\}$$

Observe that sets $\mathcal{E}$ and $\overline{\mathcal{E}}$ do not overlap: $\mathcal{E} \cap \overline{\mathcal{E}} = \varnothing$; they jointly account for each of the 16 possible pairs of the 2-bit prefixes, so that:

$$(\alpha_1\beta_1 x, \alpha_2\beta_2 y) \in E_n \iff (\alpha_1\beta_1 x, \alpha_2\beta_2 y) \notin \overline{E_n}$$

whenever $\alpha_1, \beta_1, \alpha_2, \beta_2 \in \{0,1\}$ and $x, y \in \mathcal{C}(n)$. By inspection of the two sets, we find that:

$$(\alpha_1\beta_1 x, \, \alpha_2\beta_2 y) \in E_n \iff (\alpha_2\beta_2 y, \, \alpha_1\beta_1 x) \in E_n$$

guaranteeing that the graph B_n is undirected. Furthermore, the construction specifies a self-loop for all nodes of B_n whose 2-bit prefix is equal to 00 or 01, which is disregarded, since B_n is simple and undirected.

See Figure 5.39 and Figure 5.40 for illustrations of B_1 and B_2, respectively.

To verify that B_n and $\overline{B_n}$ are indeed isomorphic, we construct a bijection $f : V_n \to V_n$, which sends every node $\alpha\beta x \in V_n$ into a node $\alpha'\beta'x \in V_n$, whose $\eta(n)$-bit suffix $x \in \mathcal{C}(n)$ is identical to that of the original node $\alpha\beta x$, but the 2-bit prefix $\alpha'\beta'$ is determined as follows.

$\alpha\beta$	$\alpha'\beta'$
00	11
01	10
10	00
11	01

By inspection of the sets $\mathcal{E}$ and $\overline{\mathcal{E}}$, we conclude that f preserves edges:

$$(\alpha_1\beta_1 x, \, \alpha_2\beta_2 y) \in E \iff (\alpha_1'\beta_1'x, \, \alpha_2'\beta_2'y) \in \overline{E} \iff (f(\alpha_1\beta_1 x), \, f(\alpha_2\beta_2 y)) \in \overline{E}$$

whenever $\alpha_1, \beta_1, \alpha_2, \beta_2 \in \{0,1\}$.

Problem 391   Construct a graph family D_n (for $n \geq 1$) of simple, undirected graphs without self-loops, such that D_n has $4n + 1$ nodes and is isomorphic to its complement. Demonstrate the isomorphism.

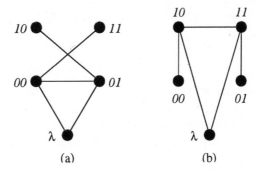

Figure 5.41: Isomorphism with the complement: (a) D_1; (b) $\overline{D_1}$

Answer: Recall that the graph $B_n = (V_n, E_n)$, constructed in the answer to Problem 390, has $4n$ nodes, for any positive integer n. To construct $D_n = (W_n, F_n)$, we augment B_n with an additional node named λ, and with $2n$ additional edges adjacent to the new node λ. Precisely, let a node set V_n and an edge set E_n be defined as in the answer to Problem 390. The node set W_n and the edge set F_n of D_n are defined as follows.

$$W_n = V_n \cup \{\lambda\}$$
$$F_n = E_n \cup \{(\lambda, 00x), (\lambda, 01x) \mid x \in \mathcal{C}(n)\}$$

yielding:

$$\overline{F_n} = \overline{E_n} \cup \{(\lambda, 10x), (\lambda, 11x) \mid x \in \mathcal{C}(n)\}$$

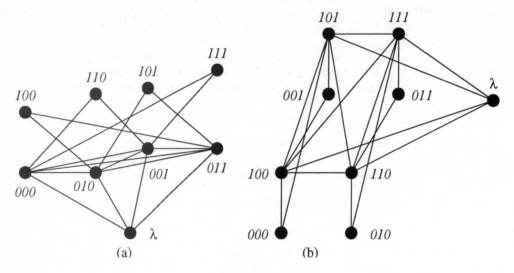

Figure 5.42: Isomorphism with the complement: (a) D_2; (b) $\overline{D_2}$

See Figure 5.41 and Figure 5.42 for illustrations of D_1 and D_2, respectively.

The isomorphism bijection $g : W_n \to W_n$ is obtained by an extension of the bijection $f : V_n \to V_n$ constructed in the answer to Problem 390. The extension maps the new node λ to itself.

$$g = f \cup \{(\lambda, \lambda)\}$$

Since f and g differ only in the mapping of the new node λ, and f preserves those edges that belong to the edge set E_n, we have to verify only that g preserves those edges that are adjacent to the new node λ (and one of the old nodes). To see this, observe that the definition of the edge set F_n and the definition of the bijections f and g imply:

$$(\lambda, \alpha\beta x) \in F_n \iff (\lambda, f(\alpha\beta x)) \in \overline{F_n} \iff (g(\lambda), g(\alpha\beta x)) \in \overline{F_n}$$

for all nodes $\alpha\beta x \in V_n$, whence the claim.

Problem 392   Construct a graph that is isomorphic to its complement, but does not belong to any of the graph families defined in the Problem 390 and Problem 391. If this is impossible, explain why.

Answer: The 5-node cycle C_5 is isomorphic to its complement, as is evident from the illustration given on Figure 5.43. However, C_5 is not isomorphic to the 5-node instance D_1 of the graph family constructed in the Problem 391, as is readily verified, say by a comparison between degrees of individual nodes.

Note that a family of graphs not isomorphic to D_n is obtained from B_n in a way analogous to the construction of D_n given in the answer to Problem 391, but with a modification in the set of edges adjacent to the new node. Precisely, the following modified definition of the new edge set F_n yields a graph family whose every member is isomorphic to its complement, and whose first instance is isomorphic to the 5-node cycle.

$$F_n = E_n \cup \{(\lambda, 10x), (\lambda, 11x) \mid x \in C(n)\}$$
$$\overline{F_n} = \overline{E_n} \cup \{(\lambda, 00x), (\lambda, 01x) \mid x \in C(n)\}$$

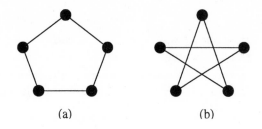

(a) (b)

Figure 5.43: C_5 is isomorphic to $\overline{C_5}$, but not to D_1: (a) C_5; (b) $\overline{C_5}$

Problem 393 Recall that a converse of a directed graph $G = (V, A)$ is the directed graph $G^{-1} = (V, A^{-1})$, whose arc set A^{-1} comprises exactly the arcs reverse to those in the arc set A. Precisely, $(x, y) \in A \Longleftrightarrow (y, x) \in A^{-1}$.

(a) Prove (by construction) that for every pair of positive integers n, e such that $0 \leq e \leq n(n-1)/2$, there exists a simple, directed graph $G(n, e)$ without self-loops, such that $G(n, e)$ is isomorphic to its converse, but does not contain any pair of arcs with identical endpoints and opposite directions.

Answer: Observe that the the only possible value of e in the case when $n = 1$ is $e = 0$, allowing $G(1, 0)$ to be an isolated node, which is vacuously isomorphic to its converse, by virtue of its empty arcs set. When $n = 2$, two values of e are possible: $e = 0$ and $e = 1$, and only the latter is nontrivial. Indeed, a two-node directed path is isomorphic to its converse.

Assuming that $n \geq 3$, let the node set of $G(n, e)$ be the set of all positive integers not exceeding n: $V_n = \{1, 2, \ldots, n\}$. Each of the $n(n-1)/2$ pairs of distinct elements of V_n is a candidate for an arc of $G(n, e)$, and our task is to select and direct e arcs so that $G(n, e)$ is isomorphic to its converse. To this end, we organize the set of all possible arcs into pairs of the form:

$$\langle (k, \ell), (n - \ell + 1, n - k + 1) \rangle \text{ where } 1 \leq k < \ell \leq n$$

In other words, every potential arc, say (a, b) is directed from the lower-numbered endpoint a to the higher-numbered endpoint b, and paired up with an arc (c, d) such that:

$$a + d = b + c = n + 1$$

Since this condition defines a symmetric relationship on the set of all potential arcs, we conclude that the arc (a, b) appears together with the arc (c, d) in two pairs: once as the first component and once as the second component. Moreover, since the function $k \mapsto n - k + 1$ (for a fixed n) is a bijection, neither of the two arcs appears in any other pair. This means that our pool of eligible arc pairs should include exactly one out of any two arc pairs of the following form.

$$\langle (a, b), (c, d) \rangle \text{ or } \langle (c, d), (a, b) \rangle$$

After this is done, each one of the $n(n-1)/2$ possible arcs appears in exactly one arc pair. However, some pairs contain two distinct arcs, while some arcs appear paired up with themselves. This occurs once for each pair (k, ℓ) such that $1 \leq k < \ell \leq n$ and:

$$k = n - \ell + 1$$
$$\ell = n - k + 1$$

or, equivalently:

$$k + \ell = n + 1 \text{ and } 1 \leq k < \ell \leq n$$

The number of such pairs (k, ℓ) is equal to $\left\lfloor \dfrac{n}{2} \right\rfloor$, meaning that exactly that many among our selected arc pairs specify the same arc as both the first and the second component. For convenience, term these pairs as odd. Since $n \geq 3$, at least one such odd pair always exists.

Finally, the actual arcs of $G(n, e)$ are selected in pairs, according to our pairing scheme. A selection of an arc pair selects both of its arcs simultaneously (unless the pair is odd.) We distinguish between two cases.

If e is even, select as many pairs as is needed to attain the required number of e arcs.

If e is odd, reserve one odd pair, select as many pairs as is needed to attain the required number of $(e - 1)$ arcs, and then adjoin the reserved odd pair as the eth arc.

The isomorphism bijection that maps the nodes of the graph $G(n, e) = (V_n, E)$ to the nodes of its converse $G^{-1}(n, e) = (V_n, E'_n)$ is a function $f : V_n \to V_n$, defined as follows.

$$f(k) = n - k + 1 \text{ for all } k \in V_n$$

To verify that f preserves arcs, consider an arbitrary arc $(k, \ell) \in E$. By our selection rules, this arc is included into E simultaneously with the arc that is paired with it:

$$(k, \ell) \in E \iff (n-\ell+1, n-k+1) \in E \iff (f(\ell), f(k)) \in E \iff (f(k), f(\ell)) \in E'$$

(b) Illustrate your construction presented in the answer to the part **(a)** on an instance of a graph with six nodes and nine arcs.

Answer: See Figure 5.44 for an illustration of $G(6, 9)$ and its converse, with node names that indicate the isomorphism bijection. The converse is obtained by the following selection of arcs. The odd pair is:

$$\langle (2, 5), (2, 5) \rangle$$

while the first eight arcs come from the following pairs.

$$\langle (1, 2), (5, 6) \rangle \, , \ \langle (1, 3), (4, 6) \rangle \, , \ \langle (1, 4), (3, 6) \rangle \, , \ \langle (2, 4), (3, 5) \rangle$$

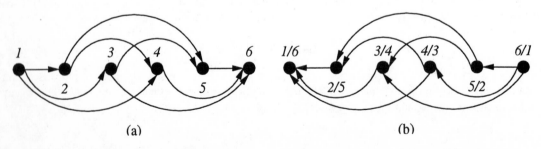

(a)   (b)

Figure 5.44: Isomorphism with the converse: (a) $G(6, 9)$; (b) $G^{-1}(6, 9)$

5.3 Recursion in Graphs

Problem 394 G_n is a family of undirected graphs defined recursively as follows.

G_0 is a two-node path between nodes 0 and 1. For $n \geq 1$, G_n is obtained by adjoining to G_{n-1} two new nodes: $2n$ and $2n + 1$, with an edge between each new node and node $2n - 1$.

(a) Write a formal recursive definition of the graph family G_n.

Answer: Let $G_n = (V_n, E_n)$, for $n \geq 0$.

$$V_0 = \{0, 1\}$$
$$E_0 = \{(0, 1)\}$$

$$V_n = V_{n-1} \cup \{2n, 2n + 1\}$$
$$E_n = E_{n-1} \cup \{(2n - 1, 2n), (2n - 1, 2n + 1)\} \text{ for } n \geq 1$$

(b) Draw G_1, G_2 and G_3.

Answer: See Figure 5.45.

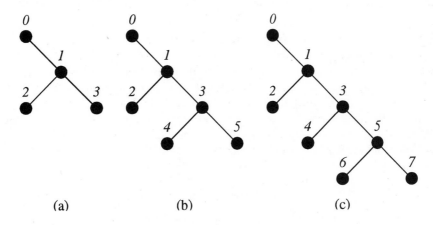

(a) (b) (c)

Figure 5.45: (a) G_1; (b) G_2; (c) G_3

(c) What is the length of the longest path in the graph G_n? Prove your answer.

Answer: The length of the longest path in G_n is $n+1$, since G_n is a caterpillar whose odd numbered nodes (together with node 0) form a path, while each even numbered node is connected by an edge to an internal node of this path.

First, we use induction to show that G_n contains as a subgraph a path with $n + 2$ nodes, whose node sequence is: $0, 1, 3, \ldots, 2n + 1$.

In the base case, by definition, G_0 indeed is a two-node path from node 0 to node 1.

For the inductive step, assume that G_n contains a length-$(n + 1)$ path whose nodes are:

$$0, 1, 3, \ldots, 2n + 1$$

in that order. We need to prove that G_{n+1} contains a length-$(n + 2)$ path whose nodes are:

$$0, 1, 3, \ldots, 2n + 1, 2(n + 1) + 1$$

in that order. Indeed, by the recursive step of the graph definition, G_{n+1} contains all the edges of G_n, including the length-$(n+1)$ path from 0 to $2n+1$. Additionally, G_{n+1} contains the edge $(2n+1, 2n+3)$, thereby extending the path from its previous endpoint $2n+1$ to the new endpoint $2(n+1)+1$.

Next, we use induction to show that each even numbered node (different from 0) is connected by an edge to some (odd-numbered) internal node of the caterpillar path.

In the base case, the only nodes of G_0 are 0 and 1, and the claim holds vacuously. A consideration of G_1 offers some more insight, and the claim is verified in this case by inspection of the graph, as is illustrated on Figure 5.45 (a).

For the inductive step, assume that each even numbered node in G_n is connected by an edge to some odd numbered node, different from node $2n+1$ (the endpoint of the caterpillar path.) We need to show that each even numbered node in G_{n+1} is connected by an edge to some odd numbered node, different from node $2n+3$.

We conclude from the recursive step of the graph definition that the edge set of G_{n+1} preserves all the edges of G_n. Hence, by the inductive hypothesis, each even numbered node that exists in G_n is still connected by an edge (in G_{n+1}) to an odd numbered node different from $2n+1$. Additionally, G_{n+1} has a new even numbered node. By the recursive step of the graph definition, a new edge indeed connects the new even numbered node $2n+2$ to the odd numbered node $2n+1$, and to no other nodes.

To complete the proof, we have to verify that G_n cannot contain as a subgraph any path longer than the demonstrated caterpillar path, which includes all odd numbered nodes and the node 0. If there existed a longer path, it would have to contain even numbered nodes other than 0. However, the degree of every even numbered node, say $2x$, is equal to 1, forcing the node $2x$ to be an endpoint of any path that contains it. Since node $2x$ is adjacent to node $2x-1$, the initial edge $(2x, 2x-1)$ could be eliminated from the candidate longest path in favor of a segment of length not less than 1 between node $2x-1$ and a suitable endpoint of the caterpillar path—recall that node $2x-1$ is not an endpoint. This would contradict the assumption that the path originating at node $2x$ is longer than the caterpillar path.

Problem 395 G_n is a family of directed graphs defined recursively as follows.

G_1 is a length-2 path containing nodes $2, 0, 1$ in that order. For $n \geq 2$, G_n is obtained by adjoining to G_{n-1} two new nodes named $2n$ and $2n-1$, with an arc from node $2n$ to node $2n-2$, and from node $2n-3$ to node $2n-1$.

(a) Write a formal recursive definition of the graph family G_n.

Answer: Let $G_n = (V_n, A_n)$, for $n \geq 1$.

$$V_1 = \{0, 1, 2\}$$
$$A_1 = \{(0,1), (2,0)\}$$

$$V_n = V_{n-1} \cup \{2n, 2n-1\}$$
$$A_n = A_{n-1} \cup \{(2n-3, 2n-1), (2n, 2n-2)\} \text{ for } n \geq 2$$

(b) Draw G_4.

Answer: See Figure 5.46.

(c) How many nodes does G_n have? Explain your answer.

Answer: G_n has $2n+1$ nodes. This is readily proved by induction—observe that G_1 has 3 nodes, and every higher instance has two nodes more than its predecessor.

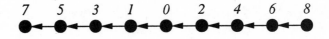

Figure 5.46: G_4

(d) How many arcs does G_n have? Explain your answer.

Answer: G_n has $2n$ arcs. This is readily proved by induction—observe that G_1 has 2 arcs, and every higher instance has two arcs more than its predecessor.

(e) What is the diameter of the undirected graph obtained by turning every arc of G_n into an edge? Explain your answer.

Answer: The undirected version of the graph G_n is a path from node $2n$ to node $2n-1$, the length of which is equal to $2n$, which is the diameter of the graph. This is readily proved by induction. For the base case, observe that G_1 is a length-2 path between nodes 2 and 1. Inductively, every higher instance extends this path by adding one arc before its initial end-point and one arc after its final end-point.

Problem 396 G_n is a family of directed graphs defined recursively as follows.

G_1 is a length-1 path from node 0 to node 1. For $n \geq 2$, G_n is obtained by adjoining to G_{n-1} a new node named n, with arcs from nodes 0 and $(n-1)$ to the new node.

(a) Write a formal recursive definition of the graph family G_n.

Answer: Let $G_n = (V_n, A_n)$, for $n \geq 1$.

$$V_1 = \{0, 1\}$$
$$A_1 = \{(0, 1)\}$$

$$V_n = V_{n-1} \cup \{n\}$$
$$A_n = A_{n-1} \cup \{(0, n), (n-1, n)\} \text{ for } n \geq 2$$

(b) Draw G_5.

Answer: See Figure 5.47.

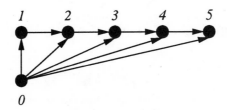

Figure 5.47: G_5

(c) Calculate the maximum in-degree and maximum out-degree of G_n as a function of n. Explain your answer.

Answer: For $n = 1$, the maximum in-degree and the maximum out-degree are equal to 1, as the graph G_1 has only one arc. For $n \geq 2$, the maximum in-degree is equal to 2, and all nodes $\{2, \ldots n\}$ have this in-degree. Node 0 has in-degree equal to 0. The maximum out-degree of G_n is equal to n, and node 0 has this out-degree. Furthermore, the out-degree of nodes $\{1, \ldots n-1\}$ is equal to 1, while the out-degree of node n is is equal to 0.

To prove all of these claims simultaneously, note that the base case is verified by inspection of G_5, given on Figure 5.47, which contains as subgraphs graphs G_n for all n such that $1 \leq n \leq 4$.

For the inductive step, assume that all the claims are true for G_{n-1} for some $n \geq 3$. Note that graph G_n contains G_{n-1} as a subgraph. In G_n, the new arcs do not alter the degree of any of the nodes $\{1, \ldots n-2\}$, which satisfy the claim by the inductive hypothesis. One of the new arcs increments the out-degree of node 0 from $n-1$ to n, while the other increments the out-degree of node $n-1$ from 0 to 1. The out-degree of all other nodes remains intact. In particular, the out-degree of the new node n is equal to 0. Since the two new arcs are incident into the new node n, the in-degree of this node becomes equal to 2, while the in-degree of all other nodes remains intact.

(d) Calculate the number of arcs of G_n as a function of n. Explain your answer.

Answer: $|A_n| = 2n - 1$. This is readily proved by induction, if we observe that G_1 has exactly one arc, and that every subsequent instance introduces two new arcs.

Problem 397 G_n is a family of directed graphs defined recursively as follows.

G_0 has three nodes: 0, 1 and 2, and a pair of arcs incident out of node 0, each one connecting node 0 with one of the two remaining nodes. For $n \geq 1$, G_n is obtained by adjoining to G_{n-1} three new nodes, named $3n$, $3n + 1$, and $3n + 2$, along with an arc from node $3n - 2$ to node $3n + 1$, from node $3n - 1$ to node $3n + 2$, and from node $3n$ to node 0.

(a) Write a formal recursive definition of the graph family G_n.

Answer: Let $G_n = (V_n, A_n)$, for $n \geq 0$.

$$V_0 = \{0, 1, 2\}$$
$$A_0 = \{(0, 1), (0, 2)\}$$

$$V_n = V_{n-1} \cup \{3n, 3n + 1, 3n + 2\}$$
$$A_n = A_{n-1} \cup \{(3n - 2, 3n + 1), (3n - 1, 3n + 2), (3n, 0)\} \text{ for } n \geq 1$$

(b) Draw G_3.

Answer: See Figure 5.48.

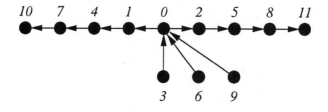

Figure 5.48: G_3

(c) What is the maximum in-degree of G_n, for an arbitrary $n > 0$, and which nodes have it?

Answer: The maximum in-degree of G_n is equal to n when $n > 0$, and node 0 has such in-degree.

(d) What is the maximum out-degree of G_n, for an arbitrary $n > 0$, and which nodes have it?

Answer: The maximum out-degree of G_n is equal to 2, and node 0 has such out-degree.

(e) What is the length of the longest path that is a subgraph of G_n, for an arbitrary $n > 0$, and which nodes are end-points of this path?

Answer: G_n contains $2n$ paths of length $n + 2$, for any $n \geq 1$. Each path originates at node $3k$, for any k such that $1 \leq k \leq n$; it passes through node 0, and continues either through node 1 or through node 2, ending at node $3n + 1$ or $3n + 2$, respectively.

(f) Is graph G_n strongly connected? Prove your answer.

Answer: Graph G_n is not strongly connected—for instance, there is no path from node 1 to node 0.

Problem 398 G_n is a family of undirected graphs defined recursively as follows.

G_2 is a four-node cycle containing nodes $1, 2, 3, 4$, in that order. For $n > 2$, G_n is obtained by adjoining to G_{n-1} two new nodes, $2n - 1$ and $2n$, and connecting each new node with nodes n and $n + 1$.

(a) Write a formal recursive definition of the graph family G_n.

Answer: Let $G_n = (V_n, E_n)$, for $n \geq 2$.

$$V_2 = \{1, 2, 3, 4\}$$
$$E_2 = \{(1, 2), (2, 3), (3, 4), (4, 1)\}$$

$$V_n = V_{n-1} \cup \{2n - 1, 2n\}$$
$$E_n = E_{n-1} \cup \{(n, 2n - 1), (n + 1, 2n - 1), (n, 2n), (n + 1, 2n)\} \text{ for } n \geq 3$$

(b) Draw G_3, G_4 and G_5.

Answer: See Figure 5.49.

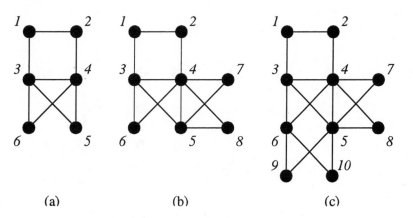

(a)   (b)   (c)

Figure 5.49: (a) G_3; (b) G_4; (c) G_5

(c) Prove that G_n has an Euler cycle for every $n \geq 2$.

Answer: We prove, by induction on n, that the degree of every node of G_n is even, which is a necessary and sufficient condition for the existence of an Euler cycle in G_n.

The base case of the recursive definition of the graph family G_n implies that G_2 is a simple cycle of length 4, which means that the degree of every node of G_2 is equal to 2.

For the inductive step, assume that all nodes of G_n have an even degree. We have to prove that all nodes of G_{n+1} have an even degree.

By the recursive step of the graph definition, G_{n+1} contains G_n as a subgraph. Hence, the only way for a node of G_{n+1} that appears in G_n to acquire an odd degree is to be an endpoint of one of the new edges. By the recursive step of the graph definition, the new edges appear in pairs: one pair is incident to node $n+1$, while the other pair is incident to node $n+2$. By the inductive hypothesis, the degree of each of the nodes $n+1$ and $n+2$ in G_n is even. Since the new edges augment the already even degree of each of these nodes by exactly 2, we conclude that the degree of nodes $n+1$ and $n+2$ remains even in G_{n+1}. To verify that the degree of the new nodes, $2n+1$ and $2n+2$, which appear in G_{n+1} but not in G_n, is even, observe that each of these nodes is an endpoint of exactly two edges: one incident to node $n+1$, and the other incident to node $n+2$. Hence, the degree of each of the new nodes in G_{n+1} is equal to 2, whence the claim.

Problem 399 G_n is a family of undirected graphs defined recursively as follows.

G_0 consists of two isolated nodes: 0 and 1.

For $n > 1$, G_n is obtained by adjoining to G_{n-1} two new nodes, $2n$ and $2n+1$, with an edge between nodes $2n$ and $2n-2$.

(a) Write a formal recursive definition of the graph family G_n.

Answer: Let $G_n = (V_n, E_n)$, for $n \geq 0$.

$$V_0 = \{0, 1\}$$
$$E_0 = \emptyset$$

$$V_n = V_{n-1} \cup \{2n, 2n+1\}$$
$$E_n = E_{n-1} \cup \{(2n, 2n-2)\} \ \text{ for } n \geq 1$$

(b) Draw G_0, G_1, G_2, and G_3.

Answer: See Figure 5.50.

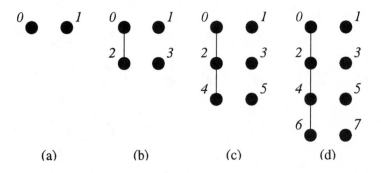

Figure 5.50: (a) G_0; (b) G_1; (c) G_2; (d) G_3

(c) How many connected components does G_n have, as a function of n? Prove your answer.

Answer: We prove, by induction on n, that G_n has $n+2$ connected components, one of which is a path of length n connecting all even-numbered nodes in the order of their numbers, while each of the remaining $n+1$ connected components is an isolated, odd-numbered node.

In the base case, G_0 indeed contains two isolated nodes, each of them constituting a connected component. The (even-numbered) node 0 is indeed a length-0 path, while the (odd-numbered) node 1 is indeed one connected component containing an isolated node.

For the inductive step, assume that the graph G_n consists of a length-n path whose node sequence is $0, 2, 4, \ldots, 2n$, together with $n + 1$ isolated nodes: $1, 3, 5, \ldots, 2n + 1$. We have to prove that G_{n+1} consists of a length-$(n + 1)$ path whose node sequence is $0, 2, 4, \ldots, 2n, 2n + 2$, together with $n + 2$ isolated nodes: $1, 3, 5, \ldots, 2n + 1, 2n + 3$. To verify this claim, observe that G_{n+1} contains G_n as a subgraph. Hence, by the inductive hypothesis, it contains the path $0, 2, 4, \ldots, 2n$ and $n + 1$ isolated odd-numbered nodes. By the recursive step of the graph definition, the inherited length-n path is extended in G_{n+1} by the only new edge $(2(n + 1), 2(n + 1) - 2)$ to include the new even numbered node $2n + 2$, thereby reaching the required length of $n + 1$. Since no other edges are introduced, the $n + 1$ odd-numbered nodes, isolated in G_n by the inductive hypothesis, remain isolated in G_{n+1}, while the new odd-numbered node $2n + 3$, which has no incident edges, forms a new connected component by itself, thereby incrementing the number of connected components.

Problem 400 G_n and H_n are families of undirected graphs, defined recursively as follows.

G_0 and H_0 consist of two isolated nodes: 0 and 1.

For $n > 1$, G_n is obtained by adjoining to G_{n-1} two new nodes, $2n$ and $2n + 1$, with an edge connecting each new node, say node x, with every node of G_{n-1}, except $x - 1$.

For $n > 1$, H_n is obtained by adjoining to H_{n-1} two new nodes, $2n$ and $2n + 1$, with an edge connecting each new node, say node x, with every other node of H_n, except $x - 2$.

(a) Write a formal recursive definition of the graph families G_n and H_n.

Answer: Let $G_n = (V_n, E'_n)$, $H_n = (V_n, E''_n)$, for $n \geq 0$.

$$V_0 = \{0, 1\}$$
$$E'_0 = E''_0 = \varnothing$$

For $n \geq 1$:

$$V_n = V_{n-1} \cup \{2n, 2n + 1\}$$

$$E'_n = E'_{n-1} \bigcup \left(\bigcup_{0 \leq k \leq 2n-2} \{(k, 2n)\} \right) \bigcup \left(\bigcup_{0 \leq k \leq 2n-1} \{(k, 2n + 1)\} \right)$$

$$E''_n = E''_{n-1} \bigcup \left(\bigcup_{0 \leq k \leq 2n-3} \{(k, 2n)\} \right) \bigcup \left(\bigcup_{0 \leq k \leq 2n-2} \{(k, 2n + 1)\} \right) \bigcup \{(2n - 1, 2n), (2n, 2n + 1)\}$$

(b) Draw G_3 and H_3.

Answer: See Figure 5.51.

Problem 401 Consider the families G_n and H_n of undirected graphs defined in Problem 400. Determine whether G_n and H_n are isomorphic and prove your answer.

Answer: Graphs G_n and H_n are isomorphic for every $n \geq 0$. To prove this, we show that their complements $\overline{G_n}$ and $\overline{H_n}$ are isomorphic to a path of length equal to $2n + 1$. Precisely, we use induction to prove that $\overline{G_n}$ is a path whose node sequence is: $0, 1, \ldots, 2n, 2n + 1$, while $\overline{H_n}$ is a path whose node sequence is: $2n, 2n - 2, \ldots, 0, 1, \ldots, 2n - 1, 2n + 1$. In other words, $\overline{G_n}$ is a path in which the $2n + 2$ nodes numbered from 0 to $2n + 1$ appear in the increasing order; $\overline{H_n}$ is a path in which the $2n + 2$ nodes numbered from 0 to $2n + 1$ appear so that all the even-numbered nodes appear first, in the decreasing order, followed by all the odd-numbered nodes, in the increasing order. See Figure 5.52 for an illustration of $\overline{G_3}$ and $\overline{H_3}$.

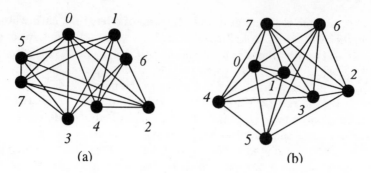

Figure 5.51: A tedious case of isomorphism verification: (a) G_3; (b) H_3

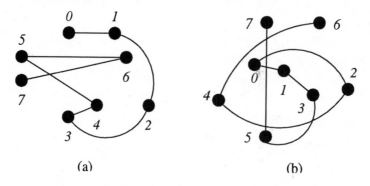

Figure 5.52: Evidently isomorphic: (a) $\overline{G_3}$; (b) $\overline{H_3}$

The base case follows from the base case of the recursive definition of G_n and H_n, which leaves the edge sets of G_0 and H_0 empty, forcing $\overline{G_0}$ and $\overline{H_0}$ to be isomorphic to a length-1 path between nodes 0 and 1.

For the inductive step, consider G_n first and assume that G_n contains an edge between any pair of its nodes, except for the edges that belong to the path $0, 1, \ldots, 2n, 2n + 1$, which comprise the edge set of $\overline{G_n}$. We have to prove that G_{n+1} contains an edge between any pair of its nodes, except for the edges that belong to the path $0, 1, \ldots, 2n, 2n + 1, 2n + 2, 2n + 3$. Indeed, by the recursive step of the definition of G, graph G_{n+1} contains G_n as a subgraph, and thereby contains all of its edges. Additionally, G_{n+1} contains two new nodes and exactly those edges that connect these nodes to every other node of G_{n+1}, except for the edges $(2n + 1, 2n + 2)$ and $(2n + 2, 2n + 3)$, whence the claim.

To prove the inductive step of the claim for H_n, assume that H_n contains an edge between any pair of its nodes, except for the edges that belong to the path $2n, 2n - 2, \ldots, 0, 1, \ldots, 2n - 1, 2n + 1$, which comprise the edge set of $\overline{H_n}$. We have to prove that H_{n+1} contains an edge between any pair of its nodes, except for the edges of the path $2n + 2, 2n, 2n - 2, \ldots, 0, 1, \ldots, 2n - 1, 2n + 1, 2n + 3$. Indeed, by the recursive step of the definition of H, graph H_{n+1} contains H_n as a subgraph, and thereby contains all of its edges. Additionally, H_{n+1} contains two new nodes and exactly those edges that connect these nodes to every other node of H_{n+1}, except for the edges $(2n + 2, 2n)$ and $(2n + 1, 2n + 3)$, whence the claim.

Problem 402 G_n is a family of undirected graphs defined recursively as follows.

G_1 is a 3-node cycle whose nodes are: 0, 1, and 2.

For $n \geq 1$, G_{n+1} is obtained from G_n as follows. First, remove the edge $(2n-1, 2n)$ from G_n. Next, adjoin to G_n two new nodes: $2n+1$ and $2n+2$, as well as the edge $(2n-1, 2n+1)$ and the cycle $(2n, 2n+1, 2n+2)$.

(a) Write a formal recursive definition of the graph family G_n.

Answer: Let $G_n = (V_n, E_n)$, for $n \geq 1$.

$$V_1 = \{0, 1, 2\}$$
$$E_1 = \{(0,1), (1,2), (2,0)\}$$

For $n \geq 1$:

$$V_{n+1} = V_n \cup \{2n+1, 2n+2\}$$
$$E_{n+1} = (E_n \setminus \{(2n-1, 2n)\}) \cup \{(2n-1, 2n+1), (2n, 2n+1), (2n+1, 2n+2), (2n+2, 2n)\}$$

(b) Draw G_1, G_2 and G_3.

Answer: See Figure 5.53.

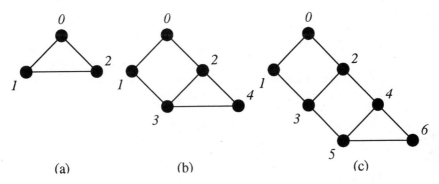

Figure 5.53: (a) G_1; (b) G_2; (c) G_3

Problem 403   Consider the family G_n of undirected graphs defined in Problem 402.

Prove that the k-node cycle C_k is a subgraph of graph G_n for every k such that $3 \leq k \leq 2n+1$.

Answer: We use induction to prove the claim with an additional constraint. Precisely, our claim is as follows. Whenever k is odd, G_n contains as a subgraph a k-node cycle that contains edge $(2n-1, 2n)$; whenever k is even, G_n contains as a subgraph a k-node cycle that contains edge $(2n-2, 2n-1)$ but does not contain edge $(2n-2, 2n)$, for any k such that $3 \leq k \leq 2n+1$.

Since G_1 is itself a 3-cycle (evidently compliant with the claim), better insight can be gained by considering G_2 for the base case. Observe that G_2 contains the 3-cycle $(2,3,4)$, and the 5-cycle $(0,1,3,4,2)$; each of these odd-length cycles indeed contains the edge $(3,4)$. Furthermore, G_2 contains the 4-cycle $(0,2,3,1)$, which indeed contains edge $(2,3)$, but not $(2,4)$.

For the inductive step, assume that for some $n \geq 2$, graph G_n contains as a subgraph a cycle of every length (not exceeding $2n+1$), such that every such designated odd-length cycle contains edge $(2n-1, 2n)$, while every such designated even-length cycle contains edge $(2n-2, 2n-1)$ but not $(2n-2, 2n)$. We have to prove that whenever k is odd, G_{n+1} contains as a subgraph a k-node cycle that contains edge $(2n + 1, 2n + 2)$; whenever k is even, G_{n+1} contains as a subgraph a k-node cycle that contains edge $(2n, 2n + 1)$ but not $(2n, 2n + 2)$, for any k such that $3 \leq k \leq 2n + 3$.

Consider the odd-length cycles first. Let C_k be an odd-length cycle in G_n that contains edge $(2n - 1, 2n)$. Note that, by the recursive step of the graph definition, G_{n+1} contains all the edges of G_n, except for the edge $(2n - 1, 2n)$. Hence, G_{n+1} contains the cycle C_k, except for the edge $(2n - 1, 2n)$, and C_k is open in G_{n+1} at nodes $2n - 1$ and $2n$. Construct a cycle C_{k+2} in G_{n+1} from the $k - 1$ inherited edges of C_k, closing the cycle with the length-3 path $(2n - 1, 2n + 1, 2n + 2, 2n)$. This path exists, by the recursive step of the graph definition, and none of its three edges may already appear in C_k, since each of these edges is incident to at least one new node. The resulting cycle C_{k+2} indeed contains edge $(2n + 1, 2n + 2)$, as claimed. Moreover, since k is odd, $k + 2$ is also odd. We conclude that our construction accounts for all cycles of odd length $k + 2$ where $3 \leq k \leq 2n + 1$, or equivalently, for all cycles of odd length k such that $5 \leq k \leq 2n + 3$. To complete the proof of the claim, we have to verify the existence of a cycle C_3 that contains the required edge $(2n + 1, 2n + 2)$. This cycle is indeed formed by the edges $(2n, 2n + 1)$, $(2n + 1, 2n + 2)$, $(2n + 2, 2n)$, present in G_{n+1} by the recursive step of the graph definition.

The argument for even-length cycles is analogous to that given for odd-length cycles. Each such even-length cycle C_k in G_n, where $4 \leq k \leq 2n$, generates an even-length cycle C_{k+2} in G_{n+1}. To construct the cycle C_{k+2}, first eliminate from C_k edge $(2n - 2, 2n - 1)$, which is contained in C_k by the inductive hypothesis. Next, close the cycle by the length-3 path $(2n - 1, 2n + 1, 2n, 2n - 2)$. This path exists, by the recursive step of the graph definition (applied to G_{n+1} and G_n.) Moreover, none of the three new edges may already appear in C_k: edge $(2n - 2, 2n)$ is excluded by the inductive hypothesis, while each of the other two edges is incident to the new node $2n + 1$. Note that the new cycle contains the required edge $(2n, 2n + 1)$, but not $(2n, 2n + 2)$. Having accounted for all the cycles of length 6 or more, we complete the proof by an observation that the edge eliminated in the construction of C_{k+2} and the length-3 path employed in its place together form a 4-cycle: $(2n - 2, 2n - 1, 2n + 1, 2n)$, which indeed contains the required edge $(2n, 2n + 1)$.

Problem 404   Construct a family of undirected rooted trees T_h, such that, for every $h \geq 0$, the height of the tree T_h is equal to h, and the number of nodes $N_T(h)$ of T_h is $\Omega\left(2^h\right)$. Prove that your construction is correct.

Answer: The complete binary tree of height h is defined recursively as follows. The nodes of T_h are binary strings of length not exceeding h. In the base case, for $h = 0$, tree T_0 is an isolated node named by the empty string. For the recursive step, to obtain T_{h+1}, form a union of two copies of T_h, append the bit 0 as a prefix to the name of each node in one of the copies, and append the bit 1 as a prefix to the name of each node in the other copy. Finally, introduce a new node, named by the empty string, as the root of T_{h+1}, and introduce two new edges connecting the root to nodes 0 and 1 (the former roots of the two copies of T_h.) See Figure 5.54 for an illustration of the recursive construction of the family of complete binary trees.

We have to prove that T_h is indeed a height-h tree, with the required number of nodes. Precisely, we use induction to prove that, for every $h \geq 0$, T_h is a connected and acyclic graph with $N_T(h) = 2^{h+1} - 1$ nodes, in which the maximum distance between the root and any other node is equal to h.

The base case occurs when $h = 0$. By the base case of the graph definition, T_0 comprises an isolated node, and thereby vacuously has no cycles, no disconnected pairs of nodes, and no distances greater than 0. Furthermore, its number of nodes is:

$$N_T(0) = 1 = 2^{0+1} - 1$$

whence the claim for the base case.

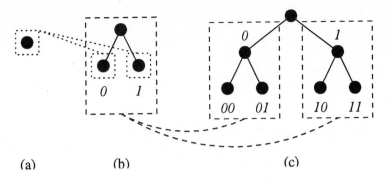

Figure 5.54: (a) T_0; (b) T_1; (c) T_2

For the recursive step, assume that for some $h \geq 0$, T_h is a connected and acyclic graph with $N_T(h) = 2^{h+1} - 1$ nodes and the maximum distance equal to h between the root and any other node. We have to prove that T_{h+1} is a connected and acyclic graph with $2^{h+2} - 1$ nodes and the maximum distance equal to $h + 1$ between the root and any other node.

To prove that T_{h+1} is connected, consider two arbitrary nodes of T_{h+1}, say x and y. We have to show that there exists a path in T_{h+1} between x and y. Recall that, by the recursive step of the graph definition, T_{h+1} is obtained by connecting two copies of T_h with a new root node. If nodes x and y belong to the same copy of T_h, then they are connected by a path within that copy, by the inductive hypothesis. If x and y belong to different copies of T_h, then, by the inductive hypothesis, each of them is connected to the former root of its copy, while the two former roots are connected in T_{h+1} via the new root and its adjacent edges, by the recursive step of the graph definition. If one of the nodes x, y, say x, is the new root, then the path between x and y consists of the new edge between x and the former root of the copy of T_h which contains y, followed by the path between this former root and y, which exists in T_h, by the inductive hypothesis.

To prove that T_{h+1} is acyclic, assume that there exists a cycle in T_{h+1}. By the inductive hypothesis, this cycle cannot be entirely contained within any one of the two copies of T_h. However, by the recursive step of the graph definition, each of the two new edges is a cut edge of T_{h+1}. Hence, any cycle that would exit one copy of T_h would have no way to re-enter it.

To prove that the height of T_h is equal to h, consider an arbitrary node x (other than root) of T_{h+1}. By the recursive step of the graph definition, x has to belong to one of the two copies of T_h. By the inductive hypothesis, there exists a path of length not exceeding h from x to the former root of its copy of T_h. The new edge incident to this former root extends the path to the (new) root of T_{h+1}, incrementing its length by one. Hence, the maximum length of such a path does not exceed $h + 1$.

Finally, by the recursive step of the graph definition, the node set of T_{h+1} contains two node sets of T_h and an additional node. By the inductive hypothesis, each copy of T_h has $N_T(h) = 2^{h+1} - 1$ nodes, whence the count for the number of nodes of T_{h+1}:

$$N_T(h + 1) = 2N_T(h) + 1 = 2\left(2^{h+1} - 1\right) + 1 = 2^{h+2} - 2 + 1 = 2^{h+2} - 1$$

Indeed, $N_T(h) = \Omega\left(2^h\right)$

Problem 405   Construct a family of undirected rooted trees P_h, such that, for every $h \geq 0$, the height of the tree P_h is equal to h, and the number of nodes $N_P(h)$ of P_h is $O(h)$. Prove that your construction is correct.

Answer: These requirements are satisfied by a length-h path. If one endpoint of the path is designated as the root, then the height of this tree is equal to h, while the number of its nodes is equal to $N_P(h) = h + 1$. See Figure 5.55.

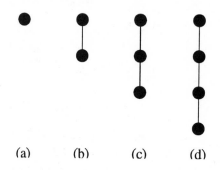

(a)   (b)   (c)   (d)

Figure 5.55: (a) P_0; (b) P_1; (c) P_2; (d) P_3

Problem 406   Construct a family of undirected rooted trees S_h, such that, for every $h \geq 0$, the height of the tree S_h is equal to h, and the number of nodes $N_S(h)$ of S_h is $\Theta(h^2)$. Prove that your construction is correct.

Answer: The family S_h is defined recursively as follows. In the base case, for $h = 0$, tree S_0 is an isolated node. For the recursive step, to obtain S_{h+1}, form a union of S_h and P_h, where P_h is the family of trees defined in the answer to Problem 405. Finally, introduce a new node as the root of S_{h+1}, and introduce two new edges connecting the root to the former roots of S_h and P_h. See Figure 5.56.

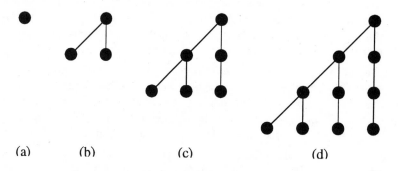

(a)   (b)   (c)   (d)

Figure 5.56: (a) S_0; (b) S_1; (c) S_2; (d) S_3

We have to prove that S_h is indeed a height-h tree, with the required number of nodes. Precisely, we prove that, for every $h \geq 0$, S_h is a connected and acyclic graph with $N_S(h) = (h+1)(h+2)/2$ nodes, in which the maximum distance between the root and any other node is equal to h.

The proof of the claim that S_h is a tree (connected and acyclic graph) of height equal to h is identical to the proof of the corresponding claims given in the answer to Problem 404, after adjusting the names of the component graphs. We use induction to prove the claim for the number of nodes $N_S(h)$ of the tree S_h.

The base case occurs when $h = 0$. By the base case of the graph definition, S_0 comprises an isolated node. Hence, its number of nodes is:

$$N_S(0) = 1 = \frac{(0+1)(0+2)}{2}$$

whence the claim for the base case.

For the recursive step, assume that for some $h \geq 0$, S_h has $N_S(h) = (h+1)(h+2)/2$ nodes. We have to prove that S_{h+1} has $(h+2)(h+3)/2$ nodes.

By the recursive step of the graph definition, the node set of S_{h+1} contains the node set of S_h, the node set of P_h, and an additional node. By the inductive hypothesis, the copy of S_h has $N_S(h) = (h+1)(h+2)/2$ nodes. Moreover, the path P_h contains $N_P(h) = h+1$ nodes, whence the count for the number of nodes of S_{h+1}:

$$\begin{aligned} N_S(h+1) \quad &= N_S(h) + N_P(h) + 1 = \frac{(h+1)(h+2)}{2} + (h+1) + 1 = \frac{(h+1)(h+2)}{2} + (h+2) \\ &= \frac{(h+1)(h+2) + 2(h+2)}{2} = \frac{(h+2)(h+1+2)}{2} = \frac{(h+2)(h+3)}{2} \end{aligned}$$

Indeed, $N_S(h) = \Theta\left(h^2\right)$

Problem 407 Construct a family of undirected rooted trees R_h, such that, for every $h \geq 0$, the height of the tree R_h is equal to h, and the number of nodes $N_R(h)$ of R_h is $\Theta\left(h^3\right)$. Prove that your construction is correct.

Answer: The family R_h is defined recursively as follows. In the base case, for $h = 0$, tree R_0 is an isolated node. For the recursive step, to obtain R_{h+1}, form a union of R_h and S_h, where S_h is the family of trees defined in the answer to Problem 406. Finally, introduce a new node as the root of R_{h+1}, and introduce two new edges connecting the root to the former roots of R_h and S_h. See Figure 5.57.

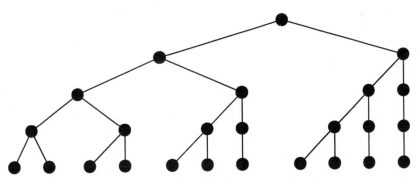

Figure 5.57: R_4

We have to prove that R_h is indeed a height-h tree, with the required number of nodes. The proof of the claim that R_h is a tree (connected and acyclic graph) of height equal to h is identical to the proof of the corresponding claims given in the answer to Problem 404, after adjusting the names of the component graphs.

To calculate the number of nodes $N_R(h)$ of R_h, observe that the node set of R_{h+1} contains the node set of R_h, the node set of S_h, and an additional node. Hence, $N_R(h)$ is governed by the following recurrence.

$$N_R(0) = 1$$
$$N_R(h+1) = N_R(h) + 1 + N_S(h)$$

This recurrence leads to the following closed form.

$$
\begin{aligned}
N_R(h+1) \; &= N_R(0) + \sum_{k=0}^{h}(N_S(k)+1) = 1 + \sum_{k=0}^{h}\left(\frac{(k+1)(k+2)}{2}+1\right) \\
&= 1 + \sum_{k=0}^{h}\left(\frac{k^2}{2}+\frac{3k}{2}+\frac{2}{2}+1\right) = 1 + \frac{1}{2}\left(\sum_{k=0}^{h}k^2\right)+\frac{3}{2}\left(\sum_{k=0}^{h}k\right)+\sum_{k=0}^{h}2 \\
&= 1 + \left(\frac{h(h+1)(2h+1)}{12}\right)+\left(\frac{3h(h+1)}{4}\right)+2(h+1) \\
&= 1 + \left(\frac{h(h+1)}{12}\right)(2h+1+9)+2(h+1) = 3+2h+\frac{h(h+1)(h+5)}{6}
\end{aligned}
$$

Indeed, $N_R(h) = \Theta\left(h^3\right)$.

Problem 408 $A_n = (V_n, E_n)$ is a family of undirected graphs defined recursively as follows. Nodes of A_n are tuples of length n whose components are integers:

$$V_n \subseteq Z^n$$

where $Z = \{\ldots, -2, -1, 0, 1, 2, \ldots\}$ is the set of integers.

A_1 is an isolated node, labeled with $\langle 0 \rangle$.

For $n > 1$, A_n is constructed as a union of two instances of A_{n-1} and a path with $2n+1$ nodes as follows. Let $A'_{n-1} = (V'_{n-1}, E'_{n-1})$ and $A''_{n-1} = (V''_{n-1}, E''_{n-1})$ be obtained from A_{n-1} by a node renaming that appends a number at the head of each tuple as follows. Whenever:

$$\langle x_{n-1}, \ldots, x_1 \rangle \in V_{n-1}$$

we have:

$$\langle -n, x_{n-1}, \ldots, x_1 \rangle \in V'_{n-1} \wedge \langle n, x_{n-1}, \ldots, x_1 \rangle \in V''_{n-1}$$

while the edges are preserved. Precisely, whenever:

$$(\langle x_{n-1}, \ldots, x_1 \rangle, \langle y_{n-1}, \ldots, y_1 \rangle) \in E_{n-1}$$

we have:

$$(\langle -n, x_{n-1}, \ldots, x_1 \rangle, \langle -n, y_{n-1}, \ldots, y_1 \rangle) \in E'_{n-1}$$
$$\wedge$$
$$(\langle n, x_{n-1}, \ldots, x_1 \rangle, \langle n, y_{n-1}, \ldots, y_1 \rangle) \in E''_{n-1}$$

Additionally, let S_n be a set of $2n-1$ nodes labeled with n-tuples whose last $n-1$ components are all equal to 0 and the first component is an integer, as follows.

$$S_n = \{\, \langle \pm k, 0^{n-1} \rangle \mid 0 \le k < n \,\}$$

Let M_n be the edge set of a path that connects the node $\langle -n, 0^{n-1} \rangle$ of A'_{n-1} with the node $\langle n, 0^{n-1} \rangle$ of A''_{n-1} via the nodes of S_n, encountered in the increasing order of first components:

$$M_n = \{(\langle k, 0^{n-1} \rangle, \langle k+1, 0^{n-1} \rangle) \mid -n \le k < n\}$$

Finally, the node set V_n of A_n is:

$$V_n = V'_{n-1} \cup V''_{n-1} \cup S_n$$

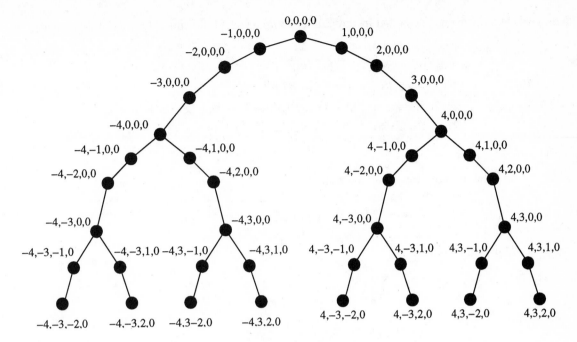

Figure 5.58: A_4

while its edge set contains the edges inherited in the two copies of A_{n-1}, as well as the path that connects them:

$$E_n = E'_{n-1} \cup E''_{n-1} \cup M_n$$

Draw A_4.

Answer: See Figure 5.58.

Problem 409 Consider the family $A_n = (V_n, E_n)$ of undirected graphs defined in Problem 408. Let v_n be the number of nodes of A_n.

(a) Write a recurrence relation with the initial conditions for v_n.

Answer: The recursive definition of the graph family yields the following recurrence.

$$
\begin{aligned}
v_1 &= 1 \\
v_n &= 2v_{n-1} + 2n - 1, \quad \text{for } n > 1
\end{aligned}
$$

(b) Recall the following identity:

$$\sum_{k=0}^{\ell} k \cdot 2^k = (\ell - 1) \cdot 2^{\ell+1} + 2$$

and employ it to obtain the closed form for v_n.

Answer: The recurrence obtained in the part **(a)** implies the following sequence of recurrences:

$$
\begin{aligned}
v_n = \ & 2v_{n-1} + 2(n-0) - 1 = \\
& 2\left(2v_{n-2} + 2(n-1) - 1\right) + (2(n-0) - 1) = \\
& 2^2 v_{n-2} + 2^1(2(n-1) - 1) + 2^0(2(n-0) - 1) = \\
& 2^2\left(2v_{n-3} + 2(n-2) - 1\right) + 2^1(2(n-1) - 1) + 2^0(2(n-0) - 1) = \\
& 2^3 v_{n-3} + 2^2(2(n-2) - 1) + 2^1(2(n-1) - 1) + 2^0(2(n-0) - 1) = \\
& \vdots
\end{aligned}
$$

which implies the following expression for v_n:

$$
v_n = 2^{n-1} \cdot v_1 + \sum_{k=0}^{n-2} \left(2^k \cdot (2(n-k) - 1)\right)
$$

which is evaluated as follows.

$$
\begin{aligned}
v_n = \ & 2^{n-1} \cdot 1 + \sum_{k=0}^{n-2} \left(2^k \cdot (2(n-k) - 1)\right) = \\
& 2^{n-1} + \sum_{k=0}^{n-2} \left(n \cdot 2^{k+1} - k \cdot 2^{k+1} - 2^k\right) = \\
& 2^{n-1} + 2n \left(\sum_{k=0}^{n-2} 2^k\right) - 2\left(\sum_{k=0}^{n-2} (k \cdot 2^k)\right) - \sum_{k=0}^{n-2} 2^k = \\
& 2^{n-1} + 2n\left(2^{n-1} - 1\right) - 2\left((n-3)\cdot 2^{n-1} + 2\right) - \left(2^{n-1} - 1\right) = \\
& 2^{n-1} + n \cdot 2^n - 2n - n \cdot 2^n + 6 \cdot 2^{n-1} - 4 - 2^{n-1} + 1 = \\
& 2^{n+1} + 2^n - 2n - 3
\end{aligned}
$$

(c) Prove that A_n is a tree.

Answer: We prove that A_n is connected by induction on n. The claim is true for A_1 vacuously. Inductively, assuming that A_{n-1} is connected for some $n > 1$, we conclude that A_n is also connected, since it is obtained by connecting two copies of A_{n-1}, as specified in the recursive step of the graph definition.

Recall that a connected graph is a tree if the number of its edges is one less than the number of its nodes. We use induction to prove that the number of edges of A_n, say e_n, is one less than the number of its nodes v_n :

$$
e_n = v_n - 1 = 2^{n+1} + 2^n - 2n - 4
$$

The base case occurs for $n = 1$. By definition, A_1 has no edges. Indeed:

$$
e_1 = 2^{1+1} + 2^1 - 2 \cdot 1 - 4 = 4 + 2 - 2 - 4 = 0
$$

Inductively, assume that the number of edges of A_n is given by e_n, for some $n > 0$. We have to prove that the number of edges of A_{n+1} is equal to e_{n+1}. By the recursive step of the graph definition, A_{n+1} inherits all the edges of two copies of A_n, in addition to the new edges of a $(2n+3)$-node path—precisely $2n + 2$ of them. Hence, the number of edges of A_{n+1} is equal to:

$$
\begin{aligned}
2e_n + (2n+2) \ &= 2\left(2^{n+1} + 2^n - 2n - 4\right) + (2n+2) = 2^{n+2} + 2^{n+1} - 4n - 8 + 2n + 2 \\
&= 2^{(n+1)+1} + 2^{n+1} - 2(n+1) - 4 \\
&= e_{n+1}
\end{aligned}
$$

Problem 410   Consider the family $A_n = (V_n, E_n)$ of undirected trees defined in Problem 408 and analyzed in Problem 409.

(a) Calculate the radius of A_n as a function of n.

Answer: We use induction to prove that the radius of A_n is given by the sequence:

$$r_n = \frac{n(n+1)}{2} - 1$$

for all $n \geq 1$, with node $\langle 0^n \rangle$ as a center. Precisely, we show that there exists a path in A_n of length not exceeding $(n(n+1)/2) - 1$ between the node $\langle 0^n \rangle$ and any other node. Moreover, the distance from any of the two nodes $\langle \pm n, n-1, n-2, \ldots, 2, 0 \rangle$ to the node $\langle 0^n \rangle$ is exactly equal to $(n(n+1)/2) - 1$.

The radius of A_1 is equal to 0, vacuously, which is in turn equal to $r_0 = (1 \cdot (1+1)/2) - 1$. Better insight may be gained by considering A_2 as the base case. The claim is verified for A_2 by inspection (Figure 5.58), which reveals that all nodes of A_2 are found within a distance not exceeding 2 from the node $\langle 0, 0 \rangle$; moreover, the distance of 2 is attained by nodes $\langle -2, 0 \rangle$ and $\langle 2, 0 \rangle$. Indeed:

$$r_2 = \frac{2(2+1)}{2} - 1 = 2$$

Inductively, assume that all nodes of A_{n-1} are found within a distance not exceeding r_{n-1} from the node $\langle 0^{n-1} \rangle$, and that the distance of r_{n-1} is attained by nodes $\langle \pm(n-1), n-2, n-3, \ldots, 2, 0 \rangle$, for some $n > 2$. We have to prove that all nodes of A_n are found within a distance not exceeding r_n from the node $\langle 0^n \rangle$, and that the distance of r_n is attained by nodes $\langle \pm n, n-1, n-2, n-3, \ldots, 2, 0 \rangle$.

Recall the construction of A_n, and consider a node, say $x \in V'_{n-1} \cup V''_{n-1}$, which belongs to any of the two copies of A_{n-1} inherited by A_n. By the inductive hypothesis, node x is connected to one of the nodes $\langle \pm n, 0^{n-1} \rangle$ of A_n (the former node $\langle 0^{n-1} \rangle$ of the corresponding copy of A_{n-1}) by a path whose length does not exceed r_{n-1}. Furthermore, the node $\langle \pm n, 0^{n-1} \rangle$ is connected to the node $\langle 0^n \rangle$ by the path of length equal to n, consisting of new edges of A_n. Assuming, for definiteness, that $x \in V'_{n-1}$, we conclude that there exists a path:

$$x \to \langle -n, 0^{n-1} \rangle \to \langle 0^n \rangle$$

whose length is not greater than:

$$r_{n-1} + n = \left(\frac{(n-1) \cdot n}{2} - 1 \right) + n = \frac{n(n-1) - 2 + 2n}{2} = \frac{n^2 + n - 2}{2} = \frac{n(n+1)}{2} - 1 = r_n$$

By the inductive hypothesis, the length of r_{n-1} is attained in the first section of the path by nodes $\langle \pm n, \pm(n-1), n-2, n-3, \ldots, 2, 0 \rangle$. Hence, the entire length of r_n is attained by the same nodes.

To verify that node $\langle 0^n \rangle$ is the only center of A_n, consider the symmetry of the graph A_n. Precisely, observe that for any legal choice of tuples $y, z \in Z^{n-1}$, the only path between nodes $s = \langle -n, y \rangle$ and $t = \langle n, z \rangle$ passes through the node $\langle 0^n \rangle$. Consider an arbitrary candidate for a center, say node c of A_n. For definiteness, assume that $c = \langle -n, w \rangle$, for some $w \in Z^{n-1}$. Now, the only path between nodes t and c passes through the node $\langle 0^n \rangle$, and is longer than the distance between nodes t and $\langle 0^n \rangle$ by the distance between nodes $\langle 0^n \rangle$ and c. If t is chosen as one of the nodes for which the distance of r_n is attained, then the distance from t to c is greater than r_n, contradicting the claim that c is a center.

(b) Consider A_n as a tree rooted at its center, which you discovered in your answer to the part **(a)**. Determine the asymptotic relationship between the height of A_n and the number of its nodes.

Answer: Since A_n is rooted at its center, node $\langle 0^n \rangle$, the height of A_n is equal to its radius. The answers given in the part **(a)** of this problem and the part **(b)** of Problem 409 provide the following closed forms for the height r_n and the number of nodes v_n of A_n.

$$r_n = \frac{n(n+1)}{2} - 1$$
$$v_n = 2^{n+1} + 2^n - 2n - 3$$

These expressions imply that:

$$r_n = \Theta(n^2) \text{ and } v_n = \Theta(2^n)$$

whence the conclusion:

$$r_n = \Theta(\lg^2 v_n)$$

Problem 411   Consider the family $A_n = (V_n, E_n)$ of undirected trees defined in Problem 408 and analyzed in Problem 409 and Problem 410.

Describe an algorithm that constructs the names of all the nodes adjacent to an arbitrary node $x = \langle x_n, x_{n-1}, \ldots, x_2, x_1 \rangle$ of A_n.

Answer: Consider the graph A_n as a tree rooted at node $\langle 0^n \rangle$.

The root $\langle 0^n \rangle$ has two children: $\langle \pm 1, 0^{n-1} \rangle$.

Given a node $x = \langle x_n, x_{n-1}, \ldots, x_2, x_1 \rangle$, such that $x \neq \langle 0^n \rangle$, let ℓ be the index of the rightmost component of x which is not equal to zero:

$$x = \langle x_n, x_{n-1}, \ldots, x_\ell, 0^{\ell-1} \rangle, \text{ where } x_\ell \neq 0$$

The nodes adjacent to $\langle x_n, x_{n-1}, \ldots, x_\ell, 0^{\ell-1} \rangle$ are determined as follows:

The parent of node $\langle x_n, x_{n-1}, \ldots, x_\ell, 0^{\ell-1} \rangle$ is node:
$\langle x_n, x_{n-1}, \ldots, x'_\ell, 0^{\ell-1} \rangle$,
where $x'_\ell = x_\ell + 1$ if $x_\ell < 0$, and $x'_\ell = x_\ell - 1$ if $x_\ell > 0$.

If $|x_\ell| < \ell$, node $\langle x_n, x_{n-1}, \ldots, x_\ell, 0^{\ell-1} \rangle$ has one child:
$\langle x_n, x_{n-1}, \ldots, x''_\ell, 0^{\ell-1} \rangle$,
where $x'_\ell = x_\ell - 1$ if $x_\ell < 0$, and $x'_\ell = x_\ell + 1$ if $x_\ell > 0$.

If $|x_\ell| = \ell$ and $\ell > 2$, node $\langle x_n, x_{n-1}, \ldots, x_\ell, 0^{\ell-1} \rangle$ has two children:
$\langle x_n, x_{n-1}, \ldots, x_\ell, 1, 0^{\ell-2} \rangle$, and $\langle x_n, x_{n-1}, \ldots, x_\ell, -1, 0^{\ell-2} \rangle$

If $|x_\ell| = \ell$ and $\ell = 2$, node $\langle x_n, x_{n-1}, \ldots, x_\ell, 0^{\ell-1} \rangle$ is a leaf.

Problem 412 $B_n = (V_n, E_n)$ is a family of undirected graphs defined recursively as follows.

Nodes of B_n are binary strings of length not exceeding n. B_0 is an isolated node, labeled with the empty string.

For $n \geq 1$, B_n is constructed as a union of two instances of B_{n-1}, an additional node, and three additional edges as follows. Let $B'_{n-1} = (V'_{n-1}, E'_{n-1})$ and $B''_{n-1} = (V''_{n-1}, E''_{n-1})$ be obtained from B_{n-1} by appending a bit at the head of each name string as follows. Whenever $x \in V_{n-1}$, we have:

$0x \in V'_{n-1}$ and $1x \in V''_{n-1}$. The renaming preserves edges. Precisely, whenever: $(x, y) \in E_{n-1}$, we have: $(0x, 0y) \in E'_{n-1}$ and $(1x, 1y) \in E''_{n-1}$.

Additionally, let λ be a new node labeled with the empty string. Let M_n be the following set of three new edges.

$$M_n = \{(0, 1), (0, \lambda), (1, \lambda)\}$$

Finally, the node set V_n and the edge set E_n of B_n are defined as follows.

$$V_n = V'_{n-1} \cup V''_{n-1} \cup \{\lambda\}$$
$$E_n = E'_{n-1} \cup E''_{n-1} \cup M_n$$

Draw B_4.

Answer: See Figure 5.59.

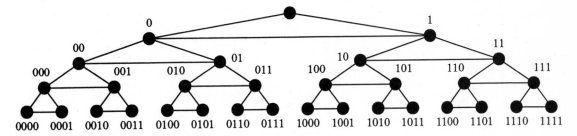

Figure 5.59: B_4

Problem 413 Consider the family $B_n = (V_n, E_n)$ of undirected graphs defined in Problem 412. Assign directions to the edges of B_n so that the resulting directed graph is strongly connected. Prove that you construction is correct.

Answer: We provide a recursive construction of the required assignment of edge directions. Our goal is to direct the edges so that there exists a directed path between any pair of nodes.

Figure 5.60: An orientation of B_1

The base case occurs when $n = 0$, and the construction is vacuously implemented by an empty assignment, since B_0 does not have any edges. Better insight may be gained by considering B_1 as the base-case instance. Since B_1 is a length-3 cycle, it can be oriented as on Figure 5.60, which corresponds to the arc set $\{(\lambda, 0), (0, 1), (1, \lambda)\}$, which indeed provides for a directed path from any of the three nodes to any other.

For the recursive step, assume that for some $n \geq 1$, the graph B_n has an assignment of edge directions that renders it strongly connected. We have to prove that B_{n+1} has an assignment of edge directions that renders it strongly connected.

Recall that B_{n+1} contains two copies of B_n, say the copy 0 and the copy 1, along with three additional edges. Let both copies of B_n retain their orientation in B_{n+1}, and let the three new edges be directed so as to form the following arcs:

$$(\lambda, 0), (0, 1), (1, \lambda)$$

To verify the assignment, consider two arbitrary nodes of B_{n+1}, say x and y. We have to show that there exists a directed path from x to y. Several cases are possible.

If x and y are in the same copy of B_n, then, by the inductive hypothesis, there exists a directed path from x to y consisting solely of the edges of that copy of B_n.

If x is in the copy 0 of B_n and y is in the copy 1, then there exists a directed path $x \to 0 \to 1 \to y$, whose initial and final sections are found within the corresponding copies of B_n by the inductive hypothesis, while the middle section consists of a new edge, directed from 0 to 1.

If x is in the copy 1 of B_n and y is in the copy 0, then there exists a directed path $x \to 1 \to \lambda \to 0 \to y$, whose initial and final sections are found within the corresponding copies of B_n by the inductive hypothesis, while the middle section consists of two new directed edges: $(1, \lambda)$ and $(\lambda, 0)$

If either x or y is the new node λ, then the required path is a subpath of a path constructed in one of the two previous cases.

Problem 414   Consider the family $B_n = (V_n, E_n)$ of undirected graphs defined in Problem 412. Assign directions to the edges of B_n so that the resulting directed graph is acyclic. Prove that your construction is correct.

Answer: We provide a recursive construction of the required assignment of edge directions. Our goal is to direct the edges of B_n so that B_n admits a topological sorting. In other words, we direct the edges of B_n so that all the nodes of B_n can be drawn on a straight horizontal line in such a way that each of its arcs is directed from left to right.

Figure 5.61: B_1: (a) Edge directions; (b) Topological sorting

The base case occurs when $n = 0$, and the construction is vacuously implemented by an empty assignment, since B_0 does not have any edges. Better insight may be gained by considering B_1 as the base-case instance. To be acyclic, B_1 should be assigned edge directions as on Figure 5.61, which corresponds to the arc set $\{(\lambda, 0), (0, 1), (\lambda, 1)\}$, which indeed admits a topological sorting, illustrated on the same figure.

For the recursive step, assume that for some $n \geq 1$, the graph B_n has a direction assignment that admits a topological sorting. We have to prove that B_{n+1} has a direction assignment that admits a topological sorting. Recall that B_{n+1} contains two copies of B_n, say the copy 0 and the copy 1, along with three additional edges. Let both copies of B_n retain their edge directions in B_{n+1}, and let the three new edges be directed so as to form the following arcs:

$$(\lambda, 0), \ (0, 1), \ (\lambda, 1)$$

To verify the assignment, construct the topological sorting of B_{n+1} as follows. (See Figure 5.62.)

1. Place the new node λ on a horizontal line.

2. Place the entire copy 0 of B_n in B_{n+1} to the right of the new node λ, according to the topological sorting of B_n, which exists by the inductive hypothesis.

3. Place the entire copy 1 of B_n in B_{n+1} to the right of the layout of copy 0, according to the topological sorting of B_n, which exists by the inductive hypothesis.

4. Lay out the three new arcs: $\{(\lambda, 0), (0, 1), (\lambda, 1)\}$.

Since the second and third steps are correct by the inductive hypothesis, we have only to argue that the three arcs laid out in the last step are directed from the left to the right. Indeed:

Node 0 is placed to the right of the new node λ, since the entire copy 0 is placed to the right of λ, whence the claim for the arc $(\lambda, 0)$.

Node 1 is placed to the right of node 0, since the entire copy 1 is placed to the right of copy 0, whence the claim for the arc $(0, 1)$.

Node 1 is placed to the right of the new node λ, since the entire copy 1 is placed to the right of λ, whence the claim for the arc $(\lambda, 1)$.

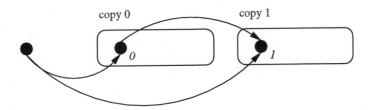

Figure 5.62: Topological sorting of B_{n+1}: two copies of B_n and the new arcs

Problem 415   Consider the family $B_n = (V_n, E_n)$ of undirected graphs defined in Problem 412. Construct an Euler tour in B_n.

Answer: We provide a recursive construction of an Euler tour in B_n. The base case occurs when $n = 0$, and holds vacuously, since B_0 does not have any edges. The next instance, B_1, is a 3-node cycle, which itself is an Euler tour. One possible sequence of nodes along the Euler tour in B_1 is:

$$R = \lambda, \ 0, \ 1, \ \lambda$$

For the recursive step, assume that B_n has an Euler tour for some $n \geq 1$. We have to construct an Euler tour in B_{n+1}.

Recall that B_{n+1} contains two copies of B_n, say the copy 0 and the copy 1, along with three additional edges. By the inductive hypothesis, each of the two copies have an Euler tour, say R_0 and R_1, respectively. List R_0 so that it begins and ends at node 0, and list R_1 so that it begins and ends at node 1. Then, the following tour is an Euler tour in B_{n+1}, which begins and ends at node λ.

$$(\lambda, 0) \to R_0 \to (0, 1) \to R_1 \to (1, \lambda)$$

Problem 416   Consider the family $B_n = (V_n, E_n)$ of undirected graphs defined in Problem 412.

(a) Prove that B_n does not have a Hamilton cycle for any n such that $n > 1$.

Answer: Assume that B_n has a Hamilton cycle, for some $n > 1$. Since the cycle has to visit both copies of B_{n-1} in B_n, there have to be two disjoint paths connecting the two copies. However, the only connection between the two copies consists of the new edges, which are incident to only one node in each copy of B_{n-1}, precisely to nodes 0 and 1. Hence, the only simple cycle that visits both copies is the cycle $\lambda, 0, 1$, which is not the Hamilton cycle of the graph B_n unless nodes 0 and 1 are the only nodes in their respective copies of B_n, which happens exactly when $n = 1$.

(b) Prove that B_n does not contain as a subgraph any (simple) cycle of length other than 3.

Answer: We prove the claim by induction on n. The base case occurs when $n = 0$, and the claim is vacuously true in this case, since B_0 does not have any edges.

For the recursive step, assume that B_{n-1} does not contain any cycles of length greater then 3. We have to prove that B_n does not contain any cycles of length greater then 3. Assume that B_n contains such a cycle, say C. By the inductive hypothesis, C is not confined to either of the two copies of B_{n-1} in B_n. Hence, C has to visit both of the two copies, or at least the new node λ, which in turn is adjacent only to nodes 0 and 1, and can belong to a cycle only together with these two nodes. However, each of the two copies of B_{n-1} is connected to the other via a single node, 0 or 1. Hence, no simple cycle that contains nodes 0 and 1 can contain any other nodes from either of the two copies of B_{n-1}. This leaves the cycle $(\lambda, 0, 1)$, formed by the new edges, as the only cycle that visits both copies of B_{n-1}. However, the length of this cycle is equal to 3, contradicting the assumption.

Problem 417 $D_n = (V_n, E_n)$ is a family of undirected graphs defined recursively as follows.

Nodes of D_n are binary strings of length not exceeding n. D_1 is a 3-node cycle, whose nodes are labeled with 0, 1, and the empty string.

For $n > 1$, D_n is constructed as a union of two instances of D_{n-1}, an additional node, and three additional edges as follows. Let $D'_{n-1} = (V'_{n-1}, E'_{n-1})$ and $D''_{n-1} = (V''_{n-1}, E''_{n-1})$ be obtained from D_{n-1} by appending one bit at the head of each name string as follows. Whenever $x \in V_{n-1}$, we have: $0x \in V'_{n-1}$ and $1x \in V''_{n-1}$. The renaming preserves edges. Precisely, whenever: $(x, y) \in E_{n-1}$, we have: $(0x, 0y) \in E'_{n-1}$ and $(1x, 1y) \in E''_{n-1}$.

Additionally, let λ be a new node, labeled with the empty string. Let M_n be the following set of three new edges.

$$M_n = \{(01, 10), (0, \lambda), (1, \lambda)\}$$

Finally, the node set V_n and the edge set E_n of D_n are defined as follows.

$$V_n = V'_{n-1} \cup V''_{n-1} \cup \{\lambda\}$$
$$E_n = E'_{n-1} \cup E''_{n-1} \cup M_n$$

Draw D_4.

Answer: See Figure 5.63.

Problem 418   Consider the family $D_n = (V_n, E_n)$ of undirected graphs defined in Problem 417.

Prove that D_n does not have an Euler path for any n such that $n > 1$.

Answer: If D_n had an Euler path, than it would have exactly two nodes whose degree is odd. However, D_2 has as many as four nodes whose degree is equal to 3, as is verified by inspection (see Figure 5.63.) The node names of two of these four nodes, precisely nodes 01 and 10, are bit strings

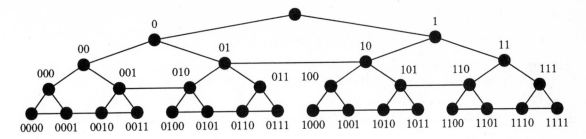

Figure 5.63: D_4

of length equal to 2. In the recursive construction of D_3 from two copies of D_2, the degree of these two nodes is not modified. To see this, recall that these two nodes are renamed in D_3 into bit strings of length equal to 3, while the new edges are incident only to nodes named as 01 and 10, leaving the degree of all other nodes intact. Hence, there are at least fours nodes in D_3, two from each copy of D_2, that have a degree equal to 3 and a name string of length equal to 3. This argument justifies the base case of the claim that D_n contains at least 2^{n-1} nodes of degree equal to 3 whose names are bit strings of length equal to n. The inductive step follows straightforwardly from the recursive construction of D_n, which puts together two identical copies of D_{n-1}. Whenever $n > 2$, the nodes whose name length is equal to n are unaffected by the new edges. Hence, D_n has at least twice as many nodes with the required degree and name length as a single copy of D_{n-1}. Since $2 \cdot 2^{n-2} = 2^{n-1}$, the claim follows.

Problem 419   Consider the family $D_n = (V_n, E_n)$ of undirected graphs defined in Problem 417. Construct a Hamilton cycle in D_n.

Answer: We provide a recursive construction of a Hamilton cycle in D_n, for all $n \geq 1$.

The base case occurs when $n = 1$. D_1, being a length-3 cycle, is identical to its Hamilton cycle.

Recursively, assume that D_{n-1} has a Hamilton cycle. Observe that this Hamilton cycle has to include both of the two edges $(\lambda, 0)$ and $(\lambda, 1)$, since this is the only way for the cycle to visit the node λ. Consequently, the copy 0 of D_{n-1} in D_n has a Hamilton cycle that contains the edge $(0, 01)$; let C_0 be the path obtained from this Hamilton cycle by removing the edge $(0, 01)$; the endpoints of C_0 are nodes 0 and 01. Likewise, the copy 1 of D_{n-1} in D_n has a Hamilton cycle that contains the edge $(1, 10)$; let C_1 be the path obtained from this Hamilton cycle by removing the edge $(1, 10)$; the endpoints of C_1 are nodes 1 and 10. Finally, connect the two paths C_0 and C_1 with the new edges, to complete the Hamilton cycle in D_n, as follows.

$$\lambda \to 0 \to C_0 \to 01 \to 10 \to C_1 \to 1 \to \lambda$$

Problem 420   Consider the family $D_n = (V_n, E_n)$ of undirected graphs defined in Problem 417. Prove that a simple cycle of any length other than 4 is a subgraph of any graph D_n that is large enough to hold it.

Answer: We use induction to prove the claim for cycles of length 5 or more—since D_1 is a 3-node cycle, inherited by all higher instances of D, it is evidently true that the length-3 cycle is a subgraph of the graph D_n for any $n \geq 1$.

Observe that the number of nodes of the graph D_n is $2^{n+1} - 1$ (identical to the number of nodes of a height-n complete binary tree.) Given any integer ℓ, define $\nu(\ell)$ as the index of the smallest instance of D that has no fewer than ℓ nodes:

$$2^{\nu(\ell)} \le \ell \le 2^{\nu(\ell)+1} - 1$$

For any $\ell \ge 5$, we provide a recursive construction of an ℓ-node cycle in $D_{\nu(\ell)}$. By the recursive definition of the graph family, all higher instances of D inherit this cycle.

For the base case, consider the explicit construction given on Figure 5.64 and Figure 5.65, for the cycles of length $\ell = 5, 6, 7$, and $\ell = 8, 9, 10$, respectively.

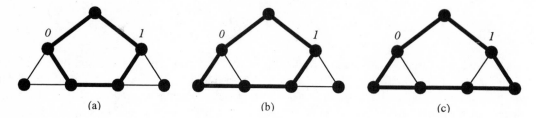

Figure 5.64: Cycles of length ℓ in D_2: (a) $\ell = 5$, (b) $\ell = 6$, (c) $\ell = 7$

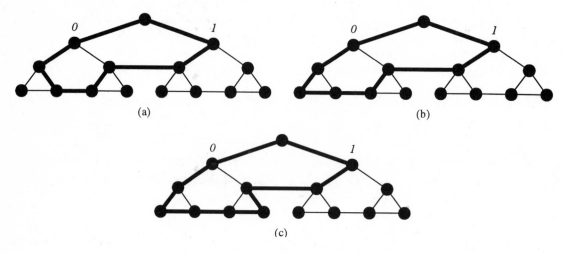

Figure 5.65: Cycles of length ℓ in D_3: (a) $\ell = 8$, (b) $\ell = 9$, (c) $\ell = 10$

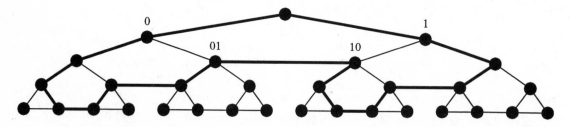

Figure 5.66: The cycle of length $17 = 8 + 8 + 1$ in D_4

For the recursive step, assume that a cycle of length m is a subgraph of the graph $D_{\nu(m)}$, for every m such that $5 \le m \le \ell$, for some $\ell \ge 10$. We have to show that the cycle of length $\ell + 1$ is a subgraph of the graph $D_{\nu(\ell+1)}$. We recognize two cases, and construct the required cycle for the two cases separately.

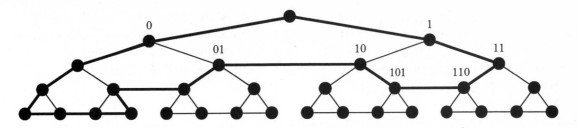

Figure 5.67: The cycle of length $16 = (2^4 - 6) + 6$ in D_4

Case (1): $\ell \neq 2^k - 1$, for any integer k.

In this case, there exists an integer $k = \nu(\ell)$ such that:

$$\ell = 2^k - p \text{ where } 1 < p \leq 2^{k-1}$$

for some integer p. Hence:

$$2^{k-1} < \ell + 1 \leq 2^k - 1$$

yielding:

$$\nu(\ell + 1) = k - 1 = \nu(\ell)$$

meaning that the cycle $C_{\ell+1}$ should be constructed within the same instance D_{k-1} that contains C_ℓ. Define ℓ_1 and ℓ_2 as follows.

$$\ell_1 = \left\lceil \frac{\ell}{2} \right\rceil = 2^{k-1} - \left\lfloor \frac{p}{2} \right\rfloor \text{ where } 1 \leq \left\lfloor \frac{p}{2} \right\rfloor \leq 2^{k-2}$$

$$\ell_2 = \left\lfloor \frac{\ell}{2} \right\rfloor = 2^{k-1} - \left\lceil \frac{p}{2} \right\rceil \text{ where } 1 \leq \left\lceil \frac{p}{2} \right\rceil \leq 2^{k-2}$$

meaning:

$$\nu(\ell_1) = \nu(\ell_2) = k - 2 = \nu(\ell + 1) - 1$$

Since $\ell + 1 = \ell_1 + \ell_2 + 1$, we proceed to construct a cycle of length $\ell + 1$ in D_{k-1}, by composing two cycles of length ℓ_1 and ℓ_2 in the copies 0 and 1 of D_{k-2} in D_{k-1}, respectively, and including the new node λ. By the inductive hypothesis, such cycles, say C_0 in the copy 0, and C_1 in the copy 1 exist. Furthermore, each of these cycles visits the node λ in its copy of D_{k-2}, or else there would be a smaller instance of D containing these cycles, contradicting the choice of k. After the renaming for D_{k-1}, we conclude that C_0 visits the path $(00, 0, 01)$, while C_1 visits the path $(10, 1, 11)$. Let P_0 be the ℓ_1-node path obtained from C_0 by removing the edge $(0, 01)$; let P_1 be the ℓ_2-node path obtained from C_1 by removing the edge $(1, 10)$. Then a cycle of length $\ell + 1 = \ell_1 + \ell_2 + 1$ is obtained as follows:

$$\lambda \to 0 \to P_0 \to 01 \to 10 \to P_1 \to 1 \to \lambda$$

See Figure 5.66 for an illustration of the construction in a case when $k = 4$.

Case (2): $\ell = 2^k - 1$, for some integer k.

In this case, $\ell + 1 = 2^k$, implying that $\nu(\ell + 1) = k > k - 1 = \nu(\ell)$, and the cycle $C_{\ell+1}$ has to be constructed in the instance D_k—the next larger instance after D_{k-1}, which contains C_ℓ. Furthermore, the construction employed in the previous case is not applicable here. To see this, observe that the previous construction would start with two cycles of length $\ell_1 = 2^{k-1}$ and $\ell_2 = 2^{k-1} - 1$, each within one copy of D_{k-1}. However, $\nu(\ell_2) = k - 2$, meaning that a cycle of length ℓ_2 can be found in D_{k-2}, whereas our previous construction essentially relies on the assumption that each of the two component cycles is a subgraph of the smallest instance of D that contains it.

Define ℓ_1 and ℓ_2 as follows.

$$\ell_1 = 2^k - 6,$$
$$\ell_2 = 5$$

Since $\ell \geq 10$, it is possible that $\ell = 2^k - 1$ only if $k \geq 4$, which implies that $2^{k-1} \geq 8$. Hence:

$$\ell_1 = 2^k - 6 < 2^{(k-1)+1} - 1$$
$$\ell_1 = 2^k - 6 = 2 \cdot 2^{k-1} - 6 = 2^{k-1} + (2^{k-1} - 6) \geq 2^{k-1} + (8 - 6) > 2^{k-1}$$

yielding:

$$\nu(\ell_1) = k - 1$$

Furthermore, $k \geq 4$ implies that:

$$\ell_1 = 2^k - 6 \geq 16 - 6 = 10$$

Hence, the inductive hypothesis applies to ℓ_1, guaranteeing that D_{k-1} contains as a subgraph a cycle of length ℓ_1. Furthermore, this cycle visits both copies of D_{k-2} in D_{k-1}—otherwise, an ℓ_1-node cycle could be found in a smaller instance of D, contradicting the conclusion that $\nu(\ell_1) = k - 1$.

To construct a cycle of length $\ell + 1$ in D_k, employ two copies of D_{k-1}. Let C_0 be the cycle of length ℓ_1, constructed in the copy 0 of D_{k-1}, by the inductive hypothesis. Recall that C_0 visits node λ in its copy of D_{k-1}; after the renaming for D_k, we conclude that C_0 visits the path $(00, 0, 01)$. Let P_0 be the ℓ_1-node path obtained from C_0 by removing the edge $(0, 01)$. Since $\ell + 1 = \ell_1 + 6$, the construction is completed by extending the path P_0 into the copy 1 of D_{k-1} in D_k, so as to close it via six additional nodes as follows.

$$0 \rightarrow \lambda \rightarrow 1 \rightarrow 11 \rightarrow 110 \rightarrow 101 \rightarrow 10 \rightarrow 01 \rightarrow P_0 \rightarrow 0$$

See Figure 5.67 for an illustration of the construction in the case when $k = 4$.

Problem 421   Construct an Euler tour in the hypercube, or prove that this is impossible.

Answer: The family Q_n of n-dimensional hypercubes is defined recursively as follows. The nodes of Q_n are binary strings of length equal to n. In the base case, for $n = 0$, the 0-dimensional hypercube Q_0 is an isolated node named by the empty string. For the recursive step, to obtain Q_{n+1}, form a union of two copies of Q_n and introduce an edge between each node in one copy of Q_n and the node with the same name in the other copy. Finally, append the bit 0 as a prefix to the name of each node in one of the copies, and append the bit 1 as a prefix to the name of each node in the other copy.

Since the n-dimensional hypercube Q_n is a regular graph of degree n, it has an Euler tour exactly when n is even. We provide a recursive construction of an Euler tour in the hypercube Q_{2m} for all $m \geq 1$.

The base case occurs when $m = 1$. The 2-dimensional hypercube Q_2 is isomorphic to a 4-node cycle, which itself is an Euler tour. One possible sequence of nodes along the Euler tour in Q_2 is:

$$R = 00, 01, 11, 10, 00$$

For the recursive step, assume that for some $m \geq 1$, the $2m$-dimensional hypercube Q_{2m} has an Euler tour with the node sequence:

$$x_1, x_2, \ldots, x_e, x_1$$

where $e = m \cdot 2^{2m}$ is the number of edges of the hypercube Q_{2m}, and each x_i, for $1 \leq i \leq e$, is a bit string of length equal to $2m$. We have to prove that the $(2m + 2)$-dimensional hypercube Q_{2m+2} has an Euler tour.

By the recursive step of the graph definition, the hypercube Q_{2m+2} is isomorphic to the union of four copies of the hypercube Q_{2m}, with additional edges that connect into a cycle each quartet of equal-named nodes in the four copies. Precisely, the node names of Q_{2m+2} are obtained by appending a distinct 2-bit prefix to all nodes in each of the four copies of Q_{2m}. For each bit string x of length equal to $2m$ (or, equivalently, for each node in any of the four copies of Q_{2m}) the additional edges form the 4-node cycle:

$$C(x) = 10x, 00x, 01x, 11x, 10x$$

By the inductive hypothesis, each of the four copies of Q_{2m} has an Euler tour:

$$R_{\alpha\beta} = \alpha\beta x_1, \alpha\beta x_2, \alpha\beta x_3, \ldots, \alpha\beta x_e, \alpha\beta x_1$$

where $\alpha\beta$ is the 2-bit prefix appended to that copy of Q_{2m}.

To construct an Euler tour R' in Q_{2m+2}, we compose the four Euler tours which, by the inductive hypothesis, exist in the four copies of Q_{2m} and modify (any) one of these tours to include the 4-node cycle C_x of additional edges, upon visiting every node x of that copy Q_{2m}. Precisely, the Euler tour in Q_{2m+2} is defined by the following node sequence:

$$
\begin{aligned}
R' \;=\; & 00x_1, \quad 00x_2, 00x_3, \ldots, 00x_{e-1}, 00x_e, 00x_1, \\
& 01x_1, \quad 01x_2, 01x_3, \ldots, 01x_{e-1}, 01x_e, 01x_1, \\
& 11x_1, \quad 11x_2, 11x_3, \ldots, 11x_{e-1}, 11x_e, 11x_1, \\
& 10x_1, \quad C(x_2), C(x_3), \ldots, C(x_{e-1}), C(x_e), 10x_1, \\
& 00x_1 \\
=\; & 00x_1, \quad 00x_2, 00x_3, \ldots, 00x_{e-1}, 00x_e, 00x_1, \\
& 01x_1, \quad 01x_2, 01x_3, \ldots, 01x_{e-1}, 01x_e, 01x_1, \\
& 11x_1, \quad 11x_2, 11x_3, \ldots, 11x_{e-1}, 11x_e, 11x_1, \\
& 10x_1, \quad (10x_2, 00x_2, 01x_2, 11x_2, 10x_2), \\
& \qquad\quad\ (10x_3, 00x_3, 01x_3, 11x_3, 10x_3), \ldots, \\
& \qquad\quad\ (10x_{e-1}, 00x_{e-1}, 01x_{e-1}, 11x_{e-1}, 10x_{e-1}), \\
& \qquad\quad\ (10x_e, 00x_e, 01x_e, 11x_e, 10x_e), 10x_1, \\
& 00x_1
\end{aligned}
$$